CELL AND TISSUE DESTRUCTION

CELL AND TISSUE DESTRUCTION

MECHANISMS, PROTECTION, DISORDERS

JÜRGEN ARNHOLD
Associate Professor, Institute for Medical Physics and Biophysics, Leipzig University, Leipzig, Germany

Academic Press is an imprint of Elsevier
125 London Wall, London EC2Y 5AS, United Kingdom
525 B Street, Suite 1650, San Diego, CA 92101, United States
50 Hampshire Street, 5th Floor, Cambridge, MA 02139, United States
The Boulevard, Langford Lane, Kidlington, Oxford OX5 1GB, United Kingdom

Library of Congress Cataloging-in-Publication Data
A catalog record for this book is available from the Library of Congress

British Library Cataloguing-in-Publication Data
A catalogue record for this book is available from the British Library

ISBN: 978-0-12-816388-7

For information on all Academic Press publications visit our website at
https://www.elsevier.com/books-and-journals

Publisher: Andre Gerhard Wolff
Acquisition Editor: Glyn Jones
Editorial Project Manager: Sandra Harron
Production Project Manager: Sreejith Viswanathan
Cover Designer: Vicky Pearson

Typeset by TNQ Technologies

Contents

6. Immune Response and Tissue Damage

7. Acute-Phase Proteins and Additional Protective Systems

III

AGING PROCESSES AND DEVELOPMENT OF PATHOLOGICAL STATES

8. Aging in Complex Multicellular Organisms

9. Cell and Tissue Destruction in Selected Disorders

10. Organ Damage and Failure

11. Conclusions

Preface

Living organisms fascinate by their multifaceted complexity, enormous variability, and versatile functionality. The all-encompassing progress in life sciences and medicine during the last years allows now a deep penetration into molecular details of enzyme reactions and receptor-mediated signaling, and it enlarges considerably our knowledge about cell physiological and regulatory processes. Modern analysis methods of the genome, transcriptome, proteome, lipidome, metabolome, and others add a tremendous amount of data about living processes that can be overlooked only by few specialists. Despite sophisticated methods, ingenious simulations, and novel hypothesis, there are large gaps in the understanding of regulatory mechanisms and deviations from normal functions in living structures. Often we can read that the cause of a disease is unknown, details of pathogenesis mechanisms are unclear, and so on. That concerns widespread, very common disorders affecting most of all people of advanced age. The same is valid for the thorough understanding of aging processes. We can describe many facets of how we age, and numerous theories about aging exists, but an all-encompassing explanation for aging phenomena is not given.

These antagonisms reflect two opposing sides of living systems. On the one hand, there are clearly defined routes for the synthesis of complex biomolecules, which are encoded in the genes, and the arrangement of these molecules to ordered structures. On the other hand, numerous unwanted reactions exist, which disturb the integrity and high order in cells and tissues. Biomolecules are not only involved in specified functional processes, they are also subjected to structural and chemical modifications. Hence, numerous disturbances of physiological processes result. It would be too easy to categorize chemical reactions in living systems as good and bad ones. Destructions of biomolecules and therefore functional disturbances are an indispensable part of life. To minimize damaging reactions and functional constraints, numerous protective mechanisms are employed by cells, tissues, and organisms.

Inspired from the enduring interplay between the maintenance of homeostatic conditions and their disturbances, the analysis of destructive events in complex organisms and protection against these deteriorations starts with a thermodynamic approach to have a rational explanation for this endless struggle. This book gives an overview about main routes of destructive chemical conversions in proteins, lipids, carbohydrates, and nucleic acids. Key players of these modifications and protective mechanisms are described too. Cell and tissue destruction are further regarded in relation to physiological processes of energy production and storage, mechanism of cell death, and the whole field of innate and acquired immunity. Finally, the contribution of damaging mechanisms is analyzed during aging processes and in the pathogenesis of chronic inflammatory diseases. A last chapter is devoted to tissue destructions in the kidney and liver as well as to sepsis. These wide-ranging subjects are exemplified by detailed descriptions, but it was impossible for me to go into each detail and to analyze all possible disturbances and disease scenarios. A unifying concept is given for the development of chronic inflammatory states and applied to aging processes and chronic diseases.

Subjects presented in this book are the result of more than 40 years of intense scientific work around topics such as lipid peroxidation, immune cell activation, reactions of heme peroxidases, redox biology, hemolysis, pathogenesis mechanisms of diseases, and some others that are all related to the field of cell and tissue destruction.

This book would be impossible without numerous inspirations, suggestions, and ideas received by other scientists. Therefore, it is a high pleasure for me to thank all my colleagues, whom I met in their laboratories and on conferences, for their fruitful discussions, thoughts, and critical remarks. I am grateful to acknowledge all my coworkers from other institutions for intense, long-lasting collaborations and joint publications. A special gratitude is devoted to all previous and actual members of my working group at the Leipzig University. Their contribution is important for this book. I am also very thankful to my wife for her immense patience, when I was working on the computer. Last but not least, I am very grateful to Sandra Harron, Glyn Jones, Sreejith Viswanathan, and other coworkers from Academic Press for their great assistance in writing this book.

Leipzig, April 2019

PART I

PHYSICAL AND CHEMICAL REASONS FOR DESTRUCTION IN LIVING SYSTEMS

CHAPTER

1

Cells and Organisms as Open Systems

1.1 MAIN PROPERTIES OF CELLS

Living organisms comprise a large variety of forms reaching from unicellular (e.g., bacteria) to complex multicellular species including human beings. Life is divided into three domains, bacteria, archaea, and eukaryotes, on the basis of genetic similarities as classification factor. Higher forms of life such as animals, plants, fungi, and others belong to the domain of eukaryotes. Living systems fascinate as natural creation by their fabulous range of manifestation. They cumulate a concentrated amount of structure and energy on smallest space and execute at once discrete functions.

In all living species, cells represent the basic structural elements. All information about structure and function of a given cell type and the species to whom this cell belongs to is encoded in nucleic acids and genes. By this genetic program, cells are able to synthesize consistently novel proteins, other cellular constituents, and if necessary extracellular matrix material.

The permanent renewal of living material is a key property in all domains of life. Cells can continually divide and increase their biomass. As long as space and nutrients are unlimitedly available, a nearly exponential growth can be observed in bacterial cultures. In growing multicellular animal organisms, where all cells are closely associated to each other, cells become functionally differentiated according to their location. These specialized cells fulfill particular functions highly necessary for the purpose of the whole organism.

Biological material is composed of highly ordered molecules such as proteins, carbohydrates, lipids, nucleic acids, and others. Energy is consumed for their synthesis. Cells are also equipped with systems for energy production and storage. In plants, some algae, and others, solar radiation energy is absorbed by chlorophylls and used for the synthesis of glucose, a process known as photosynthesis. Some bacterial species are able to exploit energy from inorganic chemical sources. Most animals utilize plant and/or animal material and produce energy from digestion of this organic food. Adenosine triphosphate (ATP) is a universal intracellular energy substrate molecule that contains high-energy phosphate bonds. If required, this energy can be applied for multiple cellular functions on hydrolysis of ATP [1]. This molecule links energy-producing and energy-utilizing processes in cells.

Taken together, living cells are equipped with systems to use the information from their genetic program, to renew permanently their constituents, and to supply sufficient energy for realization of all functional tasks. These general properties ensure the existence of life.

Cell and Tissue Destruction
https://doi.org/10.1016/B978-0-12-816388-7.00001-2

1.2 THERMODYNAMIC BASIS OF LIFE

1.2.1 Energy Flow to Living Systems

Living systems are nothing else as a special state of matter. The same natural laws are valid for living systems and for nonliving matter. As for all matter, the first and second laws of thermodynamics are also highly important for the description of physical states in cells and organisms.Although the first principle of thermodynamics represents another formulation of the principle of conservation of energy under inclusion of heat as a kind of energy, the second principle is related to the direction of natural processes. According to the latter principle, only those processes spontaneously take place in isolated systems, whose thermodynamic entropy increases. These processes are irreversible. In Box 1.1, thermodynamic characterization of a system is given with respect to processes of energy and matter exchange. In some textbooks, the term "closed system" is used instead of "isolated system."

Cells and organisms represent open systems because they exchange energy and material with their environment. For thermodynamic description of these systems, we have to mandatorily include the interaction with their surroundings. Of course, living systems contain a high number of complex biomolecules and highly ordered structural elements. These systems produce the high order of their components by utilizing radiation energy of photons (e.g., photosynthesis in plants) or by liberating energy stored in chemical bonds of complex organic foodstuffs on metabolic processes (e.g., oxidation of glucose).

In characterization of cells and organisms from a thermodynamic point of view, we have to extent our focus on uptake of energy and material by the living system from external sources as well as processes of energy and matter release. Hence, cells and organisms represent a localized part of a much larger system necessary for the full thermodynamic description of an open system. With this extension, the second law of thermodynamics is always valid for the application to all forms of life.

The amount of energy taken up by an organism is used for cell-specific functions such as maintenance of ionic gradients, synthesis of biological material, contraction of microfibers, and others, whereby these functions are mainly triggered by ATP. To complete the energy balance, some portion of the intake of energy is stored in form of high-energetic macromolecules such as glycogen, body fat, triglycerides, and others.

BOX 1.1

THERMODYNAMIC CHARACTERIZATION OF A SYSTEM CONCERNING EXCHANGE PROCESSES WITH THE SURROUNDING

Isolated system	No exchange of energy and matter
Closed system	Exchange of energy but no exchange of matter
Open system	Exchange of energy and matter

Another part of this energy is wasted as heat. Generally, the principle of energy conservation is always fulfilled in cells and organisms.

1.2.2 The Concept of Thermodynamic Entropy

The second law of thermodynamics is closely related to the concept of entropy. In an isolated system, the total entropy S can never decrease over time. It should stay the same or increase. When the entropy remains constant, the system is in the state of a thermodynamic equilibrium.

Under the latter condition, entropy S as a state function of a system is defined over its infinitesimal increment dS, which is equal to the ratio of infinitesimal amount of heat δQ transferred to a closed system divided by the common temperature T of the system and the surroundings.

$$dS = \frac{\delta Q}{T} \quad (1.1)$$

Any processes that take place in an isolated system without any change in entropy are reversible.

As long as an irreversible process runs in an isolated system, the entropy increases. Then, Eq. (1.1) should be changed into

$$dS > \frac{\delta Q}{T} \quad (1.2)$$

Typical examples for irreversible processes are expansion of a gas into a space with lower pressure, dissolving of salt crystals in water, inelastic collision, and equalizing processes in systems with local temperature or concentration gradients. It is totally impossible to reverse these processes without energy supply. Energy equivalents must be put on the corresponding system to concentrate gas molecules in the former space, to extract salt crystals from an aqueous solution, to repair mechanical deformations, or to create local differences in temperature or concentration.

A comparable small energy is sufficient to break a glass pane into numerous pieces, as this material is very brittle. To form from these pieces a new pane of glass, a much higher amount of energy must be applied. There are numerous other examples for spontaneous processes in accordance with the second law of thermodynamics.

The second principle of thermodynamics is an empirical law. Several common formulations of this principle are given in Box 1.2.

1.2.3 Entropy Versus Gibbs Energy

Cells and organisms are composed of highly structured components. In a thermodynamic

BOX 1.2

COMMON FORMULATIONS OF THE SECOND PRINCIPLE OF THERMODYNAMICS

Entropy can never decrease over time for an isolated system.

Heat can never pass from a colder to a warmer body without some other change, connected therewith, occurring at the same time.

It is impossible to devise a cyclically operating device, the sole effect of which is to absorb energy in the form of heat from a single thermal reservoir and to deliver an equivalent amount of work.

In every neighborhood of any state S of an adiabatically enclosed system, there are states inaccessible from S.

context, they are open systems. In an open system, entropy may increase, stay the same, or decrease over time. Considering the low entropy of living forms in relation to the entropy of less structured components in the environment, Erwin Schroedinger stated in 1944 that life feeds from negative entropy [2]. Later editions of his famous book "*What Is Life—The Physical Aspect of the Living Cell*" contain additionally a short note added in 1945, where he focused on free energy as a minimizing principle as true source for life [3]. As biological processes occur at roughly constant temperature and pressure, the more correct term is the Gibbs energy that drives thermodynamic processes in living matter.

In cells and organisms, all aspects of energy metabolism are realized by enzyme-driven coupled processes, where two or more chemical reactions as well as physical alterations are closely linked with each other [4]. Such coupled processes spontaneously take place when a certain amount of energy will be released considering all partial processes. In this way, energy-consuming reactions can successfully be interconnected with energy-producing processes. A coupling can be implemented by two linked chemical reactions, ligand-driven conformational changes of proteins, transport processes of ions and metabolites through membranes, and others. These links are essential for the functioning of cells and organisms as open systems.

A known example for such coupling is the intracellular formation of the energy-rich substrate ATP from adenosine diphosphate and inorganic phosphate by ATP synthase, an energy-intensive reaction that is driven by a proton gradient at the inner membrane of mitochondria. This gradient supplies the required energy for ATP synthesis [5]. Another example for coupled processes concerns sodium-dependent glucose cotransporters. In enterocytes of the small intestine and in the proximal tubule of the nephron, these transporters link the passive transport of sodium ions with the active transport of glucose [6].

Under the condition of constant temperature and pressure, a condition that predominates in living systems, the Gibbs energy G (or more precisely its change dG) is the thermodynamic parameter of choice. A coupled process proceeds spontaneously if

$$dG < 0, \quad (1.3)$$

where dG corresponds to the total change of Gibbs energy in a coupled process. Such processes are called exergonic processes in contrast to endergonic ones with positive dG. The system is in equilibrium when dG remains unchanged. Then no changes occur in concentrations of the participants. A short characterization of Gibbs energy G as thermodynamic state parameter is given in Box 1.3.

BOX 1.3

GIBBS ENERGY AS THERMODYNAMIC STATE PARAMETER

Definition at constant temperature and pressure

$\Delta G = \Delta H - T\Delta S$ with ΔH—change in enthalpy and ΔS—change in entropy

ΔG corresponds to the maximum of nonmechanical work that can be performed by a system under isothermal and isobaric conditions

Alternative names: Gibbs free energy, Gibbs function, free enthalpy

Thus, under isothermal and isobaric conditions, Gibbs energy acts as minimizing principle. Gibbs energy replaces entropy that plays the role as maximizing principle in an isolated system. Of course, other parameters can in turn replace Gibbs energy as minimizing principle when other constraints dominate in an open system. For example, a minimum of the Helmholtz free energy A is the driving principle at constant temperature and volume.

1.2.4 Two Faces of Entropy

Again it should be emphasized that life is always in accordance with thermodynamic laws. There are numerous parameters used for characterization of physicochemical processes in open systems. Anyway, the term entropy dominates in thermodynamic characterization of living systems. There are some problems with the interpretation of life on the basis of entropy [7,8]. First, the thermodynamic term entropy S or better the change in entropy dS can only be calculated but not measured. Moreover, calculations are restricted to simple reactions and straightforward systems. Thus, this thermodynamic term is hardly clear and less demonstrative. Second, there is another way of description of entropy by linking this term to disorder. Low entropy means high degree of order and vice versa. An increase of entropy corresponds to an increase of disorder. This nonthermodynamic explanation comes from the information theory [9], where the total entropy of a system is proportional to the logarithm of the numbers of ways how this system can be realized.

Often, both interpretations of entropy are mixed together as to what causes a number of confusions. It is not the purpose of this book to give more details about this controversial discussion. As life is always based on principles of thermodynamics, the term entropy will mainly be used here in a thermodynamic sense.

Nevertheless, the link between entropy and order/disorder is more vivid especially from a philosophical point of view. A cell with its complex macromolecules and highly ordered structure is characterized by a comparable low value of entropy. Neglecting energy and matter exchange, a cell resembles an isolated system. Then, over time, the entropy will increase in this living system.

1.2.5 A Short View on Heat Engines as Open Systems

In the 19th century, thermodynamic principles were primarily formulated to better understand key physical processes in engines, most of all in heat engines. An engine can be regarded as an open system that converts the applied energy into mechanical work. In case of a heat engine, thermal energy from an external source is transferred to the working substance in the engine. During its action, an engine undergoes several working cycles. Under assumption of ideal conditions, the engine is after each cycle in the same thermodynamic state as before. In a cyclic process, energy transfer and conversion are sequentially linked. There are a number of idealized thermodynamic cycles. Common textbooks mostly mention the Carnot cycle [10] to describe thermodynamic properties of heat engines.

In thermodynamic cycles, the main focus is directed on evaluation of the degree of efficiency of a working engine. For a heat engine, this important property is defined as the amount of useful work produced per unit of supplied heat energy. The theoretical maximum value of efficiency is derived under idealized conditions, assuming in case of a heat engine unlimited large hot and cold reservoirs and no changes in the temperature of these reservoirs during their contact with the working substance. In this description, only processes of energy transfer and conversion are viewed. The outer envelope of an engine and the internal arrangement of structural elements are assumed to remain unchanged after each cycle.

1.2.6 Steady State Condition in Open Systems

In system theory, an open system is regarded to be in a steady state when input and output processes of energy and matter are nearly consistent to each other. Similar to a thermodynamic (especially a chemical) equilibrium, the steady state represents a time-independent state where the system as a whole remains unchanged. Although chemical equilibria are reversible, steady states are irreversible and maintained by fluxes into and from the system. All variables describing the steady state remain constant. A demonstrative example for a steady state condition is the water flow through a cascade of successively arranged basins.

The concept of steady state is widely common in different fields of science including economics, engineering sciences, biology, cybernetics, and others. It was first applied by Ludwig von Bertalanffy, one of the founders of the general system theory [11,12], to characterization of cells and organisms [13,14].

An open system does not accumulate mass or energy as long as it is in a steady state. According to this concept, no changes in structure, composition, and internal arrangement of structural elements of the system should occur over the time of interest. In this concept, the main focus is directed on flow processes of energy and matter but less on the structure and properties of the system itself. The latter appears as an inert black box interacting by different fluxes with its surroundings (Fig. 1.1).

To apply thermodynamic description to irreversible processes and open systems, Prigogine divided the total change of entropy dS in an open system into two terms [15,16]:

$$dS = d_eS + d_iS \tag{1.4}$$

where d_eS and d_iS denote the change of entropy by import to the system and the production of

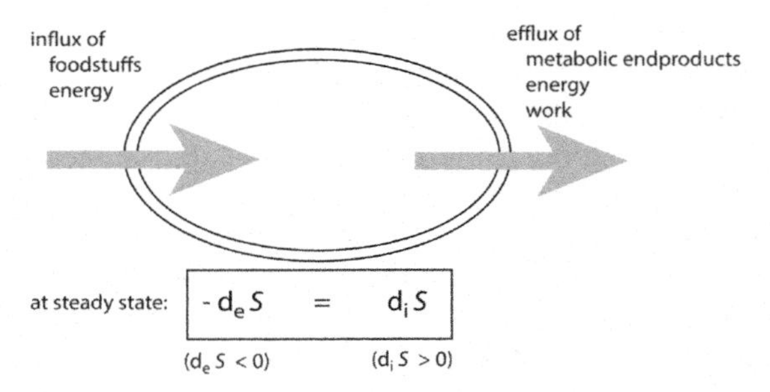

FIGURE 1.1 A cell as an open system under steady state condition. d_eS denotes changes in entropy of the cell due to the influx of matter and energy from external sources. d_iS represents changes in entropy of the cell associated with internal metabolic processes.

entropy by irreversible processes in the system, respectively. Chemical reactions, diffusion, and heat transport were listed as examples for the latter processes. While the term d_iS is always positive in accordance with the second law of thermodynamics, the term d_eS may be negative as well as positive. Thus, in dependence on the values of these two entropy terms, the total entropy either decreases or remains unchanged or increases in an open system with time. The unchanged state corresponds to the steady state condition.

Although quite different in their function, there is one striking concurrent feature in thermodynamic description of heat engines and open systems in a steady state. Assuming in both cases idealized conditions, variables characterizing their thermodynamic state including the total entropy remain unchanged over time.

$$dS = 0 \tag{1.5}$$

This is the case in a cyclic working engine and in an open system, in special case in a cell, under steady state conditions. In all these cases, the main focus is directed on processes of energy flow.

1.2.7 Apparent Thermodynamic Stability of Living Systems

Life exists for more than 4 billion years on earth undergoing a permanent evolution of its forms. If we focus our attention on small time periods and neglect evolutionary and genetic variations of living species, there is a constant, nearly identical reproduction of all living forms with time.

Bacteria grow under defined laboratory conditions, whereupon one generation follows the previous one and so on. Usually a bacterium taken from a later reproduction cycle resembles closely in appearance and function to a bacterium from a former cycle. In other words, there are no changes in general properties between these two microorganisms taken at different points in time. All thermodynamic variables describing the state of a single individual remain nearly constant in comparison with other single individuals of this species from a former reproduction cycle. In contrast to open systems regarded in the preceding subchapter, the consistency of thermodynamic state variables here is not related to the same unchanged system but to another newly reproduced cell.

The same conclusion can principally be drawn for complex multicellular organisms. In this case, it is necessary to compare with each other species of the same age, i.e. of the same state in their reproduction cycle. Of course, time scales for reproduction cycles are quite different between bacteria and higher animals, with minutes to hours for the former ones and months to years for the latter species. Despite greater individual variations in complex animal organisms in comparison to unicellular ones, it is better to state that in animals long-term average values are the same for thermodynamic state variables.

If we depict entropy as state variable for the long-term characterization of life forms without considering genetic or evolutionary changes, it follows that

$$dS = 0 \tag{1.6}$$

The same conclusion was drawn before for the long-term stability of an open system being an idealized engine or for the condition of a steady state. However, both statements differ in that aspect that Eq. (1.5) is applied to an open system holding the system's structure and properties unchanged over time, whereas Eq. (1.6) is based on the close similarity of individuals from different reproduction cycles (i.e., from different systems). Thus, in the latter case, no individual identity is given for the system parameters.

Before we further analyze the thermodynamic properties in living systems, we will throw a glance at common requirements and consequences ensuring the long-term existence of cells and organisms as open system.

1.3 FUNCTIONING OF LIFE AS OPEN SYSTEM

1.3.1 Membranes and Covers

Living structures are separated by a barrier from their surroundings. The outer surface of this barrier confines all components of the corresponding living system, giving them a more or less defined shape. This barrier also functions as a kind of filter controlling and regulating the exchange processes of energy and matter with the environment.

In a cell, this barrier is the plasma membrane. Moreover, cells contain numerous organelles with vesicular and cylindrical structures separated by a membrane from the cytoplasm. Of course, the structure of membranes and composition of their elements varies widely with cell type and function. A membrane separates different compartments from each other. It is a barrier that limits the diffusion of solutes and water. Membranes are equipped with special transport molecules, carriers, receptors that bind specific ligands, ion channels, and other regulatory agents that control the influx of components into and efflux from a cell. Metabolites can either permeate passively across a membrane, i.e. along a concentration gradient, or actively against a concentration difference. For active transport mechanisms, energy is required, which is supplied by hydrolysis of ATP, electrochemical gradients, or cotransporters. Additionally, electrical potentials can be generated at some membranes. Taken together, membrane elements help to uptake those chemical components from the surrounding medium needed for cellular metabolic processes. They contribute to maintenance of a given internal milieu within cells and cell organelles.

Animals are covered by a skin. Although some gases and fluids can freely permeate through skin, there are special organs in higher animals for uptake and release of material. The exchange of dioxygen and carbon dioxide occurs primarily via lungs or gills. These organs are structured to enlarge the area for gas exchange.

Foodstuffs are mainly taken up through mucous surfaces of the intestine. Multiple villi highly enlarge the area of intestine's surface and facilitate the uptake of foodstuffs. There are also special elements in intestinal cells promoting the transfer of important food molecules from gut into the organism. Liver with gallbladder, kidney, and colon are special organs for waste production and release.

Single cells will have the same temperature as their proximate environment. Otherwise, higher animals are able to thermoregulate. Across their outer skin cover, large temperature gradients may arise, allowing the survival of some endothermic species in cool regions.

Taken together, plasma membranes, skin covers, and mucous surfaces separate the living systems with their high-ordered components from the surroundings. As a result, living systems are characterized by low entropy due to the accumulation of energy-rich components in their cells and their arrangement to internal structures.

1.3.2 Ionic Gradients Across Membranes

In higher animals, the composition of cytoplasm of a living cell differs from the composition in the extracellular medium. Cytoplasm is usually rich in potassium ions and poor in sodium ions, whereas the reserve situation is observed in serum and interstitial fluid. In most excitable cells, for example, the potassium ion concentration is 20–50 times greater in cytoplasm than that in the external medium, whereas sodium ions are 3–15 times more enriched in the outside than in the cell [17]. Even higher concentration differences of about four orders of magnitude are found for calcium ions. Concentration of Ca^{2+} is about $1{-}2 \times 10^{-7}$ mol L^{-1} in cytoplasm of resting cells and slightly higher than 10^{-3} mol L^{-1} in the extracellular fluid [18].

These ionic gradients are caused by the presence of ionic pumps in the plasma membrane and membranes of subcellular structures. The ionic pumps utilize ATP as energy substrate and transport the aforementioned ions against the existing concentration gradient through the membranes. In this way, a special internal ionic milieu inside cells and gradients for these ions across the cell membrane is created. Cells use these conditions for realization of specific functions.

Differences in sodium ions are exploited by special cotransporters that couple the passive influx of sodium ions into the cell with the active transport of another substrate against the concentration gradient either inside or outside the cell. Known examples are the sodium-glucose transporter in epithelium cells of intestine [19] or the sodium-iodide transporter in the thyroid gland [20]. Differences in sodium and potassium ion distribution also play a great role in formation of resting and action potentials in nerve and muscle cells. The named functional responses are accompanied by small deviations of the original distribution of sodium and potassium ions. These deviations will immediately be balanced by a short-term activity of the Na^+-K^+-ATPase.

In neurons, immune, muscle, and other cells, the short-term liberation of calcium ions from internal stores or by controlled influx into cells through Ca^{2+}-channels triggers numerous processes of cell activation such as contraction of muscle fibers [21], release of neurotransmitters [22], degranulation in immune cells [23,24], and others. The transient increase of cytoplasmic Ca^{2+} activates Ca^{2+}-ATPases that pump these ions back into the extracellular space or into internal stores.

Within some cell organelles of eukaryotes, different values for proton concentration are also found. The internal compartment of mitochondria, the mitochondrial matrix, has an alkaline pH value around 7.8 [25], while inside lysosomes acidic pH values predominate [26]. These pH gradients are highly important for functional processes of these organelles. They are also the result of energy-intensive processes.

The aforementioned ATP-driven and other active transport processes ensure a special internal ion composition inside cells and cell compartments. Importantly, these mechanisms restore after each cellular activity the initial ion composition typical of the resting state in the cell. This provides the basis for numerous metabolic, regulatory, and cell-specific functions in complex multicellular organisms.

Of course, the special ionic milieu in cells can only be maintained when the barrier function of the corresponding membrane is intact. A slight increase in the passive permeability of the membrane for these ions can be compensated by higher activities of ATPases and can lead to partial depletion of ATP. However, ionic gradients are highly disturbed by more serious membrane leakage and by loss of membrane integrity.

1.3.3 Expenditure of Energy for Physiological Processes

Numerous physiological processes require energy from hydrolysis of ATP. Using standard metabolic conditions of the generated ATP, about 19%–28% is used by the Na^+-K^+-ATPase, 4%–8% by the Ca^{2+}-ATPase, and 2%–8% by the actinomyosin ATPase. Synthesis processes utilize 28% of ATP for protein synthesis, 7%–10% for gluconeogenesis, 3% for ureagenesis, and the remaining part is used for mRNA synthesis, substrate cycling, and others [27]. Of course, these rough values can vary in different tissues and in dependence on the activity state.

A particular large contribution of Na^+-K^+-ATPase is found in the brain and kidney [28–30]. Ca^{2+}-ATPase activity is important in contracting skeletal and heart muscles as well as brain [31].

In the resting human brain, the rate of ATP consumption is threefold higher in gray matter than in white matter [32]. Gray matter comprises a much higher ratio of neurons to nonneuronal cells than white matter [33]. Further, it has been estimated that a nonneuronal brain cell utilizes about 3% of ATP energy consumed by a neuron [32]. Indeed, the majority of mitochondria is located in neurons in dendrites and synapses [27]. These data prove extensive energy expenditure for neurotransmission and interneuron signaling. Another analysis estimates equal rates for energy expenditure for maintenance of resting and action potentials in cerebral cortex, while about three times more energy is required to maintain resting potential in relation to action potential in the cerebellar cortex [30].

Taken together, these analyses indicate that a sufficient part of ATP is utilized for the maintenance of ionic gradients, especially in metabolic active cells. These gradients drive important functional responses such as muscle contraction, propagation of action potentials, neurotransmission, and cell activation. Slight changes in ion concentrations during physiological responses, which lead to slight decrease in ionic gradients, activate ATPases to reestablish to former concentration difference for the corresponding ion.

1.3.4 Homeostasis in Organisms

The maintenance of a special internal milieu with characteristic physical and chemical properties is crucial for complex animal organisms. This is realized by numerous regulatory circuits consisting of a sensor unit, a control center, and an effector unit [34]. The given parameter is regulated by a negative feedback mechanism. In the organism, transport of gases, nutrients, and waste products is conducted by blood, connecting all tissues and organs. A number of essential blood parameters are maintained unchanged in certain limits. Body core temperature, arterial blood pressure, electrolyte composition, pH value, osmolarity, calcium level, and concentrations of glucose, dioxygen, and carbon dioxide are among these values, whose homeostasis will be ensured. Needless to say, their actual values can vary from individual to individual. They will also be changed by enhanced activities and under stress situations.

Irritability and autonomous activities can cause disturbances of homeostatic physiological parameters in organisms [14]. In the first case, the system returns to its steady state after a short-term deviation of the desired value. Autonomous activities consist in periodic fluctuations of parameters necessary for functional processes.

Endothermic organisms hold their body temperature more or less constant. Intensive metabolic processes in the liver, heart, brain, skeletal muscles, and some other organs produce heat, which is distributed by blood circulation to other parts of the organism. Otherwise, the organism can lose heat by conduction, convection, radiation, and evaporation. Hypothalamus is the control center for thermoregulation in the human organism [35,36]. This center gets the information from thermoreceptors scattered throughout the whole organism. Core temperature can be maintained constant by different mechanisms such as regulation of the blood flow through regions adjacent to skin, exudation, and effects on cell metabolism. Of course, humans are able to have an active effect on thermoregulation by choosing appropriate clothing and adapting the temperature in their near environment.

Several cell-based receptor systems regularly determine the actual state of blood parameters. Among these specialized cells are baroreceptors, osmoreceptors, chemoreceptors, receptors for dioxygen content in the kidney, sensors for calcium ions in the parathyroid glands, and others.

In case of deviations from desired values, a counterregulation will be initiated via the autonomous nervous system and/or by release of endocrine hormones. In kidneys, the renin–angiotensin system is an important hormone system involved in regulation of blood pressure and fluid balance [37]. These feedback control systems provide optimum conditions for a long-term functioning of our organism.

Homeostatic conditions are not only important for normal functioning of the whole organism but also for single organs, such as the liver and kidney.

1.3.5 Comparison With Other Open Systems

Thus, cells and organisms are characterized by a more or less stable internal milieu and the presence of regulatory mechanisms ensuring the stability of this setting. For maintenance of this inner milieu, there exist numerous gradients for chemical constituents, especially for ions, across membranes, and on systemic level, there also exist gradients for physical parameters. Living systems utilize energy from external sources to hold their high degree of order and to fulfill specific functions. For these purposes, complex processes of response to the environment and metabolism are realized by cells and organisms. In addition, heredity of traits, reproduction, and growth are typical features of all forms of life.

When we further apply the concept of steady state to a living system, we have a striking similarity to an engine focusing on the time dependence of thermodynamic parameters. In both cases, state functions of these systems remain constant over the time interval of interest. Both living systems and engines utilize energy from external sources for their specific functions and ongoing processes.

Otherwise, living systems differ clearly from engines in the following aspect. In case of a functional disturbance of any part of an engine, it is necessary to replace this part by a new one. A more or less extensive repair has to be performed. By a more strong impairment, the whole device must be replaced by a new one. In contrast, cells and organisms are self-repairing systems. All necessary information for repair processes and synthesis of novel material is encoded in nucleic acids and genes.

1.4 HIGH ORDER OF BIOLOGICAL MATERIAL VERSUS DESTRUCTION

1.4.1 Biological Material Under Real Conditions

Till now, a number of idealizations have been made to thermodynamically describe cells and organisms as open systems. If we depict a living system at a given time point, it is characterized by a certain internal structure based on the arrangement of their macromolecules and components. Any changes in chemical and physical properties of these ingredients are ignored when we direct our focus only on processes of energy metabolism such as uptake of energy-rich organic material, energy conversion and storage, and cell-specific energy-consuming processes. The same holds if we apply the concept of steady state to life forms. In these idealizations, a cell is nothing else but a miniaturized constantly and endlessly working engine.

Although some biological materials are very stable and robust, many others are prone to chemical degradation. Proteins, lipids, carbohydrates, and other cellular constituents permanently undergo chemical destructive processes. An unwanted change in the physical state of biological macromolecules is another source for destructions. Examples are the partial unfolding of proteins or the breakdown of secondary bonds. In living systems, chemical and physical destructive processes take always place, mostly on a very low level. Under these conditions, it is hardly to determine which defined destructive process was responsible for any damage of biological material.

Numerous different impacts are known, which enhance destructions in living systems. Among them are mechanical stress, osmotic stress, altered composition of ions in the external milieu, lack of nutrients and dioxygen, the presence of toxic agents, all kinds of radiation, effect of higher or lower temperatures, the presence of harmful microorganisms, and many others. On cellular level, results of destructive processes are denaturation of proteins, oxidation of biological material, formation of reactive oxygen and other species, enhanced consumption of energy, disturbance of barrier functions of membranes, impairment of cell organelles especially of mitochondria, and some other effects. In the worst case, stressed cells die. Of course, there are also numerous impacts of destructive processes on physiological functions in organisms.

When we regard cells and organisms as open systems, it is useful to divide all occurring chemical reactions into two main groups (Fig. 1.2). The first group is composed of reactions necessary to synthesize and arrange all structural elements of cells and to realize all processes of uptake, conversion, storage, and consumption

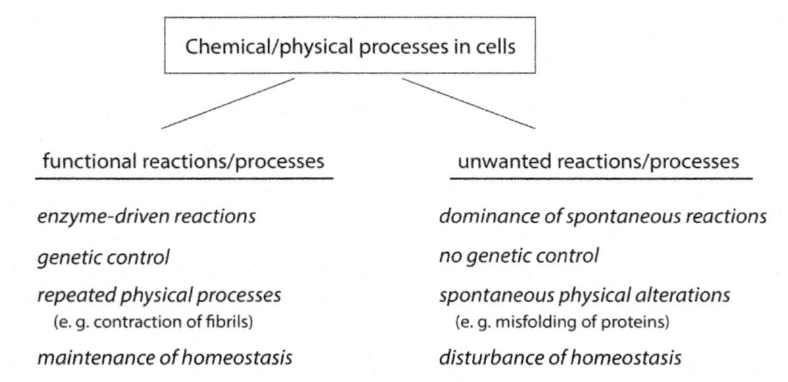

FIGURE 1.2 Characterization of ongoing chemical and physical processes in living systems.

of energy including cell-specific functional processes. Numerous structural and functional proteins are involved in these processes. Their amino acid sequences are encoded by nucleic acids. Thus, reactions of the first group are driven by the genetic program. They ensure the high internal structural arrangement and functioning of living systems. Their presence is indispensable for the existence of cells and organisms as open system.

The second group of reactions comprises a high number of unwanted, destructive reactions that occur independent of heritable information. These reactions are spontaneous. As a result physicochemical properties of biological material are changed over time, which disturbs the structural and functional integrity of macromolecules and, thus, the high order in living systems.

In some cases, destructive chemical reactions can be actively reversed by cells. A known example is the turning back of the spontaneous formation of methemoglobin from ferrous hemoglobin by methemoglobin reductase in erythrocytes [38]. In such a way, a high level of ferrous hemoglobin is maintained in red blood cells.

A similar subdivision in wanted and unwanted processes is also useful for changes in the physical state of biological macromolecules. Contractions of fibrils of the cytoskeleton and muscles fibers, the functioning of transporters, and conformational changes of receptors on binding to their agonists are examples of reversible physical alterations of proteins highly important for specific cell functions. Again information about these proteins, signaling molecules, and regulatory elements is included in genes. Among unwanted physical destructions are unfolding and denaturation of proteins, loss of electrical potentials, loss of local concentration gradients, mechanic injury, and others.

Often both chemical and physical destructions simultaneously take place.

1.4.2 General Adaptations of Living Systems to Destructions

It is quite impossible for a living system to prevent all kinds of potential damage of biological material. It is also useless to make arrangements against very strong and seldom occurring deteriorations. Thus, protective mechanisms of cells and organisms against impairment of their material and functional inactivation are directed to avoid and repair often occurring nonlethal destructions and to hold the number of damaged molecules per mass unit on a low level. Main protective strategies against destructions of biological material can be divided into three groups (Fig. 1.3).

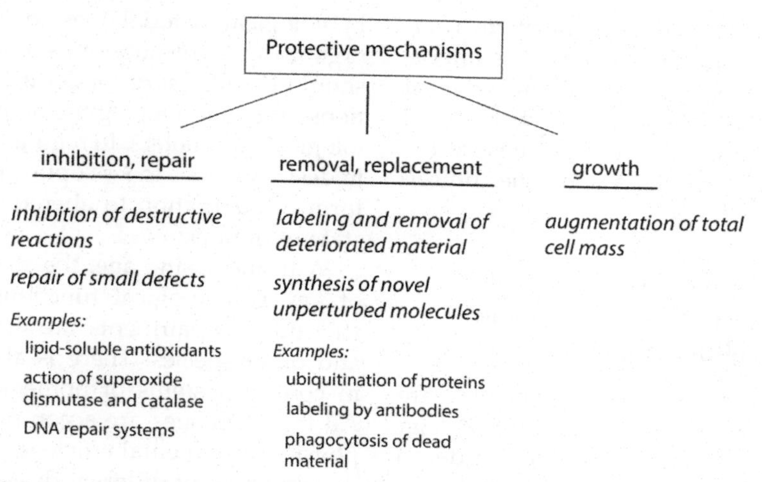

FIGURE 1.3 Main strategies of protective mechanisms against destructions in living systems.

The first group of protective mechanisms comprises systems that inhibit destructive reactions or remove damaged material. Among them are water- and lipid-soluble antioxidants, proteins that scavenge and inactivate reactive oxygen species, systems that bind, transfer, and store transition metal ions, DNA repair mechanisms, and systems that recognize incorrectly folded proteins. Cells are also equipped with stress response systems, and with their help they can upregulate beneficial proteins. Important protective functions are provided by cells and mediators of the immune system in tissues, organs, and organisms.

The second line of adaptation is based on replacement of damaged material by newly synthesized molecules. In cells and tissues, there is a permanent turnover for many kinds of biological material. Moreover, unwanted, functionally inactive, or degenerated cells can be recognized and removed in complex organisms.

The existence of the aforementioned protective systems is highly essential for cells and organisms. With their help, the number of destructive processes and impairing consequences to living systems can be reduced. However, these mechanisms do not generally avoid the appearance of destructive reactions. If protective mechanisms would solely be based on these strategies, life would be unimaginable.

Despite these mechanisms, the accumulation of structurally and functionally deteriorated biological material can only be avoided by a sufficient high rate of growth processes in combination with cell divisions. Increase of total cell mass by permanent synthesis of novel unperturbed molecules keeps the relative number of damaged molecules per mass unit on a low level. In this way, biological material effectively escapes being strongly damaged by destructive processes.

Growth processes and permanent cell divisions contribute, thus, to the survival of unicellular species. Under strong idealized conditions of unlimited space and nutrients, a nearly exponential growth of bacteria and other unicellular species is expected [39]. Mostly, there are limitations in space or in both properties and problems with accumulated waste products. Consequently, not all microorganisms have optimal conditions for growth processes and cell divisions. Only a part of them will survive.

The reproduction rate of unicellular organisms should be high enough to avoid a significant accumulation of destructive loci.

Otherwise, enhanced reproduction of microorganisms under optimum prerequisites is important to compensate a loss of living cells at less appropriate conditions. Thus, permanent growth and reproduction processes ensure the long-term survival of bacteria and other unicellular species.

1.4.3 Peculiarities of Adaptations in Complex Multicellular Organisms

All multicellular animal organisms are derived from a single cell, the fertilized egg cell. During their development, they undergo repeated cycles of cell division. In the growing cell cluster, cells start to differentiate in dependence on their location and form finally specialized tissues and organs. With increasing age, the growth rate of organisms reduces usually step by step. At adult age, there is commonly a stop in further growth processes even though cells are continuing to divide. For a given species, physiological functions are well adapted to the final size of the organism.

In higher animals, there is a permanent division and reproduction of male and female germ cells under the influence of sex hormones. From these cells, novel organisms are formed. This multiplies the number of individuals and ensures the long-term survival of a given species.

As long as a multicellular organism grows at sufficient rate, there is a good change to compensate any accumulated destructions by augmentation of the total cell mass. In adult and elderly organisms, after the slowdown of growth processes, the other aforementioned strategies dominate now in the battle against destructions. Many physiological parameters are held at the same level due to manifold regulatory mechanisms in complex organisms. To perform a counterregulation of these parameters in case of deviations from the desired value, a certain amount of energy should be applied. Assuming a permanent increase of defects and destructions in biological material over time in a nongrowing organism, more and more energy equivalents should be necessary for compensation of deviations and guaranteeing the homeostasis of physiological parameters. In regulatory circuits, shifts in desired values to less optimum ones are also a form of adaptation to altered conditions in the state of organisms.

With increasing age, the stepwise worsening of key physiological functions and numerous structural impairments are evident in humans and other species. There is also a gradual loss in control systems of homeostatic parameters. All these changes are accompanied by reduced physical and mental efficiency as well as by the development of different disease scenarios.

1.4.4 Thermodynamic Consequences From the Existence of Destructive Processes

The accidentally and spontaneously occurring destructive processes alter the thermodynamic state of living material. In order words, biological material tends to increase their low entropy. When we regard a single cell or a given complex living organism as an open system and neglect growth processes, i.e. the total mass of this system remains nearly unchanged, the following expression should be valid for the alteration of the total entropy over time.

$$dS > 0 \tag{1.7}$$

Although being in line with the second law of thermodynamics, the statement given in Eq. (1.7) contradicts at first view to the long-term existence of living forms as open systems. A constancy of entropy was stated in Eq. (1.6) for the latter conditions, when state variables are compared belonging with individuals from different reproduction cycles of a species. Thus, Eq. (1.7) is related to time-dependent alterations in entropy of an individual living system, while in Eq. (1.6), the main focus was directed on comparison of different individuals existing at different time points. Hence, in the latter equation, growth processes and cell divisions are additionally involved.

The aforementioned three main strategies to prevent and control destructions in living systems (see Section 1.4.2) differ in their impact on the destruction-induced increase of entropy. All actions that minimize the extent of damage of cell components will diminish the increase in entropy to a low level. In a similar way, replacement of damaged macromolecules or destroyed cells by novel ones will also abate this entropy increase. Of course, assuming idealized conditions, these both strategies provoke apparently unchanged values of entropy over time.

The only strategy that prevents the time-dependent increase in entropy and efficiently diminishes the total entropy per mass unit of living matter is based on growth processes in combination with cell divisions. In contrast to the other two procedures, this strategy enables to reduce the relative number of damaged molecules below the previous level. In summary, to prevent negative impacts of destructions in living systems, the growth rate should be higher than the rate of destructive processes.

1.5 DESTRUCTIONS AND THEIR PREVENTION IN COMPLEX ORGANISMS

1.5.1 External Sources for Destructions

Destruction of biological material and disturbances of homeostatic parameters are an indispensable part of life. Thus, protecting and repairing mechanisms are highly necessary to keep the integrity of cells and tissues under control. Otherwise, life cannot exist.

A large variety of external impacts can disturb the integrity in living systems. In this book, the main focus will be directed on prominent destruction pathways in cells and tissues of complex organisms, especially in human beings. Those processes that do not completely destroy the organism but cause comparable minor alterations in living material and where inhibitory reactions, repair mechanisms, inflammatory events, and other surviving strategies were induced in response to damaging factors will be depicted. Without any doubt, such destructions can lead to more fatal consequences for the whole organism, to disease states, and to death, when they emerge to a great extent and when the protecting systems fail.

Potential external sources for damaging events that cause destructions in cells and organisms can roughly be classified into four main categories:

1. Destructions as a result of the lack of energy substrates and nutrients;
2. Physical sources for destructions such as ionizing radiation, mechanical stress, traumata, strong deviation in outer temperature, and pressure;
3. Defined chemical sources for destructions such as the action of toxic components, proteolytic agents, lipolytic agents, denaturating agents, allergic components, or reactive species;
4. Action of viruses, bacteria, fungi, or parasites.

Selected examples for primary attacks of chemical and physical sources on a cell are summarized in Fig. 1.4. These impacts can lead to formation of reactive species and enhanced oxidation of cell components. Further different signaling pathways can be induced in stressed cells. As a result, the synthesis of proteins can be changed. These activities also imply alterations in surface-exposed and secreted proteins. On a (patho)physiological level, different inflammatory responses, induction of growth processes, cell differentiation, and processes of cell death are observed. Altogether, these mechanisms can contribute to pathogenesis of diseases.

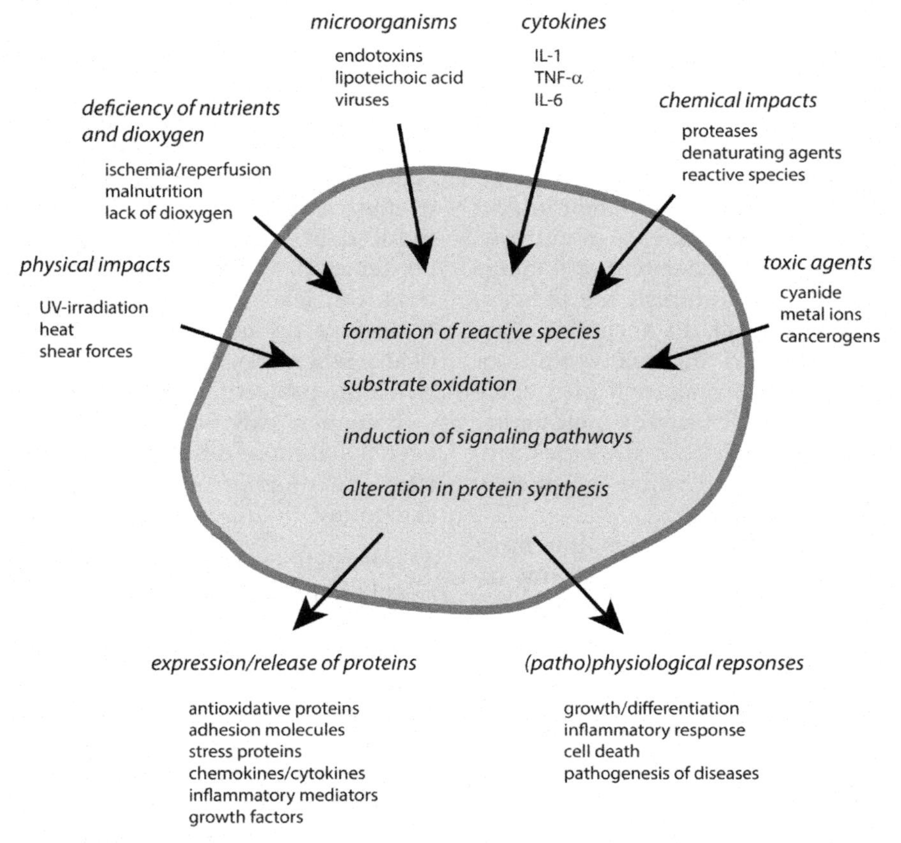

FIGURE 1.4 Typical examples of factors affecting the integrity of a cell and potential responses of this cell. *IL*, interleukin, *TNF-α*, tumor necrosis factor alpha.

Of course, it is impossible to predict which of these impacts or other not mentioned influences will act at a given time on the organism and how strong they will affect the integrity of cells and tissues. In some cases, it can be possible to define a certain probability for exposure of organism to a given impact. In many other cases, it is quite senseless to speculate about the likelihood of snakebites, exposure to carbon monoxide, or other very seldom but possible impacts that might highly disturb our health status. Such deteriorations sometimes occur unfortunately and will occur in future. Their appearance depends on many unpredictable random coincidences and succumb stochastic rules.

1.5.2 Frequency of External Damaging Events

A key aspect in analyzing damaging events and their effects on the living organism is the frequency of the appearance of these impacts. From the aforementioned influences, our organism is always exposed to low level radiation and to the presence of a variety of microorganisms, which colonize most notably on skin and mucous surfaces, and sometimes in tissues.

A temporary lack of energy substrates and nutrients arises, for example, as a result of food deficiency, intense exercise, enhanced stress, or under disease conditions. Elderly persons and persons with diminished immune response are more prone to this source for destructions. These groups of people are also more affected by microorganisms settling inside the tissues.

Another group of influences concerns the temporary exposure of organisms to more or less toxic components in foodstuffs and drinks, the inhalation of cigarette smoke and toxic gases, or the exposure to allergic agents. Of course, the degree and duration of exposure to these agents varies considerable among individuals. Moreover, persons respond very differently to toxic components in dependence on their constitution and health status.

All other impacts act very seldom or never during the lifetime of a given organism.

1.5.3 Internal Reasons for Destructions

In animals and humans, processes of cell and tissue destruction are also initiated in close association with functional activities. For example, reactive species are produced in mitochondria and during activation of immune cells such as granulocytes, monocytes, and macrophages. During phagocytosis of bacteria, neutrophils release additionally proteolytic enzymes and other hydrolyzing agents that can harm surrounding tissue. Another example concerns the release of damaging agents from necrotic cells.

Usually, these functional responses are strictly controlled to avoid substantial damage of cells and tissues. However, control mechanisms can be limited or fail at inflammatory sites, by massive immune response, and under some disease conditions.

1.5.4 Consequences to Damaging Attacks

Among the consequences to primary events, which are dangerous to cells and tissues, are oxidation of cell material, formation of reactive species, loss of barrier function of membranes, induction of cellular signaling pathways, release of cell constituents, activation of immune cells, discharge of mediators, and induction of cell death mechanisms. At very strong actions and in case of insufficient protection, these primary events can cause further injuries in adjacent molecules, cells, and tissues and induce, thus, cascades of deteriorating processes highly affecting the integrity in tissues, organs, and the whole organism.

The formation of reactive species plays a prominent role in nearly all chemical pathways of destructive events. They are not only produced as a result of ionizing radiation but also as an important hallmark in many of the listed damaging processes such as lack of energy substrates and nutrients, the induction of immune reactions, and in the response to toxic components.

1.5.5 Two-Stage Organization of Protective Systems

It makes no sense for a complex organism to make dispositions against all possible influences. The existing protective systems against potential damaging events are directed to impede or at least attenuate very often occurring and temporary acting impacts. This is normally possible without serious impairment of functional responses as long as these impacts operate at low level. Protective systems are also directed to maintain homeostasis for important cellular metabolites and physiological parameters. Under more serious impacts, these systems will restore homeostatic values.

Both on the cellular and systemic level, the existing protective mechanisms are organized in a two-stage way (Fig. 1.5). The first stage includes all mechanisms that are always present in the living system. These mechanisms can immediately respond to an appropriate destructive action. They are ready to use. These systems are directed against very often occurring

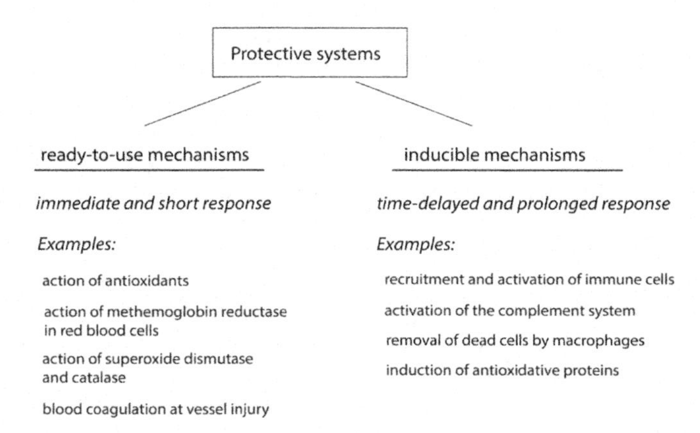

FIGURE 1.5 General properties of protective systems against destructions.

disturbances. Examples are the presence of water- and lipid-soluble antioxidants, the presence of superoxide dismutase and catalase inactivating superoxide anion radicals and hydrogen peroxide, respectively, the action of methemoglobin reductase in red blood cells avoiding the accumulation of methemoglobin, and the presence of acute-phase proteins in serum inactivating potentially toxic constituents released from erythrocytes or immune cells. The coagulation system belongs also to the first-stage protective mechanisms. It starts to act immediately after vessel damage and prevents the loss of blood as long as the degree of injury is low.

The second level comprises all mechanisms that are additionally activated in cells and induced in the organism in response to damaging events. This response is time-delayed and prolonged, but it can be much more pronounced than the answer from components of the first level. In cells, the activation of signaling pathways leads, for example, to the release of mediators, to enhanced transcription of protecting proteins, or to induction of apoptosis. On the level of the whole organism, the recruitment and activation of cells of the innate and acquired immune system, the activation of repair mechanisms, and renewal of tissues are selected examples for these processes. These inducible mechanisms are part of an inflammatory response to combat and inactivate different destructive exposures.

In summary, protecting mechanisms are highly important for guaranteeing the long-term integrity in cells and tissues and ensuring homeostasis of cellular and systemic parameters.

1.6 ON THE STRUCTURE OF THIS BOOK

Considering the significance of reactive species in destructions of living matter, the formation of different kinds of these species including protective mechanisms against them will be highlighted at first in Chapter 2. Basic mechanisms concerning oxidative processes in cell constituents and their prevention are the subject of Chapter 3.

Chapters 4–7 comprise the second main part of this book. These chapters are devoted to cell and tissue destructions and their prevention occurring on a more systemic level. Disturbances in the supply of energy substrates and energy production will be highlighted. Mechanisms of cell death, activation of the immune system, and the useful role of additional systems such as the coagulation system, acute phase proteins, or the complement system are further key aspects.

Last chapters are devoted to the role of destructive processes during aging, the development of diseases scenarios closely associated with accumulation of deteriorated material and disturbed homeostatic parameters, and with the failure of essential organs.

1.7 SUMMARY

In a thermodynamic sense, the long-term existence of life forms as open systems is only possible on the basis of a permanent energy flow from external sources and consumption of parts of this energy by cells and organisms. Utilized energy contributes to synthesis of biological material and maintenance of key ionic gradients and physiological parameters as a paramount requirement to ensure the high degree of order in cells, tissues, and organisms and to guarantee an optimum functioning of living systems. Under isothermal and isobaric conditions, the predominant conditions on earth surface, minimizing Gibbs energy is the driving thermodynamic principle for any changes in life forms over time.

The biological material is exposed to numerous accidentally and spontaneously occurring processes that disturb the chemical and physical integrity of macromolecules. Living systems exploit several strategies to avoid and minimize often occurring nonlethal destructions. Important protective mechanisms are the inhibition of destructive processes, recognition and removal of damaged material, de novo synthesis of macromolecules, removal of unwanted, inactive, or degenerated cells, and a well-functioning immune system.

Growth processes in combination with cell divisions can efficiently hold the number of damaged material per mass unit on a low level. This ensures the long-term survival of bacteria and other unicellular organisms. In humans and complex animals, after reaching an optimum size, a stepwise worsening of physiological functions takes place with increasing age. In addition, physical and mental properties are reduced, and many age-related disease scenarios result.

References

[1] S.C.L. Kamerlin, A. Warshel, On the energetics of ATP hydrolysis in solution, J. Phys. Chem. B 113 (2009) 15692–15698.

[2] E. Schroedinger, What Is Life? the Physical Aspect of the Living Cell, Cambridge University Press, 1944.

[3] E. Schroedinger, What Is Life? and Mind and Matter, Cambridge University Press, 1967, pp. 74–75.

[4] D.E. Green, H.D. Vande Zande, Universal energy principle of biological systems and the unity of bioenergetics, Proc. Natl. Acad. Sci. U.S.A. 78 (1981) 5344–5347.

[5] M. Yoshida, E. Muneyuki, T. Hisabori, ATP synthase – a marvellous rotary engine of the cell, Nat. Rev. Mol. Cell Biol. 2 (2001) 669–677.

[6] E.M. Wright, B.A. Hirayama, D.F. Loo, Active sugar transport in health and disease, J. Intern. Med. 261 (2007) 32–43.

[7] C.J. Smith, Problems with entropy in biology, Biosystems 7 (1975) 259–265.

[8] Y. Mitrokhin, Two faces of entropy and information in biological systems, J. Theor. Biol. 359 (2014) 192–198.

[9] C.E. Shannon, A mathematical theory of communication, Bell System Techn. J. 27 (1948) 379–423.

[10] S. Carnot, Réflexions sur la puissance motrice du feu et sur les machines propres à développer cette puissance, Annales scientifiques de l'école normale suprieme, 2e série 1 (1872) 393–457.

[11] L. von Bertalanffy, Zu einer allgemeinen Systemlehre, Biologia Generalis 19 (1949) 114–129.

[12] L. von Bertalanffy, General System Theory: Foundations, Development, Applications, George Braziller Incorp, 1976.

[13] L. von Bertalanffy, Der Organismus als physikalisches System betrachtet, Naturwissenschaften 28 (1940) 521–531.

[14] L. von Bertalanffy, The theory of open systems in physics and biology, Science 111 (1950) 23–29.

[15] I. Prigogine, J.M. Wiame, Biologie et thermodynamique des phénomènes irréversibles, Experientia 2 (1946) 451–453.

[16] I. Prigogine, Etude thermodynamique des phénomènes irréversibles, Durres, Paris, 1947.

[17] A.L. Hodgkin, The ionic basis of electrical activity in nerve and muscle, Biol. Rev. 26 (1951) 339–409.

[18] E. Carafoli, Calcium pump of the plasma membrane, Physiol. Rev. 71 (1991) 129–153.

[19] M. Palazzo, S. Garibaldi, L. Zanobbio, S. Selleri, G.F. Dusio, V. Mauro, A. Rossini, A. Balsari, C. Rumio, Sodium-dependent glucose transporter-1 as a novel immunological player in the intestinal mucosa, J. Immunol. 181 (2008) 3126–3136.

[20] O. Dohán, A. de la Vieja, V. Paroder, C. Riedel, M. Artain, M. Reed, C.S. Ginter, N. Carrasco, The sodium/iodide symporter (NIS): characterization, regulation, and medical significance, Endocr. Rev. 24 (2003) 48–77.

[21] A.G. Szent-Györgyi, Calcium regulation of muscle contraction, Biophys. J. 15 (1975) 707–723.

[22] Y. Goda, T.C. Südhof, Calcium regulation of neurotransmitter release: reliably unreliable? Curr. Opin. Cell Biol. 9 (1997) 513–518.

[23] J.T. O'Flaherty, C.L. Swendsen, C.J. Lees, C.E. McCall, Role of extracellular calcium in neutrophil degranulation responses to 1-O-alkyl-2-O-acetyl-sn-glycero-3-phosphocholine, Am. J. Pathol. 105 (1981) 107–113.

[24] E. Neher, The influence of intracellular calcium concentration on degranulation of dialysed mast cells from rat peritoneum, J. Physiol. 395 (1988) 193–214.

[25] A.M. Porcelli, A. Ghelli, C. Zanna, P. Pinton, R. Rizzuto, M. Rugolo, pH difference across the outer mitochondrial membrane measured with a green fluorescent protein mutant, Biochem. Biophys. Res. Commun. 326 (2005) 799–804.

[26] J.A. Mindell, Lysosomal acidification mechanisms, Annu. Rev. Physiol. 74 (2012) 69–86.

[27] D.F.S. Rolfe, G.C. Brown, Cellular energy utilization and molecular origin of standard metabolic rate in mammals, Physiol. Rev. 77 (1997) 731–758.

[28] J. Astrup, P.M. Sorensen, H.R. Sorensen, Oxygen and glucose consumption related to Na^{+}-K^{+} transport in canine brain, Stroke 12 (1981) 726–730.

[29] S.P. Soltoff, ATP and the regulation of renal cell function, Annu. Rev. Physiol. 48 (1986) 9–31.

[30] C. Howarth, P. Gleeson, D. Attwell, Updated energy budgets for neural computation in the neocortex and cerebellum, J. Cereb. Blood Flow Metab. 32 (2012) 1222–1232.

[31] T. Clausen, C. van Hardeveld, M.E. Everts, Significance of cation transport in control of energy metabolism and thermogenesis, Physiol. Rev. 71 (1991) 733–774.

[32] X.-H. Zhu, H. Qiao, F. Du, Q. Xiong, X. Liu, X. Zhang, K. Ugurbil, W. Chen, Quantitative imaging of energy expenditure in human brain, Neuroimage 60 (2012) 2107–2117.

[33] F.A. Azevedo, L.R. Carvalho, L.T. Grinberg, J.M. Farfel, R.E. Ferretti, R.E. Leite, W. Jacob Filho, R. Lent, S. Herculano-Herzel, Equal numbers of neuronal and nonneuronal cells make the human brain an isometrically scaled-up primate brain, J. Comp. Neurol. 513 (2009) 532–541.

[34] N. Wiener, Time, communication, and the nervous system, Ann. N. Y. Acad. Sci. 50 (1948) 197–220.

[35] A.A. Romanovsky, Thermoregulation: some concepts have changed. Functional architecture of the thermoregulatory system, Am. J. Physiol. Regul. Integr. Comp. Physiol. 292 (2007) R37–R46.

[36] K. Kanosue, L.I. Crawshaw, K. Nagashima, T. Yoda, Concepts to utilize in describing thermoregulation and neurophysiological evidence for how the system works, Eur. J. Appl. Physiol. 109 (2010) 5–11.

[37] H. Kobori, M. Nangaku, L.G. Navar, A. Nishiyama, The intrarenal renin-angiotensin system: from physiology to the pathobiology of hypertension and kidney disease, Pharmacol. Rev. 59 (2007) 251–287.

[38] D. Choury, A. Leroux, J.C. Kaplan, Membrane-bound cytochrome b_5 reductase (methemoglobin reductase) in human erythrocytes. Study in normal and methemoglobinemic subjects, J. Clin. Investig. 67 (1981) 149–155.

[39] S.J. Hagen, Exponential growth of bacteria: constant multiplication through division, Am. J. Phys. 78 (2010) 1290–1296.

CHAPTER

2

Role of Reactive Species in Destructions

2.1 SHORT CHARACTERIZATION OF REACTIVE SPECIES

Chemical conversions of biological material are often induced and potentiated by small reactive species. In many cases, these conversions are oxidative reactions, but reductive processes of substrate molecules also take place. Some basics about redox reactions in living systems are given in Appendix 1.

In resting cells and under normal metabolic conditions, the concentration of small reactive species is rather low and barely detectable. Enhanced formation of these species might play a certain role in activated cells, under stress situations, and under pathological conditions. In particular, immune cells are able to produce large amounts of reactive species by special enzyme reactions and use them for defined purposes. Thus, destructions in cells and tissues caused by reactive species result from stochastic processes induced by external and internal sources as well as from body's own production during immune response.

The term reactive species is not clearly defined. Its specification highly depends on involved interactions and regarded systems. Moreover, reactivity of these species can vary in a wide range as some species might induce very unspecific, nearly diffusion-controlled reactions (e.g., hydroxyl radicals), while others act specifically on defined targets. Some species are hardly detectable as they are very short-living. Other agents have a low ability for injury, but they can serve as prestage for more reactive species. There are also various denominations for these reactive species such as reactive oxygen species, reactive oxygen intermediates, reactive nitrogen species, free radicals, oxidants, and others. I prefer here the short-term reactive species, as this term best covers all possible representatives of this heterogeneous group.

In living systems, primary sources for reactive species are all kinds of ionizing radiation, one-electron reduction of dioxygen, and defined enzyme-catalyzed reactions. Once formed, there are multiple other subsequent reactions as a result of which further reactive species might be produced.

Concerning chemical sources and key constituents, reactive species can roughly be categorized into four main groups: species derived from dioxygen, nitrogen-based reactive species, transition metal ion–based species, and (pseudo)halogen-based species. In Fig. 2.1, important reactive species are listed according to this classification. This listing focusses on most common reactive species discussed in scientific literature in relation to destructions of organic material in animals and humans. Of course, numerous other reactive species are described in scientific literature. They are either related to special experimental conditions or concern specific cell cultures or mentioned only

Cell and Tissue Destruction
https://doi.org/10.1016/B978-0-12-816388-7.00002-4

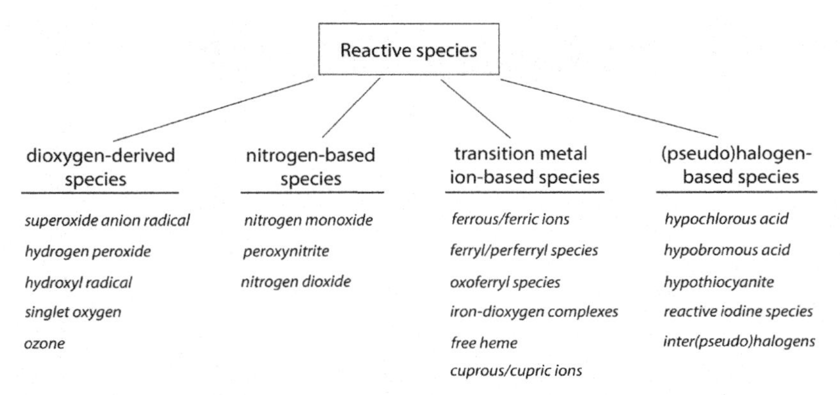

FIGURE 2.1 Classification of reactive species according to key constituents and chemical sources. In case of species with varying hydration states, the name is given for the dominating species at physiological pH value. Further explanations are given in the text.

in a few papers. It is not the purpose of this book to give a comprehensive overview about all kinds of reactive species and the involvement of these species in processes of cell and tissue destruction. An all-encompassing survey about this topic is given in the excellent book of Barry Halliwell and John M. C. Gutteridge "*Free Radicals in Biology and Medicine*" [1].

2.2 DIOXYGEN-DERIVED REACTIVE SPECIES

2.2.1 Electronic Configurations

The stepwise reduction of dioxygen to water generates the following intermediate reactive species: superoxide anion radical, hydrogen peroxide, and hydroxyl radical (Fig. 2.2).

To better understand the properties of these species, a short view on their electronic configurations is useful (Fig. 2.3). Atomic oxygen bears eight electrons, whereby the four 2p electrons are energetically equivalent. They are distributed to the p_x (two electrons), p_y, and p_z atomic orbitals. In dioxygen O_2, molecular orbitals are split into bonding and antibonding orbitals. Ground state dioxygen is in the triplet state as two unpaired electrons are located in a different π^* antibonding orbital. The σ^* antibonding orbital is empty. This electron configuration is in accordance with the quantum mechanical rules for allocation of electrons to orbitals. Hence, dioxygen is a less-reactive biradical and has paramagnetic property.

In superoxide anion radical, one of the two π^* antibonding orbitals is filled, and the other one is half-filled. In hydrogen peroxide, both π^* antibonding orbitals are fully occupied by electrons [2]. Electronic configurations of $O_2^{\bullet-}$ and O_2^{2-}, which corresponds to hydrogen peroxide, are given in Fig. 2.3.

In hydroxyl radical $HO^{\bullet}$, the five 2p electrons occupy the σ bonding orbital, one of the nonbonding orbitals, and the half of the second

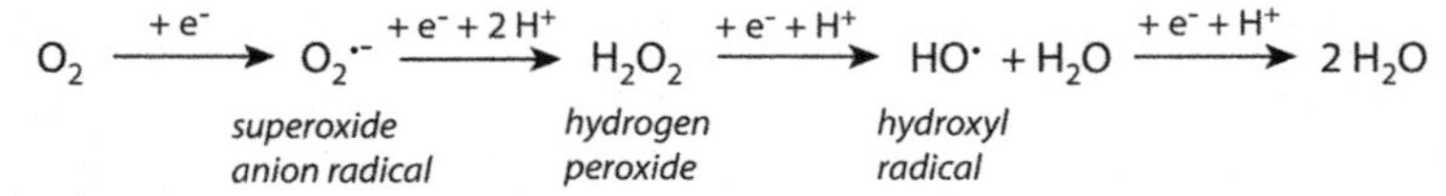

$$O_2 \xrightarrow{+e^-} O_2^{\bullet-} \xrightarrow{+e^- + 2H^+} H_2O_2 \xrightarrow{+e^- + H^+} HO^{\bullet} + H_2O \xrightarrow{+e^- + H^+} 2\,H_2O$$

superoxide anion radical — *hydrogen peroxide* — *hydroxyl radical*

FIGURE 2.2 Stepwise one-electron reductions of dioxygen to water. Formulas are given for those intermediate species that dominate at physiological pH value.

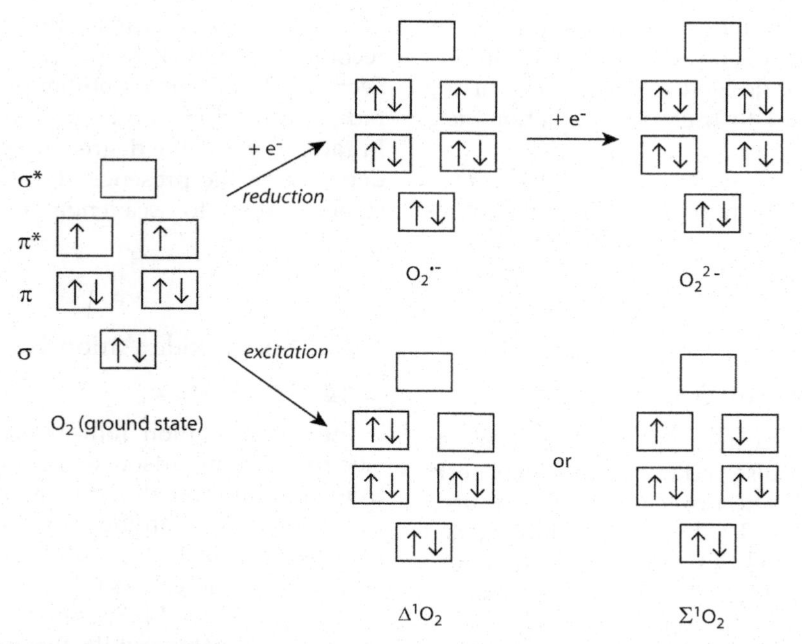

FIGURE 2.3 Configuration of 2p electrons in dioxygen (ground state), superoxide anion radical ($O_2^{\bullet -}$), hydrogen peroxide (represented as O_2^{2-}), and two forms of singlet oxygen (Δ^1O_2, Σ^1O_2). Further explanations are given in the text.

nonbonding orbital. The σ^* antibonding orbital is empty [3].

Electronic excitation of dioxygen yields singlet oxygen $^1O_2$ that exists in two forms as Δ^1O_2 and Σ^1O_2. Their electronic configurations are also illustrated in Fig. 2.3. In the dominating Δ form, both electrons of π^* antibonding orbitals are localized in one orbital. Nevertheless, this species is paramagnetic due to the resulting net orbital electronic angular momentum [4].

A further dioxygen-based product is ozone O_3 consisting of three oxygen atoms (see Section 2.2.8).

2.2.2 Ionizing Radiation

Ionizing radiation is a collective name for all kinds of radiations that carry enough energy to abstract electrons from atoms or molecules. Substrate ionization results either from high-energetic particles (e.g., alpha particles, beta particles, neutrons) or from high-energy electromagnetic waves (gamma rays, X-rays, short-wavelength ultraviolet illumination).

As a result of radiation, reactive ions, free radicals, and free electrons are formed, which in turn can induce further chemical destructive events. In living systems, radiolysis of water is of utmost importance for formation of reactive species such as solvated electron, superoxide anion radical, hydroxyl radical, hydrogen atom, oxygen atom, and hydrogen peroxide.

In our daily life, ionizing radiation is always present at very low level. Main sources of exposure to ionizing radiation are the uptake of radioactive material with food and drinks, inhalation of the radioactive noble gas radon, cosmological and terrestrial radiation, and application of medical procedures in radiology [5]. Higher doses of radiation can result from prolonged occupational exposure, frequent long-distance flights, improper use of radiation sources, accidents with radioactive material, disaster in nuclear reactors, and application of nuclear weapons. Enhanced and prolonged exposure to ionizing radiation can lead to mutation, radiation sickness, cancer, and death.

Otherwise, there exist powerful techniques with γ-rays, X-rays, radioisotopes, and protons to destroy cancer tissue by local induction of reactive species resulting from water radiolysis [6,7].

2.2.3 Radiolysis of Water

Radiolysis of water has been extensively investigated by pulse radiolysis and other techniques. On application of short electron pulses, solvated electrons, H_2O^+ cations, and excited water molecules (H_2O^*) result [8,9]. Main primary reactions of these species are summarized in Fig. 2.4. Water in an excited state can decompose to species such as hydroxyl radicals ($HO^\bullet$), hydrogen atoms ($H^\bullet$), and oxygen atoms ($O^\bullet$). The H_2O^+ cation itself can interact with a water molecule under formation of a hydrated proton (H_3O^+) and a hydroxyl radical. The solvated electron can react with H^+, yielding a hydrogen atom, or with O_2 to a superoxide anion radical.

The highly reactive hydroxyl radical can abstract an electron from any substrate (see Section 2.2.6). Further products result from radical recombination such as hydrogen peroxide (see Section 2.2.5). There is another profile of aqueous radiation products in the presence of dioxygen. In radiotherapy, the degree of tissue destruction depends on the presence of dioxygen and substances able to scavenge radical products [10–12].

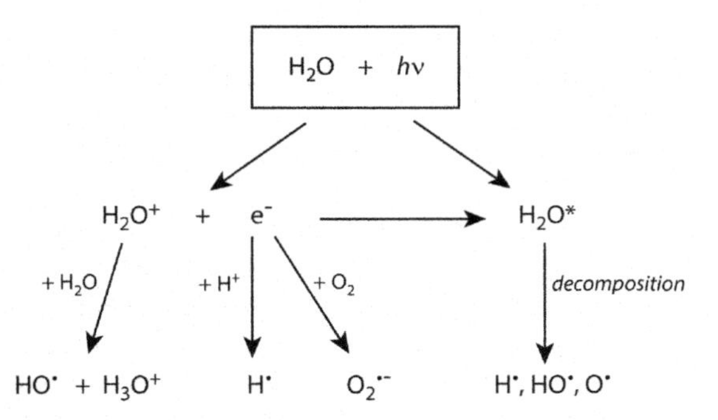

FIGURE 2.4 Main primary reactions and products of water radiolysis. On application of energy (*hν*) water either decomposes or becomes excited. Further explanations are given in the text.

2.2.4 Superoxide Anion Radicals

2.2.4.1 *Properties*

Superoxide anion radical has an additional electron in comparison with dioxygen. Its electron configuration allows orbital overlapping of the half-filled π^* antibonding orbital (Fig. 2.3) in electron transfer reactions between electron donators and acceptors [2].

At physiological pH, the anionic form of this species dominates, while the protonated form $HO_2^\bullet$, shortly known as superoxide, becomes evident at more acidic conditions. The pK_a value of $HO_2^\bullet$ is 4.81 [13].

2.2.4.2 *Formation*

In living systems, main route for formation of superoxide anion radical is the one-electron reduction of dioxygen.

$$O_2 + e^- \longrightarrow O_2^{\bullet-}$$

The standard reduction potential of the couple $O_2(g)/O_2^{\bullet-}$ is equal to −0.33 V at pH 7 [14]. Considering the maximum concentration of dissolved dioxygen of 1.17×10^{-3} mol L^{-1} [15,16] at 25°C, 101.3 kPa, and ionic strength of 150 mmol L^{-1}, a standard reduction potential of −0.16 V results for the couple $O_2(aq)/O_2^{\bullet-}$ at pH 7.

Therefore, only a few possibilities exist for redox processes to be linked with the one-electron reduction of dioxygen in living systems. Important ways for realization of this reduction are listed in Table 2.1.

TABLE 2.1 Routes for the Formation of Superoxide Anion Radical From Reduction of Dioxygen.

Linked Redox Reaction as Reduction Process	Occurrence	References
$NADP^+ + 2\,e^- + H^+ \rightarrow NADPH$	NADPH oxidase, neutrophils, macrophages, and other cells	[17,18]
Hypoxanthine + 2 e^- → xanthine Xanthine + 2 e^- → urate	Xanthine oxidase	[19,20]
Myoglobin-Fe(III) + O_2 + e^- → myoglobin-Fe(II)–O_2	Autoxidation of myoglobin, striated muscles	[21,22]
Hemoglobin-Fe(III) + O_2 + e^- → hemoglobin-Fe(II)–O_2	Autoxidation of hemoglobin, red blood cells	[21,22]
Adriamycin + e– → adriamycin•	Xenobiotica	[23]

NADPH oxidase is found to a large extent in neutrophils, eosinophils, monocytes, and macrophages. This enzyme consists of several components that assemble to the active enzyme on cell activation [24]. Then sufficient amount of dioxygen is reduced to superoxide anion radicals resulting in a burst of reactive species. Within subunits of NADPH oxidase, electrons are transferred from NADPH via flavin adenine dinucleotide (FAD) and cytochrome b_{558} to dioxygen. As a result, 2 moles O_2 are reduced by oxidation of 1 mole NADPH. The reduction potential of the couple $NADP^+/NADPH$ is −0.32 V at pH 7 [25]. In phagocytes, the catalytic subunit of NADPH oxidase is also known as NOX2.

Other locations of NADPH oxidase are cells of the blood vessel wall, respiratory tract, gastrointestinal tract, and thyroid glands [26–28]. These enzymes contain other NOX subunits than NOX2 or DUOX subunits. They are also able to reduce dioxygen, but to lesser degree than phagocytes.

Xanthine oxidase is mainly distributed in liver and small intestine. In other human organs, only trace amounts of xanthine oxidase are found [29]. Under disease conditions with injured liver and/or intestine, this enzyme can leak from these organs and show enhanced activity at other locations [29,30]. This enzyme is part of the purine catabolism. It catalyzes the electron transfer from hypoxanthine to xanthine and from xanthine to urate, which is coupled with the reduction of dioxygen to superoxide anion radical. In the active center, electrons are transferred thereby via molybdenum, FAD, and an iron–sulfur cluster, all having increasing values of their reduction potentials.

Several mitochondrial enzymes are able to reduce dioxygen to superoxide anion radical [31,32]. These mechanisms are described in detail in Section 4.5.3.

Transport of dioxygen is realized by binding to hemoglobin and myoglobin in red blood cells and striated muscle tissue, respectively. On binding of dioxygen to these proteins, a small portion of dioxygen is one-electronically reduced (Section 4.2.2).

Several semiquinone radicals (X•) of xenobiotics (X) promote the formation of superoxide anion radical from dioxygen, as their standard reduction potentials $E'^{o}(X/X^{\bullet})$ are slightly lower than those of the couple $O_2(aq)/O_2^{\bullet-}$. These semiquinone radicals can be formed by a cytochrome P450 driven process during detoxification reaction. This redox recycling allows generation of sufficient amounts of superoxide anion radicals [33,34]. Among them are adriamycin [23], alloxan [35], paraquat [36], menadione [14], daunomycin [23], epirubicin [37], and cisplatin [34].

A similar redox recycling with enhanced formation of $O_2^{\bullet-}$ is known for catecholic breakdown products of estrogens [38].

In principle, the one-electron reduction of hydrogen peroxide also results in formation of superoxide anion radical. For example, this reaction can occur on reaction of H_2O_2 with Compound I of myeloperoxidase (MPO) [39].

2.2.4.3 Reactions

Superoxide anion radicals undergo, in the presence of their protonated form, a spontaneous dismutation to hydrogen peroxide and dioxygen according to

$$O_2^{\bullet-} + HO_2^{\bullet} + H^+ \longrightarrow H_2O_2 + O_2 \text{ and}$$

$$HO_2^{\bullet} + HO_2^{\bullet} \longrightarrow H_2O_2 + O_2$$

A dismutation between two superoxide anion radicals is unlikely [40]. The highest rate of spontaneous dismutation of superoxide anion radicals is found with $3 \times 10^7\ M^{-1}s^{-1}$ at pH 4.81 [13], the pK_a value of superoxide. At slightly acidic, neutral, and alkaline pH values, the rate of this reaction decreases by one order of magnitude per unit increasing pH. This rate is $2 \times 10^5\ M^{-1}s^{-1}$ at pH 7.4 [13].

It was postulated that dioxygen is formed in the singlet state during spontaneous dismutation of superoxide anion radicals [41,42]. However, there are only scarce experimental evidences for formation of this species [43]. One reason for failure to detect singlet oxygen in this reaction is the efficient quenching of singlet oxygen by superoxide anion radicals [44,45].

Besides spontaneous dismutation, other reactions of superoxide are the interaction with iron–sulfur clusters in mitochondria (Section 4.5.4), the formation of peroxynitrite in its reaction with NO (see Section 2.3.2), and the formation of Compound III with heme peroxidases (Section 2.5.1).

2.2.4.4 Protection

Superoxide anion radicals are controlled by superoxide dismutases (SODs) and, in serum, by ceruloplasmin. These proteins enhance the rate of dismutation of superoxide anion radical by several orders of magnitude in neutral and slightly alkaline media.

SODs use the permanent redox switch between Cu^+/Cu^{2+}, Mn^{2+}/Mn^{3+}, Fe^{2+}/Fe^{3+}, or Ni^{2+}/Ni^{3+} at active site. In case of Cu–Zn dismutase, the reaction sequence is given by

$$\text{SOD-}Cu^{2+} + O_2^{\bullet-} \longrightarrow \text{SOD-}Cu^{+} + O_2$$

$$\text{SOD-}Cu^{+} + O_2^{\bullet-} + 2\,H^+ \longrightarrow \text{SOD-}Cu^{2+} + H_2O_2$$

In mammals, there are three types of SODs (Table 2.2). They differ in their fine structure at active site and preferred location.

These isoforms cover all main locations where superoxide anion radicals are formed during metabolic activity. Mitochondria contain two isoforms of SOD in different compartments. This is an adaptation to the preferred formation of $O_2^{\bullet-}$ in mitochondria and the low

TABLE 2.2 Characterization of Human Superoxide Dismutases (SODs).

Isoform	Active Site	Location	Distribution	References
SOD 1	Cu^+/Cu^{2+}, zinc	Cytoplasm, intermembrane space of mitochondria, nucleus	Ubiquitous	[46,47]
SOD 2	Mn^{2+}/Mn^{3+}	Mitochondrial matrix	Ubiquitous	[48]
SOD 3	Cu^+/Cu^{2+}, zinc	Extracellular space, attached to cell surfaces and extracellular matrix	Blood vessel wall, lungs	[49]

permeability of superoxide anion radicals through membranes.

Ceruloplasmin is a liver-derived multifunctional copper-containing serum protein that also exhibits a SOD activity [50]. Dismutation of superoxide anion radicals is realized by the extremely fast copper ion–based redox switch. In ceruloplasmin, six copper ions are tightly associated with the protein, and a further copper ion can be weakly bound [51].

2.2.5 Hydrogen Peroxide

2.2.5.1 *Properties*

Hydrogen peroxide is involved in both reduction and oxidation processes. Standard reduction potentials for these redox conversions are given in Table 2.3.

Hydrogen peroxide has a lower reactivity in comparison with superoxide anion radical. The reason is its electron configuration with full occupied π^* antibonding orbitals (Fig. 2.3). This largely restrains an overlap of its empty antibonding σ^* orbital between orbitals of electron donors and acceptors in redox reactions. However, an overlap between d orbitals of iron or copper and the empty antibonding σ^* orbital of H_2O_2 is possible [2]. Otherwise, H_2O_2 is able to oxidize selenocysteine residues of selected proteins and cysteine residues to a lesser degree [53]. Standard reduction potentials of the redox couples –Se-S–/–SeH, –SH and –Se-Se-/2 SeH are lower at pH 7 by 120 and 180 mV, respectively, in comparison to the corresponding value of the redox couple oxidized glutathione/reduced glutathione (GS-SG/2 GSH) [54].

Thus, redox conversions of H_2O_2 are mainly restricted to transition metal ions, complexes of these ions, and some proteins containing selenocysteine (or cysteines) at active site.

2.2.5.2 *Formation*

The main route of formation of hydrogen peroxide is the spontaneous and catalyzed dismutation of superoxide anion radicals (Section 2.2.4).

In peroxisomes, direct reduction of dioxygen to hydrogen peroxide is catalyzed by several acyl-CoA oxidases during β-oxidation of fatty acids [55]. Other peroxisomal flavoenzymes that directly produce hydrogen peroxide are D-aspartate oxidase, pipecolic oxidase, and L-alpha-hydroxy acid oxidase. Urate oxidase, which converts urate to allantoin and reduces thereby dioxygen to hydrogen peroxide, is found in rodents but not in humans [56,57]. The same holds for L-gulonolactone oxidase that also converts dioxygen to hydrogen peroxide and catalyzes the last step of ascorbic acid synthesis from glucose. This enzyme is deficient in humans, other primates, and guinea pigs, but it works well in mice and rats [58,59].

2.2.5.3 *Reactions*

Hydrogen peroxide can accumulate around locations where it is formed because it is less reactive. It can be utilized in reactions with transition metal ions under formation of highly reactive hydroxyl radicals (Section 2.2.6) and metal ion–based reactive species (Section 2.4.3).

Different heme proteins are also a target for H_2O_2. The resulting activated heme proteins

TABLE 2.3 Standard Reduction Potentials E'° of Redox Conversions Under Participation of Hydrogen Peroxide.

Half Reaction as Reduction Process	Redox Couple	E'° (at pH 7)	References
$H_2O_2 + 2\,H^+ + 2\,e^- \rightarrow 2\,H_2O$	$H_2O_2/2\,H_2O$	1.32 V	[40]
$H_2O_2 + H^+ + e^- \rightarrow HO^{\bullet} + H_2O$	$H_2O_2/HO^{\bullet}$, H_2O	0.32 V	[40]
$O_2^{\bullet-} + 2\,H^+ + e^- \rightarrow H_2O_2$	$O_2^{\bullet-}/H_2O_2$	0.94 V	[40]
$O_2 + 2\,H^+ + 2\,e^- \rightarrow H_2O_2$	$O_2(aq)/H_2O_2$	0.37 V	[52]

Catalase

Glutathione peroxidase

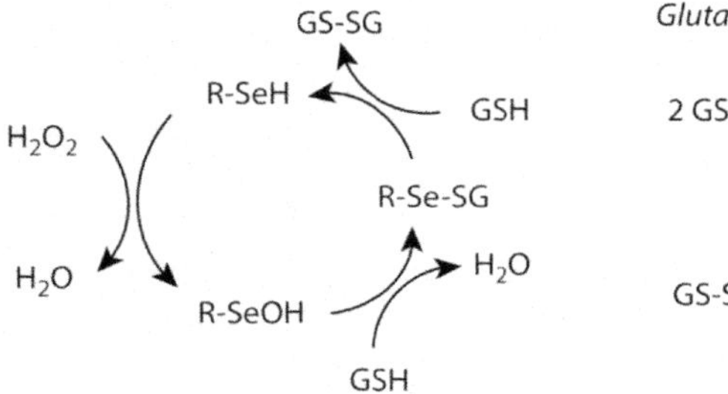

Peroxiredoxin

FIGURE 2.5 Removal of hydrogen peroxide by catalase, glutathione peroxidase, and peroxiredoxin. In catalase, Por denotes the porphyrin ring, Por-Fe^{3+} the ferric resting enzyme, and $^{\bullet+}$Por-Fe^{4+}=O Compound I. GSH and GS-SG are reduced and oxidized glutathione, respectively. PRDX stands for peroxiredoxin, and TRX for thioredoxin.

are involved in tissue degradation, substrate hydroxylation, and defense reactions against pathogens. Selected reactions of H_2O_2 with heme proteins concern reactions with catalase (Section 2.2.5.4), hemoglobin and myoglobin, cytochromes, and heme peroxidases (Section 2.5.1).

Further targets for H_2O_2 are glutathione peroxidases (GPX) with a selenocysteine or cysteine residue at active site [60].

2.2.5.4 *Protection*

In cells and tissues, control over hydrogen peroxide is realized by catalases, GPX, and peroxiredoxins (PRDX). Some isoforms of the two latter proteins are ubiquitously distributed in cells. They are responsible to remove comparable low levels of H_2O_2, peroxynitrite, and organic hydroperoxides. Mechanisms of removal of H_2O_2 by these proteins are given in Fig. 2.5.

Catalase is important in cells, where higher amounts of H_2O_2 can be produced. High yield of catalase is found in the red blood cells, kidney cells, and liver [61,62]. Within cells, catalase is mainly located, if present, in peroxisomes [55]. Catalase is one of the most efficient enzymes having an enormous V_{max} value. Thus, it can efficiently remove high load of hydrogen peroxide [63].

The heme enzyme catalase converts hydrogen peroxide into water and dioxygen by the following mechanism.

$$Por\text{-}Fe^{3+} + H_2O_2 \longrightarrow {}^{\bullet+}Por\text{-}Fe^{4+}{=}O + H_2O$$

$${}^{\bullet+}Por\text{-}Fe^{4+}{=}O + H_2O_2 \longrightarrow Por\text{-}Fe^{3+} + O_2 + H_2O$$

$$\text{in summary}: \quad 2\,H_2O_2 \rightarrow 2\,H_2O + O_2$$

During catalase-mediated H_2O_2 degradation, heme switches permanently between the ferric form (resting enzyme, $Por\text{-}Fe^{3+}$, Por denotes the porphyrin ring) and Compound I (${}^{\bullet+}Por\text{-}Fe^{4+}{=}O$) being an oxo-ferryl heme species with a radical function on the porphyrin ring [64].

In humans, eight different isoforms of GPX are known [60]. In most cells, the isoforms GPX1 and GPX4 are important for protection against H_2O_2, peroxynitrite, and organic hydroperoxides [65,66]. GPX4 has a high preference for lipid hydroperoxides. This isoform also acts as thiol peroxidase. GPX2 and GPX3 are secretory enzymes and act in the extracellular space and plasma. The other GPXs are involved in tissue-specific functions.

In human GPX1−GPX4, and GPX6, a selenocysteine group is present at active site. The selenol moiety of selenocysteine residue (R−SeH) is oxidized by H_2O_2 to a selenenic acid group (R−SeOH).

$$R - SeH + H_2O_2 \rightarrow R - SeOH + H_2O$$

This species interacts consecutively with two molecules of glutathione (GSH) under recovery of the selenol group and formation of oxidized glutathione (GS-SG).

$$R - SeOH + GSH \rightarrow R - Se - SG + H_2O$$

$$R - Se - SG + GSH \rightarrow R - SeH + GS - SG$$

The summary reaction of GPX catalysis is

$$2\,GSH + H_2O_2 \rightarrow GS - SG + H_2O$$

Glutathione is usually recovered from GS-SG by glutathione reductase according to

$$GS - SG + NaDPH + H^+ \rightarrow 2\,GSH + NADP^+$$

Six isoforms of human PRDX are known so far. In red blood cells, PRDX2 is one of the most abundant proteins [67]. PRDX2 and some other isoforms reduce H_2O_2 to water under oxidation of two neighbored cysteines to a disulfide residue. The redox cascade to convert oxidized PRDX back into the reduced form includes thioredoxin, thioredoxin reductase, and the NADPH/NADP$^+$ redox couple [68,69].

2.2.6 Hydroxyl Radicals

2.2.6.1 Properties

Hydroxyl radicals have the highest reactivity among reactive species. They easily abstract an H atom from any substrate (RH) under formation of a substrate radical ($R^\bullet$).

$$RH + HO^\bullet \longrightarrow R^\bullet + H_2O$$

In this process, the hydroxyl radical is reduced to water.

$$HO^\bullet + e^- + H^+ \longrightarrow H_2O$$

The standard reduction potential of the couple ($HO^\bullet/H_2O$) was determined to be 2.31 V at pH 7 [40]. This value is one of highest reduction potentials known in chemistry and the highest value for biologically relevant reductions. In consequence, a hydroxyl radical reacts very unspecific and nearly diffusion-controlled with most substrates in its direct proximity.

2.2.6.2 Formation

Besides radiolysis of water (Section 2.2.3), hydroxyl radicals can be formed as a result of the Fenton reaction, i.e. the transition metal ion−catalyzed breakdown of hydrogen peroxide.

$$H_2O_2 + Fe^{2+} \longrightarrow HO^\bullet + HO^- + Fe^{3+}$$

This reaction runs also with Cu^{+} that is oxidized to Cu^{2+} [70].

$$H_2O_2 + Cu^{+} \longrightarrow HO^{\bullet} + HO^{-} + Cu^{2+}$$

Although the Fenton reaction has been widely accepted, real nature of the powerful oxidizing product remains puzzling. Alternative reactive products of this reaction are excited iron–oxygen complexes such as ferryl or perferryl compounds [71–75].

Formation of HO• was also described in reactions of the MPO product hypochlorous acid (HOCl) with superoxide anion radical [76–78] or Fe^{2+} [79,80].

$$HOCl + O_2^{\bullet -} \longrightarrow HO^{\bullet} + O_2 + Cl^{-}$$

$$HOCl + Fe^{2+} \longrightarrow HO^{\bullet} + Cl^{-} + Fe^{3+}$$

2.2.6.3 *Reactions*

Two types of reactions of hydroxyl radicals are known. Besides the aforementioned abstraction of an H-atom from a substrate, $HO^{\bullet}$ can be added to an unsaturated system under formation of a hydroxylated product with a radical function [81]. Both reaction types lead to formation of substrate radicals that are involved in manifold further reactions.

2.2.6.4 *Protection*

Because of the nearly diffusion-controlled reaction rates of hydroxyl radicals, there are no enzymatic systems to scavenge this species. Therefore, main strategy of living matter to avoid any HO• formation is the efficient control of transition free metals. As copper and iron ions are involved as essential cofactor in a number of proteins, their transport, uptake by cells, intracellular storage, and metabolic processes are strictly regulated in the organism. Further details about the involvement of transition metal ions in damaging reactions are given in Section 2.4.

In some experiments, millimolar concentrations of carbohydrates, in particular mannitol, and others are applied as scavenger of $HO^{\bullet}$ [82]. Good scavengers of $HO^{\bullet}$ are able to bind transition metal ions.

2.2.7 Singlet Oxygen

2.2.7.1 *Properties*

Singlet oxygen is the electronically activated form of dioxygen. There are two types of singlet oxygen, (delta and sigma form) which differ not only in their activation energies but also in electronic configuration (Section 2.2.1). The energy levels of Δ^1O_2 and Σ^1O_2 are 94.2 kJ mol^{-1} and 157 kJ mol^{-1}, respectively, above ground level. As the sigma form decays easily into the delta form [83], most data about biological activity concern the delta form.

Despite the short lifetime of $^1O_2$ in water of about 4 μs [84,85], this time is quite sufficient for diffusion and reactions with different targets.

2.2.7.2 *Formation*

Absorption of ultraviolet or visible light can induce photooxidative processes in biological material. Electronically excited molecules either oxidize other substrate molecules or transfer their energy to dioxygen under formation of singlet oxygen. Flavines, quinones, porphyrins, and heme proteins play a prominent role as photosensitizer [86].

Production of singlet oxygen has also been reported for certain physiologically relevant reactions such as spontaneous dismutation of superoxide anion radicals [41,42], reaction between HOCl or HOBr and H_2O_2 [43,87], and as a result of reaction of peroxynitrite with H_2O_2 [88,89].

2.2.7.3 *Reactions*

Singlet oxygen reacts with aliphatic double bonds under formation of hydroperoxides. Endoperoxides result in case of reactions of $^1O_2$

with conjugated double bonds [90,91]. Double bonds with a rigid configuration or those lacking hydrogen atoms are transformed into dioxethanes by singlet oxygen [92,93].

DNA bases, especially guanine, are a preferred target for $^1O_2$ [94,95]. Among the main reaction products with guanine are 8-oxo-7,8-dihydro-2′-deoxyguanosine (8-oxodG), the 5-hydroperoxide of 8-oxodG, and the stable decomposition product oxaluric acid [95]. The latter species is found in single-stranded DNA on exposure to $^1O_2$ [96]. Oxidation of guanine may induce G to T transversions in DNA.

2.2.7.4 *Protection*

During its short lifetime, singlet oxygen can contribute to the aforementioned reactions. Physical quenching is important in deactivation of singlet oxygen. A nearly diffusion-controlled quenching of singlet oxygen was found for various carotenoids including lycopene, α-carotene, β-carotene, γ-carotene, astaxanthine, and others [89,97,98]. Interestingly, carotenes like vitamin A are located predominantly in the eye and skin in human organism, i.e. in those organs and tissues that are most of all exposed to light.

The polyamines spermine and spermidine are physical quenchers of $^1O_2$. It is assumed that they are able to prevent $^1O_2$-mediated DNA damage at millimolar concentrations under in vivo conditions [99,100].

2.2.8 Ozone

Ozone (O_3) is an inorganic, gaseous molecule consisting of three atoms of oxygen. It is formed from dioxygen on the action of ultraviolet light. In the presence of NO_2 and some other air pollutants, the formation of O_3 increases.

As a powerful oxidizing agent, ozone is cytotoxic against the respiratory system [101,102]. It reacts readily with double bonds under formation of ozonides and cytotoxic aldehydes [103]. Components of the lining fluid such as ascorbate, GSH, and urate are able to scavenge O_3 [104,105].

2.3 NITROGEN-BASED REACTIVE SPECIES

2.3.1 Nitrogen Monoxide

Reactive species of this group are derived from nitrogen monoxide (NO). In humans, nitrogen monoxide is produced by three different nitrogen monoxide synthases (Table 2.4). These homodimeric heme enzymes produce NO during conversion of L-arginine to citrulline. The reaction mechanism is rather complex and includes dioxygen as substrate and several cofactors such as NADPH, FAD, flavin mononucleotide (FMN), and (*6R*-)5,6,7,8-tetrahydrobiopterin [112].

Although NO produced by neuronal NOS plays a certain role in some neuronal functions,

TABLE 2.4 Isoforms of Nitrogen Monoxide Synthase in Humans.

Isoforms	Expression	Distribution	Functions	References
Neuronal NOS, nNOS, NOS1	Constitutive	Specific neurons of the brain	Synaptic plasticity, blood pressure regulation, atypical neurotransmitter	[106–108]
Inducible NOS, iNOS, NOS2	Inducible	Macrophages, microglia, and other cells	Probably antimicrobial effect	[109]
Endothelial NOS, eNOS, NOS3	Constitutive	Endothelial cells, cardiac myocytes, platelets, and others	Vasodilation, blood pressure regulation, inhibition of platelet functions, antiatherosclerotic principle	[110,111]

most attention is directed on endothelial NOS. Nitric monoxide derived by the latter isoform is a potent physiological vasodilator. In endothelium, NO diffuses to smooth muscle cells and induces formation of cyclic GMP by soluble guanylate cyclase. This signaling molecule is involved in blood pressure regulation and numerous essential cardiovascular functions [113,114].

In humans, the biological significance of inducible NOS remains unresolved [109]. Cultured human macrophages show little or no expression of iNOS [115]. Otherwise, macrophages from patients with infectious and inflammatory diseases express this isoform [116]. There is no expression of iNOS induced by lipopolysaccharide and interferon γ in human macrophages in contrast to the mouse system, where iNOS is upregulated by several orders of magnitude by these agents [59].

Nitrogen monoxide is a free radical species. In addition to its beneficial physiological functions, nitrogen monoxide can react with other radicals, heme proteins, and iron–sulfur clusters. Reactive nitrogen-based species produced as a result of these reactions contribute to cell and tissue damage.

In aqueous media, autoxidation of NO was also reported [117]. This slow reaction proceeds with a third order rate constant of $4k_{aq} = 8 \times 10^6\ M^{-2}s^{-1}$ [117]. This rate law is second order with respect to NO concentration.

2.3.2 Peroxynitrite

At inflammatory sites, $NO^{\bullet}$ interacts very rapidly with $O_2^{\bullet-}$ to yield to the powerful oxidant peroxynitrite $ONOO^-$.

$$O_2^{\bullet-} + NO^{\bullet} \longrightarrow ONOO^-$$

A nearly diffusion-controlled rate constant of $1.9 \times 10^{10}\ M^{-1}s^{-1}$ was determined for this reaction [118]. Other authors determined rate constants that are 3–4 times lower [119–121]. Under physiological conditions, peroxynitrite is in an acid–base equilibrium with peroxynitrous acid ONOOH. This weak acid has a pK_a value of 6.5 at mM phosphate [118]. However, its exact pKa value highly depends on medium composition [118]. Although $ONOO^-$ is a more or less stable molecule, ONOOH isomerizes spontaneously to nitrite with a rate of $1.3\ s^{-1}$ at 25°C [122].

Peroxynitrous acid is also assumed to decay homolytically into $NO_2^{\bullet}$ and $HO^{\bullet}$. However, this decay reaction is thermodynamically not feasible [123]. Careful analysis revealed the formation of $HO^{\bullet}$ when extraordinary high concentrations of total $ONOO^-/ONOOH$ (higher than 10^{-4} M) were applied. In this case, $HO^{\bullet}$ results from decay of an intermediary complex between the protonated and nonprotonated form of the reagent [118]. Hence, the $ONOO^-/ONOOH$-mediated formation of $HO^{\bullet}$ is unlikely in cells and tissues.

Peroxynitrite is a powerful oxidant. Standard reduction potentials of the couples $ONOO^-/NO_2^{\bullet}$, H_2O (one-electron reduction) and $ONOO^-/NO_2^-$, H_2O (two-electron reduction) were determined to be 1.6 and 1.3 V, respectively, at pH 7 [123]. Peroxynitrite mediates the formation of thiyl radicals from sulfhydryls [124], mediates nitration of tyrosine residues [125], and initiates lipid peroxidation reactions [126,127]. Another reaction of peroxynitrite is the degradation of hyaluronan [128]. In contrast, heparin and heparin sulfate are not affected by peroxynitrite, apparently due to electrostatic repulsion of peroxynitrite by sulfate residues [129].

Peroxynitrous acid also interacts easily with CO_2 under formation of nitrosoperoxycarbonate $ONOOCO_2^-$. The latter species decomposes into $CO_3^{\bullet-}$ and $NO_2^{\bullet}$ or into CO_2 and NO_3^- [130,131].

2.3.3 Nitrite and Nitrate

Nitrite and nitrate are produced by the aforementioned metabolic conversions of NO and peroxynitrite. Otherwise NO_2^- and NO_3^- are taken up with our food and absorbed in the gut. Moreover, some oral and intestinal bacteria are able to reduce nitrate to nitrite [132,133].

In healthy persons, blood levels of NO_2^- are reported to be 0.5–3.6 μM [134,135]. Higher values of nitrite can be found in respiratory tract lining fluids, in saliva, in gastric juice, and in serum under inflammatory conditions [136]. Serum NO_3^- value highly depends on diet and is around 20 μM in healthy persons [137].

Nitrite is taken up by red blood cells presumably via channel three proteins. It is oxidized by oxyhemoglobin to nitrate under formation of methemoglobin [138,139]. Intoxication by nitrite or nitrite-related compounds may lead to enhanced methemoglobin level [140]. In a similar reaction, oxymyoglobin is oxidized to metmyoglobin. The deoxygenated form of hemoglobin reduces nitrite to NO and thus cause vasodilation when O_2 level is falling [141].

The formation of toxic nitrosamines results from the reaction of nitrite with secondary amines [142,143]. This reaction is promoted during cooking or under strong acidic conditions in the stomach. In the latter case, the responsible species for this reaction is nitrous acid (HNO_2).

Nitrite and nitrate are also used as preservative for meat and other nutrients as they inhibit the growth of many bacteria [144].

At inflammatory sites, the enhanced formation of peroxynitrite and decay products is likely. The heme protein MPO that is released from activated neutrophils may act as a sink for peroxynitrite. This species converts ferric MPO directly to Compound II [145–147]. Among MPO-dependent reactions are the rapid one-electron oxidation of NO by Compounds I and II [148,149]. The resulting NO^+ is extremely unstable and rapidly hydrolyzes to NO_2^- [150].

2.3.4 Nitrogen Dioxide and Other Nitrogen-Based Reactive Species

Nitrite itself is also oxidized by activated MPO under formation of the nitrating species nitrogen dioxide ($NO_2^\bullet$) [151]. This species is also formed during one-electron reduction of peroxynitrite by ferric heme peroxidases [145], and during the decay of nitrosoperoxycarbonate [130,131].

Nitrogen dioxide is a strong one-electron oxidant. The reduction potential of the couple $NO_2^\bullet/NO_2^-$ is 1.04 V at pH 7 [152]. This species is involved in protein nitration [153,154] and nitrite-dependent lipid peroxidation [155–158].

Other nitrogen-based reactive species are nitryl chloride (NO_2Cl) [159], and peroxynitrate (O_2NOO^-) [160]. However, their physiological relevance remains unclear.

2.4 TRANSITION METAL ION-BASED SPECIES

2.4.1 Control Over Transition Metal Ions

Iron and copper ions function as essential components in a number of proteins. Copper is included in ceruloplasmin, Cu–Zn dismutase, cytochrome *c* oxidase, hephaestin, lysyl oxidase, tyrosinase and others [161]. Iron as a cofactor has a wider distribution than copper. Iron is of paramount importance for heme proteins, iron–sulfur clusters in mitochondria, lipoxygenases, and some other biological materials [162].

In proteins and cofactors, iron exists predominantly either in the ferrous (Fe^{2+}) or ferric (Fe^{3+}) form. In some proteins, the redox state of iron switches continuously. In activated heme proteins, iron can be also present in the oxoferryl state, where the iron valency is +4. In copper-containing proteins, the dominating redox states are the cuprous (Cu^+) and cupric (Cu^{2+}) ions.

All aspects of the metabolism of these ions are strictly controlled to avoid an increase of any free metal ions [163,164]. Intestinal absorption of iron is regulated by hepcidin [165]. In blood, iron is transported bound to transferrin [166]. The transferrin receptor is responsible for the iron uptake to cells [167]. Within cells, iron is stored in ferritin representing a hollow protein sphere with an iron mineral core [168].

Similar to iron, uptake, transport, and storage of copper are also strictly regulated by several import and efflux transporters and copper chaperones [169]. Disturbances in copper metabolism can lead to Wilson disease [170].

Thus, the concentration of free copper and iron ions is extremely low in living systems. Hard data are very scarce about the involvement of these ions in cell and tissue destruction. There are numerous speculations and assumptions above the species involved.

2.4.2 Labile Iron Pool

In cells and biological fluids, free iron is weakly chelated by low molecular weight substrates and forms the so-called "labile iron pool." Among these iron binding molecules are citrate, phosphate, nucleotides, nucleosides, carbohydrates, carboxylates, functional residues of proteins, and lipids. Thus, this pool of redox active iron is scarcely defined [171,172]. Under stress situations, iron concentration in this pool is enhanced and related to oxidation of biological material [173,174]. This pool can temporary increase on release of iron from iron–sulfur clusters, as a result of enhanced necrosis of cells, on destruction of free heme, during attack of hemolyzing microbes on red blood cells, and under septic conditions. Increased values for transition metal ions are also possible in inflamed extracellular fluids.

The release of iron from free hemoglobin or free myoglobin is addressed in Sections 4.2.6 and 4.3.2.

2.4.3 Iron- and Copper-Mediated Oxidative Processes

Among reactive species, the group of transition metal ion–based species is less specified. This concerns both the real nature of active species and gaps in knowledge of fine mechanisms of their action. Moreover, little is known about copper-mediated processes. Most data are given for redox processes with iron ions. In summary, there are numerous unjustified assumptions and speculations in redox chemistry of copper and iron.

Most references about iron- and copper-mediated damage of biological material are related to the Fenton reaction. Of course, Cu^+ and Fe^{2+} are oxidized by hydrogen peroxide and by organic hydroperoxides. As mentioned in Section 2.2.6, the fine mechanism and the nature of the powerful oxidant (hydroxyl radical vs. ferryl/perferryl species) formed in the reaction with H_2O_2 is a matter of debate [71–75]. The significance of the Fenton reaction is overstressed in a way that it seems that no other reactions of free iron and copper ions contribute to tissue damage. This is not the case.

Hydroxylation of aromatic molecules was observed by weakly phosphate of bicarbonate bound Fe^{2+} even in the absence of hydrogen peroxide [175,176]. These ions induce also strand breaks in DNA plasmids without participation of H_2O_2 [177]. It has been assumed that activated Fe^{2+}-dioxygen and Fe^{2+}-dioxygen-Fe^{3+} complexes are responsible for oxidation of biological material [33,178,179]. Other authors also favor this pathway in iron- and copper-mediated tissue degradation [180–182]. Apparently, autoxidation of Fe^{2+} is accelerated by biological buffer components most of all by phosphate [177,181]. A second order rate constant of $1.770\ M^{-1}s^{-1}$ was calculated for chelation of Fe^{2+} by phosphate and/or bicarbonate [183]. Taken together, these data support an involvement of transition metal ions in tissue destruction beyond Fenton chemistry.

Despite uncertainties of reactive species and fine mechanisms, free copper and iron ions are involved in numerous tissue-damaging reactions. There are no specific targets for their action. Fragmentation of nucleic acids, carbohydrates, components of extracellular matrix, and oxidation of lipids and proteins belong to reactions catalyzed by these ions. These reactions are addressed in Chapter 3.

A third mechanism of iron-mediated destruction is associated with heme metabolism and heme degradation. The heme group is an essential component in numerous proteins involved in transport of dioxygen, electron transfer reactions, and immune defense. Heme proteins are usually located at defined loci within cells. Free heme (ferric protoporphyrin IX) is highly cytotoxic notably to the kidney, liver, and spleen, causes hemolysis of unperturbed red blood cells, incorporates into biological membranes and lipoproteins, and activates epithelial cells on binding to toll-like receptor IV (see Section 4.4.2). Apparently, the sole presence of free heme in biological fluids is sufficient for its tremendous activities.

2.5 (PSEUDO)HALOGEN-BASED REACTIVE SPECIES

2.5.1 Heme Peroxidases: Halogenation and Peroxidase Cycles

In higher animals and humans, the oxidation of halides and thiocyanate is catalyzed by MPO, eosinophil peroxidase (EPO), and lactoperoxidase (LPO). In these heme peroxidases, the heme is covalently linked to the apoprotein giving them a slightly bow-shaped form and extraordinary redox properties [184–187].

MPO is present in azurophilic granules of neutrophils and to a lesser extent in monocytes [188]. EPO is a component of eosinophils [189], whereas LPO is found in secretions such as tears, milk, saliva, and upper airway's fluids [190]. These enzymes and their products are involved in key immune defense reactions. Otherwise, heme peroxidase–catalyzed reactions can also contribute to damage of host's tissues at inflammatory sites.

A scheme of main reaction sequences of heme peroxidases is given in Fig. 2.6. These enzymes utilize hydrogen peroxide to convert the ferric heme into so-called Compound I, an oxo-ferryl iron species with a second oxidizing equivalent on the porphyrin ring [191]. Compound I is reduced to ferric enzyme by abstracting two electrons from a (pseudo)halide, which is oxidized to the corresponding (pseudo)hypohalous acid. Reactions 1 and 2 comprise the halogenation cycle in Fig. 2.6. All three enzymes oxidize thiocyanate (^{-}SCN) to hypothiocyanite ($^{-}OSCN$). The oxidation of bromide to hypobromous acid (HOBr) is possible by MPO and EPO, whereas chloride is only converted by MPO to hypochlorous acid (HOCl). Human heme peroxidases are also able to oxidize iodide under formation of hypoiodous acid (HOI) that decays to an array of oxidized iodine species.

The formation of HOCl and HOBr by MPO and EPO is thermodynamically feasible only in a certain pH range [192]. Standard reduction potentials for the two-electron oxidation of (pseudo)halides and pK_a values of the resulting (pseudo)halous acids are given in Table 2.5.

Compound I of heme peroxidases can also be reduced to the ferric enzyme form by two consecutive one-electron steps via formation of Compound II having a ferryl iron–oxygen moiety in the heme center. In these reactions, an electron is abstracted from different small substrate molecules under formation of a substrate radical. Among substrates are sulfhydryls, tyrosine, tryptophan, phenol and indole derivatives, nitrite, hydrogen peroxide, flavonoids, and others [39,151,202–204]. In Compound I of LPO, the radical moiety is spontaneously shifted from porphyrin ring to apoprotein in the absence of substrates [205].

Reactions 1, 3, and 4 (Fig. 2.6) comprise the peroxidase cycle. In Table 2.6, the known standard reduction potentials are given for all redox couples of human heme peroxidases involved in the peroxidase cycle.

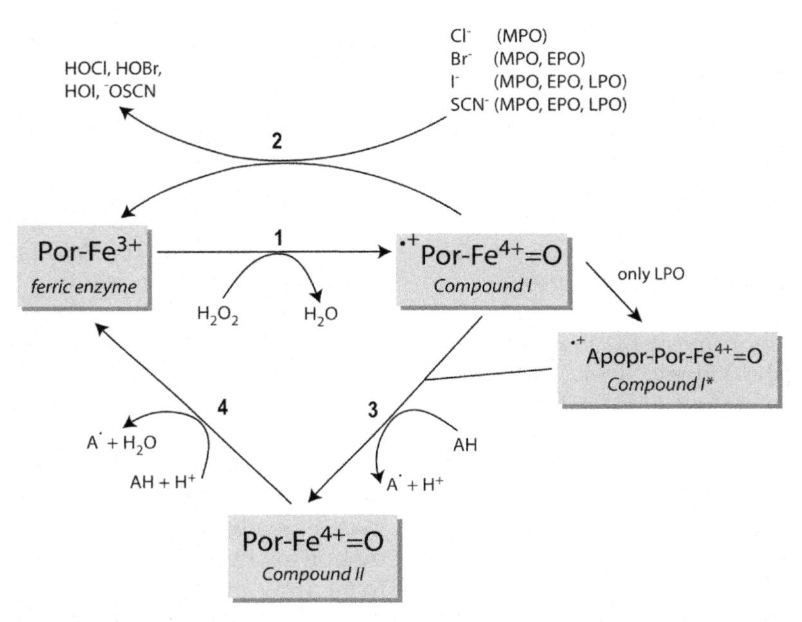

FIGURE 2.6 Halogenation and peroxidase cycles of human heme peroxidases (MPO, myeloperoxidase; EPO, eosinophil peroxidase; LPO, lactoperoxidase). Reactions 1 and 2 comprise the halogenation cycle, and reactions 1, 3, and 4 the peroxidase cycle. Por denotes the porphyrin ring, and Apopr the apoprotein. AH is an oxidizable substrate, and $A^{\bullet}$ the resulting substrate radical. Further explanations are given in the text.

TABLE 2.5 Standard Reduction Potentials of (Pseudo)halous Acids.

Half Reaction	$E'^{o}(HOX/X^-, H_2O)^a$, pH 7	pK_a Value (HOX)
$HOCl + H^+ + 2e^- \rightarrow Cl^- + H_2O$	1.28 V [193]	7.53 [194]
$HOBr + H^+ + 2e^- \rightarrow Br^- + H_2O$	1.13 V [195]	8.8 [196]
$HOI + H^+ + 2e^- \rightarrow I^- + H_2O$	0.78 V [197]	10.0 [198]
$^-OSCN + 2\,H^+ + 2e^- \rightarrow SCN^- + H_2O$	0.56 V [199]	5.3 [200], 4,85 [201]

[a] *X denotes Cl, Br, I, or SCN.*

TABLE 2.6 Standard Reduction Potentials (E'^{o}) of Redox Couples in Heme Peroxidases.

	E'^{o} (at pH 7)		
	Compound I/Ferric Enzmye	**Compound I/Compound II**	**Compound II/Ferric Enzyme**
Myeloperoxidase	1.16 V [206]	1.35 V [207]	0.97 V [207]
Eosinophil peroxidase	1.10 V [206]	n.d.[a]	n.d.
Lactoperoxidase	1.09 V [208]	1.14 V [208]	1.04 V [208]

[a] *not determined.*

A further state of heme peroxidases is Compound III that results from reaction of superoxide anion radical with the ferric enzyme [209].

2.5.2 Heme Peroxidases: Modulation of (Pseudo)Halogenating Activity

Halides and thiocyanate compete with different one-electron substrates including hydrogen peroxide for Compound I of human heme peroxidases directing the reaction route in either halogenation or peroxidase cycle. In the peroxidase cycle, often an arrest at Compound II occurs as the interconversion of Compound II to the ferric enzyme is the rate-limiting reaction in this cycle and only a few substrates react fast enough with Compound II.

The following substrates are known to enhance the chlorinating activity of MPO by overcoming the arrest of inactive Compound II: ascorbic acid [210], urate [211], tyrosine [212], superoxide anion radicals [213], acetaminophen [214], 5-aminosalicylic acid [215], and (−)-epicatechin [216,217]. The by far highest rate with MPO Compound II at pH 7 was determined for (−)-epicatechin with $4.5 \times 10^6 \, M^{-1}s^{-1}$ [218] followed by serotonin [219].

In a similar way, the hypothiocyanite production by LPO is enhanced in the presence of substrates reacting rapidly with LPO Compound II. A strong LPO-reactivating effect was found for the standardized olive leaf dry extract (Ph. Eur.) and for main components of this extract such as oleuropein, oleacein, hydroxytyrosol, caffeic acid, and luteolin [220]. These components contain a 3,4-dihydroxyphenyl moiety. Another plant extract, the motherwort (*Leonurus cardiaca* L.) extract, also increased the thiocyanate oxidation by LPO. In this extract, the active components such as caffeic acid derivatives and phenylethanoids also have a 3,4-dihydroxyphenyl partial structure [221]. The importance of this structural element in selected aromatic substrates, flavonoids, and tannins was further evaluated in enhancing hypothiocyanite production by LPO [222–224] and evidenced by docking studies and molecular orbital analysis [223,224]. Polyphenols also support the reactivation of peroxidase activity in human saliva [225].

2.5.3 Hypochlorous Acid

At neutral pH values, only MPO is able to produce HOCl. The chlorinating activity of this enzyme increases with more acidic pH values [192,226]. The formation of taurine chloramine by MPO was reported at pH 7.4 [214,227], while other authors failed to detect HOCl production at this pH in the absence of taurine [192,228,229]. Taurine directly interacts with an intermediary high-spin complex between Cl^- and Compound I under formation of taurine chloramine and resting MPO [230,231].

Formation of HOCl by MPO is also highly affected by the presence of SCN^- and Br^-. Although less abundant in serum, the latter ions react more efficiently than Cl^- with Compound I of MPO [186]. On the basis of specificity constants, it was concluded that the oxidation of 100 mM Cl^- and 100 μM SCN^- by compound I of MPO produces nearly equal amounts of HOCl and hypothiocyanite [232]. In serum, SCN^- concentration varies from 10 to 120 μM [233,234]. Much higher SCN^- values are found in secretions, especially in saliva with 0.5–4 mM [235–237]. Bromide concentrations are reported to be 20–100 μM in blood [238]. At neutral pH values, the MPO-H_2O_2 system is more prone to oxidize Br^- than Cl^- even in the presence of a mixture of both ions at physiological relevant concentrations [239,240]. Thermodynamic considerations revealed the existence of a pH threshold value in the neutral pH region above them and the formation of HOCl by the MPO–H_2O_2–Cl^- system is unlikely [192,199].

Moreover, HOCl reacts with both Br^- and SCN^- to yield HOBr or ^-OSCN [241,242].

$$HOCl + Br^- \rightarrow Cl^- + HOBr$$

$$HOCl + SCN^- \rightarrow Cl^- + H^+ + {}^-OSCN$$

Taken together, chlorination of substrates by MPO dominates at acidic pH values. In the neutral range, chlorination is only expected having low concentrations of SCN^- and Br^- in the medium.

Numerous reactions of the reagent HOCl with different substrates are reported at neutral pH values [243,244]. HOCl reacts efficiently with methionine and cysteine residues of proteins. Other targets of HOCl on proteins are cystine, histidine, tryptophan, lysine, and α-amino groups.

2.5.4 Hypobromous Acid

Similar to HOCl, hypobromous acid reacts also with a wide variety of targets [245]. For most protein residues, HOBr reacts 30–100-fold faster than HOCl. Cysteine and methionine residues are oxidized by HOBr about 10 times less efficient than by HOCl. In contrast, ring halogenation of tyrosine occurs approximately 5000-fold faster with HOBr compared with HOCl.

In bronchoalveolar lavage fluid from patients with asthma, the level of brominated tyrosine residues increases 35-fold after allergen challenge in contrast to a control group. Additionally, the eosinophil count raises more than 150-fold in these patients [246].

2.5.5 Hypothiocyanite

Thiocyanate is oxidized by MPO, EPO, and LPO to hypothiocyanite (^-OSCN) being in equilibrium with the membrane permeable form hypothiocyanous acid (HOSCN). The pK_a of HOSCN is 5.3 [200] or 4.85 [201]. Most production of ^-OSCN is expected in secretions due to the presence of LPO and high micromolar or even low millimolar concentrations of SCN^- [247,248]. As a weak oxidant, ^-OSCN/HOSCN attacks mainly intracellular free and protein-bound sulfhydryls and selenium-containing amino acid residues [249,250]. These properties allow the penetration of HOSCN into microorganisms of biofilms and to disturb specifically the redox metabolism in these cells [251,252].

Further oxidation products of SCN^- are under discussion to be formed by heme peroxidases. Among them are thiocyanogen ($(SCN)_2$) and thiocarbamate-(*S*)-oxide [253]. Hypothiocyanite is known to decay into sulfate, hydrocyanic acid, and cyanate with cyanosulforous acid (HO_2SCN) and cyanosulfuric acid (HO_3SCN) as transient intermediates [254–256].

2.5.6 Oxidized Iodine Species

Iodide oxidation by heme peroxidases leads to hypoiodous acid (HOI), iodine I_2, I_3^-, and some other reactive iodine species [257,258]. There are numerous interconversions between oxidized iodine species in dependence on pH and iodide concentration [259]. For example, hypoiodous acid dominates only at pH 8–9 [259]. Because of the low abundance of iodide in biological fluids (about 0.1 μM [260]), iodide oxidation by these enzymes is unlikely under in vivo conditions. However, iodide accumulates in the thyroid glands where, as part of thyroid hormone synthesis, it is oxidized and incorporated into tyrosine residues of thyroglobulin by thyroid peroxidase [261].

2.5.7 Inter(Pseudo)Halogens

The interhalogen bromine chloride (BrCl) was postulated as a potential product of halogenating activity of MPO [262]. The formation of cyanogen iodide (ICN) was shown by the LPO-hydrogen peroxide system in the presence of SCN^- and I^- when I^- is applied in excess over SCN^- [263,264]. Other related inter(pseudo) halogens are cyanogen chloride (ClCN) and cyanogen bromide (BrCN). They are formed on reaction of HOCl (or HOBr) with cyanide [265,266] or HOCl with cyanocobalamin [267].

2.6 REACTIVE SPECIES IN CELL AND TISSUE DESTRUCTION

2.6.1 Low- and High-Reactive Species

In living systems, reactive species are produced as a result of numerous external impacts, physiological processes, immune reactions, and under disease conditions. Primary formed reactive species are $O_2^{\bullet-}$, H_2O_2, $NO^{\bullet}$, and $^1O_2$. All other reactive species are derived from them by consecutive reactions of these species either by direct interaction or under catalysis by enzymes or transition metal ion–based reactive species.

Once generated, superoxide anion radicals trigger the formation of hydrogen peroxide as a product of spontaneous and catalyzed dismutation of $O_2^{\bullet-}$. Thus, both species appear almost concurrently in activated cells and inflamed loci. Hydrogen peroxide has a longer lifetime than superoxide anion radical due to its restricted reactivity. Both species are characterized by a comparable low potential for tissue damage. The same holds for $NO^{\bullet}$. However, high-reactive species such as $HO^{\bullet}$, HOCl, HOBr, $ONOO^-$, and $NO_2^{\bullet}$ are derived from reactions, where $O_2^{\bullet-}$ and H_2O_2 participate. Thus, an efficient control over the latter species diminishes deleterious effects by high-reactive species.

In deleterious processes of cell and tissue damage, transition metal ion–based reactive species also play a major role. In contrast to reactive species of other groups, reactions of iron- and copper-based reactive species are less well-characterized, mainly because several iron–oxygen complexes are badly defined. For example, the real nature of the catalyst in Fenton reaction is unknown till now [71–75]. Thus, numerous speculations about iron- and copper-based reactive species exist. To avoid destructive processes under participation of transition metal ion–based reactive species, iron and copper metabolism is strictly controlled in cells and tissues. Moreover, iron and copper ions are tightly sequestered by transport and storage proteins.

Serious damage by highly reactive species occurs only under extensive stress, in inflamed loci, and in association with some disease scenarios. Under these conditions, protective mechanisms are either exhausted or there is an uncontrolled increase in formation of reactive species and catalytically active transition metal ions.

2.6.2 Redox Recycling

In some cases, production of $O_2^{\bullet-}$ from dioxygen is greatly amplified by redox recycling, when the corresponding reductant is present in great excess or renewed permanently by a linked redox process. For example, NADPH serves as reductant for production of $O_2^{\bullet-}$ by NADPH oxidase in activated neutrophils and other cells [17,18]. Cytoplasmic level of NADPH is efficiently maintained by glucose-6-phosphate dehydrogenase and 6-phosphogluconate dehydrogenase, two enzymes of the oxidative pentose phosphate pathway of glucose [268].

Sufficient amounts of $O_2^{\bullet-}$ can also be formed from semiquinonic xenobiotics [14,23,33–37] or estrogen breakdown products [38,269] by redox recycling due to linkage with cytochrome P450 reductase–driven reactions. These mechanisms are involved in intoxication by xenobiotics or in estrogen-driven breast cancer pathology. Cytochrome P450 reductase is usually responsible for electron transfer from NADPH to cytochrome P450, a large group of proteins inserting as monooxygenase an oxygen atom into organic substrates [270]. In cytochrome P450 reductase, electron transfer occurs from NADPH via FAD and FMN to cytochrome P450 heme [271,272]. FMN subunit of cytochrome P450 reductase can directly interact with quinone intermediates of xenobiotics or estrogen breakdown products.

Selected examples for redox recycling of $O_2^{\bullet-}$ by NADPH oxidase and cytochrome P450 reductase–driven reactions are illustrated in Fig. 2.7.

Transition metal ions also participate in redox recycling. An often-cited example is the iron ion-catalyzed Haber–Weiss reaction [273,274]. The Fenton reaction is part of this sequence.

$$Fe^{3+} + O_2^{\bullet-} \longrightarrow Fe^{2+} + O_2$$

$$Fe^{2+} + H_2O_2 \longrightarrow Fe^{3+} + HO^{\bullet} + HO^-$$

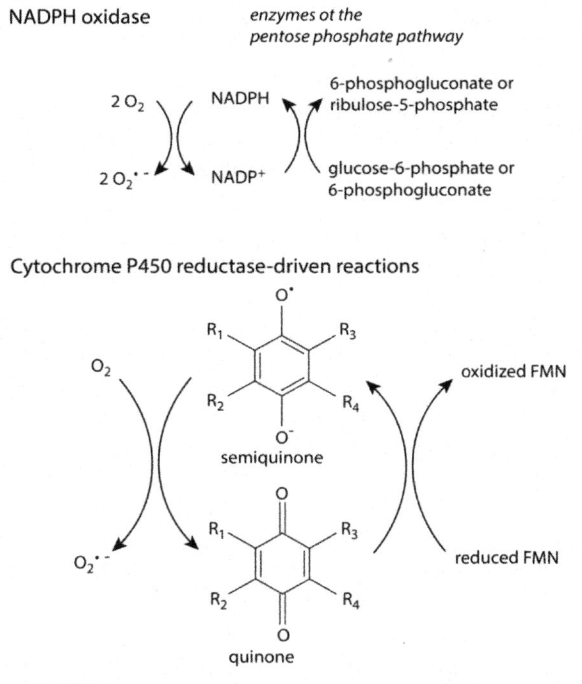

FIGURE 2.7 Redox recycling of the formation of superoxide anion radicals from dioxygen by NADPH oxidase in combination with enzymes of the pentose phosphate pathway or by cytochrome p450 reductase-driven processes. From the whole electron transfer machinery of cytochrome P450 reductase, only the terminal step, the oxidation of reduced FMN, is shown. R_1, R_2, R_3, and R_4 are functional residues of xenobiotics or estrogen breakdown products.

In summary, this reaction can shortly be written as

$$H_2O_2 + O_2^{\bullet-} \longrightarrow HO^{\bullet} + HO^- + O_2$$

and is known as Haber–Weiss reaction [275]. In aqueous solutions, the direct reaction between $O_2^{\bullet-}$ and H_2O_2 is, however, unlikely as its rate is close to zero [276].

Redox recycling of transition metal ions is also possible under involvement of small redox sensitive molecules such as ascorbic acid and others [277].

2.6.3 Beneficial and Harmful Actions of Reactive Species

Although the main focus of this book is directed on processes of cell and tissue destruction, we do not forget that reactive species are involved in humans and other organisms in numerous beneficial functions. Most of all neutrophils, eosinophils, and other immune cells apply reactive species to combat against invading microorganisms. Usually, these mechanisms are well regulated to minimize any destructions of host's own tissue. Activation of neutrophils and other immune cells is accompanied by a massive production of superoxide anion radicals and hydrogen peroxide and highly reactive species derived from subsequent reactions. These processes of cell activation are part of immune response against different agents (Chapter 6). In severe inflammations, these reactive species can cause cell and tissue damage.

Mitochondria and peroxisomes are a further source of reactive species in cells. Normally, these organelles produce a low amount of reactive species that are efficiently inactivated by protecting systems. Under stress situations, however, their production increases and contributes now to cell alterations (see Section 4.5).

Local ischemic events followed by reperfusion are another source for enhanced production of reactive species. During short-term ischemia, some products accumulate at these loci, which give rise to toxic metabolites at the beginning of reperfusion phase due to enhanced dioxygen supply (Section 4.6.5).

As long as the exposure to tissue by reactive species is enhanced but not too strong, cells will adapt to the altered redox environment by enhanced transcription of antioxidant proteins and activating antiinflammatory pathways. Under inflammatory conditions, the transcription factors NF-κB and activator protein 1 (AP-1) are activated and contribute to transcription of multiple genes that promote the further course of inflammation [278,279]. Among them are genes that encode proteins for recruitment of novel immune cells and for enhancing protective mechanisms. Important antioxidant genes with NF-κB responsive elements in their promoter regions are genes for HIF-1α, MnSOD, ferritin heavy chain, thioredoxins 1 and 2 (Trx1, Trx2), glutathione S-transferase, metallothionein, and acute-phase proteins [280].

Numerous genes of antioxidant proteins are regulated by the redox-sensitive transcription factor nuclear factor erythroid 2-related factor 2 (Nrf2), which forms an inhibitory complex with Kelch-like ECH-associated protein (Keap1) and the ubiquitin ligase cullin- 3 in unperturbed cells [281–283]. Under stress conditions, Nrf2 dissociates from this complex and activates antioxidant genes by binding to the antioxidant response element in the promoter region of these genes [284,285]. Examples for these genes are genes for heme oxygenase 1, NADPH quinone oxidoreductase, glutamate–cysteine ligase, glutathione S-transferase, and thioredoxin reductase 1.

Higher exposure of cells and tissues to reactive species disturbs the cell integrity and induces apoptotic cell death or acts necrotic. It is hardly to say under which conditions an apoptotic mechanism becomes necrotic. More about mechanisms on how cells die is given in Chapter 5. Here, I will mention a problem that is usually overlooked in experiments with isolated cells. Often experiments are performed and conclusions are drawn under conditions of well oxygenation, whereas under real in vivo conditions, oxygenation of cells and tissues is lower [286,287]. Normally cells and tissues are well adapted to survive low levels of stress including slightly diminished levels of dioxygen without significant damage.

Enhanced levels of reactive species are mostly accompanied by other alterations in activated cells and inflamed tissues. The release of bioactive molecules, cytokines, hydrolases, damage-associated molecular patterns, and many others contribute to tissue damage at these loci. Of course, highly reactive species are important for damaging reactions. But their activity should be always regarded in the whole context with other mechanisms of destruction. This is a great challenge to all scientists in this field.

2.7 INSTEAD OF A SUMMARY: REACTIVE SPECIES VERSUS GENERAL PROTECTIVE MECHANISMS

Most protective systems against reactive species are already present in cells and can immediately act. Some others will be additionally induced during inflammatory response and stress-mediated adaptations. In general, protective mechanisms are designed to control potential deleterious effects of reactive species. They are directed to avoid the formation of reactive species, to deactivate them, or to form inactive complexes with them.

Protection against reactive species is embedded in the whole arsenal of strategies of our organisms to combat with external and internal impacts. Injured cells are prone to apoptosis or other kinds of programmed cell death or become necrotic. Macrophages regularly eliminate any undergoing organic material such as apoptotic bodies, cell debris, and necrotic tissues. Last but not least, induction of proliferative processes can synthesize novel cells and restore tissues. Taken together, our organism disposes of several stages of protection against destructive reactions to ensure and reestablish homeostatic conditions in cells and tissues. More details about the involvement of physiological and immunological systems to manage with cell and tissue damage and their implication in deleterious activities are given in Chapters 5–7.

Nevertheless, the question arises on which kind of reactive species is mainly responsible for tissue destruction. It is impossible to give an adequate answer to this warrantable question. First of all, in case of severe destructions, usually an array of reactive species is involved. Because of numerous relationships between different reactive species, the coexistence of low and highly reactive species, and the presence of processes of redox recycling, it is often hard to capture which kinds of reactive species did really participate in vivo in cell and tissue destruction.

Secondly, reactive species do seldom act alone in injuring biological material. They act together with proteolytic enzymes and if present with toxic products of microorganisms. Numerous antibodies including autoantibodies can also participate in the whole process of immune activation and subsequent tissue destruction. In many cases, an inflammatory response is induced with massive infiltration of immune cells. Efficiency of immune response depends not only on the severity of the impairing impact but also on the personal immune status.

Thirdly, yes some impacts exist, where an enhanced formation of reactive species is evident as driving source of tissue destruction. Examples are processes of water radiolysis and breakdown of organic molecules due to exposure to X-rays, radioactive radiation, and ionizing particles. Illumination with UV light with short wavelengths contributes to enhanced formation of singlet oxygen and induces several photo processes.

Fourthly, the formation of reactive species can also be home-made. This is for example the case in pathological conditions associated with massive intravascular hemolysis and/or rhabdomyolysis. After exhaustion of naturally protecting systems, free hemoglobin, free myoglobin, and in particular free heme act as highly cytotoxic agents (Sections 4.4.1 and 4.4.2).

Under normal health conditions, our organism is well protected against many impacts of low and middle intensity by well-organized protective systems. Very seldom occurring strong incidences can be problematic as protective mechanisms are limited and become exhausted. With increasing age and under disease conditions, the capacity of protection is considerably diminished. Then low intense impacts can be very dangerous.

References

[1] B. Halliwell, J.M.C. Gutteridge, Free Radicals in Biology and Medicine, fifth ed., Oxford Press, 2015.

[2] W.H. Koppenol, J. Butler, Mechanisms of reactions involving singlet oxygen and the superoxide anion, FEBS Lett. 83 (1977) 1–6.

[3] K. Maeda, M.L. Wall, L.D. Carr, Hyperfine structure of the hydroxyl free radical (OH) in electric and magnetic fields, New J. Phys. 17 (2015) 045014, https://doi.org/10.1088/1367-2630/17/4/045014.

[4] M. Ruzzi, E. Sartori, A. Moscatelli, I.V. Khudyakov, N.J. Turro, Time-resolved EPR study of singlet oxygen in the gas phase, J. Phys. Chem. 117 (2013) 5232–5240.

[5] Sources and effects of ionizing radiation, United Nations Scientific Committee on the Effects of Atomic Radiation UNSCEAR 2008. Report to the General Assembly with Scientific Annexes, vol. I, United Nations, New York, 2010.

[6] R. Baskar, J. Dei, N. Wenlong, R. Yeo, K.-W. Yeoh, Biological response of cancer cells to radiation treatment, Front. Mol. Biosci. 1 (2014), Article 24, 1–9, https://doi.org/10.3389/fmolb.2014.00024.

[7] A.J. Privett, E.S. Teixeira, C. Stopera, J.A. Morales, Exploring water radiolysis in proton cancer therapy: time-dependent, non-adiabatic simulations of $H^+ + (H_2O)_{1-6}$, PLoS One 12 (2017), e0174456, 1–22, https://doi.org/10.1371/journal.pone.0174456.

[8] P. Wardman, Application of pulse radiolysis methods to study the reactions and structure of biomolecules, Rep. Prog. Phys. 41 (1978) 259–302.

[9] C. von Sonntag, H.-P. Schuchmann, Pulse radiolysis, Methods Enzymol. 233 (1994) 3–56.

[10] V.A. Baraboi, P.V. Beloshiskii, P.V. Krasiuk, V.I. Korkach, Oxygen dependent processes in the irradiate organisms, Fiziol. Zh. (Kiev) 40 (1994) 116–128.
[11] V.M. Biakov, S.V. Stepanov, Mechanism of primary radiobiologic action, Radiat. Biol. Radioecol. 37 (1997) 469–474.
[12] C.K.K. Nair, D.K. Parida, T. Nomura, Radioprotectors in radiotherapy, J. Radiat. Res. 42 (2001) 21–37.
[13] B.H.J. Bielski, D.E. Cabelli, R.L. Arudi, Reactivity of $HO_2/O_2^{\bullet -}$ radicals in aqueous solution, J. Phys. Chem. Ref. Data 14 (1985) 1041–1100.
[14] Y.A. Ilan, G. Czapski, D. Meisel, The one-electron transfer redox potentials of free radicals. I. The oxygen/superoxide system, Biochim. Biophys. Acta 430 (1976) 209–224.
[15] J.H. Carpenter, New measurements of oxygen solubility in pure and natural water, Limnol. Oceanogr. 11 (1966) 264–277.
[16] W.H. Koppenol, J. Butler, Energetics of interconversion reactions of oxyradicals, Adv. Free Radical Biol. Med. 1 (1985) 91–131.
[17] A.W. Segal, O.T.G. Jones, Novel cytochrome b system in phagocytic vacuoles from human granulocytes, Nature 276 (1978) 515–517.
[18] F.B. Wientjes, A.W. Segal, NADPH oxidase and the respiratory burst, Semin. Cell Biol. 6 (1995) 357–365.
[19] J.S. Olson, D.P. Ballou, G. Palmer, V. Massey, The reaction of xanthine oxidase with molecular oxygen, J. Biol. Chem. 249 (1974) 4350–4362.
[20] R.F. Anderson, R. Hille, V. Massey, The radical chemistry of milk xanthine oxidase as studied by radiation chemistry technique, J. Biol. Chem. 261 (1986) 15870–15876.
[21] R.W. Carrell, C.C. Winterbourn, E.A. Rachmilewitz, Activated oxygen and hemolysis, Br. J. Hematol. 30 (1975) 259–264.
[22] S. Harel, J. Kanner, Hemoglobin and myoglobin as inhibitors of hydroxyl radical generation in a model system of "iron redox" cycle, Free Radic. Res. Commun. 6 (1989) 1–10.
[23] E.J. Land, T. Mukherfee, A.J. Swallow, J.M. Bruce, One-electron reduction of adriamycin: properties of the semiquinone, Arch. Biochem. Biophys. 225 (1983) 116–121.
[24] R.M. Smith, M.J. Curnutte, Molecular basis of chronic granulomatous disease, Blood 77 (1991) 673–686.
[25] F.L. Rodkey, Oxidation-reduction potentials of the triphosphopyridine nucleotide system, J. Biol. Chem. 213 (1955) 777–786.
[26] G. Cheng, Z. Cao, X. Xu, E.G. van Meir, J.D. Lambeth, Homologs of gp91phox: cloning and tissue expression of Nox3, Nox4, and Nox5, Gene 269 (2001) 131–140.
[27] M. Geiszt, J. Witta, J. Baffi, K. Lekstrom, T.L. Leto, Dual oxidases represent novel hydrogen peroxide sources supporting mucosal host defense, FASEB J. 17 (2003) 1502–1504.
[28] R.W. Harper, C. Xu, J.P. Eiserich, Y. Chen, C.Y. Kao, P. Thai, H. Setiadi, R. Wu, Differential regulation of dual NADPH/peroxidases, Duox1 and Duox2, by Th1 and Th2 cytokines in the respiratory tract epithelium, FEBS Lett. 579 (2005) 4911–4917.
[29] A. Kooij, M. Schijns, W.M. Frederiks, C.J.F. van Noorden, J. James, Distribution of xanthine oxidoreductase activity in human tissues – a histochemical and biochemical study, Virchows Arch. B Cell Pathol. 63 (1993) 17–23.
[30] T.A. Krenitsky, J.V. Tuttle, E.L. Cattau, P. Wang, A comparison of the distribution and electron acceptor specificities of xanthine oxidase and aldehyde oxidase, Comp. Biochem. Physiol. B 49 (1974) 687–703.
[31] J.F. Turrens, Mitochondrial formation of reactive oxygen species, J. Physiol. 552 (2003) 335–344.
[32] S. Dröse, U. Brandt, The mechanism of mitochondrial superoxide production by the cytochrome bc_1 complex, J. Biol. Chem. 283 (2008) 21649–21654.
[33] T.P. Ryan, S.A. Aust, The role of iron in oxygen-mediated toxicities, Crit. Rev. Toxicol. 22 (1992) 119–141.
[34] S.L. Camhi, P. Lee, A.M.K. Choi, The oxidative stress response, New Horizons 3 (1995) 170–182.
[35] C. Houee, M. Grandies, J. Pucheault, C. Ferradini, Radical chemistry of alloxan-dialuric acid: role of the superoxide radical, Clin. Respir. Physiol. 17 (1981) 43–48.
[36] J.A. Farrington, M. Ebert, E.J. Land, Bipyridylium quaternary salt and related compounds, J. Chem. Soc. Faraday 74 (1978) 665–675.
[37] J.H. Peters, G.R. Gordon, D. Kashiwase, J.W. Lown, S.F. Yen, J.A. Plambeck, Redox activities of antitumor anthracyclins determined by microsomal oxygen consumption and assays for superoxide anion and hydroxyl radical generation, Biochem. Pharmacol. 35 (1986) 1309–1323.
[38] V. Okoh, A. Deoraj, D. Roy, Estrogen-induced reactive oxygen species-mediated signaling contribute to breast cancer, Biochim. Biophys. Acta 1815 (2011) 115–133.
[39] P.G. Furtmüller, U. Burner, W. Jantschko, G. Regelsberger, C. Obinger, Two-electron reduction and one-electron oxidation of organic hydroperoxides by human myeloperoxidase, FEBS Lett. 484 (2000) 139–143.
[40] W.H. Koppenol, Thermodynamics of reactions involving oxyradicals and hydrogen peroxide, Bioelectrochem. Bioenerg. 18 (1987) 3–11.

[41] A.U. Khan, Direct spectral evidence of the generation of singlet molecular oxygen ($^1\Delta_g$) in the reaction of potassium superoxide with water, J. Am. Chem. Soc. 103 (1981) 6516–6517.
[42] E.J. Corey, M.M. Mehrota, A.U. Khan, Water induced dismutation of superoxide generates singlet molecular oxygen, Biochem. Biophys. Res. Commun. 145 (1987) 842–846.
[43] J.R. Kanofsky, Singlet oxygen production in biological systems, Chem. Biol. Interact. 70 (1989) 1–28.
[44] H.J. Guiraud, C.S. Foote, Chemistry of superoxide ion: III. Quenching of singlet oxygen, J. Am. Chem. Soc. 98 (1976) 1984–1986.
[45] A.U. Khan, Theory of electron transfer and quenching of singlet oxygen [$^1\Sigma_g^+$ and $^1\Delta_g$] by superoxide anion. The role of water in the dismutation of $O_2^{\bullet -}$, J. Am. Chem. Soc. 99 (1977) 370–371.
[46] J. McCord, I. Fridovich, Superoxide dismutase: an enzymic function for erythrocuprein (hemocuprein), J. Biol. Chem. 224 (1960) 6049–6055.
[47] L.Y. Chang, J.W. Slot, H.J. Geuze, J.D. Crapo, Molecular immunocytochemistry of the CuZn superoxide dismutase in rat hepatocytes, J. Cell Biol. 107 (1988) 2169–2179.
[48] R.A. Weisiger, I. Fridovich, Mitochondrial superoxide dismutase. Site of synthesis and intramitochondrial localization, J. Biol. Chem. 248 (1973) 4793–4796.
[49] S.V. Antonyuk, R.W. Strange, S.L. Marklund, S.S. Hasnain, The structure of human extracellular copper-zinc superoxide dismutase at 1.7 Å resolution: insights into heparin and collagen binding, J. Mol. Biol. 388 (2009) 310–326.
[50] V.B. Vasilyev, A.M. Kachurin, N.V. Soronka, Dismutation of superoxide anion radicals by ceruloplasmin. Details of the mechanism, Biokhimija 53 (1988) 2051–2058.
[51] N.E. Hellman, J.D. Gitlin, Ceruloplasmin metabolism and function, Annu. Rev. Nutr. 22 (2002) 439–458.
[52] P.M. Wood, The potential diagram for oxygen at pH 7, Biochem. J. 253 (1988) 287–289.
[53] H.J. Reich, R.J. Hondal, Why nature chose selenium, ACS Chem. Biol. 11 (2016) 821–841.
[54] D. Besse, F. Siedler, T. Diercks, H. Kessler, L. Moroder, The redox potential of selenocystine in unconstrained cyclic peptides, Angew Chem. Int. Ed. Engl. 36 (1997) 883–885.
[55] M. Schrader, H.D. Fahimi, Peroxisomes and oxidative stress, Biochim. Biophys. Acta 1763 (2006) 1755–1766.
[56] X.W. Wu, C.C. Lee, D.M. Muzny, C.T. Caskey, Urate oxidase: primary structure and evolutionary implications, Proc. Natl. Acad. Sci. U.S.A. 86 (1989) 9412–9416.
[57] I. Ramazzina, C. Folli, A. Secchi, R. Berni, R. Perculani, Completing the uric acid degradation pathway through phylogenetic comparison of whole genomes, Nat. Chem. Biol. 2 (2006) 144–148.
[58] Y. Ohta, M. Nishikimi, Random nucleotide substitutions in primate nonfunctional gene for L-gulono-gamma-lactone oxidase, the missing enzyme in L-ascorbic acid biosynthesis, Biochim. Biophys. Acta 1472 (1999) 408–411.
[59] J. Zschaler, D. Schlorke, J. Arnhold, Differences in innate immune response between man and mouse, Crit. Rev. Immunol. 34 (2014) 433–454.
[60] R. Brigelius-Flohé, M. Maiorino, Glutathione peroxidases, Biochim. Biophys. Acta 1830 (2013) 3289–3303.
[61] S. Mueller, H.D. Riedel, W. Stremmel, Direct evidence for catalase as the predominant H_2O_2-removing enzyme in human erythrocytes, Blood 90 (1997) 4973–4978.
[62] C. Glorieux, P.B. Calderon, Catalase, a remarkable enzyme: targeting the oldest antioxidant enzyme to find a new cancer treatment approach, Biol. Chem. 398 (2017) 1095–1108.
[63] H. Aebi, Catalase in vitro, Methods Enzymol. 105 (1984) 121–126.
[64] M.M. Goyal, A. Basak, Human catalase: looking for complete identity, Prot. Cell 1 (2010) 888–897.
[65] F. Ursini, M. Maiorino, A. Roveri, Phospholipid hydroperoxide glutathione peroxidase (PHGPx): more than an antioxidant enzyme? Biomed. Environ. Sci. 10 (1997) 327–332.
[66] E. Lubos, J. Loscalzo, D.E. Handy, Glutathione peroxidase-1 in health and disease: from molecular mechanisms to therapeutic opportunities, Antioxidants Redox Signal. 15 (2011) 1957–1997.
[67] F.M. Low, M.B. Hampton, C.C. Winterbourn, Prx2 and peroxide metabolism in the erythrocyte, Antioxidants Redox Signal. 10 (2008) 1621–1630.
[68] S.G. Rhee, S.W. Kang, T.S. Chang, W. Jeong, K. Kim, Peroxiredoxin, a novel family of peroxidases, IUBMB Life 52 (2001) 35–41.
[69] C.S. Pillay, J.F. Hofmeyr, B.G. Olivier, J.L. Snoep, J.M. Rohwer, Enzymes of redox couples? The kinetics of thioredoxin and glutaredoxin reactions in a systems biology context, Biochem. J. 417 (2009) 269–275.
[70] M.R. Gunther, P.M. Hanna, R.P. Mason, M.S. Cohen, Hydroxyl radical formation from cuprous ion and hydrogen peroxide: a spin-trapping study, Arch. Biochem. Biophys. 316 (1995) 515–522.
[71] J.D. Rush, W.H. Koppenol, Oxidizing intermediates in the reaction of ferrous EDTA with hydrogen peroxide, J. Biol. Chem. 261 (1986) 6730–6735.

[72] B. Halliwell, J.M.C. Gutteridge, Iron as a biological pro-oxidant, ISI Atlas Sci. Biochem. (1988) 48–52.
[73] S. Goldstein, D. Meyerstein, G. Czapski, The Fenton reagents, Free Radic. Biol. Med. 15 (1993) 435–445.
[74] W.H. Koppenol, The centennial of the Fenton reaction, Free Radic. Biol. Med. 15 (1993) 645–651.
[75] P. Wardman, L.P. Candeias, Fenton chemistry: an introduction, Radiat. Res. 145 (1996) 523–531.
[76] C.A. Long, B.H.J. Bielski, Rate of reaction of superoxide radical with chloride-containing species, J. Phys. Chem. 84 (1980) 555–557.
[77] C.L. Ramos, S. Pou, B.E. Britigan, M.S. Cohen, G.M. Rosen, Spin trapping evidence for myeloperoxidase-dependent hydroxyl radical formation by human neutrophils and monocytes, J. Biol. Chem. 267 (1992) 8307–8312.
[78] L.P. Candeias, K.B. Patel, M.R.L. Stratford, P. Wardman, Free hydroxyl radicals are formed on reaction between the neutrophil-derived species superoxide anion and hypochlorous acid, FEBS Lett. 333 (1993) 151–153.
[79] A.N. Osipov, E.S. Yakutova, Y.A. Vladimirov, Formation of hydroxyl radicals on interaction of hypochlorite with ferrous ions, Biofizika 38 (1993) 383–388.
[80] E.S. Yakutova, E.S. Dryomina, S.A. Evgina, A.N. Osipov, V.S. Sharov, O.M. Panasenko, Y.A. Vladimirov, Free radical formation in the interaction of hypochlorite with iron(II) ions. A chemiluminescent study, Biofizika 39 (1994) 275–279.
[81] W. Bors, M. Erben-Russ, M. Saran, Fatty acid peroxyl radicals: their generation and reactivities, Bioelectrochem. Bioenerg. 18 (1987) 37–49.
[82] J.M.C. Gutteridge, Ferrous-salt-promoted damage to deoxyribose and benzoate. The increased effectiveness of hydroxyl-radical scavengers in the presence of EDTA, Biochem. J. 243 (1987) 709–714.
[83] D. Weldon, T.D. Poulsen, K.V. Mikkelsen, P.R. Ogilby, Singlet sigma: the "other" singlet oxygen in solution, Photochem. Photobiol. 70 (1999) 369–379.
[84] M.A.J. Rodgers, P.T. Snowden, Lifetime of $O_2(^1\Delta_g)$ in liquid water as determined by time-resolved infrared luminescence measurements, J. Am. Chem. Soc. 104 (1982) 5541–5543.
[85] J.R. Kanofsky, Quenching of singlet oxygen by human plasma, Photochem. Photobiol. 51 (1990) 299–303.
[86] R.M. Tyrrell, UVA (320-380 nm) radiation as an oxidative stress, in: H. Sies (Ed.), Oxidative Stress: Oxidants and Antioxidants, Academic Press, 1991, pp. 57–83.
[87] A.M. Held, D.J. Halko, J.K. Hurst, Mechanisms of chlorine oxidation of hydrogen peroxide, J. Am. Chem. Soc. 100 (1978) 5732–5740.
[88] A.A. Noronha-Dutra, M.M. Epperlein, N. Woolf, Reactions of nitric oxide with hydrogen peroxide to produce potentially cytotoxic singlet oxygen as a model for nitric oxide mediated killing, FEBS Lett. 321 (1993) 59–62.
[89] P. Di Mascio, E.J.H. Bechara, H.G. Medeiros, K. Briviba, H. Sies, Singlet molecular oxygen production in the reaction of peroxynitrite with hydrogen peroxide, FEBS Lett. 355 (1994) 287–289.
[90] D.R. Kearns, Physical and chemical properties of singlet molecular oxygen, Chem. Rev. 71 (1971) 395–427.
[91] H. Wefers, Singlet oxygen in biological systems, Bioelectrochem. Bioenerg. 18 (1987) 91–104.
[92] A.A. Frimer, The reaction of singlet oxygen with olefins: the question of mechanism, Chem. Rev. 79 (1979) 359–387.
[93] P. Di Mascio, L.H. Catalani, E.J.H. Bechara, Are dioxethanes chemiluminescent intermediates in lipid peroxidation? Free Radic. Biol. Med. 12 (1992) 471–478.
[94] J. Cadet, R. Teoule, Comparative study of oxidation of nucleic acid components by hydroxyl radicals, singlet oxygen and superoxide anion radicals, Photochem. Photobiol. 28 (1978) 661–667.
[95] L.F. Agnez-Lima, J.T.A. Melo, A.E. Silva, A.H.S. Oliveira, A.R.S. Timoteo, K.M. Lima-Bessa, G.R. Martinez, M.H.G. Medeiros, P. Di Mascio, R.S. Galhardo, C.F.M. Menck, DNA damage by singlet oxygen and cellular protective mechanisms, Mutat. Res. 751 (2012) 15–28.
[96] V. Duarte, D. Gasparutto, L.F. Yamaguchi, J.L. Ravanat, G.R. Martinez, M.H.G. Medeiros, P. Di Mascio, J. Cadet, Oxaluric acid as a major product of singlet oxygen-mediated oxidation of 8-oxo-7,8-dihydroguanine in DNA, J. Am. Chem. Soc. 122 (2000) 12622–12628.
[97] P. Di Mascio, T.P.A. Devasagayam, S. Kaiser, H. Sies, Carotenoids, tocopherols and thiols as singlet oxygen quenchers, Biochem. Soc. Trans. 18 (1990) 1054–1056.
[98] P.F. Conn, W. Schalch, T.G. Truscott, The singlet oxygen and carotenoid interaction, J. Photochem. Photobiol. B Biol. 11 (1991) 41–47.
[99] A.U. Khan, Y.-H. Mei, T. Wilson, A proposed function of spermine and spermidine: protection of replicating DANN against damage by singlet oxygen, Proc. Natl. Acad. Sci. U.S.A. 89 (1992) 11426–11427.
[100] A.U. Khan, P. Di Mascio, M.H.G. Medeiros, T. Wilson, Spermine and spermidine protection of plasmid DNA against single-strand breaks induced by singlet oxygen, Proc. Natl. Acad. Sci. U.S.A. 89 (1992) 11428–11430.
[101] R.B. Devlin, L.J. Folinsbee, F. Biscardi, G. Hatch, S. Becker, M.C. Madden, M. Robbins, H.S. Koren, Inflammation and cell damage induced by repeated exposure of humans to ozone, Inhal. Toxicol. 9 (1997) 211–234.

[102] R. Frank, M.C. Liu, E.W. Spannhake, S. Mlynarek, K. Macri, G.G. Weinmann, Repetitive ozone exposure of young adults. Evidence of persistent small airway dysfunction, Am. J. Respir. Crit. Care Med. 164 (2001) 1257–1260.
[103] W.A. Pryor, The formation of free radicals and the consequences of their reactions in vivo, Photochem. Photobiol. 28 (1978) 787–801.
[104] S. Kermani, A. Ben-Jebria, J.S. Ultman, Kinetics of ozone reaction with uric acid, ascorbic acid, and glutathione at physiologically relevant conditions, Arch. Biochem. Biophys. 451 (2006) 8–16.
[105] A.F. Behndig, A. Blomberg, R. Helleday, S.T. Duggan, F.J. Kelly, I.S. Mudway, Antioxidant responses to acute ozone challenge in the healthy human airway, Inhal. Toxicol. 21 (2009) 933–942.
[106] H. Togashi, I. Sakuma, M. Yoshioka, T. Kobayashi, H. Yasuda, A. Kitabatake, H. Saito, S.S. Gross, R. Levi, A central nervous system action of nitric oxide in blood pressure regulation, J. Pharmacol. Exp. Ther. 262 (1992) 343–347.
[107] G.A. Bohme, C. Bon, M. Lemaire, M. Reibaud, O. Piot, J.M. Stutzmann, A. Doble, J.C. Blanchard, Altered synaptic plasticity and memory formation in nitric oxide synthase inhibitor-treated rats, Proc. Natl. Acad. Sci. U.S.A. 90 (1993) 9191–9194.
[108] L. Zhou, D.Y. Zhu, Neuronal nitric oxide synthase: structure, subcellular localization, regulation, and clinical implications, Nitric Oxide 20 (2009) 223–230.
[109] C. Nathan, Role of iNOS in human host defense, Science 312 (2006) 1874–1875.
[110] D. Fulton, J.P. Gratton, T.J. McGabe, J. Fontana, Y. Fujio, K. Walsh, T.F. Franke, A. Papapetropoulos, W.C. Sessa, Regulation of endothelium-derived nitric oxide production by the protein kinase Akt, Nature 399 (1999) 597–601.
[111] I. Fleming, R. Busse, Molecular mechanisms involved in the regulation of the endothelial nitric oxide synthase, Am. J. Physiol. Regul. Integr. Comp. Physiol. 284 (2003) R1–R12.
[112] U. Förstermann, W.C. Sessa, Nitric oxide synthases: regulation and function, Eur. Heart J. 33 (2012) 829–837.
[113] L.J. Ignarro, Biosynthesis and metabolism of endothelium-derived nitric oxide, Annu. Rev. Pharmacol. Toxicol. 30 (1990) 535–560.
[114] S. Moncada, R.M.J. Palmer, E.A. Higgs, Nitric oxide: physiology, pathophysiology, and pharmacology, Pharmacol. Rev. 43 (1991) 109–142.
[115] S. Thoma-Uszynski, S. Stenger, O. Takeuchi, M.T. Ochoa, M. Engele, P.A. Sieling, P.F. Barnes, M. Rollinghoff, P.L. Bolcskei, M. Wagner, S. Akira, M.V. Norgard, J.T. Belisle, P.J. Godowski, B.R. Bloom, R.L. Molin, Induction of direct antimicrobial activity through mammalian toll-like receptors, Science 291 (2001) 1544–1547.
[116] F.C. Fang, Antimicrobial reactive oxygen and nitrogen species: concepts and controversies, Nat. Rev. Microbiol. 2 (2004) 820–832.
[117] P.C. Ford, D.A. Wink, D.M. Stanbury, Autoxidation kinetics of aqueous nitric oxide, FEBS Lett. 326 (1993) 1–3.
[118] R. Kissner, T. Nauser, P. Bugnon, P.G. Lye, W.H. Koppenol, Formation and properties of peroxynitrite as studied by laser flash photolysis, high pressure stopped-flow technique, and pulse radiolysis, Chem. Res. Toxicol. 10 (1997) 1285–1292.
[119] R.E. Huie, S. Padmaja, Reactions of $^{\bullet}NO$ and $O_2^{\bullet -}$, Free Radic. Res. Commun. 18 (1993) 195–199.
[120] S. Goldstein, G. Czapski, The reaction of $^{\bullet}NO$ with $O_2^{\bullet -}$ and $HO_2^{\bullet}$: a pulse radiolysis study, Free Radic. Biol. Med. 19 (1995) 505–510.
[121] K. Kobayashi, M. Miki, S. Tagawa, Pulse-radiolysis study of the reaction of nitric oxide with superoxide, J. Chem. Soc. Dalton Trans. (1995) 2885–2889.
[122] W.H. Koppenol, J.J. Moreno, W.A. Pryor, H. Ischiropoulos, J.S. Beckman, Peroxynitrite, a cloaked oxidant formed by nitric oxide and superoxide, Chem. Res. Toxicol. 5 (1992) 834–842.
[123] W.H. Koppenol, R. Kissner, Can 0=NOOH undergo hemolysis? Chem. Res. Toxicol. 11 (1998) 87–90.
[124] S. Goldstein, G. Czapski, Mechanism of the nitrosation of thiols and amines by oxygenated $^{\bullet}NO$ solutions: the nature of the nitrosating intermediates, J. Am. Chem. Soc. 118 (1996) 3419–3425.
[125] F.J. Schopfer, P.R.S. Baker, B.A. Freeman, NO-dependent protein nitration: a cell signaling event or an oxidative inflammatory response? Trends Biochem. Sci. 28 (2003) 646–654.
[126] R. Radi, J.S. Beckman, K.M. Bush, B.A. Freeman, Peroxynitrite-induced membrane lipid peroxidation: the cytotoxic potential of superoxide and nitric oxide, Arch. Biochem. Biophys. 288 (1991) 481–487.
[127] V.M. Darley-Usmar, N. Hogg, V.J. O'Leary, M.T. Wilson, S. Moncada, The simultaneous generation of superoxide and nitric oxide can initiate lipid peroxidation in human low density lipoprotein, Free Radic. Res. Commun. 17 (1992) 9–20.
[128] M. Li, L. Rosenfeld, R.E. Vilar, M.K. Cowman, Degradation of hyaluronan by peroxynitrite, Arch. Biochem. Biophys. 341 (1997) 245–250.
[129] R.E. Vilar, D. Ghael, M. Li, D.D. Bhagat, L.M. Arrigo, M.K. Cowman, H.S. Dweck, L. Rosenfeld, Nitric oxide degradation of heparin and heparin sulphate, Biochem. J. 324 (1997) 473–479.

[130] O. Augusto, M.G. Bonini, A.M. Amanso, E. Linares, C.C. Santos, S.L. de Menezes, Nitrogen dioxide and carbonate radical anion: two emerging radicals in biology, Free Radic. Biol. Med. 32 (2002) 841–859.

[131] G. Ferrer-Sueta, R. Radi, Chemical biology of $ONOO^-$: kinetics, diffusion, and radicals, ACS Chem. Biol. 4 (2009) 161–177.

[132] V. Kapil, S.M.A. Haydar, V. Pearl, J.O. Lundberg, E. Weitzberg, A. Ahluwalia, Free Rad. Biol. Med. 55 (2013) 93–100.

[133] M. Tiso, A.N. Schechter, Nitrate reduction to nitrite, nitric oxide and ammonia by gut bacteria under physiological conditions, PLoS One 10 (2015) e0119712, https://doi.org/10.1371/journal.pone.0119712.

[134] A.M. Leone, P.L. Francis, P. Rhodes, S. Moncada, A rapid and simple method for the measurement of nitrite and nitrate in plasma by high performance capillary electrophoresis, Biochem. Biophys. Res. Commun. 200 (1994) 951–957.

[135] T. Ueda, T. Maekawa, D. Sadmitsu, S. Oshita, K. Ogino, K. Nakamura, The determination of nitrite and nitrate in human blood plasma by capillary zone electrophoresis, Electrophoresis 16 (1995) 1002–1004.

[136] A. Van der Vliet, J.P. Eiserich, B. Halliwell, C.E. Cross, Formation of reactive nitrogen species during peroxidase-catalyzed oxidation of nitrite, J. Biol. Chem. 272 (1997) 7617–7625.

[137] H. Moshage, B. Kok, J.R. Huizenga, P.L. Jansen, Nitrite and nitrate determinations in plasma: a critical evaluation, Clin. Chem. 41 (1995) 892–896.

[138] R.C. Smith, V. Nunn, Prevention by 3-N-ribosyluric acid of the oxidation of bovine hemoglobin by sodium nitrite, Arch. Biochem. Biophys. 232 (1984) 348–353.

[139] M.D. Coleman, P.J. Hayes, D.P. Jacobus, Methemoglobin formation due to nitrite, disulfiram, 4-aminophenol and monoacetyldapson hydroxylamine in diabetic and non-diabetic human erythrocytes in vitro, Environ. Toxicol. Pharmacol. 5 (1998) 61–67.

[140] C.L. French, S.S. Yaun, L.A. Baldwin, D.A. Leonhard, X.Q. Zhao, E.J. Calabrese, Potency ranking of methemoglobin-forming agents, J. Appl. Toxicol. 15 (1995) 167–174.

[141] J.H. Crawford, T.S. Isbell, Z. Huang, S. Shiva, B.K. Chacko, A.N. Schechter, V.M. Darley-Usmar, J.D. Kerby, J.D. Lang Jr., D. Kraus, C. Ho, M.T. Gladwin, R.P. Patel, Hypoxia, red blood cells, and nitrite regulate NO-dependent hypoxic vasodilation, Blood 107 (2006) 566–574.

[142] R.A. Scanlan, Formation and occurrence of nitrosamines in food, Cancer Res. 43 (1983) 2435s–2440s.

[143] S.S. Herrmann, K. Granby, L. Duedahl-Olesen, Formation and mitigation of *N*-nitrosamines in nitrite preserved cooked sausages, Food Chem. 174 (2015) 516–526.

[144] K.O. Honikel, The use and control of nitrate and nitrite for the processing of meat products, Meat Sci. 78 (2008) 68–78.

[145] R. Floris, S.P. Piersma, G. Yang, P. Jones, R. Wever, Interaction of myeloperoxidase with peroxynitrite. A comparison with lactoperoxidase, horseradish peroxidase and catalase, Eur. J. Biochem. 215 (1993) 767–775.

[146] P.G. Furtmüller, W. Jantschko, M. Zederbauer, M. Schwanninger, C. Jakopitsch, S. Herold, W.H. Koppenol, C. Obinger, Peroxynitrite efficiently mediates the interconversion of redox intermediates by myeloperoxidase, Biochem. Biophys. Res. Commun. 337 (2005) 944–954.

[147] C.N. Koyani, J. Flemmig, E. Malle, J. Arnhold, Myeloperoxidase scavenges peroxynitrite: a novel anti-inflammatory action of the heme enzyme, Arch. Biochem. Biophys. 571 (2015) 1–9.

[148] H.M. Abu-Soud, S.L. Hazen, Nitric oxide modulates the catalytic activity of myeloperoxidase, J. Biol. Chem. 275 (2000) 5425–5430.

[149] H.M. Abu-Soud, S.L. Hazen, Nitric oxide is a physiological substrate for mammalian peroxidases, J. Biol. Chem. 275 (2000) 37524–37532.

[150] J.S. Stamler, D.J. Singel, J. Loscalzo, Biochemistry of nitric oxide and its redox-activated forms, Science 258 (1992) 1898–1902.

[151] U. Burner, P.G. Furtmüller, A.J. Kettle, W.H. Koppenol, C. Obinger, Mechanism of reaction of myeloperoxidase with nitrite, J. Biol. Chem. 275 (2000) 20597–20601.

[152] P. Wardman, Reduction potential of one-electron couples involving free radicals in aqueous solution, J. Phys. Chem. Ref. Data 18 (1989) 1637–1645.

[153] S. Baldus, J.P. Eiserich, A. Mani, L. Castrom, M. Figueroa, P. Chumley, W. Ma, A. Tousson, R. White, D.C. Bullard, M.-L. Brennan, A.J. Lusis, K.P. Moore, B.A. Freeman, Endothelial transcytosis of myeloperoxidase confers specificity to vascular ECM proteins as targets of tyrosine nitration, J. Clin. Investig. 108 (2001) 1759–1770.

[154] S. Baldus, J.P. Eiserich, M.-L. Brennan, R.M. Jackson, C.B. Alexander, B.A. Freeman, Spatial mapping of pulmonary and vascular nitrotyrosine reveals the pivotal role of myeloperoxidase as a catalyst for tyrosine nitration in inflammatory diseases, Free Radic. Biol. Med. 33 (2002) 1010–1019.

[155] J. Buyn, D.M. Mueller, J.S. Fabjan, J.W. Heinecke, Nitrogen dioxide radical generated by the myeloperoxidase-hydrogen peroxide-nitrite system promotes lipid peroxidation of low density lipoproteins, FEBS Lett. 455 (1999) 243–246.

[156] R. Zhang, M.-L. Brennan, Z. Shen, J.C. MacPherson, D. Schmitt, C.E. Molenda, S.L. Hazen, Myeloperoxidase functions as a major catalyst for initiation of lipid peroxidation at sites of inflammation, J. Biol. Chem. 277 (2002) 46116–46122.
[157] T. Kraemer, I. Prakosay, R.A. Date, H. Sies, T. Schewe, Oxidative modification of low-density lipoprotein lipid peroxidation by myeloperoxidase in the presence of nitrite, Biol. Chem. 385 (2004) 809–818.
[158] K. Zouaoui Boudjeltia, N. Moguilevsky, I. Legssyer, S. Babar, M. Guillaume, P. Delree, M. Vanhaeverbeek, D. Brohee, J. Docobu, C. Remacle, Oxidation of low-density lipoproteins by myeloperoxidase at the surface of endothelial cells: an additional mechanism to subendothelium oxidation, Biochem. Biophys. Res. Commun. 325 (2004) 434–438.
[159] J.P. Eiserich, C.E. Cross, A.D. Jones, B. Halliwell, A. van der Vliet, Formation of chlorinating and nitrating species by reaction of NO_2^- and HOCl. A novel mechanism for nitric oxide-mediated protein modification, J. Biol. Chem. 271 (1996) 19199–19208.
[160] R.E. Huie, The reaction kinetics of $NO_2^{\bullet}$, Toxicology 89 (1994) 193–216.
[161] B.R. Stern, M. Solioz, D. Krewski, P. Aggett, T.-C. Aw, S. Baker, K. Crump, M. Dourson, L. Haber, R. Hertzberg, C. Keen, B. Meek, L. Rudenko, R. Schoeny, W. Slot, T. Starr, Copper and human health: biochemistry, genetics, and strategies for modeling dose-response relationships, J. Toxicol. Environ. Health B 10 (2007) 157–222.
[162] J. Wang, K. Pantopoulos, Regulation of cellular iron metabolism, Biochem. J. 434 (2011) 365–381.
[163] R.R. Chrichton, R.J. Ward, Iron homeostasis, Met. Ions Biol. Syst. 35 (1998) 633–665.
[164] P. Ponka, Cellular iron metabolism, Kidney Int. 55 (1999) S2–S11.
[165] N. Zhao, A.-S. Zhang, C.A. Enns, Iron regulation by hepcidin, J. Clin. Investig. 123 (2013) 2337–2343.
[166] K. Gkouvatsos, G. Papanikolaou, K. Pantopoulos, Regulation of iron transport and the role of transferrin, Biochim. Biophys. Acta 1820 (2011) 188–202.
[167] E. Gamella, P. Buratti, G. Cairo, S. Recalcati, The transferrin receptor: the cellular iron gate, Metallomics 9 (2017) 1367–1375.
[168] W.H. Massover, Ultrastructure of ferritin and apoferritin: a review, Micron 24 (1993) 389–437.
[169] J.R. Prohaska, Role of copper transporters in copper homeostasis, Am. J. Clin. Nutr. 88 (2008) 826S–829S.
[170] H. Hayashi, M. Yano, Y. Fujita, S. Wakusawa, Compound overload of copper and iron in patients with Wilson's disease, Med. Mol. Morphol. 39 (2006) 121–126.
[171] G.R. Greenberg, M.M. Wintrobe, A labile iron pool, J. Biol. Chem. 165 (1946) 397–398.
[172] A. Jacobs, An intracellular transit iron pool, Ciba Found. Symp. 51 (1977) 91–106.
[173] O. Kakhlon, Z.I. Cabantchik, The labile iron pool: characterization, measurement, and participation in cellular processes, Free Radic. Biol. Med. 33 (2002) 1037–1046.
[174] M. Kruszewski, Labile iron pool: the main determinant of cellular response to oxidative stress, Mutat. Res. 53 (2003) 81–92.
[175] D.M. Miller, G.R. Buettner, S.D. Aust, Transition metals as catalysts of "autoxidation" reactions, Free Radic. Biol. Med. 8 (1990) 95–108.
[176] J.E. Biaglow, A.V. Kachur, The generation of hydroxyl radicals in the reaction of molecular oxygen with polyphosphate complexes of ferrous ions, Radiat. Res. 148 (1997) 181–187.
[177] J. Flemmig, J. Arnhold, Ferrous ion-induced strand breaks in the DNA plasmid pBR322 are not mediated by hydrogen peroxide, Eur. Biophys. J. 36 (2007) 377–384.
[178] D.M. Miller, S.D. Aust, Studies of ascorbate-dependent, iron-catalyzed lipid peroxidation, Arch. Biochem. Biophys. 271 (1981) 113–119.
[179] G. Minotti, S.D. Aust, The role of iron in the initiation of lipid peroxidation, Chem. Phys. Lipids 44 (1987) 191–208.
[180] L.A. Reinke, J.M. Rau, P.B. McCay, Characteristics of an oxidant formed during iron(II) autoxidation, Free Radic. Biol. Med. 16 (1994) 485–492.
[181] S.Y. Qian, G.R. Buettner, Iron and dioxygen chemistry is an important route to initiation of biological free radical oxidations: an electron paramagnetic resonance spin trapping study, Free Radic. Biol. Med. 26 (1999) 1447–1456.
[182] N.K. Urbanski, A. Beresewicz, Generation of $^{\bullet}OH$ initiated by interaction of Fe^{2+} and Cu^{+} with dioxygen: comparison with the Fenton chemistry, Acta Biochim. Pol. 47 (2000) 951–962.
[183] M. Saran, C. Michel, K. Stettmaier, W. Bors, Arguments against the significance of the Fenton reaction contributing to signal pathways under in vivo conditions, Free Radic. Res. 33 (2000) 567–579.
[184] C. Oxvig, A.R. Thomson, M.T. Overgaard, E.S. Sorenson, P. Højrup, M.J. Bjerrum, G.J. Gleich, L. Sottrup-Jensen, Biochemical evidence for heme linkage through esters with Asp-93 and Glu-241 in human eosinophil peroxidase. The ester with Asp-93 is only partially formed in vivo, J. Biol. Chem. 274 (1999) 16935–16958.
[185] T.J. Fiedler, C.A. Davey, R.E. Fenna, X-ray crystal structure and characterization of halide-binding sites of human myeloperoxidase at 1.8 Å resolution, J. Biol. Chem. 275 (2000) 11964–11971.

[186] P.G. Furtmüller, M. Zederbauer, W. Jantschko, J. Helm, M. Bogner, C. Jakopitsch, C. Obinger, Active site structure and catalytic mechanisms of human peroxidases, Arch. Biochem. Biophys. 445 (2006) 199–213.
[187] A.K. Singh, N. Singh, S. Sharma, S.B. Singh, P. Kaur, A. Bhushan, A. Srinivasan, T.P. Singh, Crystal structure of lactoperoxidase at 2.4 Å resolution, J. Mol. Biol. 376 (2008) 1060–1075.
[188] J. Arnhold, J. Flemmig, Human myeloperoxidase in innate and acquired immunity, Arch. Biochem. Biophys. 500 (2010) 92–106.
[189] M.E. Rothenberg, S.P. Hogan, The eosinophil, Annu. Rev. Immunol. 24 (2006) 147–174.
[190] M.J. Davies, C.L. Hawkins, D.I. Pattison, M.D. Rees, Mammalian heme peroxidases: form molecular mechanisms to health implications, Antioxidants Redox Signal. 10 (2008) 1199–1234.
[191] D. Dolphin, A. Forman, D.C. Borg, J. Fajer, R.H. Felton, Compound I of catalase and horse radish peroxidase: π-cation radicals, Proc. Natl. Acad. Sci. U.S.A. 68 (1971) 614–618.
[192] H. Spalteholz, O.M. Panasenko, J. Arnhold, Formation of reactive halide species by myeloperoxidase and eosinophil peroxidase, Arch. Biochem. Biophys. 445 (2006) 225–234.
[193] Gmelins Handbuch der Anorganischen Chemie, Chlor, eighth ed., Verlag Chemie, Weinheim, 1968, p. 244. Supplement A.
[194] J.C. Morris, The acid ionization constant of HOCl from 5 to 35°, J. Phys. Chem. 70 (1966) 3798–3805.
[195] Gmelins Handbuch der Anorganischen Chemie, Brom, eighth ed., Verlag Chemie, Weinheim, 1931, p. 290.
[196] R.C. Troy, D.W. Margerum, Non-metal redox kinetics: hypobromite and hypobromous acid reactions with iodide and with sulfite and the hydrolysis of bromosulfate, Inorg. Chem. 30 (1991) 3538–3543.
[197] Handbook of Chemistry and Physics, fifty-sixth ed., CRC Press, Cleveland, 1975, p. D-142.
[198] J.C. Wren, S. Paquette, S. Sunder, B.L. Ford, Iodine chemistry in the +1 oxidation state. II. Raman and uv-visible spectroscopic study of the disproportionation of hypoiodite in basic solutions, Can. J. Chem. 64 (1986) 2284–2296.
[199] J. Arnhold, E. Monzani, P.G. Furtmüller, M. Zederbauer, L. Casella, C. Obinger, Kinetics and thermodynamics of halide and nitrite oxidation by mammalian hem peroxidases, Eur. J. Inorg. Chem. (2006) 3801–3811.
[200] E.L. Thomas, Lactoperoxidase-catalyzed oxidation of thiocyanate: equilibria between oxidized forms of thiocyanate, Biochemistry 20 (1981) 3273–3280.
[201] P. Nagy, G.N. Jameson, C.C. Winterbourn, Kinetics and mechanisms of the reaction of hypothiocyanous acid with 5-thio-2-nitrobenzoic acid and reduced glutathione, Chem. Res. Toxicol. 22 (2009) 1833–1840.
[202] L.A. Marquez, H.B. Dunford, Kinetics of oxidation of tyrosine and dityrosine by myeloperoxidase compounds I and II, J. Biol. Chem. 270 (1995) 30434–30440.
[203] U. Burner, W. Jantschko, C. Obinger, Kinetics of oxidation of aliphatic and aromatic thiols by myeloperoxidase compounds I and II, FEBS Lett. 443 (1999) 290–296.
[204] W. Jantschko, P.G. Furtmüller, M. Allegra, M.A. Livrea, C. Jakopitsch, G. Regelsberger, C. Obinger, Redox intermediates of plant and mammalian peroxidases: a comparative transient-kinetic study of their reactivity toward indole derivatives, Arch. Biochem. Biophys. 398 (2002) 12–22.
[205] P.G. Furtmüller, W. Jantschko, G. Regelsberger, C. Jakopitsch, J. Arnhold, C. Obinger, Reaction of lactoperoxidase compound I with halides and thiocyanate, Biochemistry 41 (2002) 11895–11900.
[206] J. Arnhold, P.G. Furtmüller, G. Regelsberger, C. Obinger, Redox properties of the couple compound I/native enzyme of myeloperoxidase and eosinophil peroxidase, Eur. J. Biochem. 268 (2001) 5142–5148.
[207] P.G. Furtmüller, J. Arnhold, W. Jantschko, H. Pichler, C. Obinger, Redox properties of the couples compound I/compound II and compound II/native enzyme of human myeloperoxidase, Biochem. Biophys. Res. Commun. 301 (2003) 551–557.
[208] P.G. Furtmüller, J. Arnhold, W. Jantschko, M. Zederbauer, C. Jakopitsch, C. Obinger, Standard reduction potentials of all couples of the peroxidase cycle of lactoperoxidase, J. Inorg. Biochem. 99 (2005) 1220–1229.
[209] H. Hoogland, A. van Kuilenburg, C. van Riel, A.O. Muijsers, R. Wever, Spectral properties of myeloperoxidase compounds II and III, Biochim. Biophys. Acta 916 (1987) 76–82.
[210] B.G.J.M. Bolscher, G.R. Zoutberg, R.A. Cuperus, R. Wever, Vitamin C stimulates the chlorinating activity of human myeloperoxidase, Biochim. Biophys. Acta 784 (1984) 189–191.
[211] A.J. Kettle, C.C. Winterbourn, Superoxide modulates the activity of myeloperoxidase and optimizes the production of hypochlorous acid, Biochem. J. 252 (1988) 529–536.
[212] I.I. Vlasova, A.V. Sokolov, J. Arnhold, The free amino acid tyrosine enhances the chlorinating activity of human myeloperoxidase, J. Inorg. Biochem. 106 (2012) 76–83.

[213] A.J. Kettle, C.A. Geyde, C.C. Winterbourn, Superoxide is an antagonist of anti-inflammatory drugs that inhibit hypochlorous acid production by myeloperoxidase, Biochem. Pharmacol. 45 (1993) 2003–2010.
[214] K.W.M. Zuurbier, A.R.J. Bakkenist, R. Wever, A.O. Muijsers, The chlorinating activity of human myeloperoxidase: high initial activity at neutral pH and activation by electron donors, Biochim. Biophys. Acta 1037 (1990) 140–146.
[215] L.A. Marquez, H.B. Dunford, Interaction of acetaminophen with myeloperoxidase intermediates: optimum stimulation of enzyme activity, Arch. Biochem. Biophys. 305 (1993) 414–420.
[216] T. Kirchner, J. Flemmig, P.G. Furtmüller, C. Obinger, J. Arnhold, (–)-Epicatechin enhances the chlorinating activity of human myeloperoxidase, Arch. Biochem. Biophys. 495 (2010) 21–27.
[217] J. Flemmig, J. Remmler, F. Röhring, J. Arnhold, (–)-Epicatechin regenerates the chlorinating activity of myeloperoxidase in vitro and in neutrophil granulocytes, J. Inorg. Biochem. 130 (2014) 84–91.
[218] H. Spalteholz, P.G. Furtmüller, C. Jakopitsch, C. Obinger, T. Schewe, H. Sies, J. Arnhold, Kinetic evidence for rapid oxidation of (–)-epicatechin by human myeloperoxidase, Biochem. Biophys. Res. Commun. 371 (2008) 810–813.
[219] H.B. Dunford, Y. Hsuanyu, Kinetics of oxidation of serotonin by myeloperoxidase, Biochem. Cell Biol. 77 (1999) 449–457.
[220] J. Flemmig, D. Rusch, M.E. Czerwińska, H.-W. Rauwald, J. Arnhold, Components of a standardized olive leaf dry extract (Ph. Eur.) promote hypothiocyanite production by lactoperoxidase, Arch. Biochem. Biophys. 549 (2014) 17–25.
[221] J. Flemmig, I. Noetzel, J. Arnhold, H.-W. Rauwald, *Leonurus cardiaca* L. herb extracts and their constituents promote lactoperoxidase activity, J. Funct. Foods 17 (2015) 328–339.
[222] J. Gau, P.G. Furtmüller, C. Obinger, J. Arnhold, J. Flemmig, Enhancing hypothiocyanite production by lactoperoxidase – mechanism and chemical properties of promoters, Biochem. Biophys. Rep. 4 (2015) 257–267.
[223] J. Gau, P.G. Furtmüller, C. Obinger, M. Prévost, P. van Antwerpen, J. Arnhold, J. Flemmig, Flavonoids as promoters of the (pseudo-)halogenating activity of lactoperoxidase and myeloperoxidase, Free Radic. Biol. Med. 97 (2016) 307–319.
[224] J. Gau, M. Prévost, P. van Antwerpen, M.-B. Sarosi, S. Rodewald, J. Arnhold, J. Flemmig, Tannins and tannin-related derivatives enhance the (pseudo-)halogenating activity of lactoperoxidase, J. Nat. Prod. 80 (2017) 1328–1338.
[225] J. Gau, J. Arnhold, J. Flemmig, Reactivation of peroxidase activity in human saliva samples by polyphenols, Arch. Oral Biol. 85 (2018) 70–78.
[226] J.M. Zgliczynski, R.J. Selvaraj, B.B. Paul, T. Stelmazynska, P.K.F. Poskitt, A.J. Sbarra, Chlorination by the myeloperoxidase-H_2O_2-Cl^- antimicrobial system at acid and neutral pH, Proc. Soc. Exp. Biol. Med. 154 (1977) 418–422.
[227] A.J. Kettle, C.C. Winterbourn, Assays for the chlorination activity of myeloperoxidase, Methods Enzymol. 233 (1994) 502–512.
[228] A. Jerlich, L. Horakova, J.S. Fabjan, A. Giessauf, G. Jürgens, R.J. Schaur, Correlation of low-density lipoprotein modification by myeloperoxidase with hypochlorous acid formation, Int. J. Clin. Lab. Res. 29 (1999) 155–161.
[229] O.M. Panasenko, H. Spalteholz, J. Schiller, J. Arnhold, Myeloperoxidase-induced formation of chlorohydrins and lysophospholipids from unsaturated phosphatidylcholines, Free Radic. Biol. Med. 34 (2003) 553–562.
[230] L.A. Marquez, H.B. Dunford, Chlorination of taurine by myeloperoxidase: kinetic evidence for an enzyme-bound intermediate, J. Biol. Chem. 269 (1994) 7950–7956.
[231] D.R. Ramos, M. Victoria Garcia, L.M. Canle, J. Arturo Santaballa, P.G. Furtmüller, C. Obinger, Myeloperoxidase-catalyzed taurine chlorination: initial versus equilibrium rate, Arch. Biochem. Biophys. 466 (2007) 221–233.
[232] C.J. van Dalen, M.W. Whitehouse, C.C. Winterbourn, A.J. Kettle, Thiocyanate and chloride as competing substrates for myeloperoxidase, Biochem. J. 327 (1997) 487–492.
[233] D.K. Dastur, E.V. Quadros, N.H. Wadia, M.M. Desai, E.P. Bharucha, Effect of vegetarianism and smoking on vitamin B12, thiocyanate, and folate levels in the blood of normal subjects, Br. Med. J. 3 (1972) 260–263.
[234] N.N. Rehak, S.A. Cecco, J.E. Niemala, R.J. Elin, Thiocyanate in smokers interferes with Nova magnesium ion-selective electrode, Clin. Chem. 43 (1997) 1595–1600.
[235] J. Tenovuo, K.K. Makinen, Concentration of thiocyanate and ionizable iodine in saliva of smokers and nonsmokers, J. Dent. Res. 55 (1976) 661–663.
[236] C.P. Schultz, M.K. Ahmed, C. Dawes, H.H. Mantsch, Thiocyanate levels in human saliva: quantitation by Fourier transform infrared spectroscopy, Anal. Biochem. 240 (1996) 7–12.
[237] J.D. Chandler, B.J. Day, Thiocyanate: a potentially useful therapeutic agent with host defense and antioxidant properties, Biochem. Pharmacol. 84 (2012) 1381–1387.

[238] E.L. Thomas, P.M. Bozeman, M.M. Jefferson, C.C. King, Oxidation of bromide by the human leukocyte enzyme myeloperoxidase and eosinophil peroxidase, J. Biol. Chem. 270 (1995) 2906–2913.
[239] H. Spalteholz, K. Wenske, J. Arnhold, Interaction of hypohalous acids and heme peroxidases with unsaturated phophatidylcholines, Biofactors 24 (2005) 67–76.
[240] P. Salavej, H. Spalteholz, J. Arnhold, Modification of amino acid residues in human serum albumin by myeloperoxidase, Free Radic. Biol. Med. 40 (2006) 516–525.
[241] K. Kumar, D.W. Margerum, Kinetics and mechanisms of general-acid-assisted oxidation of bromide by hypochlorite and hypochlorous acid, Inorg. Chem. 26 (1987) 2706–2711.
[242] M.T. Ashby, A.C. Carlson, M.J. Scott, Redox buffering of hypochlorous acid by thiocyanate in physiologic fluids, J. Am. Chem. Soc. 126 (2004) 15976–15977.
[243] D.I. Pattison, M.J. Davies, Absolute rate constants for the reaction of hypochlorous acid with protein side chains and peptide bonds, Chem. Res. Toxicol. 14 (2001) 453–464.
[244] C.L. Hawkins, D.I. Pattison, M.J. Davies, Hypochlorite-induced oxidation of amino acids, peptides and proteins, Amino Acids 25 (2003) 259–274.
[245] D.I. Pattison, M.J. Davies, Kinetic analysis of the reaction of hypobromous acid with protein components: implications for cellular damage and use of 3-bromotyrosine as a marker of oxidative stress, Biochemistry 43 (2004) 4799–4809.
[246] W. Wu, M.K. Samoszuk, S.A. Comhair, M.J. Thomassen, M.J. Farver, R.A. Dweik, M.S. Kavaru, S.C. Erzurum, S.L. Hazen, Eosinophils generate brominating oxidants in allergen-induced asthma, J. Clin. Investig. 105 (2000) 1455–1463.
[247] J.D. Chandler, B.J. Day, Biochemical mechanisms and therapeutic potential of pseudohalide thiocyanate in human health, Free Radic. Res. 49 (2015) 695–710.
[248] J. Flemmig, J. Gau, D. Schlorke, J. Arnhold, Lactoperoxidase as a potential drug target, Expert Opin. Ther. Targets 20 (2016) 447–461.
[249] O. Skaff, D.I. Pattison, M.J. Davies, Hypothiocyanous acid reactivity with low-molecular-mass and protein thiols: absolute rate constants and assessment of biological relevance, Biochem. J. 422 (2009) 111–117.
[250] O. Skaff, D.I. Pattison, P.E. Morgan, R. Bachana, V.K. Jain, I. Priyardasini, M.J. Davies, Selenium-containing amino acids are targets for myeloperoxidase-derived hypothiocyanous acid: determination of absolute rate constants and implications for biological damage, Biochem. J. 441 (2012) 305–316.
[251] C.L. Hawkins, The role of hypothiocyanous acid (HOSCN) in biological systems, Free Radic. Res. 43 (2009) 1147–1158.
[252] T.J. Barrett, C.L. Hawkins, Hypothiocyanous acid: benign or deadly? Chem. Res. Toxicol. 25 (2012) 263–273.
[253] M.T. Ashby, Hypothiocyanite, Adv. Inorg. Chem. 64 (2012) 263–303.
[254] J. Chung, J.L. Wood, Oxidation of thiocyanate to cyanide and sulfate by the lactoperoxidase-hydrogen peroxide system, Arch. Biochem. Biophys. 141 (1970) 73–78.
[255] T.M. Aune, E.L. Thomas, Accumulation of hypothiocyanite ion during peroxidase-catalyzed oxidation of thiocyanate ion, Eur. J. Biochem. 80 (1977) 209–214.
[256] M. Arlandson, T. Decker, V.A. Roongta, L. Bonilla, K.H. Mayo, J.C. MacPherson, S.L. Hazen, A. Slungaard, Eosinophil peroxidase oxidation of thiocyanate. Characterization of major reaction products and a potential sulfhydryl-targeted cytotoxicity system, J. Biol. Chem. 276 (2001) 215–224.
[257] M. Huwiler, H. Kohler, Pseudo-catalytic degradation of hydrogen peroxide in the lactoperoxidase/H_2O_2/iodide system, Eur. J. Biochem. 141 (1984) 69–74.
[258] F. Bafort, O. Parisi, J.-P. Perraudin, M.H. Jijakli, Mode of action of lactoperoxidase as related to its antimicrobial activity: a review, Enzym. Res. (2014), Article ID 517164, 1–13, https://doi.org/10.1155/2014/517164.
[259] W. Gottardi, Iodine and disinfection: theoretical study on mode of action, efficiency, stability, and analytical aspects in the aqueous system, Arch. Pharm. Pharm. Med. Chem. 332 (1999) 151–157.
[260] J. Rendl, M. Luster, C. Reiners, Serum inorganic iodide determined by paired-ion reversed-phase HPLC with electrochemical detection, J. Liq. Chromatogr. Relat. Technol. 20 (2006) 1445–1449.
[261] J. Ruf, P. Carayon, Structural and functional aspects of thyroid peroxidase, Arch. Biochem. Biophys. 445 (2006) 269–277.
[262] J.P. Henderson, J. Byun, M.V. Williams, D.M. Mueller, M.L. McCormick, J.W. Heinecke, Production of brominating intermediates by myeloperoxidase. A transhalogenation pathway for generating mutagenic nucleobases during inflammation, J. Biol. Chem. 276 (2001) 7867–7875.
[263] D. Schlorke, J. Flemmig, C. Birkemeyer, J. Arnhold, Formation of cyanogen iodide by lactoperoxidase, J. Inorg. Biochem. 154 (2016) 35–41.
[264] D. Schlorke, J. Atosuo, J. Flemmig, E.-M. Lilius, J. Arnhold, Impact of cyanogen iodide in killing of *Escherichia coli* by the lactoperoxidase-hydrogen peroxide-(pseudo)halide system, Free Radic. Res. 50 (2016) 1287–1295.

[265] C.M. Gerritsen, D.W. Margerum, Non-metal redox kinetics: hypochlorite and hypochlorous acid reactions with cyanide, Inorg. Chem. 29 (1990) 2757–2762.

[266] C.M. Gerritsen, M. Gazda, D.W. Margerum, Nonmetal redox kinetics: hypobromite and hypoiodite reactions with cyanide and the hydrolysis of cyanogen halides, Inorg. Chem. 32 (1993) 5739–5748.

[267] H.M. Abu-Soud, D. Maitra, J. Byun, C.E. Souza, J. Banajee, G.M. Saed, M.P. Diamond, P.R. Andreana, S. Pennathur, The reaction of HOCl and cyanocobalamin: corrin destruction and the liberation of cyanogen chloride, Free Radic. Biol. Med. 52 (2012) 616–625.

[268] C. Riganti, E. Gazzano, M. Polimeni, E. Aldieri, D. Ghigo, The pentose phosphate pathway: an antioxidant defense and a crossroad in tumor cell fate, Free Radic. Biol. Med. 53 (2012) 421–436.

[269] D. Roy, Q. Cai, Q. Felty, S. Narayan, Estrogen-induced generation of reactive oxygen and nitrogen species, gene damage, and estrogen-dependent cancers, J. Toxicol. Environ. Health B Crit. Rev. 10 (2007) 235–257.

[270] M.J. Coon, Cytochrome P450: nature's most versatile biological catalyst, Annu. Rev. Pharmacol. Toxicol. 45 (2005) 1–25.

[271] J.L. Vermilion, D.P. Ballou, V. Massey, M.J. Coon, Separate roles of FMN and FAD in catalysis by liver microsomal NADPH-cytochrome P-450 reductase, J. Biol. Chem. 256 (1981) 266–277.

[272] A.I. Shen, T.D. Porter, T.E. Wilson, C.B. Kasper, Structural analysis of the FMN binding domain of NADPH-cytochrome P-450 oxidoreductase by site-directed mutagenesis, J. Biol. Chem. 264 (1989) 7584–7589.

[273] J.P. Kehrer, The Haber-Weiss reaction and mechanisms of toxicity, Toxicology 149 (2000) 43–50.

[274] W.H. Koppenol, The Haber-Weiss cycle – 70 years later, Redox Rep. 6 (2001) 229–234.

[275] F. Haber, J.J. Weiss, Über die Katalyse des Hydroperoxydes, Naturwissenschaften 20 (1932) 948–950.

[276] B. Halliwell, J.M.C. Gutteridge, O_2 toxicity, O_2 radicals, transition metals and disease, Biochem. J. 219 (1984) 1–14.

[277] G.R. Buettner, B.A. Jurkiewicz, Catalytic metals, ascorbate, and free radicals: combinations to avoid, Radiat. Res. 145 (1996) 532–541.

[278] V. Oliveira-Marques, H.S. Marinho, L. Cyme, F. Antunes, Role of hydrogen peroxide in NF-κB activation: from inducer to modulator, Antioxidants Redox Signal. 11 (2009) 2223–2243.

[279] R. Brigelius-Flohé, L. Flohé, Basic principles and emerging concepts in redox control of transcription factors, Antioxidants Redox Signal. 15 (2011) 2335–2381.

[280] M.J. Morgan, Z.-g. Liu, Crosstalk of reactive oxygen species and NF-κB signaling, Cell Res. 21 (2011) 103–115.

[281] S. Petri, S. Körner, M. Kiaei, Nrf2/ARE signaling pathway: key mediator in oxidative stress and potential therapeutic target in ALS, Neurol. Res. Int. (2012), Article 878030, 1–7, https://doi.org/10.1155/2012/878030.

[282] Q. Ma, Role of Nrf2 in oxidative stress and toxicity, Annu. Rev. Pharmacol. Toxicol. 53 (2013) 401–426.

[283] E. Kansanen, S.M. Kuosmanen, H. Leinonen, A.-L. Levonen, The Keap1-Nrf2 pathway: mechanisms of activation and dysregulation in cancer, Redox Biol 1 (2013) 45–49.

[284] K. Itoh, T. Chiba, T. Takahashi, K. Ishii, Y. Igarashi, T. Katoh, T. Oyake, N. Hayashi, K. Satoh, I. Hatayama, M. Yamamoto, Y. Nabeshima, An Nrf2/small Maf heterodimer mediates the induction of phase II detoxifying enzyme genes through antioxidant response elements, Biochem. Biophys. Res. Commun. 236 (1997) 313–322.

[285] M. McMahon, K. Itoh, M. Yamamoto, J.D. Hayes, Keap1-dependent proteasomal degradation of transcription factor Nrf2 contributes to the negative regulation of antioxidant response element-driven gene expression, J. Biol. Chem. 278 (2003) 21592–21600.

[286] A. Carreau, B. El Hafny-Rahbi, A. Matejuk, C. Grillon, C. Kieda, Why is the partial oxygen pressure of human tissues a crucial parameter? Small molecules and hypoxia, J. Cell Mol. Med. 15 (2011) 1239–1253.

[287] S.R. McKeown, Defined normoxia, physoxia and hypoxia in tumours – implications for treatment response, Br. J. Radiol. 87 (2014), 20130676, 1–12, https://doi.org/10.1259/brj.20130676.

CHAPTER

3

Oxidation and Reduction of Biological Material

3.1 UNWANTED DESTRUCTION OF BIOLOGICAL MATERIAL

Unwanted destruction of biological material can highly disturb the integrity of cells and tissues. Direct reasons for these destructions are the action of reactive species and activities of decomposition enzymes from necrotic host's cells or from foreign organisms. Moreover, different carbonyls and a wide range of toxic agents can also contribute to cell and tissue damage. Altogether, destructive mechanisms are very complex and multifaceted. Serious problems arise for animals and humans, when destructive processes massively take place and when naturally protecting mechanisms are limited or exhausted.

In this chapter, the main focus is directed on processes of cell and tissue destruction caused by reactive species including a short survey about prominent mechanisms of protection against these deteriorations. In many cases, reactive species are able to oxidize certain functional target groups in biological material, thus initiating subsequent redox reactions and substrate alterations. In addition, reductive processes are also involved in deterioration of cell constituents. For example, in lipids and other materials, the reduction of organic hydroperoxide moieties by transition metal ions can initiate novel cascades of damaging reactions.

A short overview about easily oxidizable biological material is given in Appendix 1. As redox processes are often involved in damaging reactions in cells and tissues, a survey about chemistry and physics of redox processes is also included in this appendix.

Naturally occurring and dietary antioxidants protect our tissues against oxidative damage. Basic mechanisms of antioxidative defense by small molecules and antioxidative proteins are also subject of this chapter.

3.2 OXIDATION OF LIPIDS

3.2.1 Initial Events in Oxidation of Lipid Molecules: Abstraction of an H-Atom

Phospholipids and cholesterol are the most abundant lipid constituents in biological membranes. Their oxidation can lead to fatal consequences such as increased permeability and loss of barrier function. Furthermore, numerous lipid oxidation products can disturb the arrangement of lipids and other components and cause subsequent alterations in physicochemical properties in membranes. Oxidized lipid components are also present in lipoproteins.

Cell and Tissue Destruction
https://doi.org/10.1016/B978-0-12-816388-7.00003-6

Abstraction of an H-atom from a polyunsaturated fatty acid residue under formation of a C-centered radical occurs mainly at allylic sites to a double bond, especially at *bis*-allylic position between two double bonds, as these H-atoms are weaker bound than other hydrogens (Fig. 3.1). Binding enthalpies and reduction potentials for different H-atoms in fatty acid chains are given in Table 3.1. The standard reduction potentials of the couples allylic-C$^{\bullet}$/allylic-CH and *bis*-allylic-C$^{\bullet}$/*bis*-allylic-CH were calculated to be 0.7 and 0.6 V at pH 7, respectively [1]. The following reactive species can abstract these H-atoms: superoxide ($HO_2^{\bullet}$) [2], nitrogen dioxide ($NO_2^{\bullet}$) [3], thiyl radical (–S$^{\bullet}$) [4], and hydroxyl radical (HO$^{\bullet}$). Lipid peroxyl (LOO$^{\bullet}$) and alkoxyl (LO$^{\bullet}$) are also able to abstract these weakly bound H-atoms.

In saturated fatty acid chains, all other H-atoms are more tightly bound. For abstraction of these H-atoms, higher energy is required, which can be only supplied by hydroxyl radicals [5].

Carbon-centered radicals with a hydroxyl group in α-position to the radical center can also be formed in the hydrophilic part of lipid molecules. In this case, an H-atom is preferentially abstracted from a C-atom that bears a hydroxyl group and is additionally located in β-position to a phosphoester bond [6–8]. Because of

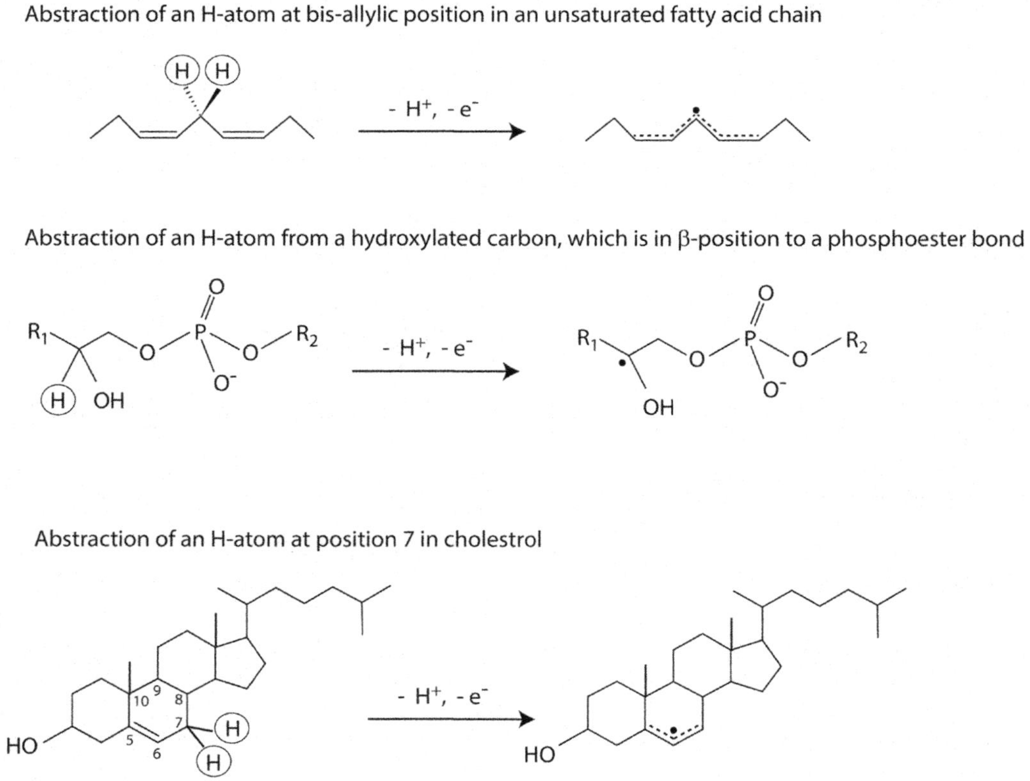

Abstraction of an H-atom at bis-allylic position in an unsaturated fatty acid chain

Abstraction of an H-atom from a hydroxylated carbon, which is in β-position to a phosphoester bond

Abstraction of an H-atom at position 7 in cholestrol

FIGURE 3.1 Scheme illustrating common locations for abstraction of an H-atom during oxidation of lipid molecules. Easily abstractive H-atoms are encircled. In cholesterol, the C-atoms of the B ring are numbered. Further explanations are given in the text.

TABLE 3.1 Binding Enthalpies and Standard Reduction Potentials for Different H-Atoms Bound to C-Atoms in Fatty Acid Chains [1].

C–H bond[a]	Binding Enthalpy	E'° at pH 7
$-CH{=}CH-\mathbf{CH_2}-CH{=}CH-$	335 kJ mol^{-1}	0.6 V
$-\mathbf{CH_2}-CH{=}CH-\mathbf{CH_2}-$	344 kJ mol^{-1}	0.7 V
$-CH_2-\mathbf{CH}{=}\mathbf{CH}-CH_2-$	372 kJ mol^{-1}	0.96 V
$-\mathbf{CH_2}-\mathbf{CH_2}-\mathbf{CH_2}-\mathbf{CH_2}-$	422–439 kJ mol^{-1}	1.5–1.6 V

[a] *Corresponding H-atoms are bold.*

electron-drawing properties of both adjacent oxygens, the C–H bond is characterized by a low binding enthalpy. This situation is given in cardiolipin, phosphatidylglycerols, phosphatidylinositols, and lysophospholipids (Fig. 3.1). A similar structural arrangement is given in glycolipids, where instead of a phosphoester bond a glycosidic bond is involved.

In cholesterol, an H-atom is predominantly abstracted by reactive species from C7 (Fig. 3.1) that is in allylic position to the C5–C6 double bond [9].

3.2.2 Initial Events in Oxidation of Lipid Molecules: Attacks on Aliphatic Double Bonds

Aliphatic double bonds in fatty acid residues can be directly attacked by reactive species such as $^1O_2$, O_3, $HO^•$, and $NO_2^•$. No radical intermediates are involved in the primary attack of $^1O_2$ to an aliphatic double bond in lipids [10]. In this "ene" addition reaction, $^1O_2$ targets one of the allylic C-atoms under formation of a hydroperoxide moiety and induces a shift of the double bond by one position in opposite direction [11]. The interaction of $^1O_2$ with conjugated double bonds leads to formation of a cyclic endoperoxide [10]. Interaction of $^1O_2$ with the C5–C6 double bond in cholesterol leads to hydroperoxides and later alcohol moieties at 5 and 6 [12].

Addition of ozone to a double bond yields a trioxide that decays into a Criegee ozonide and radical products [13].

In unsaturated fatty acid residues, both $HO^•$ and $NO_2^•$ are also known to be added to one of the C-atoms of a double bond under disappearance of this unsaturated bond and formation of a C-centered radical [3,5].

The hypohalous acids HOCl and HOBr interact with aliphatic double bonds under formation of chlorohydrins and bromohydrins [14–16].

Selected examples of targeting by reactive species to aliphatic double bonds in lipids are illustrated in Fig. 3.2.

3.2.3 Initial Events in Oxidation of Lipid Molecules: Attacks on Hydroperoxides

Lipid hydroperoxides can accumulate in biological membranes and lipoproteins as a result of nonenzymatic reactions. Lipid hydroperoxides are restricted in their reactivity as they contain fully occupied antibonding π^* orbitals. This restriction is larger expressed in comparison with H_2O_2 [17–19]. They are either utilized as substrate in peroxidase-catalyzed reactions or involved in decay processes catalyzed by transition metal ions (Fig. 3.3). Glutathione peroxidase 4 (Gpx4) converts lipid hydroperoxides (LOOH) into the corresponding alcohols (LOH). Fe^{2+} and Cu^+ generate from LOOH reactive alkoxyl radicals ($LO^•$). Latter reaction is important in uncontrolled lipid peroxidation (Section 3.2.4).

It should be noted that lipid hydroperoxides of arachidonic acid such as prostaglandin G_2, 5-HPETE, 12-HPETE, and 15-HPETE are formed in reactions catalyzed by prostaglandin H synthase and different lipoxygenases in human and animal organisms during normal metabolic activity and in greater amount during inflammatory processes. Although these hydroperoxides are further converted into physiological active prostaglandins, thromboxanes, leukotrienes,

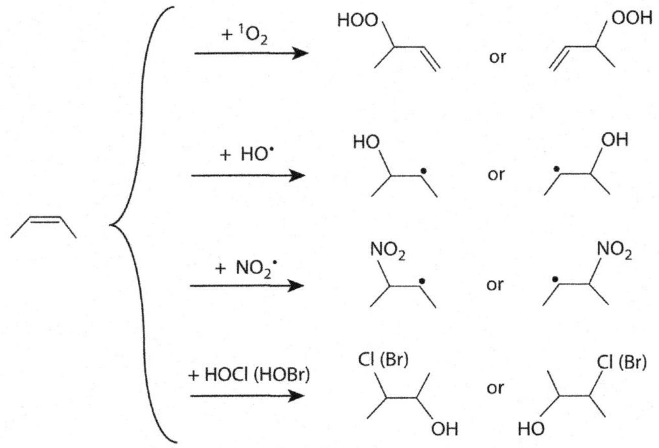

FIGURE 3.2 Common examples of direct targeting of aliphatic double bonds in lipids by reactive species.

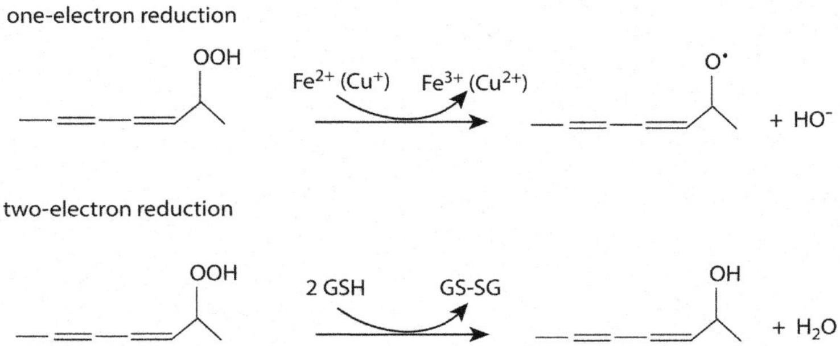

FIGURE 3.3 One- and two-electron reduction of aliphatic lipid hydroperoxides. GSH and GS-SG stand for reduced and oxidized glutathione, respectively.

and lipoxins [20], they can also interact if present with free metal ions. While prostaglandin H synthase I is ubiquitously distributed, the inducible isoform II of this enzyme plays a prominent role in immune response reactions. Different lipoxygenases occur only in leukocytes and mast cells and are mainly active under inflammatory conditions.

It has been assumed that lipid hydroperoxides can also interact with HOCl [21–23]. However, detailed investigation of this subject did not yield any evidence for this interaction [24].

3.2.4 Lipid Peroxidation as a Chain Reaction

A simplified scheme of the chain process of lipid peroxidation is given in Fig. 3.4. In most cases, the abstraction of an H-atom from a fatty acid residue (LH) represents the initial step in lipid peroxidation. This abstraction occurs mainly at a C-atom of a methylene groups between two double bonds. The resulting radical moiety ($L^{\bullet}$) is delocalized over five adjacent carbon atoms under formation of a so-called pentadienyl radical. The transient formation of this

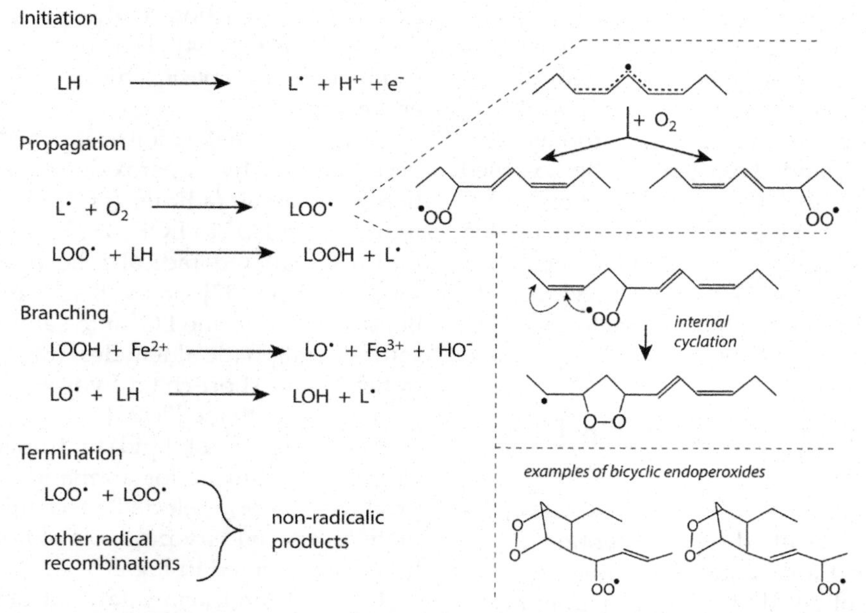

FIGURE 3.4 Simplified scheme of the lipid peroxidation as chain reaction. The inserts show the addition of dioxygen to an alkyl radical, the internal cyclation of a peroxyl radical with formation of a cyclic endoperoxide radical, and examples of bicyclic endoperoxide peroxyl radicals. *L•*, alkyl radical; *LH*, fatty acid chain with an abstractive H-atom; *LO•*, alkoxyl radical; *LOO•*, peroxyl radical; *LOOH*, lipid hydroperoxide. Further explanations are given in the text.

radical can be followed by its strong absorbance at 280 nm [25]. Moreover, pentadienyl radicals can undergo several cis–trans transitions. As a result, their reaction products are characterized by a variety of stereochemical isomers [26].

The alkyl radical ($L^•$) reacts rapidly with dioxygen to yield a peroxyl radical ($LOO^•$). As dioxygen is usually added at one of terminal C-atoms of the pentadienyl structure of $L^•$, the resulting $LOO^•$ bears two conjugated double bonds (forming a so-called diene conjugate) characterized by an intense absorbance at 234 nm [27]. Among lipid radicals, peroxyl radicals are the most stable ones. They predominantly abstract an H-atom from *bis*-allyl methylene bridges of neighbored fatty acid residues forming a lipid hydroperoxide (LOOH) and a novel alkyl radical ($L^•$) [5]. As primary products of this reaction cascade, lipid hydroperoxides result. In polyunsaturated acyl chains, the peroxyl radical moiety can also abstract an H-atom within the same acyl chain under formation of cyclic and bicyclic endoperoxides [28,29].

Lipid hydroperoxides interact with free ferrous ions, whereby the hydroperoxide moiety is reduced to an alkoxyl radical ($LO^•$) and further to an alcohol species (LOH) by abstracting an H-atom from another lipid molecule.

$$LOOH + Fe^{2+} \longrightarrow LO^• + Fe^{3+} + {}^{-}OH$$

$$LO^• + LH \longrightarrow LOH + L^•$$

Of course, ferrous ions can be replaced by Cu^+ in the first reaction. Both reactions contribute to branching of the chain reaction in lipid peroxidation.

Termination of chain process of lipid peroxidation is realized by recombination of two radicals. As peroxyl radicals predominate as radical species, usually the recombination of two $LOO^{\bullet}$ takes place under formation of a carbonyl compound, alcohol, and dioxygen by the so-called Russell mechanism [30]. In this reaction, either the carbonyl or dioxygen is formed in an excited state [31,32]. As a result of radical recombination under involvement of alkyl and alkoxyl radicals, dimeric and oligomeric products are formed [11].

3.2.5 Secondary Products of Lipid Peroxidation

Chemical processes in lipid peroxidation are very complex. Already the abovementioned processes of radical formation in unsaturated fatty acid residues followed by generation of lipid hydroperoxides lead to a variety of stereoisomeric products [26]. Further decomposition of these hydroperoxides and cyclic endoperoxides generates numerous secondary products including reactive carbonyls. It is impossible to describe here all facets of these reactions. Overviews about these reaction pathways are given in Refs. [11,13,26,33,34]. Hence, only dominating products and basic mechanisms will be enumerated.

Transition metal ion–catalyzed homolytic decay of lipid hydroperoxides results in formation of alkoxyl radicals ($LO^{\bullet}$). In addition to reduction of $LO^{\bullet}$ to LOH, $LO^{\bullet}$ can be converted into a C-centered radical followed by epoxide formation [35–37] or by β-scission of a C–C bond adjacent to the $LO^{\bullet}$ into saturated and unsaturated aliphatic aldehydes, fragmentary lipid radicals, and short-chain hydrocarbons such as pentane and ethane [26,33].

Hock scission of lipid hydroperoxides is a further mechanism for formation of carbonyl products [26,38]. Selected mechanisms for formation of secondary products of lipid peroxidation are presented in Fig. 3.5.

The heat- and acid-catalyzed decomposition of prostaglandin-like cyclic endoperoxides is the source for malondialdehyde [39], a species often used for general detection of lipid peroxidation products [34]. The significance of malondialdehyde for biological degradation is often overestimated. Most malondialdehyde is only

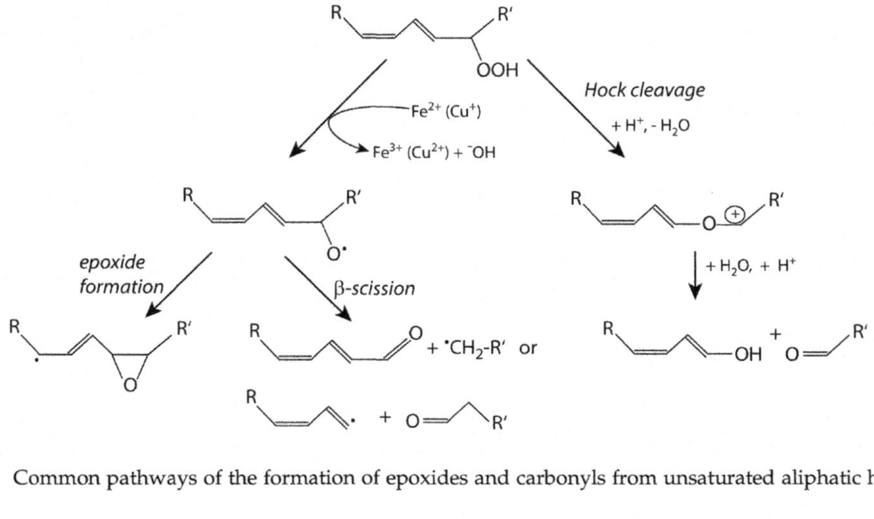

FIGURE 3.5 Common pathways of the formation of epoxides and carbonyls from unsaturated aliphatic hydroperoxides.

formed under assay conditions as in case of application of thiobarbituric acid to detect malondialdehyde.

3.2.6 Carbonyl-Based Formation of Advanced Lipoxidation End Products

During lipid peroxidation, different short chain carbonyls are formed such as α,β-unsaturated aldehydes (4-hydroxy-2-nonenal (HNE), 4-hydroxy-2-hexenal (HHE), 4-oxo-2-nonenal (ONE), 2-nonenal, acrolein) and dicarbonyls (malondialdehyde, glyoxal, methylglyoxal, isoketals, levuglandins D_2 and E_2) [40,41]. Important short chain carbonyls are shown in Fig. 3.6. In oxidized phospholipids, the carbonyl moiety can also be on a truncated fatty acyl chain [42,43].

Carbonyls are highly important for tissue destruction. All carbonyls interact easily with primary amines under formation of Schiff bases and conjugated Schiff bases [44,45]. The latter bases are formed when both carbonyl moieties of malondialdehyde react with amines. Cyclic aldehydes also result in interaction between aldehyde and amines [46,47]. Thus, reactive carbonyls can affect not only lipids but also other cellular constituents such as proteins and nucleic acids.

Special attention has been directed on reactions of HNE, HHE, and ONE. Because of the electron-withdrawing capacities of the carbonyl group and the 4-hydroxy group (or 4-oxo group), an electron deficiency results at the alkene β-carbon atom (C3) [48] that promotes at this site reactions with nucleophilic species such the sulfhydryl group of cysteine, imidazole group of histidine, or the ε-amino group of lysine. This reaction yields a Michael adduct, which can further be converted into a cyclic hemiacetal compound when a hydroxy substituent is at position 4 in the reacting aldehyde [49,50]. In formation of Michael adducts, the most reactive carbonyl species is ONE [51]. This aldehyde also forms a stable adduct with arginine [52]. By subsequent reactions and rearrangements, further products are formed. For example, HNE yields 2-pentylpyrroles, 2-hydroxy-2-pentyl-1, 2-dihydropyrrol-3-one iminium, a mixed Michael adduct and Schiff base product, and cyclic carbinolamine [53,54]. Main reaction routes of α,β-unsaturated aldehydes on the example of HNE are presented in Fig. 3.6 as well.

Taken together, peroxidation of lipids generates a large variety of products including reactive carbonyls that can further interact with other biological macromolecules and thus considerably enlarge the product diversity. In relation to enormous diversity of glycated products, that are called advanced glycation end products (AGEs), a similar term, namely advanced lipoxidation end products was introduced for products generated as a result of lipid peroxidation [55].

3.2.7 Fragmentation of Lipid Molecules in Its Hydrophilic Region

In polar head groups of lipid molecules, C-centered radicals can also be formed after abstraction of an H-atom from a carbon bearing a hydroxyl group and being in β-position to a phosphoester, glycosidic, or amide bond [6–8]. After initial formation of an α-hydroxyl carbon-centered radical, a transient five-membered ring system is formed followed by two bond ruptures in β-position in respect to the radical center. The original lipid molecule is fragmented into two smaller products. Details are given in Fig. 3.7. One of the two fragmentation products bears a radical function. This radical can abstract an H-atom from another unperturbed lipid molecule.

Initial abstraction of an H-atom is favored by the presence of transition metal ions or by γ-irradiation. The participation of hydroxyl radicals has been proposed. Importantly, the primary

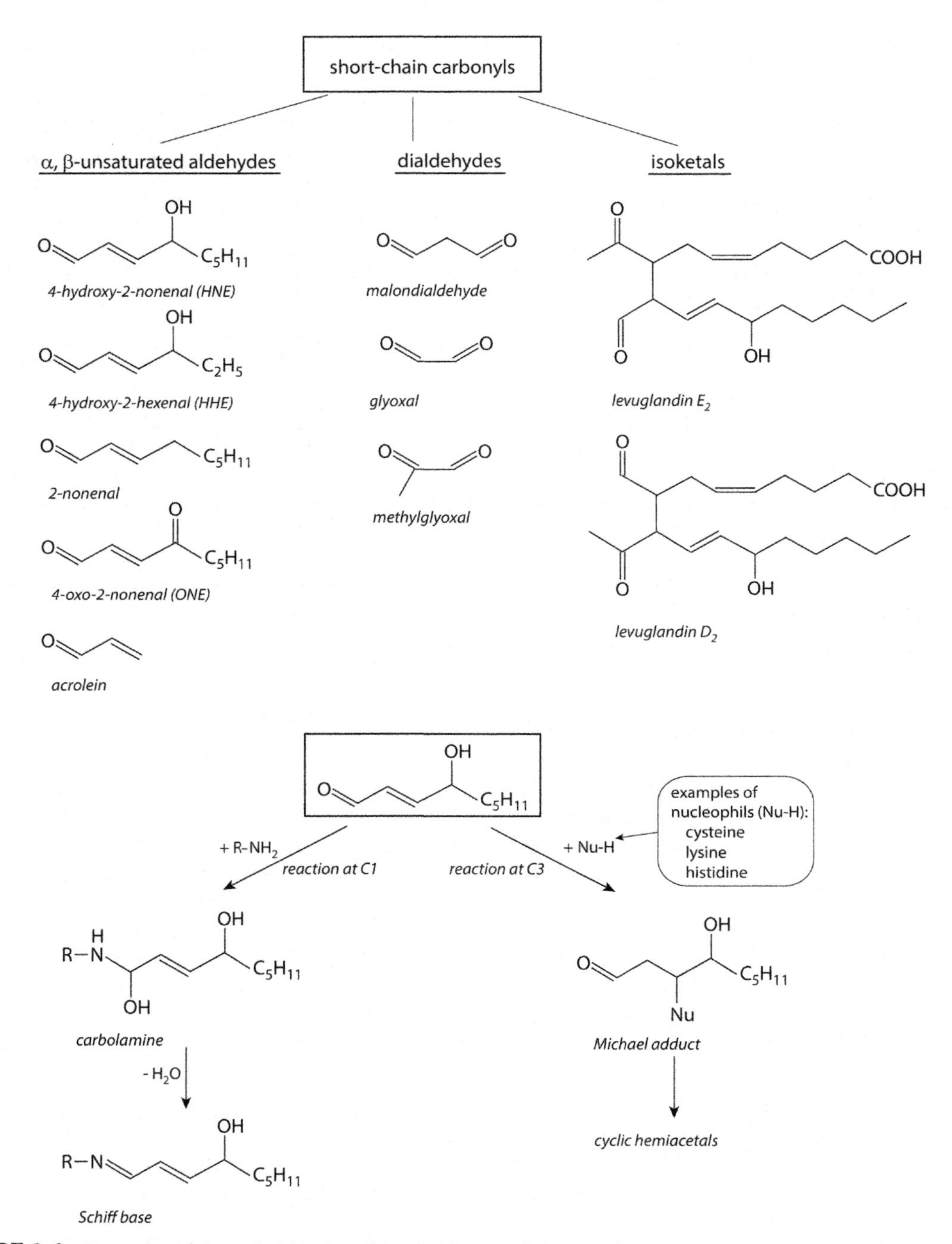

FIGURE 3.6 Examples of short-chain carbonyls as lipid peroxidation products. Main reactions routes of α,β-unsaturated aldehydes are shown for 4-hydroxy-2-nonenal in the lower part of the figure.

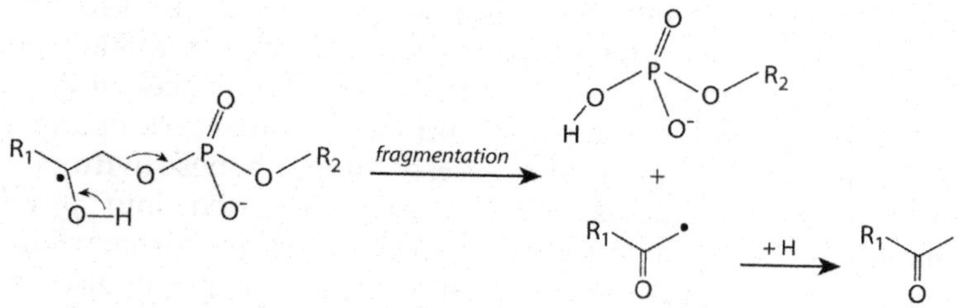

FIGURE 3.7 Fragmentation of a hydroxylated carbon-centered radical being in β-position to a phosphoester bond. This reaction pattern occurs in cardiolipin, lysophospholipids, phosphatidylglycerol, and phosphatidylinositol. Further explanations are given in the text.

radical forming attack occurs in the hydrophilic part of lipid molecules, which is exposed to the water phase.

In case of cardiolipin, phosphatidic acid and phosphatidyl hydroxyacetone are described as fragmentation products [56,57]. Phosphatidic acid was also detected as fragmentation product in experiments with phosphatidylinositol and phosphatidylglycerol [8]. Galactocerebrosides are an important constituent of the myelin sheath in the central and peripheral nervous system. Their free radical transformation yields ceramide as fragmentation product [58,59].

An important aspect of these investigations is the detection of lipid molecules as fragmentation products that may act as signaling molecules in physiological processes [8]. Of course, on oxidative processes in complex lipids, both oxidative alterations in fatty acyl chains and hydrophilic part of the molecule might occur.

3.2.8 Steroid Oxidation

Cholesterol as the major membrane steroid undergoes both nonenzymatic and enzymatic oxidation. Oxidative products of cholesterol are collectively called oxysterols.

In cholesterol, the predominant place for initial attack by one-electron oxidants is C7, which is in allylic position to the double bond [9]. Similar to peroxidation in unsaturated fatty acyl chains, at C7, a carbon-centered radical, peroxyl radical, and hydroperoxide moiety (7α- and 7β-OOH) are subsequently formed. Decomposition of hydroperoxides by free metal ions leads to 7α/β-hydroxycholesterol and 7-ketocholesterol [9,60]. Abstraction of an H-atom from side chain carbons is another major pathway in cholesterol oxidation. This occurs mainly by several enzymes of the cytochrome P450 family, by cholesterol 25-hydroxylase, a nonheme protein, or nonenzymatically. A variety of hydroxylated oxysterols is formed [9,61].

Another source for oxysterols in the human body is the uptake of these products with food. Oxysterols can be absorbed in the gut and distributed via circulating chylomicrons to tissues [9,62].

In unperturbed tissues, the amount of oxysterols is low, not exceeding 0.1% of total cholesterol. It is assumed that this low level of oxysterols has a relatively small impact on membrane properties [61]. To keep the cholesterol level constant in biomembranes and to maintain physicochemical quality in membrane bilayers, several mechanisms ensure the homeostasis of cholesterol [9]. Among these mechanisms are the action of certain oxysterols as ligand for liver X nuclear receptors, regulating genes that encode proteins involved in cholesterol synthesis and export [63]. Other oxysterol-mediated effects are the accelerated degradation of 3-hydroxy-3-methyl-glutaryl-CoA reductase, a key enzyme of cholesterol synthesis, the suppression of activation of the sterol regulatory element binding protein pathway, and the increased intracellular storage of cholesterol as esters [64].

In biological samples, cholesterol is less susceptible to nonenzymatic oxidation than unsaturated lipids [65]. Otherwise, in vitro studies with stressed cells demonstrate higher abundance of oxysterols over other oxidized lipids. Apparently, oxidized phospholipids are more efficiently recycled than oxysterols [9,66].

Under some pathophysiological conditions, oxysterol level may increase and significantly disturb membrane and lipoproteins properties. Increased amounts of these products are detected in atherosclerotic plaques [9]. Some oxysterols can pass the blood barrier in contrast to cholesterol [67]. The involvement of oxysterols in the pathogenesis of neurodegenerative disorders is under discussion [61]. Oxysterols are also associated with some kinds of cancer [68].

3.2.9 Physiological Significance of Oxidative Damage in Lipids

It is believed that processes of lipid oxidation play an important role in a number of pathologies. Despite numerous investigations, it is hard to detect an increase of peroxidation products, resulting in vivo only from nonenzymatic processes. In cells and tissues, lipids are usually well protected by antioxidants and mechanisms controlling transition metal ions and metal complexes. Moreover, any damaged material and undergoing cells are removed by professional phagocytes during metabolic processes contributing, thus, to homeostatic conditions in tissues.

Otherwise, oxidized forms of lipids, especially of arachidonic acid, are involved in numerous physiological functions. Therefore, it is challenging to make a clear separation between nonenzymatic and enzymatic processes of lipid peroxidation. Hydroperoxides of arachidonic acid are also produced by prostaglandin H synthases and lipoxygenases. These reactions are enhanced under inflammatory conditions that often accompany a pathological process. Although these hydroperoxides are usually further converted into physiologically active prostaglandins, thromboxanes, leukotrienes, and lipoxins, they can also interact with transition metal ions, when local concentration of free iron and copper is increased at inflamed sites and protection by glutathione peroxidase and glutathione fails. Thus, physiologically active hydroperoxides of arachidonic acid can also contribute to novel nonenzymatic oxidative processes in lipid material.

Taken together, the following factors favor an enhanced oxidation of lipid material:

i) increased exposure of living structures to radiation and reactive species,
ii) exhaustion of naturally occurring antioxidants including glutathione, and
iii) enhanced values of transition free metal ions.

It is most likely that these conditions are only realized at some inflammatory loci when a large number of cells die and necrotic cell material accumulates or at local loci where the blood flow is limited and an exhaustion of preventive mechanisms occurs over time.

It made no sense to regard destructive alterations in lipids without considering changes in other material of living systems. Under strong inflammatory conditions, there is a general depletion of protective mechanisms also favoring destructions in proteins, nucleic acids, and other cellular constituents and extracellular material. Moreover, reactive carbonyls from lipid peroxidation can react with amino groups of proteins and other molecules and disturb, thus, the structural arrangement of these components. Carbonyls are also associated with monomeric carbohydrates. In general, there is a complex picture of agents favoring destruction of biological material at strong inflammatory and pathologic conditions.

3.3 DESTRUCTIONS IN CARBOHYDRATES

3.3.1 Fragmentation of Hyaluronan and Sulfated Glycosaminoglycans

Polymeric carbohydrates are an essential part of extracellular matrix, mucous surfaces, and the external cover of endothelial and other cells. Their turnover is regulated by the presence of an array of carbohydrate degradation enzymes that cleave nonsulfated and sulfated glycosaminoglycans into smaller units [69,70]. Dysregulation in glycosaminoglycan and proteoglycan dynamics is associated with numerous disease states [71,72]. Here, the attention is addressed on nonenzymatic mechanisms of fragmentation of glycosaminoglycans.

Hyaluronan consists of repeating disaccharide units of D-glucuronic acid-β-(1 → 3)-*N*-acetyl-D-glucosamine linked together with β-(1 → 4) glycosidic bonds. It is a nonbranched and nonsulfated carbohydrate polymer that is able to bind enormous amounts of water [73,74]. High molecular mass hyaluronan with a mean molecular weight of several MDa is an important constituent of synovial fluids. It determines viscoelastic properties of this fluid and ensures frictionless and pain-free movement of joints [75]. Hyaluronan is the main component of the nucleus pulposus, a gel-like matter, in intervertebral discs. It is also a main component of the vitreous body of the eye [76].

Fragmentation of hyaluronan into smaller units impairs the viscoelastic properties of this polymer, a process that is enhanced in inflamed joints in different forms of arthritis. Although molecular mechanisms of degradation are largely unknown, it is well known that transition metal ions favor this fragmentation. Again, it remains unclear whether hydroxyl radicals or some transition iron–oxygen complexes are the active agent. In any case, $HO^{\bullet}$ was reported to abstract H-atoms at all C–H ring bonds within glucuronic acid and N-acetylglucosamine expect at C(2) of the latter [69]. A subsequent cleavage of glycosidic bonds is favored by radical formation at several places of these structural elements [77,78].

Other reactive species, which can be involved in degradation of hyaluronan and sulfated glycosaminoglycans, are HOCl [79,80], HOBr [81], and peroxynitrite [82].

3.3.2 Cyclic and Open-Chain Forms of Monosaccharides

D-Glucose and related monomeric carbohydrates exist predominantly in the cyclic form. The dietary monosaccharides glucose, fructose, and galactose are absorbed in the intestine from our food (Section 4.7.2). In liver, the metabolism of these three compounds is linked to each other (Sections 4.7.3 and 4.7.4). Thus, fructose and galactose can stepwise be transformed into glucose and glucose metabolites. Glucose is the most important energy fuel for metabolic processes in cells and tissues.

For all monosaccharides, several isomeric cyclic forms and an open-chain form exist at equilibrium. In case of glucose, four cyclic forms are known, the α- and β-anomer of D-glucopyranose having a hemiacetal linkage between C1 and the C5 hydroxyl group and the α- and β-anomer of D-glucofuranose where C1 is linked with the hydroxyl group of C4. The latter two anomers are only present in negligible amounts. Interconversion between cyclic forms occurs always via the transient formation of the open-chain form. As a result of this slow mutarotation, a stable ratio of 36:64 is found between the α- and β-form of D-glucose [83].

Although the cyclic forms of monosaccharides are chemically less reactive, the open form bears a reactive free aldehyde group or as in case of fructose a keto group. Among monosaccharides, D-glucose has been determined to show the lowest carbonyl content with 0.002% in aqueous solution at 20°C [84,85]. The corresponding values are 0.02% for D-galactose and 0.7% for D-fructose [84]. It has been proposed that D-glucose has been selected as major energy substrate during evolution because it is characterized by the most stable cyclic form and the lowest ability among monosaccharides for side reactions [86].

Nevertheless, glucose is prone to undergo numerous spontaneous glycation reactions due to the presence of a free aldehyde group in its open-chain form. Under hyperglycemic conditions, the probability raises for these unwanted reactions.

3.3.3 Formation of Advanced Glycation End Products

In analyzing chemical alterations in foodstuffs during storage and cooking, the formation of browned products was carefully investigated. The basis of these reactions is the so-called Maillard reaction [87], a reaction between primary amino groups of amino acids, proteins, and others with the carbonyl moiety of linear monosaccharides. The chemistry of browning reactions is rather complex, includes several stages and parallel pathways, and leads to a large variety of intermediates and products [88]. Here is not the place to review all aspects of Maillard reaction. The focus will be directed on consequences of the interaction of D-glucose with amino groups of biological constituents to cells and tissues.

The carbonyl group of the linear form of D-glucose reacts easily with an amino group present in amino acids, proteins, and other biological material in a condensation reaction. The resulting *N*-substituted glycosylamine is converted via a Schiff base intermediate into a mixture of Amadori rearrangement products including 1-amino-1-deoxy-2-ketose [89–91]. Although being more stable than the previous intermediates, the Amadori products undergo a variety of further reactions with peptides and proteins. They are involved in dehydrations, fragmentations, oxidations, and polymerizations. In these pathways, numerous reactive carbonyls and dicarbonyls are formed, which cause further multiple substrate alterations [41,92–94]. Fission reactions of Amadori products are favored by the presence of transition metal ions [95]. Finally, a very heterogeneous group of compounds results collectively known as AGEs [41,96,97]. A rough scheme of the formation of Amadori products from reaction between glucose and amines is given in Fig. 3.8. Some important products of these conversions are also listed in this figure.

Fructose also undergoes a nonenzymatic condensation reaction with primary amines leading to formation of an intermediary Schiff base product like in reaction of glucose with amines. In the fructose-mediated Schiff base, the amino moiety is linked to carbon 2 unlike to carbon 1 in glucose-driven condensation. Further, the rearrangement of the fructose-based Schiff base leads to two types of products with a carbonyl group at C1 or C3 [98]. This reaction is also known as Heyns rearrangement. However, the Heyns pathway focusses on the exclusive participation of C1 seen under the harsh conditions in experimental setup [99]. The rearrangement products derived from condensation of fructose and amines undergo similar reactions as Amadori products. Although some products such as *N*-carboxymethyllysine, carboxyethyl-lysine, or pentosidine are identical in both pathways, specific fructose-mediated AGE products should be characterized in future research [100].

Because of its higher carbonyl content, fructose is believed to be more reactive than glucose. Fructose-mediated AGE formation is likely in the intestine and liver in patients having a high intake of fructose-rich food and in cells such as red blood cells, lens cells, epithelial cells, and neurons where glucose is accumulated in an uncontrolled manner under hyperglycemic conditions, and converted into sorbitol and fructose [100].

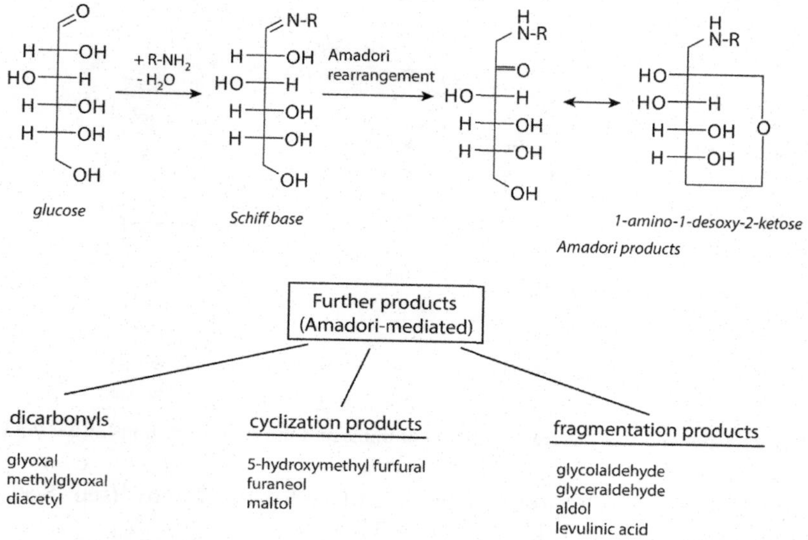

+ R-NH$_2$
- H$_2$O
Amadori rearrangement

glucose
Schiff base
1-amino-1-desoxy-2-ketose
Amadori products

Further products (Amadori-mediated)

dicarbonyls
glyoxal
methylglyoxal
diacetyl

cyclization products
5-hydroxymethyl furfural
furaneol
maltol

fragmentation products
glycolaldehyde
glyceraldehyde
aldol
levulinic acid

FIGURE 3.8 Formation of Amadori rearrangement products in the reaction between glucose and amines. Important products derived from Amadori products are also listed.

3.3.4 Biological Significance of Advanced Glycation End Products

Research has been focused on identification of special AGEs, which may have pathophysiological relevance. Among them are glycated hemoglobin [101], in particular hemoglobin A1c [102,103], carboxymethyl-lysine [104], pentosidine [105,106], 3-deoxyglucosone, methylglyoxal, glyoxal [107], glucosepane, carboxymethyl-hydroxylysine, carboxyethyl-lysine, fructose-lysine [104,107], and others.

Several receptors for AGEs have been identified [108–110]. The ubiquitary expressed receptor for AGEs (RAGE) is a multiligand, pattern recognition receptor [111]. Ligand binding to RAGE induces proinflammatory pathways via activation of the transcription factor NFκB [112].

AGEs are not only produced in vivo but also taken up by the organism with our food [113]. Excessive uptake of dietary AGEs can act cytotoxic to cells and tissues and can potentiate effects of host's own production of AGEs. In other words, the way of preparing our food and the kind of diet is also responsible for potential effects of AGEs [114,115].

In formation of AGEs and in processes of lipid peroxidation, numerous reactive carbonyls are involved. Because carbonyls interact easily with primary amines, most of all in form of proteins, both pathways have the potential to damage cells and tissues seriously. These processes play a role in development and progression of many disease scenarios.

3.4 OXIDATIONS IN PROTEINS

3.4.1 Oxidation of Sulfhydryls

In cell constituents, sulfhydryl groups are predominantly located in cysteine residues of proteins and glutathione. Because of their peculiar redox properties, these functional groups are involved in numerous physiological processes such as signal transduction, regulation of enzyme activities, control over transcription

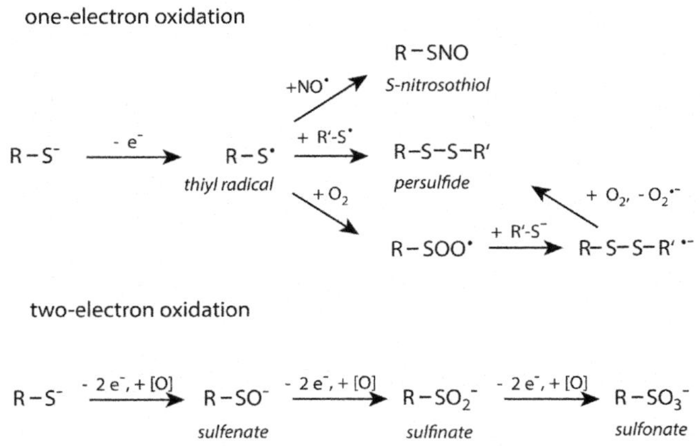

FIGURE 3.9 One- and two-electron oxidation of thiols and important reactions of the thiyl radical.

factors, functioning of ion channels, and mechanisms of redox homeostasis [116–119].

At neutral pH values, protein thiols are normally protonated as their pK_a values are around 8.5. Thiols with lower pK_a react more intensely because thiolate anions ($-S^-$) are more readily oxidized than thiols (−SH) [120].

Sulfhydryls can undergo both one- and two-electron oxidations (Fig. 3.9). One-electron oxidation of a protein sulfhydryl residue (R−SH) leads to a thiyl radical ($R-S^•$). The reduction potential of the redox couple $R-S^•$/$R-S^-$ is equal to 0.84 V [121]. Thiyl radical can add dioxygen to yield the unstable $R-SOO^•$ that is further converted into $R-SO^•$ or react more likely with a second thiol to $R-SS-R^{•-}$. The latter species interacts in an extremely fast reaction with O_2 under formation of R−SS−R and $O_2^{•-}$. The standard reduction potential of couple $R-SS-R/R-SS-R^{•-}$ is −1.5 V at pH 7, one of the lowest values in biological redox systems [121,122]. Important reactions of thiyl radicals are shown in Fig. 3.9.

In proteins, thiyl radicals are converted into disulfide residues either by forming inter- or intramolecular links between two thiols or by combining a protein thiol with glutathione. These disulfides can be reduced to the corresponding thiols by means of the thioredoxin or glutaredoxin system (see Sections 3.7.4 and 3.7.5).

The reaction between thiyl radical and NO yields S-nitrosothiols [123]. This reaction is the final step in nitrosation of biologically relevant thiols in the presence of NO, dioxygen, and reactive species. Both peroxynitrite and $NO_2^•$ are capable to abstract an electron from the thiol [123,124].

The rate of two-electron oxidation of sulfhydryls by H_2O_2 or organic hydroperoxides strongly depends on the microenvironment of thiol in proteins [125]. In peroxiredoxins, for example, the transition complex is stabilized allowing a simple substitution mechanism [126].

Other two-electron oxidation products of thiols are sulfenic acid (R−SOH), sulfinic acid ($R-SO_2H$), and sulfonic acid ($R-SO_3H$). The latter two species are only formed in the absence of suitable substrates for R−SOH. Sulfinic acids react readily with other thiols to yield disulfides or with amines under formation of sulfonamides [127].

3.4.2 Oxidation of Methionine Residues

Oxidation of methionine residues in proteins leads to formation of methionine sulfoxide and further in a slow reaction to methionine sulfone

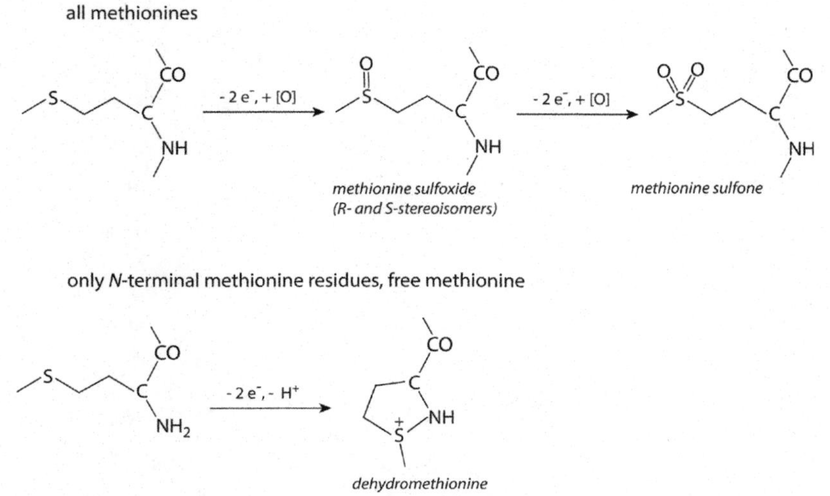

all methionines

methionine sulfoxide (R- and S-stereoisomers)

methionine sulfone

only *N*-terminal methionine residues, free methionine

dehydromethionine

FIGURE 3.10 Common pathways of methionine oxidation.

(Fig. 3.10). Methionine sulfoxide appears in two stereoisomeric forms. Hydrogen peroxide, HOCl, HOBr, taurine chloramine, peroxynitrite, ozone, $HO^{\bullet}$, and transition metal ion–based species are known to convert methionine into methionine sulfoxide [128]. Direct reaction between H_2O_2 and methionine occurs at higher concentrations of hydrogen peroxide [129,130]. In cells, methionine sulfoxide can be reduced to the corresponding thioether. L-methionine sulfone cannot be reversed.

A second pathway concerns the formation of dehydromethionine (Fig. 3.10). This pathway is only valid in free methionine and *N*-terminal methionine residues. Which of the both pathways is preferred in these thioethers depends on the oxidant's nature [131]. Hydrogen peroxide and peroxynitrite produce only methionine sulfoxide, while a mix of both products is generated by hypohalous acids. The impact of dehydromethionine increases from Cl over Br to I in these agents. Reactive iodine species generate dehydromethionine as main product. The reaction of $^1O_2$ with free methionine can also result in formation of dehydromethionine [132].

3.4.3 Oxidation of Other Amino Acid Residues

One-electron oxidation of tyrosine yields the tyrosyl radical ($Tyr^{\bullet}$) (Fig. 3.11). As the standard reduction potential of the redox couple $Tyr^{\bullet}/Tyr$ is 0.88 V at pH 7 [133], only highly reactive species and activated peroxidase complexes are able to oxidize tyrosine. In a protein or between proteins, two $Tyr^{\bullet}$ can be cross-linked to a dityrosine residue [134,135]. In biological samples, the detection of dityrosine cross-links is widely employed as a marker for oxidative stress [136–138]. $Tyr^{\bullet}$ is also involved in further oxidation steps, cyclization, and decarboxylation reactions, yielding dopamine, dopamine quinone, and 5,6-dihydroxyindol as major products [136,139]. Other oxidation products derived from L-tyrosine are trityrosine, pulcherosine, and isodityrosine [140].

Tryptophan oxidation occurs mainly on the pyrrole ring. Besides N-formylkynurenine and kynurenine (Fig. 3.11), being in the focus of many researchers, oxindolylalanine, dioxindolylalanine, 3a-hydroxy-1,2,3,3a,8,8a-hexahydro pyrrolo[2,3-b]indole-2-carboxylic acid, and

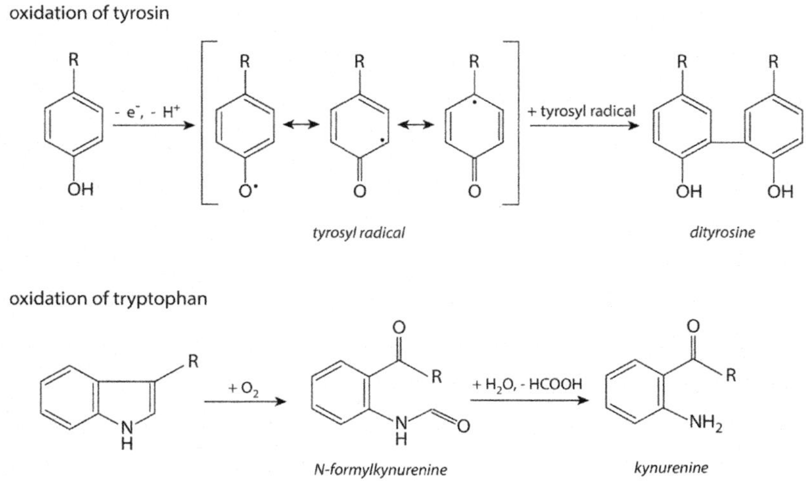

oxidation of tyrosin

tyrosyl radical

dityrosine

oxidation of tryptophan

N-formylkynurenine

kynurenine

FIGURE 3.11 Common pathways of tyrosine and tryptophan oxidation. R stands for a $-CH_2$ group attached to a carbon bearing an amino group, a carboxyl group, and a hydrogen atom.

5-hydroxytryptophan are formed as oxidation products [141].

3.5 DESTRUCTIONS IN NUCLEIC ACIDS

3.5.1 Damaging Pathways

Ionizing radiation, ultraviolet light, metal ions–catalyzed processes, carbonyls derived from oxidized lipids or glycation reactions, special enzymatic reactions, and numerous toxic agents can initiate DNA damage [142]. Of the reactive species, hydroxyl radicals, Fenton reagents, $^1O_2$, and hypohalous acids are known to attack DNA to generate DNA radicals, peroxides, and halogenated products. When DNA radicals are formed, many other reactive species can also participate in destruction. Further pathways of DNA radicals depend on the presence of reaction partners including O_2, redox properties, and near redox environment. Moreover, DNA contains with phosphate anions, nitrogen, and oxygen donor groups binding sites for transition metal ions. Chromatin proteins are also involved in binding of metal ions [143]. Thus, DNA is ever prone to low-level oxidative destructions that can be enhanced under inflammatory conditions, by the presence of necrotic cell material, and as a result of external damaging impacts.

Main structural alterations of DNA include chemical modifications in DNA bases, single and double strand scissions, cross-links between DNA and adjacent proteins, and depurination.

In living cells, numerous DNA lesions can be repaired and removed from DNA (see Section 3.8). Nevertheless, intensive damage of DNA and failure of DNA repair may lead to serious biological implications.

3.5.2 Modification of Guanine DNA Bases

Of the DNA bases, guanine is the most susceptible target to oxidation. This base has with 1.29 V at pH 7 the lowest one-electron reduction potential among DNA bases [144,145]. Hence, hydroxyl radicals and Fenton reactants dominate in formation of radicals in DNA bases.

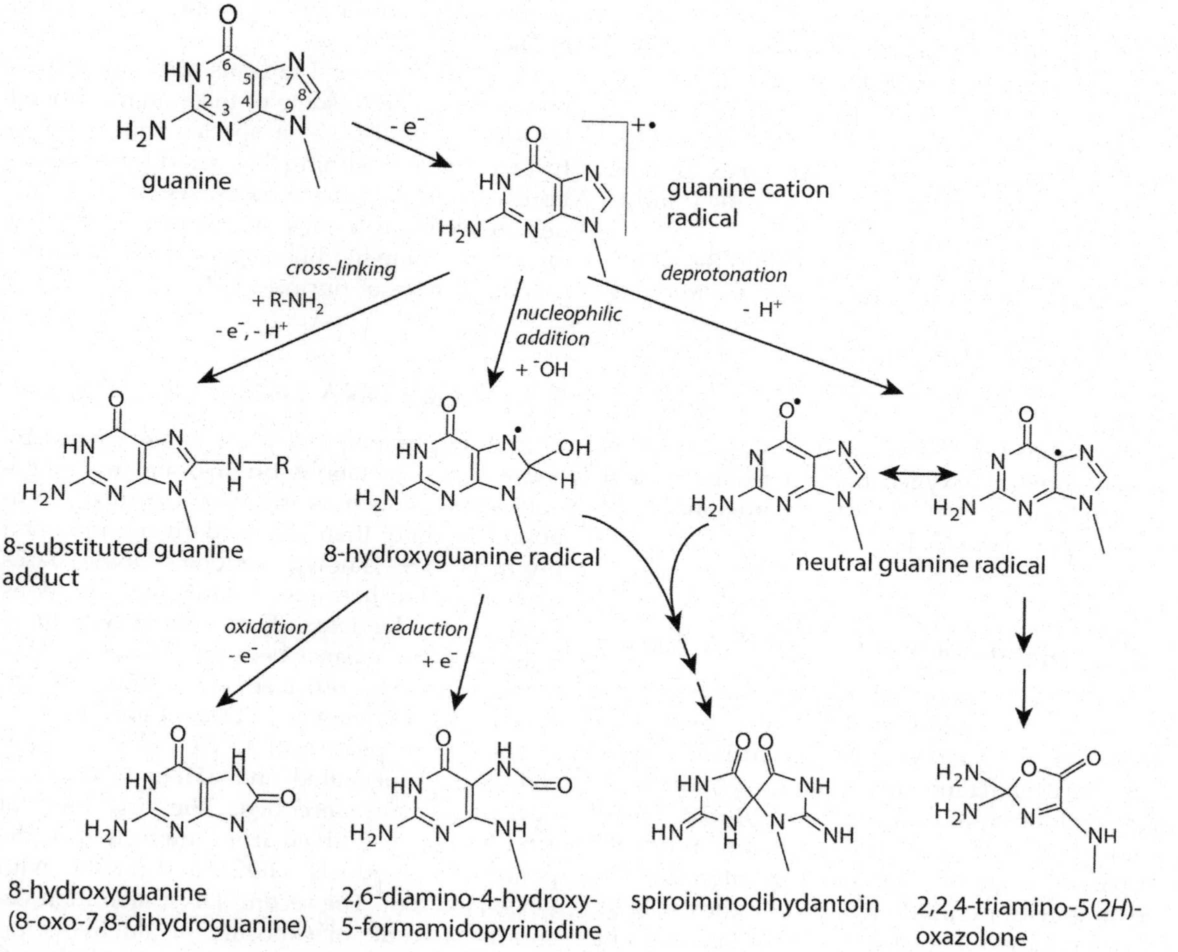

FIGURE 3.12 Transient and final products derived from the guanine cation radical at position 8 on exposure of guanine to hydroxyl radicals.

In the purine bases guanine and adenine as well as their nucleotides, an electron is primarily abstracted from carbons 4 and 8 [146,147]. Among the resulting pathways, conversions of the guanine cation radical at carbon 8 (Fig. 3.12) are intensively studied [148]. On nucleophilic addition of water, 8-oxo-7,8-dihydroguanyl radical is generated. This radical is either oxidized in the presence of O_2 to 8-oxo-7,8-dihydroguanine (also known as its tautomeric form 8-hydroxyguanine) or reduced to 2,6-diamino-4-hydroxy-5-formamidopyrimidine. Alternatively, the guanine cation radical can also interact with nucleophilic amines. This reaction can cause cross-links with other bases in the DNA strand or with adjacent proteins [149,150].

Besides nucleophilic reactions, guanine cation radical is also involved in fast deprotonation reaction to a neutral guanine radical. The latter species interacts rapidly with $O_2^{\bullet -}$ to an intermediary hydroperoxide that is further converted to

2,2,4-triamino-5(2H)-oxazolone [148]. An alternative product derived from the neutral guanine radical is spiroiminodihydantoin [151].

The abundant oxidation product 8-oxo-7,8-dihydroguanine forms a base pair with adenine [152]. The incorporation of the 8-oxo form of guanine into the genome causes G → T and C → A substitutions [153].

In cells, guanine is converted into 8-oxo-7,8-dihydroguanine by exposure to $^1O_2$ via intermediary formation of an endoperoxide and hydroperoxide [148]. This non-radical pathway is verified in many mammalian cells and human skin [154,155].

Formation of 8-oxo-7,8-dihydroguanine is also favored by oxygen transfer from a vicinal pyrimidine peroxyl radical to guanine in a DNA strand [156,157].

3.5.3 Modification of Other DNA Bases

Like in guanine, similar degradation pathways are observed in adenine oxidation. However, the yield of products such as 8-oxo-7,8-dehydroadenine or 6-amino-4-hydroxy-5-formamidopyrimidine is about 10-fold lower compared with guanine oxidation products apparently due to a less efficient interaction of adenine with pyrimidine peroxyl radicals [142].

Thymine oxidation by $HO^\bullet$ results in transient 5-hydroxy-5,6-dihydrothymine radical, 6-hydroxy-5,6-dihydrothymine radical, and 5-(uracilyl)methyl radical [158]. Cytosine is oxidized to the transient products 5-hydroxy-5,6-dihydrocytosyl radical and 6-hydroxy-5,6-dihydrocytosyl radical [159]. In thymidine oxidation, further conversions yield 5,6-dihydroxy-5,6-dihydrothymine, also called thymine glycol, 5-hydroxymethylhydantoin, 5-hydroxymethyluracil, and 5-formyluracil [160]. Main products of cytosine oxidation are 5-hydroxycytosine and 1-carbomoyl-4,5-dihydroxy-2-oxoimidazolidine [161]. Major transient and final products of one-electron oxidation of thymine and cytosine are presented in Fig. 3.13.

Pyrimidine peroxyl radicals arise after addition of dioxygen to some of the aforementioned transient radicals. These species contribute to further damage within DNA strand by interaction with vicinal purine bases. As a result, tandem nucleobase lesions are formed, consisting of formylamine-modified pyridimines and 8-oxo-derivatives of purines [157].

3.5.4 Other DNA Lesions

In living tissues, DNA lesions are caused by endogenous, metabolic origins, and by exogenous sources such as irradiation or toxic compounds. More than 20 oxidatively modified products are known among DNA bases [162,163]. Furthermore, damaging reactions concern also the deoxyribose unit. In cells, there is a permanent balance between damaging reactions and repair processes even under dominance of only endogenous DNA lesions [164].

The whole picture of DNA lesions is very complex. Halogenated and aldehyde-modified nucleobases are described. The first kind of modification is derived from interaction of the peroxidase products HOCl and HOBr with DNA [165,166]. The second DNA alteration occurs as a result of exposure of aldehydes on DNA [167,168]. Enhanced values for both modifications are detected in some pathologies in human tissues [169–172]. In both cases, the question arises: do these modifications occur in living cells or after release of DNA from necrotic cells at inflammatory site? The latter issue seems to be more likely.

Other complex DNA modifications are intrastrand base lesions, interstrand cross-links, and the formation of purine 5′,8-cyclo-2′-deoxyribonucleosides [142].

There are several repair mechanisms for single and double strand breaks in oxidatively damaged cellular DNA (see Section 3.8).

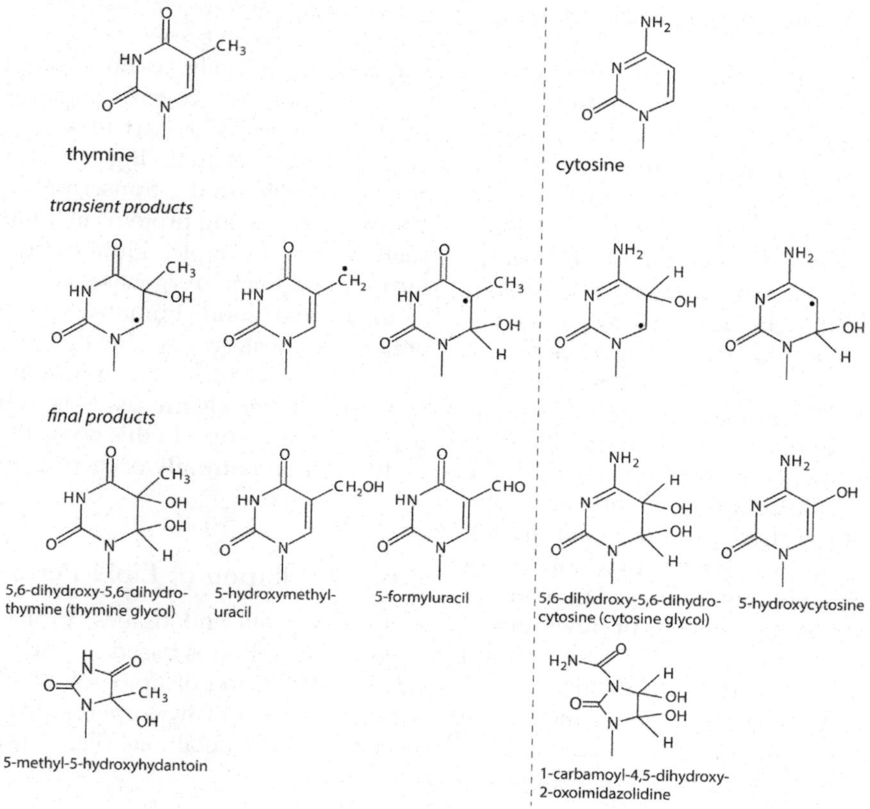

thymine

cytosine

transient products

final products

5,6-dihydroxy-5,6-dihydro-thymine (thymine glycol)

5-hydroxymethyl-uracil

5-formyluracil

5,6-dihydroxy-5,6-dihydro-cytosine (cytosine glycol)

5-hydroxycytosine

5-methyl-5-hydroxyhydantoin

1-carbamoyl-4,5-dihydroxy-2-oxoimidazolidine

FIGURE 3.13 Transient and final products on exposure of the pyrimidine bases thymine and cytosine to hydroxyl radicals.

3.6 ANTIOXIDATIVE DEFENSE BY SMALL MOLECULES

3.6.1 What Is an Antioxidant?

In biological systems, antioxidative defense is provided by small molecules and by certain proteins and enzymes. Although it is impossible to strictly separate these two basic mechanisms, in this subchapter, the main focus is directed on the role of small molecules as antioxidant. Protein- and enzyme-based antioxidative mechanisms that are crucial to maintain cellular redox homeostasis are the topic of Section 3.7.

An antioxidant is any chemical compound that inhibits an oxidative alteration of a substrate. The main problem with this term is that it is badly defined and very often used in a too broad meaning. In a pure chemical approach, the role of an antioxidant is closely related to a given oxidative process at defined conditions. Thus, either alterations of these conditions (e.g., other concentrations of reactants, other pH values) or the choice of another oxidative process highly reflects on the ability of a defined molecule to act as an antioxidant. Often, the term antioxidant is used in a more generalized sense affirming an antioxidative activity of a given molecule under many different conditions or without relating this activity to a selected oxidative process.

As antioxidants are preferentially involved in redox conversions, there are numerous examples known the certain molecules can act as an antioxidant in a given system and under other conditions as prooxidant. For example, ascorbate acts in most cases as an antioxidant. In transition metal ion–catalyzed reactions, this molecule can efficiently reduce Fe^{3+} to Fe^{2+} serving, thus, as a prooxidant.

In living systems, characterized by a large variety of oxidative reactions, antioxidants are mainly directed against long-living radicals and nonradical intermediary species in the cascade of oxidative transformation of a substrate. In this way, by decomposition of these intermediates, antioxidants prevent the formation of further reactive products and inhibit chain processes or processes of redox recycling under involvement of free radicals. Antioxidants can also scavenge free metal ions or other important components of an oxidative reaction and inhibit, thus, further redox conversions.

The action of antioxidants belongs most of all to first-stage preventive mechanisms against destruction of biological material. These molecules are always present in cells and tissues and can immediately act on induction of oxidative reactions. Of course, some antioxidative mechanisms are also part of second-stage preventive reactions in living systems. Examples are the enhanced transcription of some antioxidative-acting proteins in inflamed tissues, such as, for example, glutathione peroxidase, ferritin, and heme oxygenase.

In animals and humans, antioxidants are either synthesized by host's cells or taken up with foodstuffs. In addition, numerous synthetic chemicals are used as antioxidants in experimental setups. In this book, the main focus is directed on naturally occurring antioxidants.

3.6.2 Inhibition of Lipid Peroxidation

In living cell and tissues, protection against lipid peroxidation is based on three main mechanisms, the action of lipid-soluble antioxidants, the inactivation of hydroperoxides, and control over transition metal ions (Fig. 3.14).

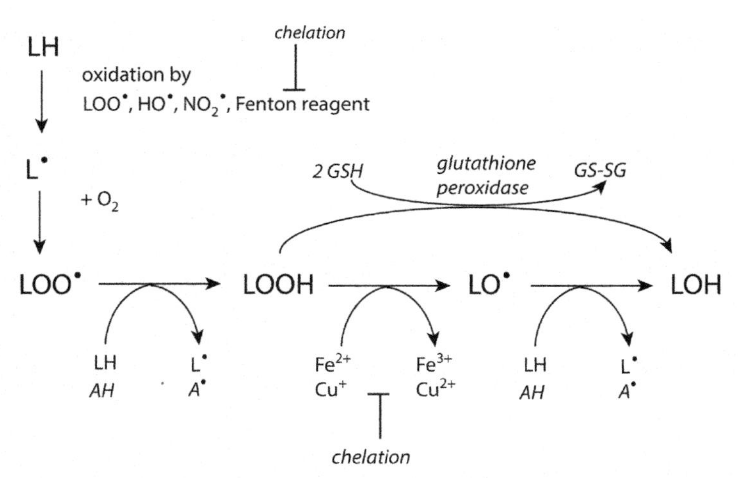

FIGURE 3.14 Common principles of inhibition of lipid peroxidation. All inhibitory mechanisms are italicized. *A•*, antioxidant radical; *AH*, lipid-soluble antioxidant; *GSH*, reduced glutathione; *GS–SG*, oxidized glutathione; *L•*, lipid alkyl radical; *LH*, lipid molecule with a weakly bound H-atom; *LO•*, lipid alkoxyl radical; *LOH*, lipid alcohol; *LOO•*, lipid peroxyl radical; *LOOH*, lipid hydroperoxide.

In biomembranes and lipoproteins, lipid-soluble antioxidants, such as α-tocopherol, γ-tocopherol, carotenoids, ubiquinol, and dihydrolipoic acid, are present. They are mainly directed to scavenge lipid peroxyl radicals. This reaction can be formulated as

$$LOO^{\bullet} + AH \longrightarrow LOOH + A^{\bullet}$$

with AH and $A^{\bullet}$ being the antioxidant and the antioxidant radical, respectively. Hence, the antioxidant competes efficiently with unperturbed unsaturated lipid molecules (LH) for $LOO^{\bullet}$. α-Tocopherol and related molecules act as chain-breaking antioxidants as they stop the oxidation of novel lipid molecules. Despite its low concentration in biological membranes, the rate constant of α-tocopherol with $LOO^{\bullet}$ is several orders of magnitude higher than those for $LOO^{\bullet}$ with LH [173,174]. Selected naturally occurring lipid-soluble antioxidants are shown in Fig. 3.15.

The resulting α-tocopheryl radical itself is a weak radical unable to abstract an H-atom from a lipid molecule. The standard reduction potential of the couple α-tocopheryl radical/α-tocopherol is 0.5 V at pH 7 [174]. Otherwise, the α-tocopheryl radical can also contribute to

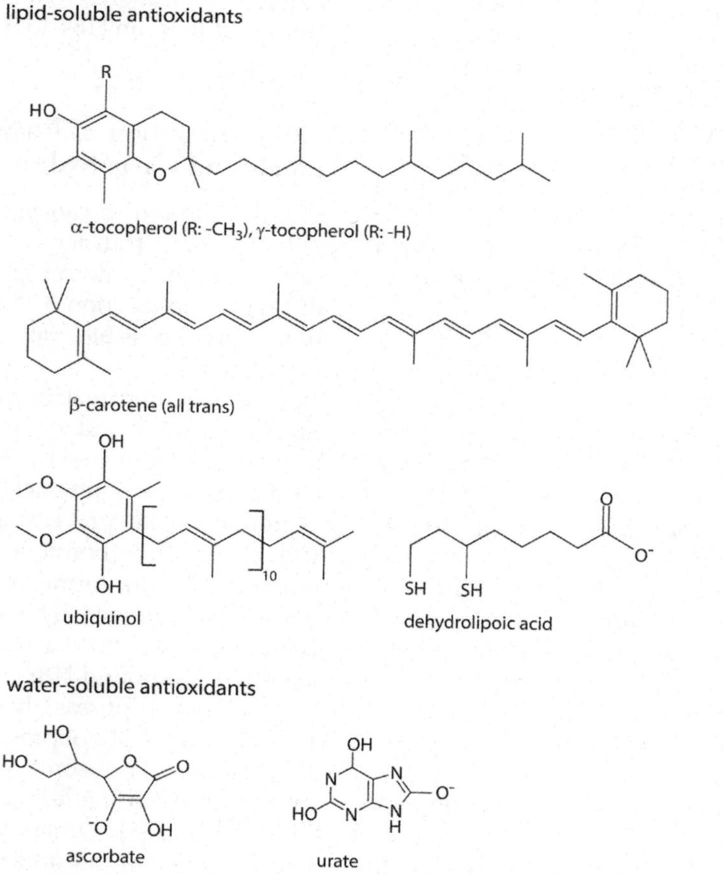

FIGURE 3.15 Selected naturally occurring lipid- and water-soluble antioxidants.

inactivation of $LOO^{\bullet}$ by radical recombination [175].

Recycling of AH from $A^{\bullet}$ is possible by interaction with other redox systems. For example, this is the case for the reaction between α-tocopheryl radical and ascorbic acid. In the latter reaction, ascorbate is oxidized to semidehydroascorbyl radical [176–178]. A reduction potential of 0.28 V at pH 7 was determined for the redox couple semidehydroascorbyl radical/ascorbate [179]. Two semidehydroascorbyl radicals can dismutate under formation of ascorbate and dehydroascorbate [174]. Taken together, there is a decrease in standard reduction potentials of the involved redox couples of antioxidants in inactivation of peroxyl radicals.

Inactivation of lipid hydroperoxides is also an important mechanism to inhibit lipid peroxidation. Here, most of all glutathione peroxidase (GPx4) is involved in combination with phospholipases [180]. By this mechanism, the hydroperoxide moiety is converted into an alcohol, and two molecules of glutathione (GSH) are oxidized to glutathione disulfide (GS-SG).

$$LOOH + 2\,GSH \rightarrow LOH + H_2O + GS - SG$$

Phospholipases cleave perturbed acyl residues and initiate, thus, processes of phospholipid recycling.

A third major antioxidative mechanism is the efficient control over transition free metal ions. Reduced forms of these ions (Fe^{2+}, Cu^{+}) and their oxidized counterparts (Fe^{3+}, Cu^{2+}) together with reducing agents are involved in abstraction of H-atoms and decomposition of lipid hydroperoxides. Details of these protecting mechanisms are already described in Section 2.4.

Despites the presence of chain-breaking antioxidants, inactivation of hydroperoxides, and thorough control of free metal ions, lipid composition of biological membranes is also an important factor for the extent of oxidized lipid products. In the liquid crystalline state of membranes, cholesterol increases the membrane rigidity and the rate of lipid peroxidation [181]. With increasing yield of peroxidation products in membrane, a self-assembling of cholesterol domains occurs [182]. In plasmalogens, H-atoms adjacent to vinyl ether bond and this double bond itself are prominent targets for peroxyl radicals and singlet oxygen [183,184]. As a result of this interaction, the propagation step of lipid peroxidation is interrupted and the yield of peroxidation products is decreased [185]. Thus, a high yield of plasmalogens protects membranes against lipid peroxidation. High amounts of 1-alkenyl-2-acyl-*sn*-glycero-3-phosphoethanolamine are found in brain [184,185].

3.6.3 Inhibition of Fragmentation in Lipids and Carbohydrates

Lipid molecules can be fragmented in their polar via transient formation of a carbon-centered radical with an adjacent hydroxyl group (Section 3.2.7). Similar fragmentations are possible in carbohydrates and nucleotides.

Fragmentation reactions in the polar part of lipid molecules and oligomeric carbohydrates as well are dampened by dioxygen [56]. Apparently, dioxygen interacts with the carbon-centered radical preventing bond ruptures. In this reaction, a carbonyl product and superoxide anion radical are formed [186,187]. Cupric ions (Cu^{2+}) are oxidized by the α-hydroxy carbon-centered radical to Cu^{+} under formation of a carbonyl compound [188].

Quinones suppress the fragmentation reaction of lysophospholipids, phosphatidylglycerols, maltose, and some other compounds by interaction with the α-hydroxy carbon-centered radical [187,188]. Other principal inhibitors of fragmentation are ascorbic acid [189], flavonoids [190], and group B vitamins [191].

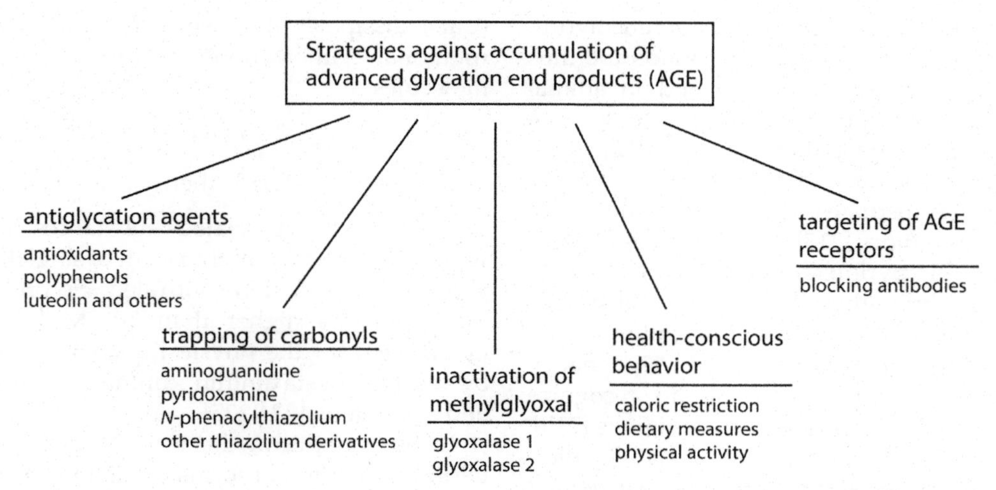

FIGURE 3.16 Important strategies against accumulation of advanced glycation end products.

3.6.4 Water-Soluble Antioxidants

In humans, the main water-soluble antioxidant is urate but not ascorbate in contrast to mice and rats [192–194]. Ascorbate is not formed by humans from glucose due to inactive L-gulonolactone oxidase [195]. It has to be taken up with food. Otherwise, urate accumulates in human cells unlike in mice as it is not further converted into (*S*)-allantoin due to the absence of active urate oxidase [196]. The significance of the dominance of urate over ascorbate in human defense reactions remains unknown. Maybe, urate also displays neuroprotective activities [197]. Moreover, urate is known to protect ascorbate against oxidation by transition metal ions [198]. Formula for ascorbate and urate are shown in Fig. 3.15.

3.6.5 Defense Against Glycation

Accumulation of AGEs seriously affects homeostatic functions and integrity of biological structures most of all in hyperglycemic patients. Otherwise, regular uptake of glucose and other sugars by food is crucial for the energy metabolism. Thus, nonenzymatic glycation reactions including formation of AGEs always take place in our organism. The question here is how these processes are controlled and how an excess accumulation of AGEs can be prevented. Important strategies against glycation are summarized in Fig. 3.16.

Caloric restriction, dietary measures, and adequate physical activity are highly recommended to stabilize metabolism. As dietary AGEs may contribute to inflammatory processes in blood vessels, a diet with low content of AGEs would beneficial to patients with diabetes and renal failure [199–201]. Numerous substances with antioxidant and metal-chelating properties are known to inhibit the degree of glycation under in vitro conditions [202–204]. Their effects indicate the close relationship between glycation processes, oxidative reactions, and participation of transition metal ions. In human subjects, systematic investigations are scarce about the action of these substances on AGE formation.

Among substances with antiglycation activity are naturally occurring stilbenes, flavonoids, and other polyphenols [205–209]. Although many of the tested flavonoids inhibited different phases of AGE formation, the most promising results were found for the flavone luteolin [205].

Aminoguanidine traps as nucleophilic hydrazine intermediary carbonyl compounds and thus limits AGE formation in vitro and in animal models [210,211]. However, clinical tests with aminoguanidine on patients failed [212]. Pyridoxamine, an isoform of vitamin B_6, scavenges carbonyl intermediates and inhibits late phase in AGE formation [213]. Clinical trials are in progress with pyridoxamine on patients with diabetic nephropathy [214]. Other substances such as dimethyl-3-phenacylthiazolium chloride, *N*-phenacylthiazolium, and *N*-phenacyl-4,5-dimethylthiazolium are able to recognize and break the Maillard reaction cross-links [215].

Methylglyoxal, an important reactive intermediate in the glycation cascade, is deactivated by a concerted action of glyoxalase 1 (GLO1) and glyoxalase 2. This enzyme system utilizes glutathione to convert 2-oxoaldehydes into the corresponding 2-hydroxyacids. In detoxification of methylglyoxal, S-D-lactoylglutathione is formed as the transient and D-lactic acid as the final product [216,217]. GLO1 is the key enzyme of glyoxalase system. Its expression is transcriptionally controlled by Nrf2 [218]. Decreased activity of GLO1 has been implicated in the pathogenesis of numerous diseases such as diabetes, cancer, and aging [111,219–221].

Targeting RAGE by inhibitors or blocking antibodies is a further potential strategy to combat against deleterious AGE affects. Promising results were obtained in inflammatory and diabetes models on laboratory animals [97,222,223]. Systemic studies are necessary to prove the relevance of this strategy for therapeutic intervention in patients.

3.6.6 Carotenoids as Singlet Dioxygen Quenchers

Carotenoids (Car) are very efficient in physical quenching of singlet dioxygen. In this process, $^1O_2$ is converted in the triplet ground state, and the resulting excited carotenoid (Car^*) dissipates its energy in vibrational and rotational interactions with surrounding molecules.

$$Car + {}^1O_2 \rightarrow Car^* + {}^3O_2$$

$$Car^* \rightarrow Car + \text{thermal energy}$$

Carotenes and related substances interact with Δ^1O_2, the dominating form of singlet dioxygen, nearly diffusion-controlled. Second order rate constants higher than $10^{10}\,M^{-1}s^{-1}$ were determined for the physical quenching of Δ^1O_2 by lycopene, astaxanthin, canthaxanthin, α-, β-, and γ-carotene [224,225].

In our organism, dietary β-carotene is converted into different forms of vitamin A. Hence, β-carotene functions as provitamin A [226]. In humans, carotinoids accumulate notably in the adipose areas, skin, intestine, liver, corpus luteum, and eye [227]. The presence of β-carotene, vitamin A, and others in skin and eye is important, as these organs are most of all exposed to light and being preferential places for $^1O_2$ production.

3.6.7 Polyphenols as Dietary Antioxidants

Polyphenols are a very heterogeneous group of moderately water-soluble compounds that contain one to several aromatic rings with joined hydroxyl groups. Often additional structural elements are incorporated into polyphenolic compounds such as sugar molecules, carboxylic and organic acids, amines, and lipids. These substances are widely distributed in plants and contained in vegetables, fruits, cereals, tea, coffee, red wine, and other beverages. Thus, we incorporate polyphenols with our food and drinks. They also play an important role as ingredients in a number of medicaments.

Major classes of dietary polyphenols are phenolic acids, lignans, stilbenes, and flavonoids [228]. Phenolic acids can be further divided into hydroxybenzoic acids and the abundant

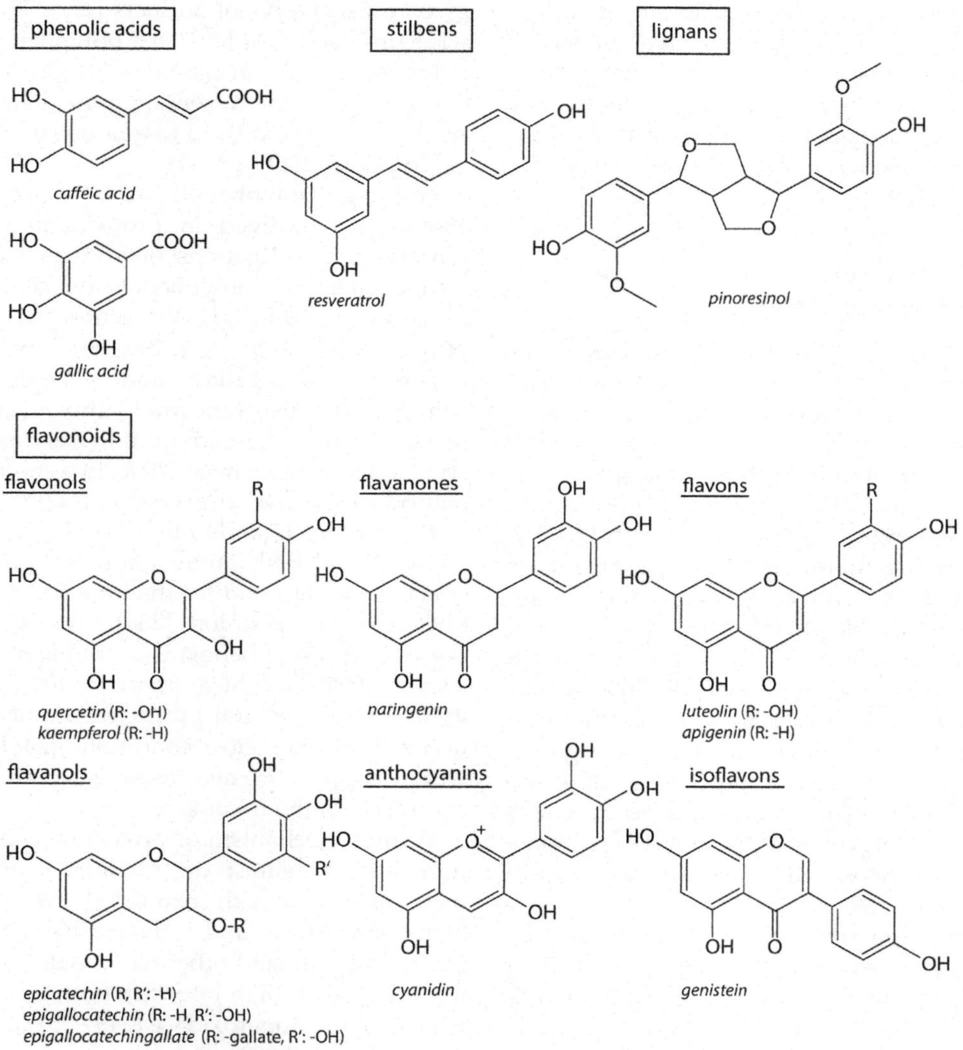

phenolic acids

caffeic acid

gallic acid

stilbens

resveratrol

lignans

pinoresinol

flavonoids

flavonols

quercetin (R: -OH)
kaempferol (R: -H)

flavanones

naringenin

flavons

luteolin (R: -OH)
apigenin (R: -H)

flavanols

epicatechin (R, R': -H)
epigallocatechin (R: -H, R': -OH)
epigallocatechingallate (R: -gallate, R': -OH)

anthocyanins

cyanidin

isoflavons

genistein

FIGURE 3.17 Selected naturally occurring polyphenols.

hydroxycinnamic acid derivatives. Lignans are dimerized phenylpropanoids. In stilbenes, two phenyl moieties are linked by a two-carbon methylene bridge. Basic structural elements in flavonoids are two aromatic rings connected together by three carbon atoms forming an oxygen containing heterocycle. According to the type of this heterocycle, several subtypes are differentiated such as flavonols, flavanones, flavanols, flavones, anthocyanins, and isoflavones. Moreover, numerous polymerized species are formed among polyphenols. Classification of polyphenols and examples for representing molecules are indicated in Fig. 3.17.

The general assumption is that polyphenols might have numerous beneficial activities including the action as antioxidant, as antiinflammatory agent, and protective substance against cancer. They provide a significant protection against development of numerous chronic diseases such as diabetes, cancer, cardiovascular diseases, infections, neurodegenerative diseases, and others [229–231].

3.6.8 Mechanisms of Antioxidative Activity by Polyphenols

Polyphenols can be oxidized by numerous reactive species under formation of the corresponding phenoxyl radicals. One-electron standard reduction potentials of tested phenolic substances are slightly higher or lower than the reduction potential of α-tocopheryl radical/α-tocopherol being 0.48 V at pH 7 [232]. For example, the standard reduction potentials are 0.43 V at pH 7 for both epigallocatechin and epigallocatechin gallate [233]. The corresponding value for catechol is 0.53 V [232]. Taken together, flavonoids and other polyphenols can scavenge peroxyl radicals, tyrosyl radicals, and other substrate radicals and thus interrupt propagation steps of oxidative chain processes. They are also a target for some nonradicalic reactive species. The resulting oxidized products of polyphenols do not have the capacity to perturb biological material.

A second important property of these substances is the sequestration of iron ions. This feature is mainly expressed in polyphenols bearing two or three neighbored hydroxyls groups on the aromatic ring like in catechol and gallol derivatives [234] or having the chelating groups (two hydroxyl group or a hydroxyl group and a keto group) on two fused rings in *peri* position [235]. For metal ion binding, the involved hydroxyl groups should be in the deprotonated form. Fe^{3+} binds stronger than Fe^{2+} to polyphenol ligands [234,236]. In the presence of O_2, the binding of Fe^{2+} to catecholate and gallate complexes promotes an autoxidation to give Fe^{3+}-polyphenol adducts [237,238]. Otherwise, on binding of Fe^{3+} to a polyphenol ligand at low pH, ferric ion can be reduced to Fe^{2+} under formation of a semiquinone ligand. The semiquinone is oxidized to quinone by a further Fe^{3+} [235,239,240].

Although polyphenols are being principally able to be involved in prooxidant activities, numerous investigations demonstrate cytoprotective effects of polyphenols on chelation of iron ions [241–244]. Protection by dietary polyphenols such as green tea constituents (−)-epicatechin-3-gallate and (−)-epigallocatechin-3-gallate also concerns pathological conditions of iron overload in β-thalassemia [245] and iron release from defective neurons in neurodegenerative diseases [246,247].

Polyphenol ligands bind strongly Cu^{2+}, but weakly Cu^{+}. Both antioxidant and prooxidant effects were reported for the interaction of polyphenols with copper ions [234]. Recycling of copper ions by polyphenols is more likely than for iron ions. This might promote prooxidant activity of copper ions in the presence of polyphenols under certain in vitro conditions [248]. Otherwise, copper homeostasis is very tightly controlled in the organism.

A third mechanism of protecting activities of polyphenols against destructions is based on their reactivity with oxo-ferryl complexes of heme peroxidases and activated heme proteins. (−)-Epicatechin and other flavonoids are known to interact with high rate with both Compounds I and II of myeloperoxidase [249–251] and lactoperoxidase [252,253]. This interaction can inhibit more harmful substrate oxidations by activated heme peroxidases. Reactivating inactive Compound II, tested flavonoids also enhance the halogenating activity of these heme peroxidases [252–255].

Some flavonoids are known to deactivate oxoferryl states in hemoglobin [256,257]. This inhibits potential substrate oxidations catalyzed by activated hemoglobin.

3.7 REDOX HOMEOSTASIS

3.7.1 NADPH as a Universal Source for Electrons in Cellular Redox Homeostasis

In cytoplasm of resting cells, a reducing milieu dominates. Homeostatic mechanisms are responsible to keep cells in this reduced state. To ensure protection against unwanted oxidation of biological material, cells are equipped with antioxidant redox-sensitive mechanisms that regularly restore these components after any attack by reactive species. Here, processes of redox recycling hold protective cellular elements in an active mode.

The major electron source for cellular redox homeostasis is NADPH [116,258]. Oxidation of this molecule to $NADP^+$ liberates two electrons. The standard reduction potential of the couple $NADP^+/NADPH$ is -0.32 V at pH 7 [259]. Because of the presence of cytosolic binding sites for $NADP^+$, NADPH dominates over $NADP^+$ providing an excellent reducing environment in cells [260,261]. In the related redox couple $NAD^+/NADH$ having also a standard reduction potential of -0.32 V at pH 7 [259], the oxidized form NAD^+ predominates in cytoplasm [116].

Moreover, $NADP^+$ is reduced to NADPH by glucose-6-phosphate dehydrogenase and 6-phosphogluconate dehydrogenase, two enzymes of the oxidative pentose phosphate pathway that runs parallel to glycolysis [262]. The former enzyme, which is the rate-limiting enzyme in this pathway, is allosterically activated by the $NADP^+/NADPH$ ratio [263,264]. Activation of the transcription factor NRF2 (nuclear factor erythroid 2–related factor 2) switches on numerous genes involved in antioxidant and xenobiotic responses including the expression of enzymes generating NADPH [265,266]. Besides enzymes of the oxidative pentose phosphate pathway, NADPH is also produced by the isoforms 1 and 2 of isocitrate dehydrogenase [267], malic enzyme [268], and methylene tetrahydrofolate dehydrogenase [269].

3.7.2 Maintenance of the Cellular Sulfhydryl Status

NADPH is required to maintain protein sulfhydryls and other cellular thiols in the reduced form. Oxidation of NADPH is coupled with reduction of oxidized glutathione (GS-SG) to two molecules glutathione (GSH), a reaction that is catalyzed by glutathione reductase [270].

$$GS - SG + NADPH + H^+ \rightarrow 2\,GSH + NADP^+$$

The standard reduction potential of the couple GS-SG/2 GSH is -0.24 V at pH 7 [116]. Moreover, the actual potential of this redox couple also depends on GSH concentration [116].

On oxidation of protein thiols, GSH can interact with different sulfhydryl oxidation products such as sulfenic acid, thiyl radical, or protein disulfides under formation of GS-SG [116]. This helps to restore the protein thiol pool. Dethiolation reaction of proteins is further favored by glutaredoxin, thioredoxin reductase, and protein disulfide isomerase [271,272].

Intracellular concentrations of glutathione are about 100–1000 times higher than extracellular ones. In the cytosol, concentrations of GSH vary from 1 to 11 mM [273]. Highest values are found in mitochondria (5–11 mM) and the nucleus [273–275]. In most compartments (cytosol, mitochondria, nucleus) of resting cells, the GSH/GS-SG ratio is typically greater than 30:1 and usually around 100:1. An exception is the endoplasmic reticulum having a GSH/GS-SG ratio of about 1:1 to 3:1 [276]. This more oxidizing state is apparently needed to synthesize proteins with disulfide bonds. A careful discussion of these dependencies is found by Schafer and Buettner [116].

It has been postulated that the redox environment of a cell might determine whether a cell will proliferate, differentiate, or die [116]. Using the midpoint potential as indicator, proliferation of cells will be favored around -0.240 V, differentiation around -0.200 V, and apoptosis around −0.170 V. In other words, there is a continuous increase in the reduction potential of the couple GS-SG/2 GSH due to decrease of GSH concentration.

3.7.3 The Glutathione System as a Sink for Intracellular Radicals

In cells and tissues, several proteins contribute to deactivation of reactive species and hydroperoxide moieties in substrate molecules. With their action, they prevent further damage of biological material. Important antioxidative proteins are superoxide dismutase (Section 2.2.4.4), catalase (Section 2.2.5.4), glutathione peroxidase (Sections 2.2.5.4 and 3.6.2), heme oxygenase (Section 4.4.3), and proteins controlling free metal ions (Section 2.4.1).

Two molecules GSH are utilized and oxidized to GS-SG by different glutathione peroxidases to remove H_2O_2, peroxynitrite, and organic hydroperoxides [277,278].

Moreover, it is believed that intracellularly formed substrate radicals are scavenged by GSH and ultimately transformed into superoxide anion radicals that disappear by dismutation [279]. This hypothesis was found to be thermodynamically favorable [122]. GSH reacts with any organic radical under formation of $GS^{\bullet}$ and reduction of the original radical to a nonradical component. $GS^{\bullet}$ can either react with another $GS^{\bullet}$ to GS-SG, in a reversible reaction with O_2 to $GSOO^{\bullet}$, or with GS^- to yield $GS - SG^{\bullet -}$. The later radical species reacts highly efficient with O_2 to $O_2^{\bullet -}$ and is thus rapidly removed from the system. Final products of this cascade are GS-SG and superoxide anion radical [279].

Thus, the GSH system contributes to maintain a very low level of any damaging radicals and hydroperoxides in cells.

3.7.4 The Thioredoxin System

In addition to the glutathione system, the thioredoxin system is also highly important for redox homeostasis in cells. This system comprises thioredoxins (Trx) and thioredoxin reductases. Thioredoxins (Trx) are small redox-active proteins that contain a disulfide/dithiol motif. In contrast to the glutathione system, the reduced state of thioredoxin (Trx(−SH, −SH)) comprises only one molecule bearing two redox-active sulfhydryls. Oxidized thioredoxin (Trx(−S−S−)) contains a disulfide bond. The following half reaction can be written for the Trx(−S−S−)/Trx(−SH, −SH) redox couple:

$$\text{Trx}(-\text{S}-\text{S}-) + 2\,\text{e}^- + 2\,\text{H}^+ \rightarrow \text{Trx}(-\text{SH},\ -\text{SH})$$

NADPH provides electrons for reduction of oxidized thioredoxin in a thioredoxin reductase−driven reaction [272,280]. Thioredoxin reductase catalyzes the following reaction:

$$\text{Trx}(-\text{S}-\text{S}-) + \text{NADPH} + \text{H}^+ \rightarrow \text{Trx}(-\text{SH},\ -\text{SH}) + \text{NADP}^+$$

Peroxiredoxins (Prx) are an important physiological substrate for Trx [281]. Interaction between reduced Trx and oxidized Prx recycles reduced Prx. A major property of Prx is the presence of a so-called peroxidatic cysteine that reacts very fast with hydrogen peroxide [282]. By this interaction, the peroxidasin−thioredoxin system efficiently removes H_2O_2 [281].

Thioredoxin 1 (Trx1) is primarily found in the cytoplasm and nucleus. It is involved in reduction of oxidized proteins such as ribonucleotide reductase, transcriptions factors, and peroxiredoxins [283−285].

Thioredoxin 2 (Trx2) acts independent of Trx1 inside mitochondria [286,287]. It eliminates mitochondria-derived reactive species and thus inhibits proinflammatory activation of NFκB and apoptosis induction. Peroxiredoxin 3 and thioredoxin reductase 2 are further mitochondria-specific proteins that are closely linked to Trx2 [288,289].

Thus, both the glutathione and thioredoxin systems contribute together with NADPH for the maintenance of cellular redox homeostasis.

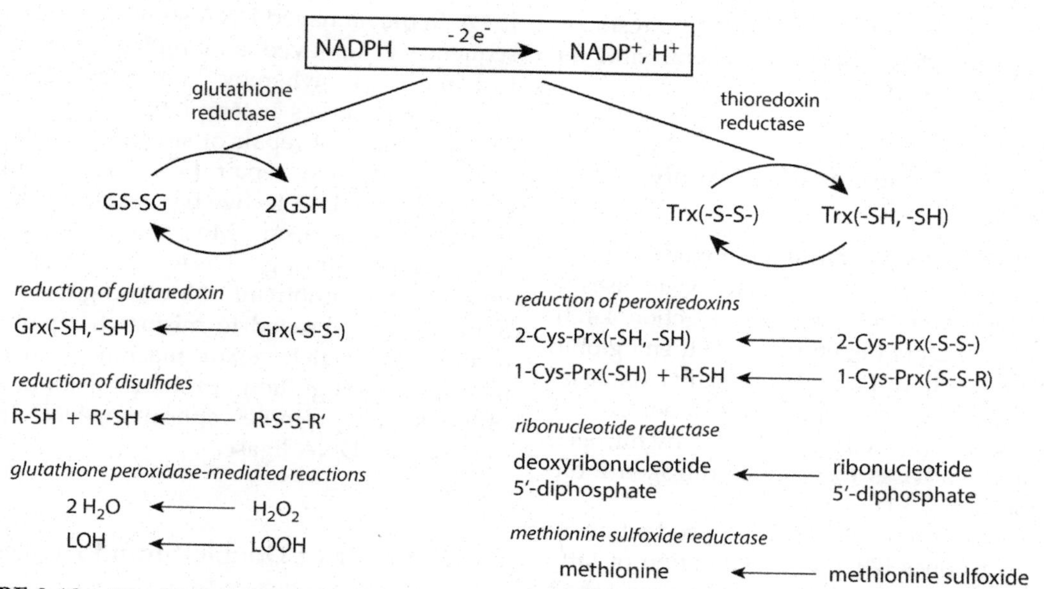

FIGURE 3.18 Glutathione and thioredoxin systems in maintenance of cellular redox homeostasis. Examples of major reactions coupled to both redox systems are indicated. Further explanations are given in the text. *Grx*, glutaredoxin; *GSH*, reduced glutathione; *GS–SG*, oxidized glutathione; *LOH*, lipid alcohol.; *LOOH*, lipid hydroperoxide; *Prx*, peroxiredoxin (with one or two active cysteine residues); *Trx*, thioredoxin.

An overview about these basic mechanisms is given in Fig. 3.18.

3.7.5 Glutaredoxins

Glutaredoxins (Grx) are small intracellular redox enzymes. Like thioredoxins, they contain in the reduced state two redox-active thiols that form a disulfide bond on oxidation. Unlike Trx, oxidized glutathione is reduced by glutathione reductase [290].

Glutaredoxins deliver electrons to ribonucleotide reductase [291]. They are involved in reversible glutathionylation of thiols and formation and reduction of mixed disulfides between protein sulfhydryls and GSH [292,293]. In human mitochondrial glutaredoxin 2, two Grx molecules are bridged by a $[2Fe{-}2S]^{2+}$ cluster keeping the enzyme in an inactive state. Oxidative stress promotes degradation of this cluster and activates Grx [294].

3.7.6 Methionine Sulfoxide Reductases

Methionine sulfoxides can be reserved by a set of methionine sulfoxide reductases. These stereospecific enzymes employ reduced thioredoxin to recover free and protein-bound methionine [295]. Oxidized thioredoxin itself is transferred to reduced thioredoxin by NADPH-dependent thioredoxin reductase.

Methionine residues can be oxidized by numerous reactive species to methionine sulfoxide. The reduction of the latter to the original thioether is provided by methionine reductase together with the thioredoxin system and

NADPH. This recycling of methionine residues plays an important role as a general sink for cellular oxidants [296,297].

3.7.7 Other Redox Systems Applying NADPH

In addition to the aforementioned systems, NADPH is also utilized by heme oxygenases to detoxify free heme in cytoplasm (Section 4.4.3), by the NADPH-dependent methemoglobin reductase (Section 4.2.3), by NADPH oxidases in neutrophils, macrophages, and other cells to reduce dioxygen (Section 2.2.4), and by nitrogen monoxide synthases to produce nitrogen monoxide (Section 2.3.1).

Cholesterol generation [298], production of steroid hormones [299], and elongation of fatty chains [300] are examples of synthesis processes dependent on the presence of NADPH. Cytochrome P450 enzymes apply NADPH for hydroxylation of aromatic compounds, alcohols, drugs, steroids, and others [301,302].

3.8 REPAIR MECHANISMS OF NUCLEIC ACIDS

3.8.1 Repair of Single-Strand Lesions

To maintain disturbances in the DNA on a very low level is crucial for an error-free transcription and translation of the genetic information. Despite common principles against oxidative damage such as control over transition metal ions and the presence of a reducing milieu in nucleus, faulty transitions can also result from replication or transcription. In nuclei, there are special enzyme systems that recognize common lesions in DNA and provide repair of these lesions.

Irradiation with UV light results in formation of pyrimidine dimers, mostly thymine dimers and some photoproducts. This kind of damage is usually reversed by nucleotide excision repair [303]. In the damaged DNA strand, the defective sequence is removed and replaced by unperturbed bases, whereby the complementary DNA strand serves as template.

Other types of repair of single-strand lesions are base excision repair [304] and mismatch repair [305,306]. Defective bases are restored by the first mechanism. Most base mismatches that occur during DNA replication are immediately proofread. Not recognized mismatches are subjected to mismatch repair. The complex mismatch repair machinery includes mismatch recognition proteins, exonuclease, proteins that promote replication, DNA polymerase, and DNA ligase.

3.8.2 Repair of Double-Strand Breaks

Repair of double-strand breaks is a challenge as the complementary DNA strand cannot serve as a template. Double-strand breaks can be repaired by three mechanisms: nonhomologous end joining [307], microhomology-mediated end joining [308], and homologous recombination [309].

In nonhomologous end joining, the repair process is guided by short homologous DNA sequences that are often present in single-stranded overhangs. Usually these lesions are accurately repaired. This mechanism is the dominating double-stranded repair process in mammalian cells [310].

Repair by microhomology-mediated end joining includes resection of the ends of double-strand breaks, annealing of microhomologies, removal of flaps, fill-in synthesis, and ligation [308,311]. This repair process is error-prone and can result in deletion mutations.

Homologous recombination repair starts with end resection followed by binding of a DNA strand—exchange protein to single-stranded DNA. This initiates a repair synthesis. Special genes and proteins involved in this repair mechanism are the breast cancer early onset genes 1 and 2 (BRCA1 and BRCA2), RAD51, and PALB2 proteins [309].

Permanent alterations in the DNA sequence are associated with mutations. Failure of DNA repair processes and the inability to induce apoptosis in affected cells can stabilize mutations in DNA. Irreparable DNA lesions like double-strand breaks and DNA cross-linkages are exceedingly dangerous. Such mutations can form the basis for the development of cancer [312,313].

3.9 SUMMARY AND OUTLOOK

Chemical alterations of biological substrates can highly disturb the integrity and functionality of these materials. In cells and tissues, all major biochemical constituents are subject of initial attacks by reactive species and other reactive components. In most cases, chemical modifications start on readily oxidizable molecules. Under normal conditions, the yield of oxidized products is low as cells are equipped with different protecting systems holding the number of damaged molecules on a minor level.

A reducing milieu predominates inside cells. This ensures to maintain critical functional groups on a reduced state. In proteins, there is a permanent recycling of oxidized sulfhydryl and methionine residues provided by the link between the glutathione and thioredoxin systems and the recovery of NADPH [314]. Glutathione and other easily oxidizable materials act additionally as a cellular, restorable sink for different reactive species, substrate radicals, and inorganic and organic peroxides. These protecting systems are completed by lipid- and water-soluble antioxidants and by special antioxidative-acting proteins such as superoxide dismutase, catalase, peroxidasins, heme oxygenase, proteins that control transition metal ions, and others. Altogether, these mechanisms are usually always present in cells and tissues. They mainly contribute to the first line defense against frequently occurring destructive agents.

On all hierarchic levels, normal cell functioning and execution of protective mechanisms highly depends on regular delivery of energy substrates and production of adenosine triphosphate. Disturbances associated with energy metabolism are outlined in Chapter 4.

Within cells, biological material is permanently recycled. The same holds for extracellular material. Moreover, many cells are replaced by novel ones. There are several mechanisms of cell death (Chapter 5) to eliminate dysfunctional and unwanted cells. These mechanisms also limit the accumulation of defective material and dysfunctional cells. The functioning of recycling systems highly depends on recognition of disturbed material. For example, defective cellular proteins are labeled by ubiquitin and further degraded by the proteasome (see Section 5.6.3).

External impacts such as traumata, toxic compounds, or invading microorganisms, as well as homemade autoantibodies can highly affect physiological functions of cells and tissues. Then activation of the immune system is very crucial to restore homeostasis in tissues, organs, and whole organisms. The significance of these protective mechanisms is the topic of Chapters 6 and 7.

References

[1] W.H. Koppenol, Oxyradicals reactions: from bond-dissociation energies to reduction potentials, FEBS Lett. 264 (1990) 165–167.

[2] B.H.J. Bielski, R.L. Arudi, M.W. Sutherland, A study of the reactivity of $HO_2/O_2^{\bullet-}$ with unsaturated fatty acids, J. Biol. Chem. 258 (1983) 4759–4761.

[3] S. Hippeli, E.F. Elstner, Oxygen radicals and air pollution, in: H. Sies (Ed.), Oxidative Stress: Oxidants and Antioxidants, Academic Press, 1991, pp. 3–55.

[4] C. Schöneich, K.-D. Asmus, Reaction of thiyl radicals with alcohols, ethers and polyunsaturated fatty acids: a possible role of thiyl free radicals in thiol mutagenesis, Radiat. Environ. Biophys. 29 (1990) 263–271.

[5] W. Bors, M. Erben-Russ, M. Saran, 1012 – Fatty acid peroxyl radicals: their generation and reactivities, Bioelectrochem. Bioenerg. 18 (1987) 37–49.
[6] I.P. Edimicheva, M.A. Kisel, O.I. Shadyro, V.P. Vlasov, I.L. Yurkova, The damage of phospholipids caused by free radical attack on glycerol and sphingosine backbone, Int. J. Radiat. Biol. 71 (1997) 555–560.
[7] O.I. Shadyro, I.L. Yurkova, M.A. Kisel, Radiation-induced peroxidation and fragmentation of lipids in a model membrane, Int. J. Radiat. Biol. 78 (2002) 211–217.
[8] O. Shadyro, I. Yurkova, M. Kisel, O. Brede, J. Arnhold, Formation of phosphatidic acid, ceramide, and diglyceride on radiolysis of lipids: identification by MALDI-FOF mass spectrometry, Free Radic. Biol. Med. 36 (2004) 1612–1624.
[9] A.J. Brown, W. Jessup, Oxysterols: sources, cellular storage and metabolism, and new insights into their role in cholesterol homeostasis, Mol. Aspect. Med. 30 (2009) 111–122.
[10] A.A. Frimer, The reaction of singlet oxygen with olefins: the question of mechanism, Chem. Rev. 79 (1979) 359–387.
[11] E.N. Frankel, Secondary products of lipid oxidation, Chem. Phys. Lipids 44 (1987) 73–85.
[12] W. Korytowski, A.W. Girotti, Singlet oxygen adducts of cholesterol: photogeneration and reductive turnover in membrane systems, Photochem. Photobiol. 70 (2000) 484–489.
[13] W.A. Pryor, The formation of free radicals and the consequences of their reactions in vivo, Photochem. Photobiol. 28 (1978) 787–801.
[14] C.C. Winterbourn, J.J.M. van den Berg, E. Roitman, F.A. Kuypers, Chlorohydrin formation from unsaturated fatty acids reacted with hypochlorous acid, Arch. Biochem. Biophys. 296 (1992) 547–555.
[15] J. Arnhold, A.N. Osipov, H. Spalteholz, O.M. Panasenko, J. Schiller, Effects of hypochlorous acid on unsaturated phosphatidylcholines, Free Radic. Biol. Med. 31 (2001) 1111–1119.
[16] H. Spalteholz, K. Wenske, O.M. Panasenko, J. Schiller, J. Arnhold, Evaluation of products upon the reaction of hypohalous acid with unsaturated phosphatidylcholines, Chem. Phys. Lipids 129 (2004) 85–96.
[17] W.H. Koppenol, J. Butler, Mechanism of reaction involving singlet oxygen and the superoxide anion, FEBS Lett. 83 (1977) 1–6.
[18] J.D. Rush, Z. Maskos, W.H. Koppenol, Reactions of iron(II) nucleotide complexes with hydrogen peroxide, FEBS Lett. 261 (1990) 121–123.
[19] J.D. Rush, W.H. Koppenol, Reactions of Fe(II)-ATP and Fe(II)-citrate complexes with *t*-butyl hydroperoxide and cumyl hydroperoxide, FEBS Lett. 275 (1990) 114–116.
[20] M.J. Stables, D.W. Gillroy, Old and new generation of lipid mediators in acute inflammation and resolution, Prog. Lipid Res. 50 (2011) 35–51.
[21] O.M. Panasenko, J. Arnhold, J. Schiller, Hypochlorite reacts with an organic hydroperoxide forming free radicals, but not singlet oxygen, and thus initiates lipid peroxidation, Biochemistry (Mosc.) 62 (1997) 951–959.
[22] O.M. Panasenko, J. Arnhold, Linoleic acid hydroperoxide favors hypochlorite- and myeloperoxidase-induced lipid peroxidation, Free Radic. Res. 30 (1999) 479–487.
[23] S. Miyamoto, G.R. Martinez, D. Rettori, O. Augusto, M.H. Medeiros, P. Di Mascio, Linoleic acid hydroperoxide reacts with hypochlorous acid, generating peroxyl radical intermediates and singlet molecular oxygen, Proc. Natl. Acad. Sci. U.S.A. 103 (2006) 293–298.
[24] J. Zschaler, J. Arnhold, The hydroperoxide moiety of aliphatic lipid hydroperoxides is not affected by hypochlorous acid, Chem. Phys. Lipids 184 (2014) 42–51.
[25] L.K. Patterson, K. Hasegawa, Pulse radiolysis in model lipid systems. The influence of aggregation on kinetic behavior of OH induced radicals in aqueous sodium linoleate, Ber. Bunsenges. Phys. Chem. 82 (1978) 951–956.
[26] E.N. Frankel, Chemistry of free radical and singlet oxidation of lipids, Prog. Lipid Res. 23 (1985) 197–221.
[27] H. Esterbauer, G. Striegl, H. Puhl, M. Rotheneder, Continuous monitoring of in vitro oxidation of human low density lipoprotein, Free Radic. Res. Commun. 6 (1989) 67–75.
[28] L.K. Dahle, E.G. Hill, R.T. Holman, The thiobarbituric acid reaction and the autoxidations of polyunsaturated fatty acid methyl esters, Arch. Biochem. Biophys. 98 (1962) 253–261.
[29] W.A. Pryor, J.P. Stanley, E. Blair, Autoxidation of polyunsaturated fatty acids: II. A suggested mechanism for the formation of TBA-reactive materials from prostaglandin-like endoperoxides, Lipids 11 (1976) 370–379.
[30] G.A. Russell, Deuterium-isotope effects in the autoxidation of aralkyl hydrocarbons. Mechanism of the interaction of peroxy radicals, J. Am. Chem. Soc. 79 (1957) 3871–3877.
[31] H. Kautsky, Quenching of luminescence by oxygen, Trans. Faraday Soc. 35 (1939) 216–219.

[32] R.E. Kellogg, Mechanism of chemiluminescence from peroxy radicals, J. Am. Chem. Soc. 91 (1969) 5433–5436.
[33] E.N. Frankel, Volatile lipid peroxidation products, Prog. Lipid Res. 22 (1982) 1–33.
[34] H. Esterbauer, R.J. Schaur, H. Zollner, Chemistry and biochemistry of 4-hydroxynoneal, malondialdehyde and related aldehydes, Free Radic. Biol. Med. 11 (1991) 81–128.
[35] T.A. Dix, L.J. Marnett, Hematin-catalyzed rearrangement of hydroperoxy-linoleic acid to epoxy alcohols via an oxygen rebound, J. Am. Chem. Soc. 105 (1983) 7001–7002.
[36] T.A. Dix, L.J. Marnett, Conversion of linoleic acid hydroperoxide to hydroxy, keto, epoxyhydroxy, and trihydroxy fatty acids by hematin, J. Biol. Chem. 260 (1985) 5351–5357.
[37] G.A. Reed, Co-oxidation of xenobiotics: lipid peroxyl derivatives as mediators of metabolism, Chem. Phys. Lipids 44 (1987) 127–148.
[38] P. DiMascio, L.H. Catalani, E.J.H. Bechara, Are dioxethanes chemiluminescent intermediates in lipid peroxidation? Free Radic. Biol. Med. 12 (1992) 471–478.
[39] A.W. Pryor, P.A. Stanley, A suggested mechanism for the production of malondialdehyde during autoxidation of polyunsaturated fatty acids. Nonenzymatic production of prostaglandin endoperoxides during autoxidation, J. Org. Chem. 40 (1975) 3615–3617.
[40] R. Pamplona, Advanced lipoxidation end-products, Chem. Biol. Interact. 192 (2011) 14–20.
[41] G. Vistoli, D. de Maddis, A. Cipak, N. Zarkovic, M. Carini, G. Aldini, Advanced glycoxidation and lipoxidation end products (AGEs and ALEs): an overview of their mechanisms of formation, Free Radic. Res. 47 (Suppl. 1) (2013) 3–27.
[42] V.N. Bochkov, O.V. Oskolkova, K.G. Birukov, A.L. Levonen, C.J. Binder, J. Stockl, Generation and biological activities of oxidized phospholipids, Antioxidants Redox Signal. 12 (2010) 1009–1059.
[43] P. Kinnunen, K. Kaarniranta, A.K. Mahalka, Protein-oxidized phospholipid interactions in cellular signaling for cell death: from biophysics to clinical correlations, Biochim. Biophys. Acta 1818 (2012) 2446–2455.
[44] D.L. Crawford, T.C. Yu, R.O. Sinnhuber, Reaction of malonaldehyde and glycine, J. Agric. Food Chem. 14 (1966) 182–184.
[45] K.S. Chio, A.L. Tappel, Synthesis and characterization of the fluorescent products derived from malonaldehyde and amino acids, Biochemistry 8 (1969) 2821–2826.
[46] K. Kikugawa, Y. Machida, M. Kida, T. Kurechi, Studies on peroxidized lipids. III. Fluorescent pigments derived from the reaction of malonaldehyde and amino acids, Chem. Pharm. Bull. 29 (1981) 3003–3011.
[47] K. Kikugawa, Fluorescent products derived from the reaction of primary amines and components in peroxidized lipids, Free Radic. Biol. Med. 2 (1986) 389–417.
[48] R.M. LoPachin, T. Gavin, D.R. Petersen, D.S. Barber, Molecular mechanisms of 4-hydroxy-2-nonenal and acrolein toxicity: nucleophilic targets and adduct formation, Chem. Res. Toxicol. 22 (2009) 1499–1508.
[49] C.V. Nadkarni, L.M. Sayre, Structural definition of early lysine and histidine adduction chemistry of 4-hydroxynonenal, Chem. Res. Toxicol. 8 (1995) 284–291.
[50] D. Lin, H.G. Lee, Q. Liu, G. Perry, M.A. Smith, L.M. Sayre, 4-Oxo-2-nonenal is both more neurotoxic and protein reactive than 4-hydroxy-2-nonenal, Chem. Res. Toxicol. 18 (2005) 1219–1231.
[51] Z.F. Liu, P.E. Minkler, L.A. Sayre, Mass spectroscopic characterization of protein modification by 4-hydroxy-2-(E)-nonenal and 4-oxo-2-(E)-nonenal, Chem. Res. Toxicol. 16 (2003) 901–911.
[52] J.A. Doorn, D.R. Petersen, Covalent adduction of nucleophilic acids by 4-hydroxynonenal and 4-oxononenal, Chem. Biol. Interact. 143 (2003) 93–100.
[53] K. Uchida, E.R. Stadtman, Covalent attachment of 4-hydroxynonenal to glyceraldehyde-3-phosphate dehydrogenase – a possible involvement of intramolecular and intermolecular cross-linking reaction, J. Biol. Chem. 268 (1993) 6388–6393.
[54] L.M. Sayre, D. Lin, Q. Yuan, X.C. Zhu, X.X. Tang, Protein adducts generated from products of lipid peroxidation: focus on HNE and ONE, Drug Metab. Rev. 38 (2006) 651–675.
[55] T. Miyata, C. van Ypersele de Strihou, K. Kurokawa, J.W. Baynes, Alterations in nonenzymatic biochemistry in uremia: origin and significance of "carbonyl stress" in long-term uremic complication, Kidney Int. 55 (1999) 389–399.
[56] O.I. Shadyro, I.L. Yurkova, M.A. Kisel, O. Brede, J. Arnhold, Radiation-induced fragmentation of cardiolipin in a model membrane, Int. J. Radiat. Biol. 80 (2004) 239–245.
[57] I.L. Yurkova, D. Huster, J. Arnhold, Free radical fragmentation of cardiolipin by cytochrome c, Chem. Phys. Lipids 158 (2009) 16–21.
[58] I. Yurkova, M. Kisel, J. Arnhold, O. Shadyro, Dopamine and iron mediated fragmentation of galctocerebroside and cardiolipin in micelles, Cent. Eur. J. Chem. 5 (2007) 970–980.

[59] I.L. Yurkova, J. Arnhold, Ceramide formation on the γ-irradiation of galactocerebroside as a constituent of micelles, High Energy Chem. 42 (2008) 261–265.
[60] A.J. Brown, W. Jessup, Oxysterols and atherosclerosis, Atherosclerosis 142 (1999) 1–28.
[61] W. Kulig, L. Cwiklik, P. Jurkiewicz, T. Rog, I. Vattulainen, Cholesterol oxidation products and their biological importance, Chem. Phys. Lipids 199 (2006) 144–160.
[62] K.L. Carpenter, Good COP, bad COP: an unsolved murder. Are dietary cholesterol oxidation products guilty of atherogenicity? Br. J. Nutr. 88 (2002) 335–338.
[63] Y. Wang, P.M. Rogers, C. Su, G. Varga, K.R. Stayrook, T.P. Burris, Regulation of cholesterologenesis by the oxysterol receptor LXRα, J. Biol. Chem. 283 (2008) 26332–26339.
[64] A.J. Brown, 24(*S*),25-Epoxycholesterol: a messenger for cholesterol homeostasis, Int. J. Biochem. Cell Biol. 41 (2009) 744–747.
[65] N. Noguchi, R. Numano, H. Kaneda, E. Niki, Oxidation of lipids in low density lipoprotein particles, Free Radic. Res. 29 (1998) 43–52.
[66] Y. Saito, Y. Yoshida, E. Niki, Cholesterol is more susceptible to oxidation than linoleates in cultured cell under oxidative stress induced by selenium deficiency and free radicals, FEBS Lett. 581 (2007) 4349–4354.
[67] I. Bjorkhem, A. Cedazo-Minguez, V. Leoni, S. Meaney, Oxysterols and neurodegenerative diseases, Mol. Aspect. Med. 30 (2009) 171–179.
[68] A. Zarrouk, A. Vejux, J. Mackrill, Y. O'Callaghan, M. Hammami, N. O'Brien, C. Lizard, Involvement of oxysterols in age-related diseases and ageing processes, Ageing Res. Rev. 18 (2014) 148–162.
[69] R. Stern, G. Kogan, M.J. Jedrzejas, L. Šoltes, The many ways to cleave hyaluronan, Biotechnol. Adv. 25 (2007) 537–557.
[70] M. Rienks, A.P. Papageorgiou, N.G. Francogiannis, S. Heymans, Myocardial extracellular matrix: an ever-changing and diverse entity, Circ. Res. 114 (2014) 872–888.
[71] C. Malavaki, S. Mizumoto, N. Karamanos, K. Sugahara, Recent advances in the structural study of functional chondroitin sulfate and dermatan sulfate in health and disease, Connect. Tissue Res. 49 (2008) 133–139.
[72] M. Sárközy, R. Gáspar, K. Gömöri, L. Dux, C. Csonka, T. Csont, Effects of proteoglycans on oxidative/nitrative stress, Curr. Org. Chem. 21 (2017) 1–8.
[73] T.C. Laurent, U.B. Laurent, J.R. Fraser, The structure and function of hyaluronan: an overview, Immunol. Cell Biol. 74 (1996) A1–A7.
[74] J. Hunger, A. Bernecker, H.J. Bakker, M. Bonn, R.P. Richter, Hydration dynamics of hyaluronan and dextran, Biophys. J. 103 (2012) L10–L12.
[75] T. Hardingham, Solution properties of hyaluronan, in: H.G. Garg, C.A. Hales (Eds.), Chemistry and Biology of Hyaluonan, Elsevier Press, Amsterdam, 2004, pp. 1–19.
[76] K. Meyer, J. Palmer, The polysaccharide of the vitreous humor, J. Biol. Chem. 107 (1934) 629–634.
[77] B.J. Parsons, S. Al-Assaf, S. Navaratnam, G.O. Philips, Comparison of the reactivity of different oxidative species (ROS) towards hyaluronan, in: J.F. Kennedy, G.O. Philips, P.A. Williams (Eds.), Hyaluronan: Chemical, Biochemical and Biological Aspects, vol. 1, Woodhead Publishing Ltd., Cambridge, 2002, pp. 141–150. V.C. Hascall, guest ed.
[78] L. Šoltes, R. Mendichi, G. Kogan, J. Schiller, M. Stankovská, J. Arnhold, Degradative action of reactive species on hyaluronan, Biomacromolecules 7 (2006) 659–668.
[79] M.D. Rees, C.L. Hawkins, M.J. Davies, Hypochlorite-mediated fragmentation of glycosaminoglycans and related *N*-acetyl glycosamines: evidence for chloramide formation, free radical transfer reactions and site-specific fragmentation, J. Am. Chem. Soc. 125 (2003) 13719–13733.
[80] M.D. Rees, M.J. Davies, Heparan sulfate degradation via reductive hemolysis of its *N*-chloro derivatives, J. Am. Chem. Soc. 128 (2006) 3085–3097.
[81] M.D. Rees, T.N. McNiven, M.J. Davies, Degradation of extracellular matrix and its components by hypobromous acid, Biochem. J. 401 (2007) 587–596.
[82] M. Li, L. Rosenfeld, R.E. Vilar, M.K. Cowman, Degradation of hyaluronan by peroxynitrite, Arch. Biochem. Biophys. 341 (1997) 245–250.
[83] F. Franks, P.J. Lillford, G. Robinson, Isomeric equilibria of monosaccharides in solution, J. Chem. Soc. Faraday Trans. I 85 (1989) 2417–2426.
[84] L.D. Hayward, S.J. Angyal, A symmetry rule for the circular dichroism of reducing sugars, and the proportion of carbonyl forms in aqueous solutions thereof, Carbohydr. Res. 53 (1977) 13–20.
[85] J.P. Dworkin, S.L. Miller, A kinetic estimate of the free aldehyde content of aldoses, Carbohydr. Res. 329 (2000) 359–365.
[86] H.F. Bunn, P.J. Higgins, Reaction of monosaccharides with proteins: possible evolutionary significance, Science 213 (1981) 222–224.

[87] L.C. Maillard, Actions des acides amines sur les sucres; formation de melanoidines par voie méthodique, Comput. Rend. 154 (1912) 66–68.
[88] J.E. Hodge, Dehydrated foods. Chemistry of browning reactions in model systems, Agricult. Food Chem. 1 (1953) 928–943.
[89] M. Amadori, The product of the condensation of glucose and p-phenetidine, Atti. Reale Accad. Nazl. Lincei 9 (1929) 68–73.
[90] J.E. Hodge, The Amadorri rearrangement, Adv Carbohydr. Chem. 10 (1955) 169–205.
[91] V.J. Yaylayan, A. Huyghues-Despointes, Chemistry of Amadori rearrangement products: analysis, synthesis, kinetics, reactions, and spectroscopic properties, Crit. Rev. Food Sci. Nutr. 34 (1994) 321–369.
[92] T. Kurata, Y. Otsuka, Amino-reductones. Formation mechanisms and structural characteristics, Adv. Exp. Med. Biol. 434 (1998) 269–276.
[93] H.Y. Wang, H. Qian, W.R. Yao, Melanoidins produced by the Maillard reaction: structure and biological activity, Food Chem. 128 (2011) 573–584.
[94] F.J. Morales, V. Somoza, V. Fogliano, Physiological relevance of dietary melanoidins, Amino Acids 42 (2012) 1097–1109.
[95] T. Davidek, F. Robert, S. Devaud, F.A. Vera, I. Blank, Sugar fragmentation in the Maillard reaction cascade: formation of short-chain carboxylic acids by a new oxidative alpha-dicarbonyl cleavage pathway, J. Agric. Food Chem. 54 (2006) 6677–6684.
[96] S.R. Thorpe, J.W. Baynes, Maillard reaction products in tissue proteins: new products and new perspectives, Amino Acids 25 (2003) 275–281.
[97] P. Gkogkolou, M. Böhm, Advanced glycation end products. Key players in skin aging? Derm. Endocrinol. 4 (2012) 259–270.
[98] G. Suárez, R. Rajaram, A.L. Oronsky, M.A. Gawinowicz, Nonenzymatic glycation of bovine serum albumin by fructose (fructation). Comparison with the Maillard reaction initiated by glucose, J. Biol. Chem. 264 (1989) 3674–3679.
[99] K. Heyns, W. Beilfuss, Ketosylamine rearrangement of D-threo-pentulose (D-xylulose) with alpha-amino acids, Chem. Ber. 103 (1970) 2873–2876.
[100] A. Gugliucci, Formation of fructose-mediated advanced glycation end products and their role in metabolic and inflammatory diseases, Adv. Nutr. 8 (2017) 54–62.
[101] H.G. Kunkel, G. Wallenius, New hemoglobin in normal adult blood, Science 122 (1955) 288.
[102] R.J. Koenig, C.M. Peterson, R.L. Jones, C. Saudek, M. Lehrmann, A. Cerami, Correlation of glucose regulation and hemoglobin A1c in diabetes mellitus, N. Engl. J. Med. 295 (1976) 417–420.
[103] H.F. Bunn, K.H. Gabbay, P.M. Gallop, The glycosylation of hemoglobin: relevance to diabetes mellitus, Science 200 (1978) 21–27.
[104] M.U. Ahmed, S.R. Thorpe, J.W. Baynes, Identification of N epsilon-carboxymethyllysine as a degradation product of fructolysine in glycated protein, J. Biol. Chem. 261 (1986) 4889–4894.
[105] D.R. Sell, V.M. Monnier, Isolation, purification and partial characterization of novel fluorophores from aging human insoluble collagen-rich tissue, Connect. Tissue Res. 19 (1989) 77–92.
[106] S.K. Grandhee, V.M. Monnier, Mechanism of formation of the Maillard protein cross-link pentosidine. Glucose, fructose and ascorbate as pentosidine precursors, J. Biol. Chem. 266 (1991) 11649–11653.
[107] P.J. Thornalley, A. Langborg, H.S. Minhas, Formation of glyoxal, methylglyoxal and 3-deoxyglucosone in the glycation of proteins by glucose, Biochem. J. 344 (1999) 109–116.
[108] M. Neeper, A.M. Schmitt, J. Brett, S.D. Yan, F. Wang, Y.C. Pan, K. Elliston, D. Stern, A. Shaw, Cloning and expression of a cell surface receptor for advanced glycosylation end products of proteins, J. Biol. Chem. 267 (1992) 14998–15004.
[109] P.J. Thornalley, Cell activation by glycated proteins. AGE receptors, receptor recognition factors and functional classification of AGEs, Cell. Mol. Biol. 44 (1998) 1013–1023.
[110] R. Ramasamy, S.J. Vannucci, S.S. Yan, K. Herold, S.F. Yan, A.M. Schmidt, Advanced glycation end products and RAGE: a common thread in aging, diabetes, neurodegeneration, and inflammation, Glycobiology 15 (2005) 16R–28R.
[111] T.H. Fleming, P.M. Humpert, P.P. Nawroth, A. Bierhaus, Reactive metabolites and AGE/RAGE-mediated cellular dysfunction affect the aging process: a mini-review, Gerontology 57 (2011) 435–443.
[112] A. Bierhaus, P.M. Humpert, M. Morcos, T. Wendt, T. Chevakis, B. Arnold, D.M. Stern, P.P. Nawroth, Understanding RAGE, the receptor for advanced glycation end products, J. Mol. Med. 83 (2005) 876–886.
[113] J. Uribarri, S. Woodruff, S. Goodman, W. Cai, X. Chen, R. Pyzik, A. Yong, G.E. Striker, H. Vlassara, Advanced glycation end products in foods and a practical guide to their reduction in the diet, J. Am. Diet. Assoc. 110 (2010) 911–916.
[114] J. O'Brien, P.A. Morrissey, Nutritional and toxicological aspects of the Maillard browning reaction in foods, Crit. Rev. Food Sci. Nutr. 28 (1989) 211–248.
[115] T. Goldberg, W. Cei, M. Peppa, V. Dardaine, B.S. Baliga, J. Uribarri, H. Vlassara, Advanced glycoxidation end products in commonly consumed foods, J. Am. Diet. Assoc. 104 (2004) 1287–1291.

[116] F.Q. Schafer, G.R. Buettner, Redox environment of the cell as viewed through the redox state of the glutathione disulfide/glutathione couple, Free Radic. Biol. Med. 30 (2001) 1191–1212.
[117] D. Barford, The role of cysteine residues as redox-sensitive regulatory switches, Curr. Opin. Struct. Biol. 14 (2004) 679–686.
[118] A. Bindoli, J.M. Fukuto, H.J. Forman, Thiol chemistry in peroxidase catalysis and redox signaling, Antioxidants Redox Signal. 10 (2008) 1549–1564.
[119] L.B. Poole, K.J. Nelson, Discovering mechanisms of signaling-mediated cysteine oxidation, Curr. Opin. Chem. Biol. 12 (2008) 18–24.
[120] C.C. Winterbourn, D. Metodiewa, Reactivity of biologically important thiol compounds with superoxide and hydrogen peroxide, Free Radic. Biol. Med. 27 (1999) 322–328.
[121] P.S. Surdhar, D.A. Armstrong, Redox potentials of some sulphur containing radicals, J. Phys. Chem. 90 (1986) 5915–5917.
[122] W.H. Koppenol, A thermodynamic appraisal of the radical sink hypothesis, Free Radic. Biol. Med. 14 (1993) 91–94.
[123] S. Goldstein, G. Czapski, Mechanism of the nitrosation of thiols and amines by oxygenated •NO solutions: the nature of the nitrosating intermediates, J. Am. Chem. Soc. 118 (1996) 3419–3425.
[124] Y. Zhang, N. Hogg, S-Nitrosothiols: cellular formation and transport, Free Radic. Biol. Med. 38 (2005) 831–838.
[125] A. Zeida, R. Babbush, M.C.G. Lebrero, M. Trujillo, R. Radi, D.A. Estrin, Molecular basis of the mechanism of thiol oxidation by hydrogen peroxide in aqueous solution: challenging the S_N2 paradigm, Chem. Res. Toxicol. 25 (2012) 741–746.
[126] A. Hall, D. Parsonage, L.B. Poole, P.A. Karplus, Structural evidence that peroxiredoxin catalytic power is based on transition-state stabilization, J. Mol. Biol. 402 (2010) 194–209.
[127] L.B. Poole, The basics of thiols and cysteines in redox biology and chemistry, Free Radic. Biol. Med. 80 (2015) 148–157.
[128] W. Vogt, Oxidation of methionyl residues in proteins: tools, targets, and reversal, Free Radic. Biol. Med. 18 (1995) 93–105.
[129] M.D. Evans, W.A. Pryor, Damage to human α1-proteinase inhibitor by aqueous cigarette tar extracts and the formation of methionine sulfoxide, Chem. Res. Toxicol. 5 (1992) 654–660.
[130] M.D. Evans, D.F. Church, W.A. Pryor, Aqueous cigarette tar extracts damage human alpha-1-proteinase inhibitor, Chem. Biol. Interact. 79 (1991) 151–164.
[131] J.L. Beal, S.B. Foster, M.T. Ashby, Hypochlorous acid reacts with the N-terminal methionines of proteins to give dehydromethionine, a potential biomarker for neutrophil-induced oxidative stress, Biochemistry 48 (2009) 11142–11148.
[132] P.K. Sysak, C.S. Foote, T.-Y. Ching, Chemistry of singlet oxygen. XXV. Photooxygenation of methionine, Photochem. Photobiol. 26 (1977) 19–27.
[133] C. Giulivi, E. Cadenas, Heme protein radicals: formation, fate, and biological consequences, Free Radic. Biol. Med. 24 (1998) 269–279.
[134] J.W. Heinecke, W. Li, H.L. Daehnke III, J.A. Goldstein, Dityrosine, a specific marker of oxidation, is synthesized by the myeloperoxidase-hydrogen peroxide system of human neutrophils and macrophages, J. Biol. Chem. 268 (1993) 4069–4077.
[135] J.W. Heinecke, W. Li, G.A. Francis, J.A. Goldstein, Tyrosyl radical generated by myeloperoxidase catalyzes the oxidative cross-linking of proteins, J. Clin. Investig. 91 (1993) 2866–2872.
[136] C. Giulivi, N.J. Traaseth, K.J.A. Davies, Tyrosine oxidation products: analysis and biological relevance, Amino Acids 25 (2003) 227–232.
[137] T. DiMarco, C. Giulivi, Current analytical methods for the detection of dityrosine, a biomarker of oxidative stress, in biological samples, Mass Spectrom. Rev. 26 (2007) 108–120.
[138] S. Mukherjee, E.A. Kapp, A. Lothian, A.M. Roberts, Y.V. Vasil'ev, B.A. Boughton, K.J. Barnham, W.M. Kok, C.A. Hutton, C.L. Masters, A.I. Bush, J.S. Beckman, S.G. Dey, B.R. Roberts, Characterization and identification of dityrosine cross-linked peptides using tandem mass spectrometry, Anal. Chem. 89 (2017) 6136–6145.
[139] C. Giulivi, K.J.A. Davies, Dityrosine and tyrosine oxidation products are endogenous markers for the selective proteolysis of oxidatively modified red blood cell hemoglobin by (the 19 S) proteasome, J. Biol. Chem. 268 (1993) 8752–8759.
[140] J.S. Jacob, D.P. Cistola, F.F. Hsu, S. Muzaffar, D.M. Mueller, S.L. Hazen, J.W. Heinecke, Human phagocytes employ the myeloperoxidase-hydrogen peroxide system to synthesize dityrosine, trityrosine, pulcherosine, and isodityrosine by a tyrosyl radical-dependent pathway, J. Biol. Chem. 271 (1996) 19950–19956.
[141] T.J. Simat, H. Steinhart, Oxidation of free tryptophan and tryptophan residues in peptides and proteins, J. Agric. Food Chem. 46 (1998) 490–498.
[142] J. Cadet, K.J.A. Davies, M.H.G. Medeiros, P. Di Mascio, J.R. Wagner, Formation and repair of oxidatively generated damage in cellular DNA, Free Radic. Biol. Med. 107 (2017) 13–34.

[143] K.S. Kasprzak, Oxidative DNA and protein damage in metal-induced toxicity and carcinogenesis, Free Radic. Biol. Med. 32 (2002) 958–967.
[144] S. Steenken, S.V. Jovanovic, How easily oxidizable is DNA? One electron reduction potentials of adenosine and guanosine radicals in aqueous solution, J. Am. Chem. Soc. 119 (1997) 617–618.
[145] S. Fukuzumi, H. Miyao, K. Ohkubo, T. Suenobu, Electron-transfer oxidation properties of DNA bases and DNA oligomers, J. Phys. Chem. A 109 (2005) 3285–3294.
[146] A.J.S.C. Vieira, S. Steenken, Pattern of OH radical reaction with adenine and its nucleosides and nucleotides. Characterization of two types of isomeric OH adduct and their unimolecular transformation reactions, J. Am. Chem. Soc. 112 (1990) 6986–6994.
[147] L.P. Candeias, S. Steenken, Reaction of $HO^{\bullet}$ with guanine derivatives in aqueous solution: formation of two different redox-active OH-adduct radicals and their unimolecular transformation reactions. Properties of $G(-H)^{\bullet}$, Chem. Eur J. 6 (2000) 475–484.
[148] J. Cadet, T. Douki, J.-L. Ravanat, Oxidatively generated damage to the guanine moiety of DNA: mechanistic aspects and formation in cells, Acc. Chem. Res. 41 (2008) 1075–1083.
[149] K. Kurbanyan, K.L. Nguyen, P. To, E.V. Rivas, A.M. Lueras, C. Kosinski, M. Steryo, M. Gonzalez, D.A. Mah, E.D. Stemp, DNA-protein cross-linking via guanine oxidation: dependence upon protein and photosensitizer, Biochemistry 42 (2003) 10269–10281.
[150] C. Crean, Y. Uvaydov, N.E. Geactinov, V. Shafirovich, Oxidation of single-stranded oligonucleotides by carbonate radical anions: generating intrastrand cross-links between guanine and thymine bases separated by cytosines, Nucleic Acids Res. 36 (2008) 742–755.
[151] W. Luo, J.G. Muller, C.J. Burrows, The pH-dependent role of superoxide in riboflavin-catalyzed photooxidation of 8-oxo-7,8-dihydroguanosine, Org. Lett. 3 (2001) 2801–2804.
[152] Y. Oda, S. Uesugi, M. Ikehara, S. Nishimura, Y. Kawase, H. Ishikawa, H. Inoue, E. Ohtsuka, NMR studies of a DNA containing 8-hydroxydeoxyguanosine, Nucleic Acids Res. 19 (1991) 1407–1412.
[153] K.C. Cheng, D.S. Cahill, H. Kasai, S. Nishimura, L.A. Loeb, 8-Hydroxyguanine, an abundant form of oxidative DNA damage, causes G → T and A → C substitutions, J. Biol. Chem. 267 (1992) 166–172.
[154] J. Cadet, E. Sage, T. Douki, Ultraviolet radiation-mediated damage to cellular DNA, Mutat. Res. 571 (2005) 3–17.
[155] S. Mouret, C. Baudouin, M. Charveron, A. Favier, J. Cadet, T. Douki, Cyclobutane pyrimidine dimers are predominant DNA lesion in whole human skin exposed to UVA radiation, Proc. Natl. Acad. Sci. U.S.A. 103 (2006) 13765–13770.
[156] H.C. Box, H.G. Freund, E.E. Budzinski, J.C. Wallace, A.E. Maccubbin, Free radical induced double base lesions, Radiat. Res. 141 (1995) 91–94.
[157] T. Douki, J. Rivière, J. Cadet, DNA tandem lesions containing 8-oxo-7,8-dihydroguanine and formamido residues arise from intramolecular addition of thymine peroxyl radical to guanine, Chem. Res. Toxicol. 15 (2002) 445–454.
[158] J. Cadet, T. Douki, J.-L. Ravanat, Oxidatively generated base damage to cellular DNA, Free Radic. Biol. Med. 49 (2010) 9–21.
[159] J.R. Wagner, J. Cadet, Oxidant reactions of cytosine DNA components by hydroxyl radical and one-electron oxidants in aerated aqueous solutions, Acc. Chem. Res. 43 (2010) 564–571.
[160] J. Cadet, J.R. Wagner, Oxidatively generated base damage to cellular DNA by hydroxyl radical and one-electron oxidants: similarities and differences, Arch. Biochem. Biophys. 557 (2014) 47–54.
[161] S. Tremblay, T. Douki, J. Cadet, J.R. Wagner, 2′-deoxycytosine glycols, a missing link in the free radical-mediated oxidation of DNA, J. Biol. Chem. 274 (1999) 20833–20838.
[162] M. Dizdaroglu, Oxidative damage to DNA in mammalian chromatin, Mutat. Res. 275 (1992) 331–342.
[163] M.S. Cooke, M.D. Evans, M. Dizdaroglu, J. Lunec, Oxidative DNA damage: mechanisms, mutation, and disease, FASEB J. 17 (2003) 1195–1214.
[164] J.A. Swenberg, K. Lu, B.C. Moeller, L. Gao, P.B. Upton, J. Nakamura, T.B. Starr, Endogenous versus exogenous DNA adducts: their role in carcinogenesis, epidemiology, and risk assessment, Toxicol. Res. 120 (2011) S130–S145.
[165] C.L. Hawkins, M.J. Davies, Hypochlorite-induced damage to DNA, RNA, and polynucleotides: formation of chloramines and nitrogen-centered radicals, Chem. Res. Toxicol. 15 (2002) 83–92.
[166] S.E. Gomez-Mejiba, Z. Zhai, M.S. Gimenez, M.T. Ashby, J. Chilakapati, K. Kitchin, R.P. Mason, D.C. Ramirez, Myeloperoxidase-induced genomic DNA-centered radicals, J. Biol. Chem. 285 (2010) 20062–20071.
[167] A.K. Chaudhary, M. Nokubo, L.J. Marnett, I.A. Blair, Analysis of the malondialdehyde-2′-deoxyguanosine adduct in rat liver DNA by gas chromatography/electron capture negative chemical ionization mass

spectrometry, Biol. Mass Spectrom. 23 (1994) 457–464.

[168] M. Wang, K. Dhingra, W.M. Hittleman, J.G. Liehr, M. de Andrade, D. Li, Lipid peroxidation-induced putative malondialdehyde-DNA adducts in human breast tissues, Cancer Epidemiol. Biomark. Prev. 5 (1996) 705–710.

[169] J.D. West, L.J. Marnett, Reactive intermediates as modulators of cell signaling and cell death, Chem. Res. Toxicol. 19 (2006) 173–194.

[170] I.A. Blair, DNA adducts with lipid peroxidation products, J. Biol. Chem. 283 (2008) 15545–15549.

[171] K. Arab, M. Pedersen, J. Nair, M. Meerang, L.E. Knudson, H. Bartsch, Typical signature of DANN damage in white blood cells: a pilot study on etheno adducts in Danish mother-newborn child pairs, Carcinogenesis 30 (2009) 282–285.

[172] B.I. Fedeles, B.D. Freudenthal, E. Yau, V. Singh, S.C. Chang, D. Li, J.C. Dealney, S.H. Wilson, J.M. Essigmann, Intrinsic mutagenic properties of 5-chlorocytosine: a mechanistic connection between chronic inflammation and cancer, Proc. Natl. Acad. Sci. U.S.A. 112 (2015) E4571–E4580.

[173] G.W. Burton, A. Joyce, K.U. Ingold, Is vitamin E the only lipid-soluble, chain-breaking antioxidant in human blood plasma and erythrocyte membrane? Arch. Biochem. Biophys. 221 (1983) 281–290.

[174] G.R. Buettner, The pecking order of free radicals and antioxidants: lipid peroxidation, α-tocopherol, and ascorbate, Arch. Biochem. Biophys. 300 (1993) 535–543.

[175] E. Niki, Antioxidants in relation to lipid peroxidation, Chem. Phys. Lipids 44 (1987) 227–253.

[176] J.E. Packer, T.F. Slater, R.L. Wilson, Direct observation of a free radical interaction between vitamin E and vitamin C, Nature 278 (1979) 737–738.

[177] M. Scarpa, A. Rigo, M. Maiorino, F. Ursini, C. Gregolin, Formation of alpha-tocopherol radical and recycling of alpha-tocopherol by ascorbate during peroxidation of phosphatidylcholine liposomes. An electron paramagnetic resonance study, Biochim. Biophys. Acta 801 (1984) 215–219.

[178] B. Frei, R. Stocker, B.N. Ames, Antioxidant defenses and lipid peroxidation in human blood plasma, Proc. Natl. Acad. Sci. U.S.A. 85 (1988) 9748–9752.

[179] N.H. Williams, J.K. Yandell, Outer-sphere electron-transfer reactions of ascorbate anions, Aust. J. Chem. 35 (1982) 1133–1144.

[180] F.J.G.M. van Kujik, A. Sevanian, G.J. Handelman, E.A. Dratz, A new role for phospholipase A_2: protection of membranes from lipid peroxidation damage, Trends Biochem. Sci. 12 (1987) 31–34.

[181] L.R. McLean, K.A. Hagaman, Effect of lipid physical state on the rate of peroxidation of liposomes, Free Radic. Biol. Med. 12 (1992) 113–119.

[182] R.F. Jacob, R.P. Mason, Lipid peroxidation induces cholesterol domain formation in model membranes, J. Biol. Chem. 280 (2005) 39380–39387.

[183] A. Broniec, R. Klosinski, M. Pawlak, M. Wrona-Krol, D. Thompson, T. Sarna, Interactions of plasmalogens and their diacyl analogs with singlet oxygen in selected model systems, Free Radic. Biol. Med. 50 (2011) 892–898.

[184] N.E. Braverman, A.B. Moser, Functions of plasmalogen lipids in health and disease, Biochim. Biophys. Acta 1822 (2012) 1442–1452.

[185] P.J. Sindelar, Z. Guan, G. Dallner, L. Ernster, The protective role of plasmalogens in iron-induced lipid peroxidation, Free Radic. Biol. Med. 26 (1999) 318–324.

[186] C. von Sonntag, The Chemical Bases of Radiation Biology, Taylor & Francis, London, Philadelphia, 1987.

[187] O.I. Shadyro, A.A. Sosnovskaya, I.P. Edimecheva, N.I. Ostrovskaya, K.M. Kazem, I.B. Hryntsevich, A.V. Alekseev, Effects of quinones on free-radical processes of oxidation and fragmentation of hydroxyl-containing organic compounds, Bioorg. Med. Chem. Lett 17 (2007) 6383–6386.

[188] O.I. Shadyro, G.K. Glushonok, T.G. Glushonok, I.P. Edimecheva, A.G. Moroz, A.A. Sosnovskaya, I.L. Yurkova, G.I. Polozov, Quinones as free-radical fragmentation inhibitors in biologically important molecules, Free Radic. Res. 36 (2002) 859–867.

[189] S.D. Brinkevich, O.I. Shadyro, The effects of ascorbic acid on homolytic processes involving α-hydroxyl-containing carbon-centered radicals, Bioorg. Med. Chem. Lett 18 (2008) 6448–6450.

[190] I.B. Hryntsevich, O.I. Shadyro, Reactions of α-hydroxyethyl radicals with flavonoids of various structures, Bioorg. Med. Chem. Lett 15 (2005) 4252–4255.

[191] P.Y. Lagustin, O.I. Shadyro, Effects of B group vitamins on reactions of various α-hydroxyl-containing organic radicals, Bioorg. Med. Chem. Lett 15 (2005) 3797–3800.

[192] P. Hochstein, L. Hatch, A. Sevanian, Uric acid: functions and determination, Methods Enzymol. 105 (1984) 162–166.

[193] K. Nyyssönen, E. Porkkala-Sarataho, J. Kaikkonen, J.T. Salonen, Ascorbate and urate are the strongest determinants of plasma antioxidative capacity and serum lipid resistance to oxidation in Finnish men, Atherosclerosis 130 (1997) 223–233.

[194] J. Zschaler, D. Schlorke, J. Arnhold, Differences in innate immune response between man and mouse, Crit. Rev. Immunol. 34 (2014) 433–454.
[195] Y. Ohta, M. Nishikimi, Random nucleotide substitutions in primate nonfunctional gene for L-gulono-gamma-lactone oxidase, the missing enzyme in L-ascorbic acid biosynthesis, Biochim. Biophys. Acta 1472 (1999) 408–411.
[196] X.W. Wu, C.C. Lee, D.M. Muzny, C.T. Caskey, Urate oxidase: primary structure and evolutionary implications, Proc. Natl. Acad. Sci. U.S.A. 86 (1989) 9412–9416.
[197] B. Alvarez-Lario, J. Macarron-Vicente, Uric acid and evolution, Rheumatology 49 (2010) 2010–2015.
[198] D.D.M. Wayner, G.W. Burton, K.U. Ingold, L.R.C. Barcley, S.J. Locke, The relative contributions of vitamin E, urate, ascorbate and proteins to the total peroxyl radical-trapping antioxidant activity of human blood plasma, Biochim. Biophys. Acta 924 (1987) 408–419.
[199] H. Vlassara, W. Cai, J. Crandall, T. Goldberg, R. Oberstein, V. Dardaine, M. Peppa, E.J. Rayfield, Inflammatory mediators are induced by dietary glycotoxins, a major risk factor for diabetic angiopathy, Proc. Natl. Acad. Sci. U.S.A. 99 (2002) 15596–15601.
[200] S. Yamagishi, S. Ueda, S. Okuda, Food-derived advanced glycation end products (AGEs): a novel therapeutic target for various disorders, Curr. Pharmaceut. Des. 13 (2007) 2832–2836.
[201] H. Vlassara, G.E. Striker, AGE restriction in diabetes mellitus: a paradigm shift, Nat. Rev. Endocrinol. 7 (2011) 526–539.
[202] D.L. Price, P.M. Rhett, S.R. Thorpe, J.W. Baynes, Chelating activity of advanced glycation end-product inhibitors, J. Biol. Chem. 276 (2001) 48967–48972.
[203] R.P. Dearlove, P. Greenspan, D.K. Hartle, R.B. Swanson, J.L. Hargrove, Inhibition of protein glycation by extracts of culinary herbs and spices, J. Med. Food 11 (2008) 275–281.
[204] K.V. Tarwadi, V.V. Agte, Effect of micronutrients on methylglyoxal-mediated in vitro glycation of albumin, Biol. Trace Elem. Res. 143 (2011) 717–725.
[205] C.H. Wu, G.C. Yen, Inhibitory effect of naturally occurring flavonoids on the formation of advanced glycation endproducts, J. Agric. Food Chem. 53 (2005) 3167–3173.
[206] Z.D. Draelos, M. Yatskayer, S. Raab, C. Oresajo, An evolution of the effect of a topical product containing C-xyloside and blueberry extract on the appearance of type II diabetic skin, J. Cosmet. Diabetol. 8 (2009) 147–151.
[207] H.A. Jung, J.J. Park, B.S. Min, H.J. Jung, M.N. Islam, J.S. Choi, Inhibition of advanced glycation endproducts formation by Korean thistle, Cirsium maackii, Asian Pac. J. Trop. Med. (2015) 1–5, https://doi.org/10.1016/S1995-7645(14)60178-4.
[208] W.-J. Yeh, S.,-H. Hsia, W.-H. Lee, C.-H. Wu, Polyphenols with antiglycation activity and mechanisms of action: a review of recent findings, J. Food Drug Anal. 25 (2017) 84–92.
[209] R. Yang, W.-X. Wang, H.-J. Chen, Z.-C. He, A.-Q. Jia, The inhibition of advanced glycation end-products by five fractions and three main flavonoids from Camellia nitidissima Chi flowers, J. Food Drug Anal. 26 (2018) 252–259.
[210] D. Edelstein, M. Brownlee, Mechanistic studies of advanced glycosylation end product inhibition by aminoguanidine, Diabetes 41 (1992) 26–29.
[211] D.R. Sell, J.F. Nelson, V.M. Monnier, Effect of chronic aminoguanidine treatment on age-related glycation, glycoxidation, and collagen cross-linking in the Fischer 344 rat, J. Gerontol. A Biol. Sci. Med. Sci. 56 (2001) B405–B411.
[212] V.P. Reddy, A. Beyaz, Inhibitors of the Maillard reaction and AGE breakers as therapeutics for multiple diseases, Drug Discov. Today 11 (2006) 646–654.
[213] P.A. Voziyan, B.G. Hudson, Pyridoxamine: the many virtues of a Maillard reaction inhibitor, Ann. N. Y. Acad. Sci. 1043 (2005) 807–816.
[214] J.P. Dwyer, B.A. Greco, K. Umanath, D. Packham, J.W. Fox, R. Peterson, B.R. Broome, L.E. Greene, M. Sika, J.B. Lewis, Pyridoxamine dihydrochloride in diabetic nephropathy (PIONEER-CSG-17): lessons learned from a pilot study, Nephron 129 (2015) 22–28.
[215] S. Vasan, P. Foiles, H. Founds, Therapeutic potential of breakers of advanced glycation end product-protein crosslinks, Arch. Biochem. Biophys. 419 (2003) 89–98.
[216] P.J. Thornalley, Glyoxalase I – structure, function and a critical role in the enzymatic defence against glycation, Biochem. Soc. Trans. 31 (2003) 1343–1348.
[217] B. Mannervik, Molecular enzymology of the glyoxalase system, Drug Metab. Drug Interact. 23 (2008) 13–27.
[218] M. Xue, N. Rabbani, H. Momiji, P. Imbasi, M.M. Anwar, N. Kitteringham, B.K. Park, T. Souma, T. Moriguchi, M. Yamamoto, P.J. Thornalley, Transcriptional control of glyoxalase 1 by Nrf2 provides a stress-responsive defence against dicarbonyl glycation, Biochem. J. 443 (2012) 213–222.
[219] M. Brownlee, Biochemistry and molecular cell biology of diabetic complications, Nature 414 (2001) 813–820.

[220] P.J. Thornalley, N. Rabbani, Glyoxalase in tumourigenesis and multidrug resistance, Semin. Cell Dev. Biol. 22 (2011) 318–325.

[221] M.G. Distler, A.A. Palmer, Role of glyoxalase 1 (Glo 1) and methylglyoxal (MG) in behavior: recent advances and mechanistic insights, Front. Genet. 3 (2012), Article 250, 1–10, https://doi.org/10.3389/fgene.2012.00250.

[222] B.I. Hudson, L.G. Bucciarelli, T. Wendt, T. Sakaguchi, E. Lalla, W. Qu, Y. Lu, D.M. Stern, Y. Naka, R. Ramasamy, S.D. Yan, S.F. Yan, V. D'Agati, A.M. Schmidt, Blockade of receptor for advanced glycation endproducts: a new target for therapeutic intervention in diabetic complications and inflammatory disorders, Arch. Biochem. Biophys. 419 (2003) 80–88.

[223] S. Bongarzone, V. Savickas, F. Luzi, A.D. Gee, Targeting the receptor for advanced glycation endproducts (RAGE): a medicinal chemistry perspective, J. Med. Chem. 60 (2017) 7213–7232.

[224] P. Di Mascio, T.P.A. Devasagayam, S. Kaiser, H. Sies, Carotenoids, tocopherols and thiols as biological singlet molecular oxygen quenchers, Biochem. Soc. Trans. 18 (1990) 1054–1056.

[225] P. Di Mascio, E.J.H. Bechara, H.G. Medeiros, K. Briviba, H. Sies, Singlet molecular oxygen production in the reaction of peroxynitrite with hydrogen peroxide, FEBS Lett. 355 (1994) 287–289.

[226] T. Grune, G. Lietz, A. Palou, C. Ross, W. Stahl, G. Tang, D. Thurnham, S. Yin, H.K. Biesalski, β-carotene is an important vitamin A source for humans, J. Nutr. 140 (2010) 2268S–2285S.

[227] A. Bendich, J.A. Olson, Biological actions of carotenoids, FASEB J. 3 (1989) 1927–1932.

[228] J.P.E. Spencer, M.M. Abd El Mohsen, A.-M. Minihane, J.C. Mathers, Biomarkers of the intake of dietary polyphenols: strengths, limitations and application in nutrition research, Br. J. Nutr. 99 (2008) 12–22.

[229] B.A. Graf, P.E. Milbury, J.B. Blumberg, Flavonols, flavones, flavanones and human health: epidemiological evidence, J. Med. Food 8 (2005) 281–290.

[230] I.C.W. Arts, P.C.H. Hollman, Polyphenols and disease risk in epidemiologic studies, Am. J. Clin. Nutr. 81 (2005) 317–325.

[231] K.B. Pandey, S.I. Rizvi, Plant polyphenols as dietary antioxidants in human health and disease, Oxid. Med. Cell. Longevity 2 (2009) 270–278.

[232] S. Steenken, P. Neta, One-electron redox potentials of phenols. Hydroxy- and aminophenols and related compounds of biological interest, J. Phys. Chem. 86 (1982) 3661–3667.

[233] S.V. Jovanovic, Y. Hara, S. Steenken, M.G. Simic, Antioxidant potential of gallocatechins. A pulse radiolysis and laser photolysis study, J. Am. Chem. Soc. 117 (1995) 9881–9888.

[234] N.R. Perron, J.L. Brumaghim, A review of the antioxidant mechanisms of polyphenol compounds related to iron binding, Cell Biochem. Biophys. 53 (2009) 75–100.

[235] R.C. Hider, Z.D. Liu, H.H. Khodr, Metal chelation of polyphenols, Methods Enzymol. 335 (2001) 190–203.

[236] H. Kipton, J. Powell, M.C. Taylor, Interactions of iron (II) and iron (III) with gallic acid and its homologues: a potentiometric and spectrophotometric study, Aust. J. Chem. 35 (1982) 739–756.

[237] M. Yoshino, K. Murakami, Interaction of iron with polyphenolic compounds: application to antioxidant characterization, Anal. Biochem. 257 (1998) 40–44.

[238] H.E. Hajii, E. Nkhili, V. Tomao, O. Dangles, Interactions of quercetin with iron and copper ions: complexation and autoxidation, Free Radic. Res. 40 (2006) 303–320.

[239] M.J. Hynes, M.O. Coinceanainn, The kinetics and mechanisms of the reaction of iron(III) with gallic acid, gallic acid methyl ester and catechin, J. Inorg. Biochem. 85 (2001) 131–142.

[240] P. Ryan, M.J. Hynes, The kinetics and mechanisms of the reaction of iron(III) with quercetin and morin, J. Inorg. Biochem. 102 (2008) 127–136.

[241] M. Tuntawiroon, N. Sritongkul, M. Brune, L. Rossander-Hulten, R. Pleehachinda, R. Suwanik, L. Hallberg, Dose-dependent inhibitory effect of phenolic compounds in foods on nonheme-iron absorption in men, Am. J. Clin. Nutr. 53 (1991) 554–557.

[242] I. Morel, G. Lescoat, P. Cillard, J. Cillard, Role of flavonoids and iron chelation in antioxidant action, Methods Enzymol. 234 (1994) 437–443.

[243] M. Ferrali, C. Signorini, B. Caciotti, L. Sugherini, L. Ciccoli, D. Giachetti, M. Comporti, Protection against oxidative damage of erythrocyte membrane by the flavonoid quercetin and its relation to iron chelating activity, FEBS Lett. 416 (1997) 123–129.

[244] P. Das, N. Raghuramulu, K.C. Rao, Effect of organic acids and polyphenols on in vitro available iron from foods, J. Food Sci. Technol. 40 (2003) 677–681.

[245] S. Srichairatanakool, S. Ounjaijean, C. Thephinlap, U. Khansuwan, C. Phisalpong, S. Fucharoen, Iron-chelating and free-radical scavenging activities of microwave processed green tea in iron overload, Hemoglobin 30 (2006) 311–327.

[246] T. Pan, J. Jankovic, W. Le, Potential therapeutic properties of green tea polyphenols in Parkin'sons disease, Drugs Aging 20 (2003) 711–721.

[247] M. Singh, M. Arseneault, T. Sanderson, V. Murthy, C. Ramassamy, Challenges for research on polyphenols from food in Alzheimer's disease: bioavailability,

metabolism, and cellular and molecular mechanisms, J. Agric. Food Chem. 56 (2008) 4855–4873.

[248] A. Rahman, Shahabuddin, S.M. Hadi, J.H. Parish, K. Ainley, Strand scission in DNA by quercetin and Cu(II): role of Cu(I) and oxygen free radicals, Carcinogenesis 10 (1989) 1833–1839.

[249] H. Spalteholz, P.G. Furtmüller, C. Jakopitsch, C. Obinger, T. Schewe, H. Sies, J. Arnhold, Kinetic evidence for rapid oxidation of (–)-epicatechin by human myeloperoxidase, Biochem. Biophys. Res. Commun. 371 (2008) 810–813.

[250] T. Kirchner, J. Flemmig, P.G. Furtmüller, C. Obinger, J. Arnhold, (–)-Epicatechin enhances the chlorinating activity of human myeloperoxidase, Arch. Biochem. Biophys. 495 (2010) 21–27.

[251] J. Flemmig, J. Remmler, F. Röhring, J. Arnhold, (–)-Epicatechin regenerates the chlorinating activity of myeloperoxidase in vitro and in neutrophil granulocytes, J. Inorg. Biochem. 130 (2014) 84–91.

[252] J. Gau, P.G. Furtmüller, C. Obinger, J. Arnhold, J. Flemmig, Enhancing hypothiocyanite production by lactoperoxidase – mechanism and chemical properties of promoters, Biochem. Biophys. Rep. 4 (2015) 257–267.

[253] J. Gau, P.G. Furtmüller, C. Obinger, M. Prévost, P. van Antwerpen, J. Arnhold, J. Flemmig, Flavonoids as promoters of the (pseudo-)halogenating activity of lactoperoxidase and myeloperoxidase, Free Radic. Biol. Med. 97 (2016) 307–319.

[254] J. Gau, M. Prévost, P. van Antwerpen, M.-B. Sarosi, S. Rodewald, J. Arnhold, J. Flemmig, Tannins and tannin-related derivatives enhance the (pseudo-)halogenating activity of lactoperoxidase, J. Nat. Prod. 80 (2017) 1328–1338.

[255] J. Gau, J. Arnhold, J. Flemmig, Reactivation of peroxidase activity in human saliva samples by polyphenols, Arch. Oral Biol. 85 (2018) 70–78.

[256] L. Gebicka, E. Banasiak, Flavonoids as reductants of ferryl hemoglobin, Acta Biochim. Pol. 56 (2009) 509–513.

[257] N.-H. Lu, C. Chen, Y.-J. He, R. Tian, Q. Xiao, Y.-Y. Peng, Effects of quercetin on hemoglobin-dependent redox reactions: relationship to iron-overload rat liver injury, J. Asian Nat. Prod. Res. 15 (2013) 1265–1276.

[258] E. Mullarky, L.C. Cantley, Diverting glycolysis to combat oxidative stress, in: K. Nakao, N. Minato, S. Uemoto (Eds.), Innovative Medicine. Basic Research and Development, Springer Open, 2015, pp. 3–24.

[259] F.L. Rodkey, Oxidation-reduction potentials of the triphosphopyridine nucleotide system, J. Biol. Chem. 213 (1955) 777–786.

[260] R.L. Veech, L.V. Eggleston, H.A. Krebs, The redox state of free nicotinamide-adenine dinucleotide phosphate in the cytoplasm of rat liver, Biochem. J. 155 (1969) 609–619.

[261] H. Sies, Nicotinamide nucleotide compartmentation, in: H. Sies (Ed.), Metabolic Compartmentation, Academic Press, 1982, pp. 205–231.

[262] C. Riganti, E. Gazzano, M. Polimeni, E. Aldieri, D. Ghigo, The pentose phosphate pathway: an antioxidant defense and a crossroad in tumor cell fate, Free Radic. Biol. Med. 53 (2012) 421–436.

[263] D. Holton, D. Proscal, H.L. Chang, Regulation of pentose phosphate pathway dehydrogenases by NADP+/NADPH ratios, Biochem. Biophys. Res. Commun. 68 (1976) 436–441.

[264] R.C. Stanton, Glucose-6-phosphate dehydrogenase, NADPH, and cell survival, IUBMB Life 64 (2012) 362–369.

[265] K.C. Wu, J.Y. Cui, C.D. Klaassen, Beneficial role of Nrf2 in regulating NADPH generation and consumption, Toxicol. Sci. 123 (2011) 590–600.

[266] Y. Mitsuishi, K. Taguchi, Y. Kawatani, T. Shibata, T. Nukiwa, H. Aburatani, M. Yamamoto, H. Motohashi, Nrf2 redirects glucose and glutamine into anabolic pathways in metabolic reprogramming, Cancer Cell 22 (2012) 66–79.

[267] X. Xu, J. Zhao, Z. Xu, B. Peng, Q. Huang, E. Arnold, J. Ding, Structures of human cytosolic NADP-dependent isocitrate dehydrogenase reveal a novel self-regulatory mechanism of activity, J. Biol. Chem. 279 (2004) 33946–33957.

[268] G. Bukato, Z. Kochan, J. Swierczyński, Purification and properties of cytosolic and mitochondrial malic enzyme isolated from human brain, Int. J. Biochem. Cell Biol. 27 (1995) 47–54.

[269] J. Fan, J. Ye, J.J. Kamphorst, T. Shlomi, C.B. Thompson, J.D. Rabinowitz, Quantitative flux analysis reveals folate-dependent NADPH production, Nature 510 (2014) 298–302.

[270] A. Meister, Glutathione metabolism and its selective modification, J. Biol. Chem. 263 (1988) 17205–17208.

[271] T. Seres, V. Ravichandran, T. Moriguchi, K. Rokutan, J.A. Thomas, R.B. Johnston, Protein-S-thiolation and dethiolation during respiratory burst in human monocytes, J. Immunol. 156 (1996) 1973–1980.

[272] D. Mustacich, G. Powis, Thioredoxin reductase, Biochem. J. 346 (2000) 1–8.

[273] C.V. Smith, D.P. Jones, T.M. Guenther, L.H. Lash, B.H. Lauterburg, Contemporary issues in toxicology. Compartmentation of glutathione: implications for the study of toxicity and disease, Toxicol. Appl. Pharmacol. 140 (1996) 1–12.

[274] A. Wahllander, S. Soboll, H. Sies, I. Linke, M. Muller, Hepatic mitochondrial and cytosolic glutathione content and the subcellular distribution of GSH-S-transferases, FEBS Lett. 97 (1979) 138–140.
[275] O. Bellomo, M. Vairetti, L. Stivala, F. Mirabelli, P. Richelmi, S. Orrenius, Demonstration of nuclear compartmentalization of glutathione in hepatocytes, Proc. Natl. Acad. Sci. U.S.A. 89 (1992) 4412–4416.
[276] C. Hwang, A.J. Sinskey, H.F. Lodish, Oxidized redox state of glutathione in the endoplasmic reticulum, Science 257 (1992) 1496–1502.
[277] F. Ursini, M. Maiorino, A. Roveri, Phospholipid hydroperoxide glutathione peroxidase (PHGPx): more than an antioxidant enzyme? Biomed. Environ. Sci. 10 (1997) 327–332.
[278] E. Lubos, J. Loscalzo, D.E. Handy, Glutathione peroxidase-1 in health and disease: from molecular mechanisms to therapeutic opportunities, Antioxidants Redox Signal. 15 (2011) 1957–1997.
[279] C.C. Winterbourn, Superoxide as an intracellular radical sink, Free Radic. Biol. Med. 14 (1993) 85–90.
[280] A. Holmgren, C. Johansson, C. Berndt, M.E. Lönn, C. Hudemann, C.H. Lillig, Thiol redox control via thioredoxin and glutaredoxin systems, Biochem. Soc. Trans. 33 (2005) 1375–1377.
[281] L.E.S. Netto, F. Antunes, The roles of peroxiredoxin and thioredoxin in hydrogen peroxide sensing and in signal transduction, Mol. Cells 39 (2016) 65–71.
[282] K.J. Nelson, S.T. Knutson, L. Soito, C. Klomsiri, L.B. Poole, J.S. Fetrow, Analysis of the peroxiredoxin family: using active-site structure and sequence information for global classification and residue analysis, Proteins 79 (2011) 947–964.
[283] J.R. Matthews, N. Wakasugi, J. Virelizier, J. Yodoi, R.T. Hay, Thioredoxin regulates the DNA binding activity of NF-kappa B by reduction of a disulphide bond involving cysteine 62, Nucleic Acids Res. 20 (1992) 3821–3830.
[284] T. Tanaka, H. Nakamura, A. Nishiyama, F. Hosoi, H. Masutani, H. Wada, J. Yodoi, Redox regulation by thioredoxin superfamily: protection against oxidative stress and aging, Free Radic. Res. 33 (2000) 851–855.
[285] G. Powis, W.R. Montfort, Properties and biological activities of thioredoxins, Annu. Rev. Pharmacol. Toxicol. 41 (2001) 261–295.
[286] J.M. Hanson, H. Zhang, D.P. Jones, Mitochondrial thioredoxin-2 has a key role in determining tumor necrosis factor-α-induced reactive oxygen species generation, NF-κB activation, and apoptosis, Toxicol. Sci. 91 (2006) 643–650.
[287] Y. Chen, J. Cai, T.J. Murphy, D.P. Jones, Overexpressed human mitochondrial thioredoxin confers resistance to oxidant-induced apoptosis in human osteosarcoma cells, J. Biol. Chem. 277 (2002) 33242–33248.
[288] A.G. Cox, A.V. Peskin, L.N. Paton, C.C. Winterbourn, M.B. Hampton, Redox potential and peroxide reactivity of human peroxiredoxin 3, Biochemistry 48 (2009) 6495–6501.
[289] B.A. Stanley, V. Sivakumaran, S. Shi, I. McDonald, D. Lloyd, W.H. Watson, M.A. Aon, N. Paolocci, Thioredoxin reductase-2 is essential for keeping low levels of H_2O_2 emission from isolated heart mitochondria, J. Biol. Chem. 286 (2011) 33669–33677.
[290] C.H. Lillig, C. Berndt, A. Holmgren, Glutatedoxin systems, Biochim. Biophys. Acta 1780 (2008) 1304–1317.
[291] P. Nordlund, P. Reichard, Ribonucleotide reductases, Annu. Rev. Biochem. 75 (2006) 681–706.
[292] S. Yoshitake, H. Nanri, M.R. Fernando, S. Minakami, Possible differences in the regenerative roles played by thioltransferase and thioredoxin or oxidatively damaged proteins, J. Biochem. 116 (1994) 42–46.
[293] M. Ruoppolo, J. Lundstrom-Ljung, F. Talamo, P. Pucci, G. Marino, Effect of glutaredoxin and protein disulfide isomerase on the glutathione-dependent folding of ribonuclease A, Biochemistry 36 (1997) 12259–12267.
[294] C.H. Lillig, C. Berndt, O. Vergnolle, M.E. Lönn, C. Hudemann, E. Bill, A. Holmgren, Characterization of human glutaredoxin 2 as iron-sulfur protein: a possible role as redox sensor, Proc. Natl. Acad. Sci. U.S.A. 102 (2005) 8168–8173.
[295] S. Boschi-Muller, G. Branlant, Methionine sulfoxide reductase: chemistry, substrate binding, recycling process and oxidase activity, Bioorg. Chem. 57 (2014) 222–230.
[296] R.L. Levine, L. Mosoni, B.S. Berlett, E.R. Stadtman, Methionine residues as endogenous antioxidants in proteins, Proc. Natl. Acad. Sci. U.S.A. 93 (1996) 15036–15040.
[297] E.R. Stadtman, J. Moskovitz, B.S. Berlett, R.L. Levine, Cyclic oxidation and reduction of protein methionine residues is an important antioxidant mechanism, Mol. Cell. Biochem. 234/235 (2002) 3–9.
[298] N.M.F.S.A. Cerqueira, E.F. Oliveira, D.S. Gesto, D. Santos-Martins, C. Moreira, H.N. Moorthy, M.J. Ramos, P.A. Fernandes, Cholesterol biosynthesis: a mechanistic overview, Biochemistry 55 (2016) 5483–5506.
[299] H.K. Ghayee, R.J. Auchus, Basic concepts and recent developments in human steroid hormone biosynthesis, Rev. Endocr. Metab. Disord. 8 (2007) 289–300.
[300] J.K. Hiltunen, M.S. Schonauer, K.J. Autio, T.M. Mittelmeier, A.J. Kastaniotis, C.L. Dieckmann, Mitochondrial fatty acid synthesis type II: more than just fatty acids, J. Biol. Chem. 284 (2009) 9011–9015.

[301] A.M. McDonnell, C.H. Dang, Basic review of the cytochrome P450 system, J. Adv. Pract.Oncol. 4 (2013) 263–268.
[302] P. Manikandan, S. Nagini, Cytochrome P450 structure, function and clinical significance: a review, Curr. Drug Targets 19 (2018) 38–54.
[303] O.D. Schärer, Nucleotide excision repair in eukaryotes, Cold Spring Harbor Perspect. Biol. 5 (2013), a012609, 1–19, https://doi.org/10.1101/cshperspect.a012609.
[304] S.S. Wallace, Base excision repair: a critical player in many games, DNA Repair 19 (2014) 14–26.
[305] G.-M. Li, New insights and challenges in mismatch repair: getting over the chromatin hurdle, DNA Repair 19 (2014) 48–54.
[306] T.A. Kunkel, D.A. Erie, Eukaryotic mismatch repair in relation to DNA replication, Annu. Rev. Genet. 49 (2015) 291–313.
[307] A.J. Davis, D.J. Chen, DNA double strand break repair via non-homologous end-joining, Transl. Cancer Res. 2 (2013) 130–143.
[308] H. Wang, X. Xu, Microhomology-mediated end joining: new players join the team, Cell Biosci. 7 (6) (2017) 1–6, https://doi.org/10.1186/s13578-017-0136-8.
[309] R. Prakash, Y. Zhang, W. Fang, M. Jasin, Homologous recombination and human health: the roles of BRCA1, BRCA2, and associated proteins, Cold Harbor Spring Perspect. Biol. 7 (2015), a016600, 1–27, https://doi.org/10.1101/cshperspect.a016600.
[310] S. Burma, B.P. Chen, D.J. Chen, Role of non-homologous end joining (NHEJ) in maintaining genomic integrity, DNA Repair 5 (2006) 1042–1048.
[311] A. Sfeir, L.S. Symington, Microhomology-mediated end joining. A back-up survival mechanism or dedicated pathway? Trends Biochem. Sci. 40 (2015) 701–714.
[312] J.R. Pon, M.A. Marra, Driver and passenger mutations in cancer, Annu. Rev. Pathol. 10 (2015) 25–50.
[313] I. Martincorena, P.J. Campbell, Somatic mutation in cancer and normal cells, Science 349 (2015) 1483–1489.
[314] R.L. Levine, J. Moskovitz, E.R. Stadtman, Oxidation of methionine in proteins: roles in antioxidant defense and cellular regulation, IUMBM Life 50 (2000) 301–307.

PART II

PROTECTION AGAINST CYTOTOXIC COMPONENTS AND DESTRUCTIONS

CHAPTER

4

Disturbances in Energy Supply

4.1 INTRODUCTION

The continuous supply of energy substrates is crucial for the long-term surveillance of humans and other multicellular organisms. Dioxygen and glucose are the main energy-related substrates regularly taken up by our body from the environment. In addition, fatty acids are also used in cellular respiration. Ketone bodies and amino acids will be utilized as energy substrate only under special pathological conditions such as diabetic ketoacidosis, prolonged exercise, fasting, or starving [1].

Cellular energy metabolism is directed to generate adenosine triphosphate (ATP) as universal store for chemical energy equivalents. This molecule is formed in cytosol during glycolysis and above all in mitochondria during oxidative phosphorylation. Although glycolysis is independent of the presence of O_2, mitochondrial processes require dioxygen. Further products of glucose utilization are the cellular redox components nicotinamide adenine dinucleotide (NADH), nicotinamide adenine dinucleotide phosphate (NADPH), and reduced flavin adenine dinucleotide ($FADH_2$).

In addition to these important pathways of energy production, dioxygen and glucose are also consumed for other metabolic processes. For example, dioxygen is utilized in peroxisomes for different biosynthesis processes (Section 2.2.5.2) and in neutrophils, eosinophils, monocytes, macrophages, and some other cells for generation of reactive species (see Sections 6.1 and 6.4.3). Intermediates of the aforementioned pathways of glucose are used for different synthesis reactions of organic material. Thus, energy metabolism of O_2 and glucose is interrelated with many other pathways in our organism.

In this chapter, important properties of utilization of dioxygen and glucose as well as limits and disturbances of these pathways will be addressed. The main focus will be directed on those disturbances of the energy supply that occur already under normal physiological conditions at low level and can be accelerated under stress situations. These disturbances concern the transport of O_2 in blood, the storage of O_2 in muscles, the release of dioxygen-carrying proteins from red blood and muscle cells, dysfunctional metabolic processes in mitochondria, and chemical side reactions of glucose. Of course, these mechanisms of disturbances can be enhanced under hypoxic and hyperglycemic conditions and generally under numerous pathological situations.

4.2 TRANSPORT OF DIOXYGEN BY RED BLOOD CELLS

4.2.1 Hemoglobin as Dioxygen Carrier

The reversible binding of dioxygen to hemoglobin in red blood cells greatly facilitates the

Cell and Tissue Destruction
https://doi.org/10.1016/B978-0-12-816388-7.00004-8

capacity of blood for the uptake and transport of dioxygen. Hemoglobin is a tetrameric protein, where each subunit contains a planar protoporphyrin IX molecule with a central iron ion. The binding of dioxygen to this prosthetic group, shortly known as heme group, is only possible if the central heme iron is in the reduced ferrous state. In its octahedral structure, ferrous heme ion is surrounded by four pyrrole nitrogen atoms forming a plane. On proximal site to this plane, the iron ion is coordinated to a nitrogen atom of adjacent histidine. The sixth position of heme Fe^{2+} can reversibly bind dioxygen. When dioxygen is not bound, a water molecule is weakly attached at the distal position. The structure of heme is schematically depicted in Fig. 4.1.

In humans, tetrameric hemoglobin comprises two α and two β subunits, which are structurally similar [2]. The tetrameric structure is also denoted as $\alpha_2\beta_2$. The four subunits are bound to each other by salt bridges, hydrogen bonds, and hydrophobic interactions, but not by covalent bonds. In fetal hemoglobin, which occurs in the developing fetus and persists up to an age of about 6 month in newborns, instead of the two β subunits two γ subunits are incorporated [3]. In adults, 0.3%–4.4% of red blood cells contains fetal hemoglobin [4]. Higher values of fetal hemoglobin and cells containing this hemoglobin form are found in some inherited and acquired disorders such as sickle cell disease and β-thalassemia [4]. The $\alpha_2\gamma_2$ form of hemoglobin has a higher affinity to dioxygen than the $\alpha_2\beta_2$ tetramer, a property that is important for dioxygen transfer from maternal to fetal blood through the placental membrane [5].

The binding of dioxygen to one subunit in deoxygenated hemoglobin facilitates the binding of further dioxygen molecules to the other subunits. As a consequence of this positive cooperative effect, a sigmoidal curve results for the binding of dioxygen to hemoglobin [6].

Hemoglobin makes up about 96% of the dry mass of red blood cells [7]. It has a binding capacity to dioxygen of 1.34 mL per gram, a value that is approximately 70-fold higher compared with dissolved dioxygen in blood [8].

4.2.2 Oxidation of Hemoglobin to Methemoglobin

Ferrous hemoglobin is prone to oxidation. The resulting ferric protein, known as methemoglobin, is unable to bind dioxygen. Autoxidation

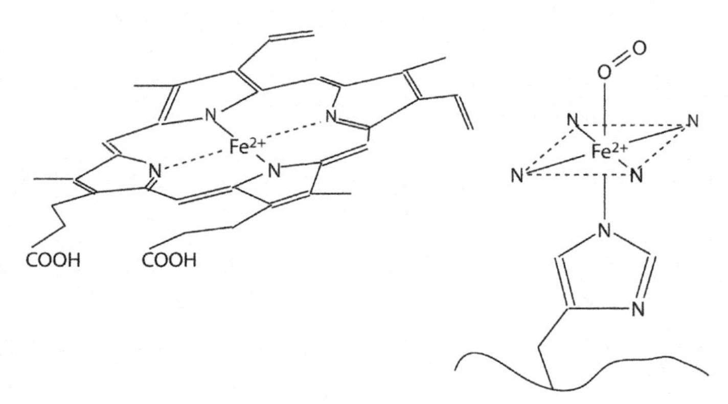

FIGURE 4.1 Schematic view of the structure of heme b and the heme component in ferrous hemoglobin. Left: heme b consisting of the porphyrin ring and a central ferrous ion; right: hexacoordinated structure of the ferrous ion in oxyhemoglobin. From the porphyrin ring, only the four pyrrole nitrogen atoms are indicated.

of the heme group in hemoglobin is a spontaneous oxidation, which always occurs in red blood cells at very low level [9].

$$\text{Por-Fe}^{2+} + O_2 \longrightarrow \text{Por-Fe}^{3+} + O_2^{\bullet-}$$

Por denotes the porphyrin ring of the heme group. In this redox reaction, dioxygen is reduced to a superoxide anion radical, whereas ferrous heme iron is oxidized to the ferric form [10,11]. Because of the presence of methemoglobin reductase in red blood cells, the amount of methemoglobin is normally maintained on a very low level. The rate of autoxidation is slower in tetrameric hemoglobin. It increases on dissociation of the tetramer into two $\alpha\beta$ dimers [9,12].

Increased levels of methemoglobin in blood seriously affect the delivery of dioxygen to tissues [13]. Although methemoglobin levels lower than 10% are usually without any symptoms, shortness of breath, cyanosis, mental status changes, fatigue, dizziness, and palpitation followed with further increasing methemoglobin. Serious complications are coma, seizure, arrhythmias, and acidosis that are observed at more than 50% methemoglobin. A level higher than 70% is lethal.

Methemoglobinemia may have a congenital cause due to deficiencies of NADH-cytochrome b5 reductase [14], pyruvate kinase [15], or glucose-6-phosphate dehydrogenase [16]. Otherwise, patients with abnormal hemoglobin variants such as hemoglobin M or hemoglobin H are also prone to methemoglobinemia [17,18].

Several small soluble agents also favor the formation of methemoglobin from ferrous hemoglobin [19]. Among them are nitrogen monoxide, nitrite, nitrate, sulfide, chlorite, and aniline dyes. These agents play a role in intoxications. A number of antibiotics and drugs are also known to cause enhanced levels of methemoglobin. Examples are trimethoprim, sulfonamides, dapsone, articaine, benzocaine, prilocaine, and lidocaine.

4.2.3 Role of Methemoglobin Reductase

To ensure a sufficient high yield of ferrous hemoglobin, red blood cells contain methemoglobin reductase that converts methemoglobin back into ferrous hemoglobin. In healthy persons, methemoglobin level in red blood cells does not exceed 1.5% of total hemoglobin.

In humans, two forms of methemoglobin reductase exist [20]. The prevailing form is the NADH-cytochrome b5 methemoglobin reductase that utilizes NADH and contains FAD as cofactors. Two electrons from NADH are transferred via FAD to two cytochrome *b*5 subunits and further to methemoglobin.

$$\text{NADH} + \text{H}^+ + 2\,\text{Fe}^{3+} - \text{cyt b5} \rightarrow \text{NAD}^+ + 2\,\text{Fe}^{2+} - \text{cyt b5}$$

$$\text{Fe}^{2+} - \text{cyt b5} + \text{Por} - \text{Fe}^{3+} \rightarrow \text{Fe}^{3+} - \text{cyt b5} + \text{Por} - \text{Fe}^{2+}$$

A second minor pathway uses NADPH as cofactor in methemoglobin reduction. The NADPH methemoglobin reductase is the main target for pharmaceutical treatment of excess methemoglobin by using methylene blue [13].

4.2.4 Binding of Nitric Monoxide and Carbon Dioxide to Hemoglobin

Inside red blood cells, hemoglobin can principally interact with nitric monoxide. However, this reaction is about 1000-fold slower than the reaction of NO with free hemoglobin [21,22]. This interaction leads to formation of methemoglobin and further to nitrosylated ferrous hemoglobin. Red blood cells are more prone to formation of nitrosylated hemoglobin in comparison with free hemoglobin [22]. The NO-binding site is the cysteine residue 93 of the β chain on hemoglobin [23]. It has been assumed that S-nitrosylated hemoglobin can deliver NO in areas of restricted blood flow and can contribute under hypoxic conditions to dilation of blood vessels [24,25]. Thus, NO mediates the formation of both nitrosylated hemoglobin and methemoglobin. The yield of these products depends on the oxygenation state of hemoglobin. The formation of the nitrosylated species is favored in deoxygenated blood [26].

Carbon dioxide binds in an allosteric fashion to terminal α-amino groups of hemoglobin [27]. This binding facilitates the release of dioxygen from hemoglobin and triggers the supply of dioxygen to tissues. Otherwise, CO_2 is more readily dissolved in deoxygenated blood. This property is important for the gaseous exchange is the lungs.

4.2.5 Heme Poisons

Carbon monoxide (CO) replaces dioxygen from ferrous hemoglobin in a competitive manner, as it is 218 times stronger bound than dioxygen to ferrous heme iron [28]. Nearly the same symptoms are observed by intoxication with carbon monoxide as by enhanced methemoglobin formation with the difference that the CO intoxication proceeds heavier, and lethal cases occur already at a yield of 60% affected hemoglobin [29].

Treatment of CO intoxication is based on the rapid elimination of CO and improvement of hypoxic state. This is possible by application of pure dioxygen and the use of hyperbaric chambers [30].

Cyanide is a poison that binds to heme proteins in the ferric state and prevents any redox conversions of these proteins. The mitochondrial proteins cytochrome *c* and cytochrome *c* oxidase are the preferred targets for cyanide [31,32]. Venous blood remains saturated with dioxygen as the cells are unable to exploit dioxygen due to the blockade of the mitochondrial electron transport chain by cyanide. The application of substances favoring a mild methemoglobinemia provokes a rapid reallocation of cyanide and facilitates mitochondrial respiration. Finally, cyanide is detoxified by rhodanese, converting cyanide into thiocyanate. The latter process can be enhanced by application of thiosulfate [33,34].

4.2.6 Intravascular Hemolysis

The term intravascular hemolysis describes the release of hemoglobin from circulating red blood cells. Normally, this release takes always place at a very low level. Cell-free plasma obtained from healthy volunteers contain about 0.2 μM total heme [35]. This value increases in plasma of patients with sickle cell disease to a medium value of 4.2 μM, whereupon values higher than 20 μM were measured in some patients. The lack of haptoglobin, a scavenger of free hemoglobin (see Section 4.2.7), in plasma of most of these patients is a further indicator for enhanced intravascular hemolysis [35].

Examples for other diseases with increased release of hemoglobin from red blood cells are thalassemia, glucose-6-phosphate dehydrogenase deficiency, paroxysmal nocturnal hemoglobinuria, hereditary spherocytosis, malaria, and some others [36–39]. Enhanced intravascular hemolysis can also be induced by physical and chemical factors such as osmotic stress, sheer stress, the presence of lytic poisons, some secreted microbial components from Gram-positive bacteria, cardiopulmonary bypass, mechanical heart valve–induced anemia, autoantibodies, elevated oxidative processes in membranes of red blood cells, burn-associated necrosis, and hemorrhagic conditions [37,40,41].

Damage of red blood cells may also take place during storage of blood for transfusion [42]. This may occur because of consumption of glucose concomitant with a decrease in 2,3-diphosphoglycerate and ATP levels and an increase in potassium ions. As a result, the integrity of plasma membrane of erythrocytes is disturbed [43].

The release and fate of hemoglobin from red blood cells is illustrated in Fig. 4.2. On release

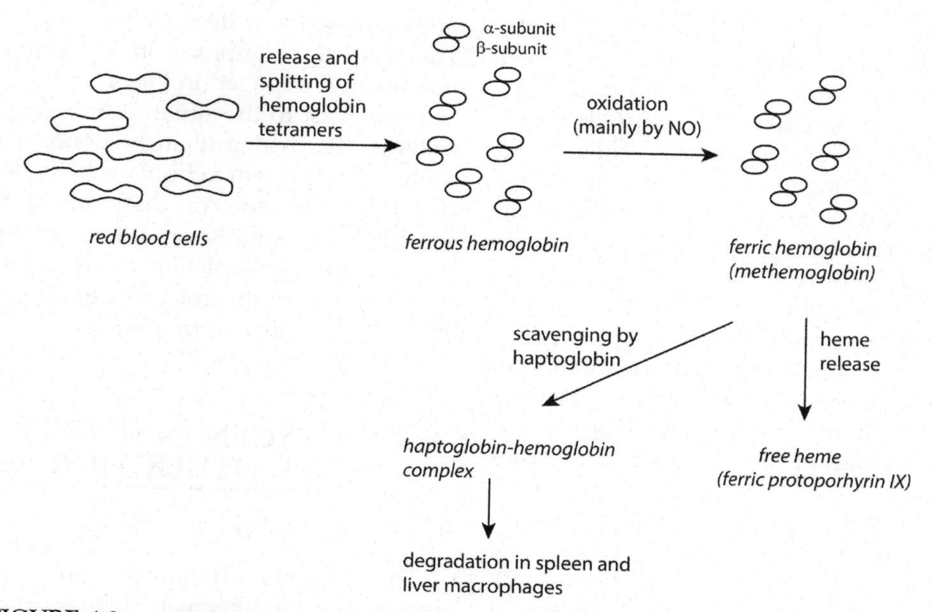

FIGURE 4.2 Release and fate of hemoglobin from red blood cells during intravascular hemolysis.

from red blood cells, tetrameric hemoglobin is split into two dimers [44]. It is easily oxidized to the ferric form most of all under the action of nitrogen monoxide [45,46]. This small molecule is permanently produced by nitrogen monoxide synthase in the blood vessel wall. It activates guanylate cyclase to produce cyclic guanosine monophosphate that causes relaxation of smooth muscle cells and dilate small blood vessels. Nitrogen monoxide can oxidize ferrous hemoglobin both inside and outside red blood cells. Because of the lack of the membrane barrier, it acts more efficiently in serum [21,22]. Massive hemolysis leads to a redistribution of NO and a vasoconstrictive activity of released hemoglobin.

Once oxidized, free methemoglobin cannot be reduced because of the lack of methemoglobin reductase in serum. In methemoglobin, the heme group is not so tightly bound as in ferrous hemoglobin [47]. Thus, free heme (ferric protoporphyrin IX) is released step by step from methemoglobin [48,49]. Both free methemoglobin and in particular free heme are highly cytotoxic (Section 4.4).

4.2.7 Protection Against Free Hemoglobin and Free Heme

To avoid cytotoxic reactions of methemoglobin, the serum protein haptoglobin forms a complex with free methemoglobin. Although free methemoglobin is subjected to glomerular filtration in the kidneys, the haptoglobin–methemoglobin complex does not permeate through the glomerular membrane. This complex is cleared from circulation via binding to CD163 by macrophages in the spleen and liver (Fig. 4.2) [41,50].

Haptoglobin also binds and inactivates free myoglobin/metmyoglobin that is released from injured muscles during rhabdomyolysis (see Section 4.3.2).

Haptoglobin is primarily synthesized in the liver. It is an acute-phase protein (see Chapter 7). Thus, its concentration may raise under inflammatory and stress condition. Otherwise, excessive intravascular hemolysis leads to an exhaustion of haptoglobin. The same holds for intense muscle damage. In consequence, not all hemoglobin released from red blood cells (and/or myoglobin from muscle tissue) can be bound by haptoglobin and cleared without further problems from circulation.

Monocytes/macrophages and neutrophils are also known to release a certain amount of haptoglobin [51]. In neutrophils, haptoglobin is stored in specific granules of neutrophils and released on cell activation [52].

After exhaustion of haptoglobin, the probability for formation of free heme increases. Free heme is also highly cytotoxic (Section 4.4.2). This flat hydrophobic molecule easily inserts in biological membranes and lipoproteins and binds to hydrophobic areas of proteins. If generated, substantial parts of free heme are associated with albumin, α_1-microglobulin, and lipoproteins [40]. The by far highest binding affinity to free heme exerts the serum protein hemopexin [53]. This protein is like haptoglobin synthesized in liver. The hemopexin-free heme complex binds via CD91 to hepatocytes and is phagocytosed and degraded by these cells [54]. The role of hemopexin in protection against free heme is depicted in Fig. 4.3.

In contrast to the situation in mice and other rodents, the liver protein hemopexin is not an acute-phase protein in humans [55]. Thus, excessive hemolytic and/or rhabdomyolytic events can lead after exhaustion of haptoglobin to partial or complete depletion of hemopexin. Then, only limited or no protection exists against the cytotoxic activities of free heme.

4.3 DIOXYGEN IN MUSCLE CELLS AND OTHER TISSUES

4.3.1 Myoglobin

The monomeric heme protein myoglobin is responsible for dioxygen delivery and storage in muscle tissue. There are several types of muscle fibers that differ in their distribution, function, and the kinds of production of ATP [56,57]. Type I muscle fibers are rich in blood capillaries, mitochondria, and myoglobin. They produce ATP by oxidative phosphorylation that makes them fatigue-resistant. Type II muscle fibers generate ATP through the glycolytic cycle. There a several subtypes of type II fibers.

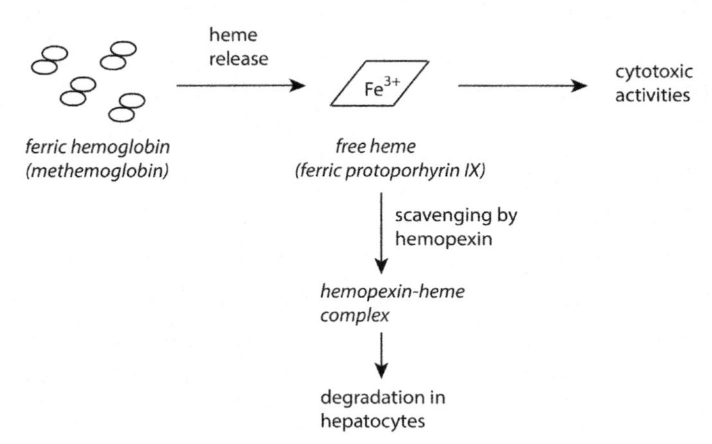

FIGURE 4.3 Protection by hemopexin to release of free heme (ferric protoporphyrin IX) from free methemoglobin.

The number of mitochondria and myoglobin is highest in type IIA and intermediate in type IIX, whereas only few mitochondria and no myoglobin are present in type IIB fibers. Muscle fibers of types I and IIA are important for long-term burden, whereas type IIB fibers have a higher contraction velocity. However, the latter fibers are not fatigue-resistant as they produce their energy from prestored glycogen.

In contrast to hemoglobin, myoglobin is a monomeric protein with only one heme group. Dioxygen is bound to ferrous myoglobin much stronger than in ferrous hemoglobin [58]. Differences and similarities between hemoglobin and myoglobin are summarized in Table 4.1. This allows the storage of dioxygen in muscles for intense exercise and serves as a reservoir of dioxygen during hypoxia. Unlike hemoglobin, myoglobin displays a hyperbolic-shaped binding curve for dioxygen that corresponds to Michaelis–Menten kinetics [58,59]. This strong binding of O_2 allows functioning of myoglobin to buffer intramuscular O_2 concentration under conditions of enhanced metabolic activity. Further myoglobin facilitates and mediates diffusion of O_2 from blood vessels to mitochondria [56,63]. For delivery of O_2 to mitochondria, a direct contact between oxygenated myoglobin and mitochondria is necessary [64].

Like hemoglobin, myoglobin is rapidly oxidized mainly by NO to metmyoglobin after its release from damaged muscle. Methemoglobin is unable to bind dioxygen. In muscles but not in blood, the ferrous form of myoglobin is maintained by the activity of an NADH-dependent metmyoglobin reductase [61,62,65].

In cardiomyocytes, myoglobin contributes also to NO homeostasis [60]. Under normoxic conditions, oxygenated myoglobin binds NO and is converted into metmyoglobin. Metmyoglobin reductase recycles ferrous myoglobin. This mechanism protects mitochondrial respiration from inhibition by NO. At reduced O_2 supply, deoxygenated myoglobin becomes the dominating form. It reduces nitrite to NO under formation of metmyoglobin. As a result, mitochondrial cytochromes are reversibly inhibited by NO, myocardial O_2 consumption is diminished, and the heart muscle is adapted to limited O_2 supply.

TABLE 4.1 Differences and Similarities Between Hemoglobin and Myoglobin.

Property	Hemoglobin	Myoglobin	References
Protein structure	Tetramer	Monomer	
Heme group	Ferrous protoporphyrin IX	Ferrous protoporphyrin IX	
Binding of O_2		Stronger than in hemoglobin	[58]
Binding curve to O_2	Sigmoidal shape	Hyperbolic shape	[6,58,59]
Cooperativity of O_2 binding	Yes	No	[6]
Autoxidation	Biphasic, more resistant as compared with myoglobin		[9]
Oxidation by NO	Yes	Yes	[45,46,60]
Reduction of the ferric heme	Methemoglobin reductase in red blood cells	Metmyoglobin reductase in muscles	[20,61,62]

4.3.2 Rhabdomyolysis

Damage of skeletal muscle cells with release of constituents into blood is called rhabdomyolysis. During this process, myoglobin and other constituents of muscle cells are liberated and appear in blood. This damage is mostly caused by intense exercise associated with overexertion and compression of muscles, alcohol and illicit drug abuses, and the use of certain medications. Other reasons for myoglobin release are infections, electrical injury, heat stroke, and prolonged immobilization [66,67].

Rhabdomyolysis causes local signs and symptoms such as muscle pain, tenderness, muscle swelling as well as systemic features like tea-colored urine, fever, malaise, and others. A serious problem is the myoglobin-induced nephrotoxicity resulting in acute renal failure [66,68]. More details about acute kidney failure are given in Section 10.2.4.

Damaged muscle tissue swells and is filled with fluid from circulating blood. Swelling may contribute to further damage either directly or by affecting the Ca^{2+} influx into muscle cells. Muscle damage is accompanied by an increase in intracellular Ca^{2+}, overexertion, destruction, and necrosis of muscle fibers [66,69]. As a result, large quantities of potassium, phosphate, myoglobin, creatine kinase, urate, and others are released into blood. Infiltration of neutrophils, their activation, and inflammatory response is also observed particularly after crush injury.

Although the myoglobin concentration is below 3 μg/100 mL in normal serum, it increases significantly because of rhabdomyolysis. Haptoglobin is able to bind free myoglobin such as hemoglobin released from red blood cells. The binding capacity of serum haptoglobin becomes saturated if more than 100 g of skeletal muscle is damaged [66]. The haptoglobin–metmyoglobin complex is cleared by macrophages of the spleen and liver from blood circulation.

Like hemoglobin released from red blood cells, free myoglobin is also rapidly oxidized primarily under the action of nitric monoxide. In addition, metmyoglobin is able to release its heme group. The same protecting systems mentioned above, namely haptoglobin and hemopexin, bind free metmyoglobin and free heme, respectively.

4.3.3 Neuroglobin and Cytoglobin

Dioxygen-binding heme proteins like neuroglobin and cytoglobin have been also described in other tissues. In comparison with hemoglobin and myoglobin, this knowledge is fragmentary as these proteins are only recently discovered on the basis of proteomic analysis [70–73].

Neuroglobin is expressed in the cells of the central and peripheral nervous system, cerebrospinal fluid, retina, and endocrine cells [70,73,74]. This monomeric heme protein reversibly binds dioxygen with an affinity higher than hemoglobin, but lower than myoglobin. Unlike hemoglobin and myoglobin, the deoxygenated neuroglobin has a hexacoordinated heme iron in the reduced and oxidized form. On binding, dioxygen displaces the sixth endogenous ligand.

Neuroglobin facilitates like myoglobin the diffusion of dioxygen to the mitochondria. It helps to provide reserves of dioxygen to neuronal cells under hypoxic-ischemic conditions [70,75,76]. An action of neuroglobin as NADH oxidase under conditions of limited dioxygen supply has been proposed [73]. The involvement of neuroglobin in NO metabolism is another hypothetic function [77].

The forth human heme protein involved in binding and storage of dioxygen is cytoglobin that is expressed in a wide range of tissues [71]. Like neuroglobin, this monomeric protein is characterized by hexacoordinated heme iron in the deoxygenated form.

Similar functional activities as described for neuroglobin are also under discussion for cytoglobin. This heme protein also exhibits a catalase activity [78].

4.4 CYTOTOXIC EFFECTS OF DIOXYGEN-BINDING HEME PROTEINS AND THEIR COMPONENTS

4.4.1 Deleterious Effects of Free Hemoglobin and Free Myoglobin

Once released from red blood cells or muscle tissue, both free hemoglobin and free myoglobin are rapidly oxidized to the ferric form. In blood, the predominant oxidant of both heme proteins is nitric monoxide (NO). Enhanced levels of intravascular hemolysis and/or rhabdomyolysis are discussed as a reason for diminished bioavailability of NO [35,37]. During hemolysis, arginase is also released from red blood cells. This enzyme converts arginine to ornithine and also diminishes the bioavailability of NO [37]. Hence, a decreased blood flow and a general increase in blood pressure are the consequences [37,42]. Other potential effects of reduced NO levels are smooth muscle dystonia, gastrointestinal contractions, clot formation, esophageal spasm, dysphagia, abdominal pain, and erectile dysfunction [37].

Free hemoglobin and free myoglobin form a complex with haptoglobin that is eliminated by spleen and liver macrophages via recognition by the macrophage receptor CD163 [41,50]. However, serum haptoglobin level decreases by massive release of these heme proteins. Unbound hemoglobin and myoglobin accumulate in kidney proximal tubular cells after glomerular filtration and reabsorption through the endocytic receptors megalin and cubilin [41,68,79,80]. Here, they contribute to cell damage and renal failure.

A third mechanism of deleterious activities of free hemoglobin and free myoglobin is the release of the free heme group in form of ferric protoporphyrin IX, shortly called free heme [48]. This mechanism can occur in blood and also in kidney after uptake by proximal tubular cells.

4.4.2 Cytotoxicity of Free Heme

Cytotoxic activity of free heme becomes evident after collapse of protecting haptoglobin and is further increased with the decline of hemopexin. The complex of hemopexin with free heme is scavenged by CD91 receptors of hepatocytes and eliminated by these cells from circulation [54]. This interaction is favored by the extremely high binding affinity of hemopexin to free heme. This property allows a reallocation of free heme from other hydrophobic targets to hemopexin.

As a highly hydrophobic molecule, free heme easily inserts into lipid bilayers of membranes, lipoproteins, and hydrophobic area of proteins and exerts numerous cytotoxic activities (Fig. 4.4).

Importantly, free heme (ferric protoporphyrin IX), which can accumulate in serum due to massive hemolysis and/or rhabdomyolysis is itself a strong inducer of hemolytic events in unperturbed red blood cells [81,82]. This very dangerous process further impairs an already existing hemolysis and contributes to long-lasting pathological effects. In interaction of free heme with red blood cells, first a massive release of potassium ions occurs followed by depletion of glutathione and ATP, swelling, and massive loss of hemoglobin [81,83]. In addition, free heme alters the conformation of cytoskeletal proteins in red blood cells [84].

Free heme also inserts into lipoproteins. It is discussed that the accompanying oxidation of lipoprotein constituents contributes to the development of atherosclerosis [85]. In low-density lipoproteins (LDL), sufficient lipid oxidation by free heme will only take place after exhaustion of antioxidants. The iron chelator deferoxamine prevents deleterious effects of heme on LDL [86].

Furthermore, free heme is highly cytotoxic to the kidney, liver, and spleen [42,87]. Accumulation of free heme in these organs causes fatal alterations in their metabolism and damage of these organs. In the human kidney cell, exposure

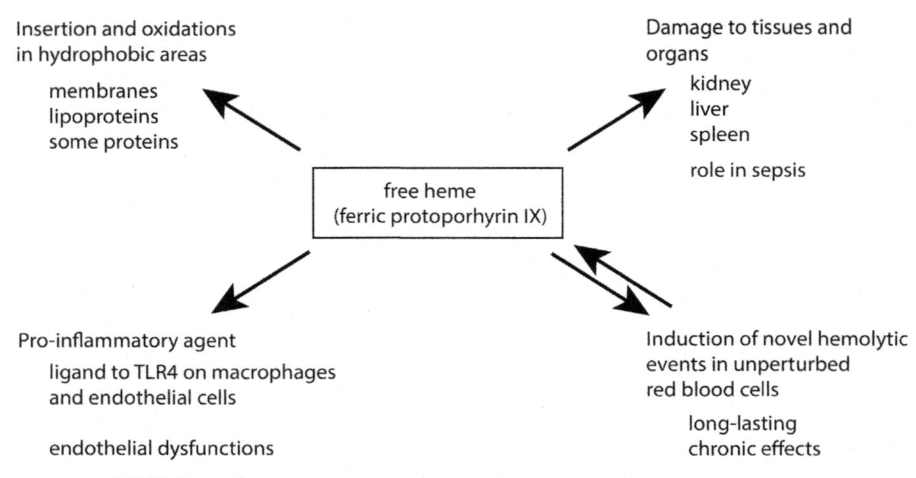

FIGURE 4.4 Cytotoxic activities of free heme (ferric protoporphyrin IX).

to low concentrations of free heme causes an adaptive response via enhanced transcription of heme oxygenase and ferritin. Higher heme amounts, however, damage these cells [88].

Free heme is a proinflammatory agent via binding to the toll-like receptor 4 [89,90]. This proinflammatory interaction activates endothelial cells, neutrophils, and other cells.

4.4.3 Role of Heme Oxygenase in Detoxification of Free Heme

Heme oxygenase (HO) intracellularly detoxifies free heme resulting from turnover of cellular heme proteins or from uptake of external heme proteins and free heme. In humans, there are two enzymatically active isoforms of HO, the constitutively expressed HO-2 and the inducible isoform HO-1, and the catalytically inactive isoform HO-3 [91].

Heme oxygenase degrades free heme into biliverdin, carbon monoxide, and Fe^{2+} (Fig. 4.5). This reaction requires the presence of NADPH and O_2 [92]. Biliverdin is further converted by biliverdin reductase into bilirubin that acts as lipid antioxidant. Fe^{2+} is sequestered by ferritin. Carbon monoxide triggers via blockade of mitochondrial cytochrome *c* oxidase cytoprotective effects [40]. Further, the binding of CO to ferrous heme proteins prevents them from oxidation and heme release [44]. Thus, heme utilization by heme oxygenase protects not only cells from cytotoxic effects of free heme but also improves the cellular defense status.

Besides HO-1, ferritin, and peroxiredoxin-1 are also upregulated under oxidative stress. Peroxiredoxin-1 is also known to bind free heme with high affinity in cytoplasm of cells [93].

Despite the upregulation of HO-1 under stress situations, the capacity of HO to detoxify free heme is limited. Heme-induced stress results from severe intravascular hemolysis or rhabdomyolysis. After saturation of HO, heme causes oxidative reactions inside the cells and binds to the 26S catalytic unit of proteasome with high affinity [94,95]. This inhibits proteasome function and activates the response to unfolding proteins. Finally, damaged proteins accumulate in affected cells and cells die [95].

Severe hemolytic and rhabdomyolysis events can acutely damage the kidney. The release of free heme from free hemoglobin or free

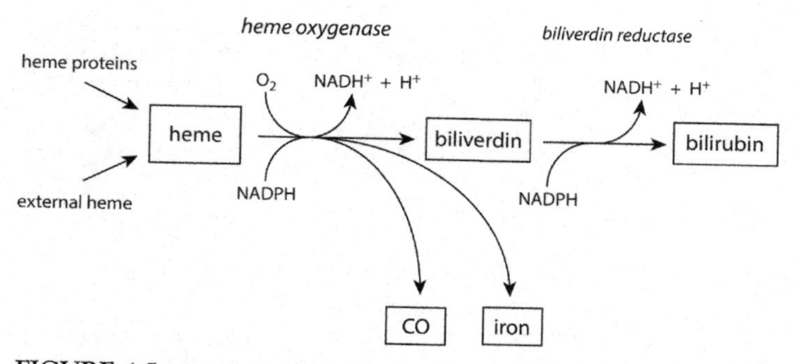

FIGURE 4.5 Intracellular destruction of free heme by heme oxygenase.

myoglobin in the tubular system promotes both oxidative damage and response to unfolded proteins [88].

4.4.4 Release of Hemoglobin and Anemia

Anemia is a common blood disorder characterized by insufficient ability of blood to supply dioxygen to peripheral tissues. General reasons of anemia can be subdivided into three main groups [96,97]:

- loss of the total amount of blood due to traumata or bleeding,
- insufficient production of hemoglobin as a result of, e.g., iron deficiency, lack of vitamin B12, or disturbances in bone marrow, and
- increased breakdown of red blood cells.

Here, the focus will be directed on enhanced intravascular hemolysis as a reason of anemia, the so-called hemolytic anemia. It is evident that this kind of anemia is caused by destruction of red blood cells.

Typical signs and symptoms of anemia are fatigue, shortness of breath, weakness, and limited ability to exercise. Special further symptoms are jaundice and an enhanced risk for gallstones and pulmonary hypertension. The later symptoms are related to the degradation of porphyrin to biliverdin/bilirubin that yields the characteristic pale color [98].

Not each increased intravascular hemolysis should result in anemia. The increased loss of cells carrying hemoglobin can be compensated by enhanced production and release into circulation of red blood cells, a process that is triggered by the kidney hormone erythropoietin [99]. This hormone stimulates erythropoiesis in the bone marrow. Hence, an increased count of reticulocytes among the total amount of red blood cells is observed.

4.5 UTILIZATION OF GLUCOSE AND DIOXYGEN IN CELLS AND MITOCHONDRIA

4.5.1 Short Overview About Aerobic Respiration

The stepwise conversion of glucose by O_2 into ATP, which stores chemical energy equivalents, is schematically illustrated in Fig. 4.6.

Inside cells, glucose is converted during processes of glycolysis by a cascade of cytosolic enzymes into pyruvate under net formation of 2 mol ATP and 2 mol NADH per mole utilized glucose. In a parallel cytosolic pathway, the pentose phosphate pathway, glucose is metabolized to pentoses, ribose 5-phosphate, and 2 moles NADPH per mole glucose. These processes do not require the presence of dioxygen.

In mitochondria, pyruvate is further converted into acetyl coenzyme A (acetyl-CoA) by

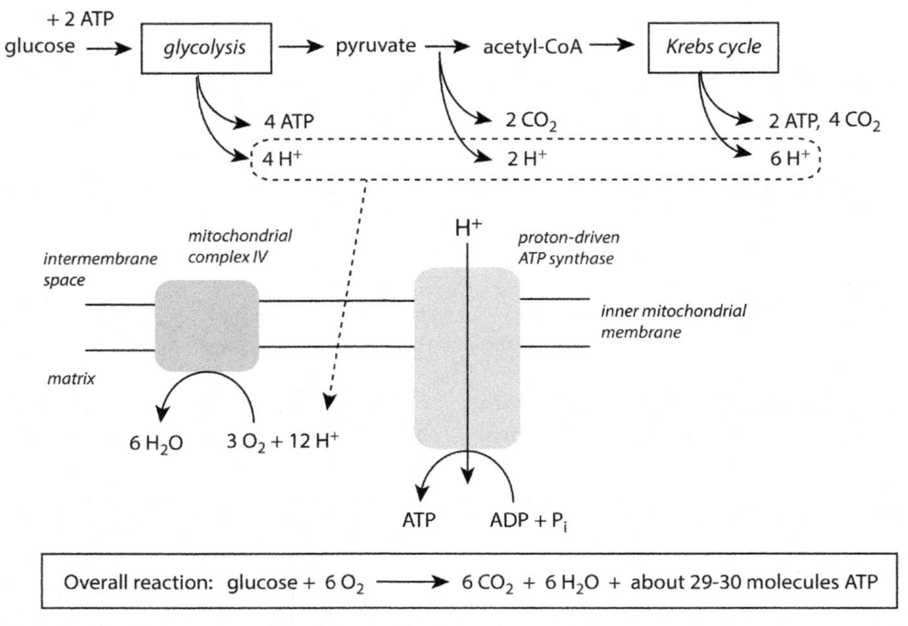

FIGURE 4.6 Scheme of the generation of ATP from glucose and dioxygen during cellular respiration. Further explanations are given in the text.

the pyruvate dehydrogenase complex under formation of 1 mol NADH and 1 mol CO_2 per mole pyruvate. Acetyl-CoA enters into the citric acid cycle (Krebs cycle) inside the mitochondrial matrix. In this cycle, 1 mol acetyl-CoA yields 2 mol CO_2, 1 mol ATP, 1 mole guanosine triphosphate, which can subsequently be converted into ATP, 3 mol NADH, and 1 mol $FADH_2$.

The final step of aerobic respiration, the oxidative phosphorylation, occurs in the mitochondrial cristae, where electron transfer reactions by four protein complexes (complexes I–IV) create a pH gradient and electrical potential across the inner mitochondrial membrane. Electrons are supplied from oxidation of NADH and $FADH_2$, which were generated in previous metabolic steps.

The final acceptor for electrons is O_2. The reduction of O_2 to H_2O is catalyzed by cytochrome *c* oxidase (complex IV). The summary of this conversion is

$$O_2 + 4\,H^+ + 4\,e^- \rightarrow 2\,H_2O.$$

The resulting electrochemical proton gradient is the driving force for the energetically unfavorable ATP synthesis from ADP and inorganic phosphate. ATP synthase is an integral protein in the inner membrane of mitochondria [100]. During aerobic respiration, about 30 mol of ATP are generated per mole glucose [101].

Long-chain fatty acids are a further source for acetyl-CoA [102,103]. During β-oxidation, the degradation of these fatty acids yields one equivalent each of NADH and $FADH_2$ per shortage of the acyl chain by two carbon atoms. Both NADH and $FADH_2$ are used to generate ATP in subsequent metabolic steps.

Although production of energy equivalents by glycolysis mainly depends on the amount of glycogen stored in cells, the greater output of ATP from oxidative phosphorylation and the possibility to run this pathway also coupled with β-oxidation of fatty acids allow a more intensive and long-lasting delivery of ATP for functional processes.

4.5.2 Mitochondria as an Indicator of Cell Viability

In cells, mitochondria are an important location for production of chemical energy equivalents in form of ATP. Redox conversions in mitochondrial complexes and the generation of a transmembrane potential for H^+ contribute to the electrical mitochondrial membrane potential ($\Delta\Psi_m$) [104]. Stable levels of intracellular ATP and $\Delta\Psi_m$ are thought to be a prerequisite for normal cell function and cellular homeostasis [105,106]. Prolonged deviations from normal level of each parameter may induce a loss of cell viability and pathological processes.

Substantial decrease in intracellular ATP level is accompanied by acidosis and enhanced degradation of cell material by proteolytic, lipolytic, and DNA-degrading enzymes, and it can lead to cell death [107,108]. Mild acidosis is also discussed to induce structural and functional reorganization of mitochondria, as a result of which ATP production increases and the cellular ATP level becomes stabilized [109].

Increased values of $\Delta\Psi_m$ can boost the formation of superoxide anion radicals, which induce several deleterious reactions (see Section 4.5.4). Otherwise, during apoptosis induction, mitochondrial membrane potential collapses because of the opening of mitochondrial permeability transition pore [110].

4.5.3 Formation of Superoxide Anion Radicals

In intact mitochondria, a very few percentage of dioxygen is reduced under formation of a superoxide anion radical ($O_2^{\bullet-}$). About 0.1%–0.5% of consumed O_2 is converted into $O_2^{\bullet-}$ by mitochondria in the resting skeletal muscle. This value decreases to about 0.01%–0.03% mimicking muscle exercise [111]. Up to now, 11 different sites of $O_2^{\bullet-}$ generation have been described, whereupon six sites are related to redox conversions associated with complex I and five others to the ubiquinone cycle (Q-cycle) transferring electrons to complex III [112].

In complex I, $O_2^{\bullet-}$ generation is favored by reverse electron transport where electrons flow in opposite direction. This situation is determined by three factors: a high proton motive force across the inner mitochondrial membrane, a highly reduced coenzyme Q pool, and a high yield of NADH [113,114]. A thorough analysis of factors affecting the $O_2^{\bullet-}$ generation by complex I–associated processes revealed that the overall thermodynamic status is mainly responsible for one-electron reduction of O_2 rather than direct actions on complex I [114].

During forward electron transfer through complex I, significant production of $O_2^{\bullet-}$ is only possible after addition of inhibitors such as rotenone or antimycin A [115–117].

The Q-cycle links both complexes I and II with complex III. Electrons transfer to complex III occurs from the reduced form ubiquinol (QH_2). In the Q-cycle, besides oxidized ubiquinone (Q) and reduced QH_2, the intermediary ubisemiquinone anion radical ($Q^{\bullet-}$) also participates [118]. Oxidation of $Q^{\bullet-}$ by O_2 is an important source for $O_2^{\bullet-}$.

$$Q^{\bullet-} + O_2 \longrightarrow Q + O_2^{\bullet-}$$

This reaction is thermodynamically possible considering the pH-independent reduction potentials of the redox couples $Q/Q^{\bullet-}$ and $O_2/O_2^{\bullet-}$ being −0.23 V [119] and −0.16 V [120], respectively. A detailed kinetic analysis of redox properties of the Q-cycle favors an enhanced formation of $O_2^{\bullet-}$ by an increase of the mitochondrial membrane potential and an increase of the pH difference across the inner mitochondrial membrane above critical values [121].

Production of $O_2^{\bullet-}$ by complex I–associated redox conversions occurs in matrix, while the Q-cycle–derived $O_2^{\bullet-}$ is mainly formed within the inner mitochondrial membrane and released into both matrix and intermembrane space [117]. A scheme illustrating the two principal locations of formation of superoxide anion radicals within the electron transfer chain in mitochondria is given in Fig. 4.7.

4.5.4 Reactions of Superoxide Anion Radicals

Of course, SOD-catalyzed and spontaneous dismutation of $O_2^{\bullet-}$ yields H_2O_2 (see Sections 2.2.4.3 and 2.2.4.4). Superoxide anion radicals are nearly impermeable and remain at sites of their formation. In contrast, H_2O_2 can cross membranes and is spread over the cytosol. Although both reactive species have only a limited potential for destruction of biological material, site reactions of $O_2^{\bullet-}$ can be dangerous to mitochondria and cells. Important reactions of mitochondrial $O_2^{\bullet-}$ are summarized in Fig. 4.8.

In mitochondria, superoxide anion radicals are known to oxidize $[4Fe{-}4S]^{2+}$ clusters present in some redox intermediates and aconitase, an enzyme of the citric cycle [122,123].

$$[4Fe\text{-}4S]^{2+} + O_2^{\bullet-} + 2H^+ \longrightarrow [4Fe\text{-}4S]^{3+} + H_2O_2$$

$$[4Fe-4S]^{3+} \rightarrow [3Fe-4S]^{+} + Fe^{2+}$$

The primary oxidation product is $[4Fe{-}4S]^{3+}$ that decays into $[3Fe{-}4S]^{+}$ under release of Fe^{2+}. Together with H_2O_2, ferrous ion can participate in Fenton-like reactions. A second-order rate constant in the range of $8 \times 10^6\ M^{-1}s^{-1}$ to $3 \times 10^7\ M^{-1}s^{-1}$ was measured for the reaction between $[4Fe{-}4S]^{2+}$ clusters and $O_2^{\bullet-}$ [124,125].

This oxidation of $[4Fe{-}4S]^{2+}$ clusters inactivates the corresponding redox elements and enzymes. Otherwise, aconitase is known to function as sensitive sensor for iron and is involved as iron-responsive element in gene regulation of cellular iron metabolism [126,127].

Superoxide anion radicals do not affect [2Fe–2S] clusters, as iron is here more tightly bound [124].

Superoxide anion radicals also interact with NO inside mitochondria under formation of the powerful oxidant peroxynitrite. The existence of a specific NO synthase in mitochondria is a matter of debate [128]. Apparently, the mitochondrial isoform of NO synthase, which is distinguished from eNOS, iNOS, and nNOS, is upregulated and active under hypoxic conditions [129,130].

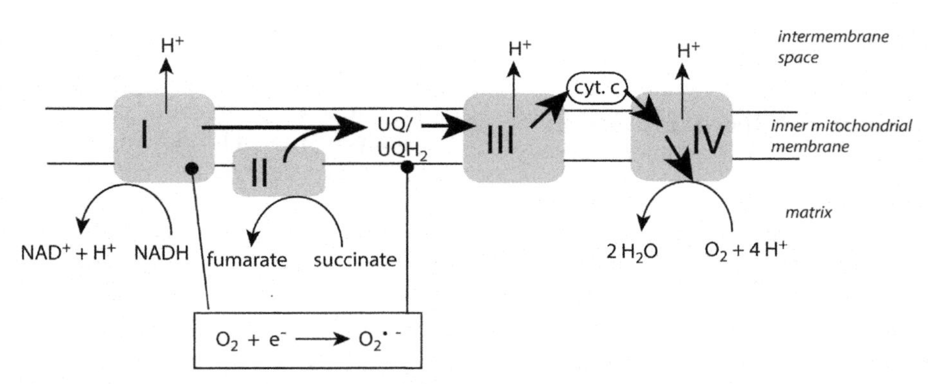

FIGURE 4.7 Potential sites for the formation of superoxide anion radicals within the electron transfer chain in mitochondria. The route for electron transfer through the mitochondrial complexes I–IV is indicated by *bold black arrows*. cyt *c*, cytochrome *c*. Further explanations are given in the text; *UQ/UQH2*, ubiquinone–ubiquinol system.

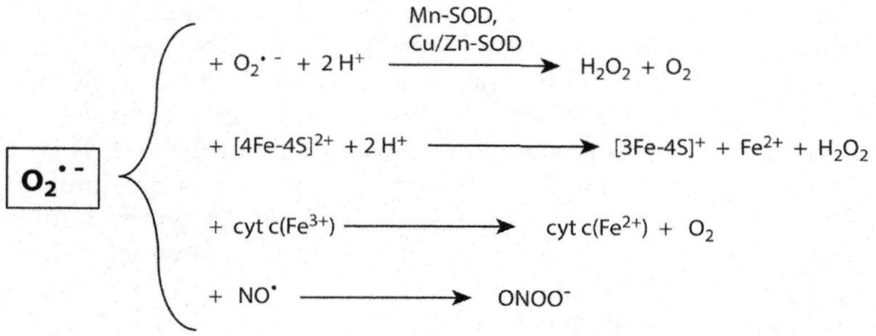

$$O_2^{\bullet -} \begin{cases} + O_2^{\bullet -} + 2H^+ \xrightarrow{\text{Mn-SOD, Cu/Zn-SOD}} H_2O_2 + O_2 \\ + [4Fe\text{-}4S]^{2+} + 2H^+ \longrightarrow [3Fe\text{-}4S]^+ + Fe^{2+} + H_2O_2 \\ + \text{cyt c}(Fe^{3+}) \longrightarrow \text{cyt c}(Fe^{2+}) + O_2 \\ + NO^{\bullet} \longrightarrow ONOO^- \end{cases}$$

FIGURE 4.8 Important reactions of superoxide anion radicals within mitochondria. *cyt c*, cytochrome *c*, *SOD*, superoxide dismutase. Further explanations are given in the text.

4.5.5 Protection Against Superoxide Anion Radicals and Hydrogen Peroxide

Mitochondrial matrix contains manganese-dependent superoxide dismutase (Mn-SOD) that is different from copper/zinc superoxide dismutase (Cu/Zn-SOD), which is present in intermembrane space and cytosol [131,132]. In mouse models, depletion of Mn-SOD has more serious effects than the knockout of Cu/Zn-SOD [133–135]. Second-order rate constants for spontaneous dismutation of superoxide anion radicals decrease with increasing pH. They are $4 \times 10^5\ M^{-1}s^{-1}$ at pH 7.2 and $7 \times 10^4\ M^{-1}s^{-1}$ at pH 8.0 [136], typical pH values of the intermembrane space and matrix, respectively. Thus, $O_2^{\bullet -}$ lives longer in the more alkaline matrix than in the intermembrane space in the absence of SOD.

Moreover, oxidized cytochrome *c* is powerful target for $O_2^{\bullet -}$ within the mitochondrial intermembrane space [137,138]. In this interconversion, cytochrome *c* is reduced and $O_2^{\bullet -}$ is oxidized to O_2.

$$\text{Cyt c}(Fe^{3+}) + O_2^{\bullet -} \longrightarrow \text{Cyt c}(Fe^{2+}) + O_2$$

This reaction occurs with a rate of about $10^7\ M^{-1}s^{-1}$ and contributes, thus, to detoxification of $O_2^{\bullet -}$ [139,140].

Defense against excess H_2O_2 is not only related to mitochondria but also to cytosol and surrounding area, as H_2O_2 permeates easily through membranes. Phospholipid-dependent glutathione peroxidase (GPx4), the thioredoxin system, peroxiredoxins, and catalase are involved in detoxification of H_2O_2.

4.5.6 Physiological and Pathophysiological Role of Mitochondrial H_2O_2

A signaling function of mitochondria-derived reactive species has been proposed. In central nervous system, mitochondrial H_2O_2 participates in rapid signaling processes on a subsecond timescale, shortly known as dynamic H_2O_2 signaling, and is implicated in intra- and intercellular regulation of neuronal properties and processes of synaptic transmission [141,142]. Other H_2O_2-mediated physiological processes ranging from embryonic development to cell death occur on timescales of minutes to hours [143–145].

In dorsolateral striatum, glutamate regulates the release of dopamine via dynamic H_2O_2 signaling. Glutamate induces $O_2^{\bullet -}$ and H_2O_2 production by mitochondrial complex I. Hydrogen peroxide inhibits the release of dopamine via activation of ATP-dependent K^+ channels [141]. This signaling is supported by the close location of both glutaminergic and dopaminergic synapses on dendritic spines of striatum neurons [146].

Hydrogen peroxide also acts at transient receptor potential channels and excites, thus, nondopaminergic neurons in the striatum and in the substantia nigra pars reticulata [147–149]. Other effects of H_2O_2 concern neuron–glia signaling in the hippocampus [150] and improvement of spatial memory in aged mice [151].

However, elevated H_2O_2 production by dysfunctional mitochondria can lead to false signals and mediate pathophysiological signaling. Disturbances of mitochondrial functions are discussed as a contributing factor to nigrostriatal degradation in Parkinson's disease [152].

4.6 DIOXYGEN IN TISSUES UNDER NORMAL AND PATHOLOGICAL CONDITIONS

4.6.1 Normal Tissue Values for Dioxygen

In tissues, the partial pressure of dioxygen (pO_2) is much lower than in air. At low altitudes, pO_2 in air is 20%–21% of the sea level standard atmospheric pressure (101.325 kPa). This value corresponds to about 152–160 mmHg.

Overviews about normal values of pO_2 in various human tissues are given in Refs. [153,154]. Median values of pO_2 range from 3.4%–6.8% (26–52 mmHg) in many organs and tissues [154]. In this publication, a total range is given for normal pO_2 values from 3% to 7.4% [154]. Functionally less active tissues such as outer layers of skin contain lower concentrations of O_2.

Selected data about pO_2 in tissues under normal and pathological conditions are presented in Fig. 4.9.

Of course, the indicated data for pO_2 are rough approximate values that can vary in a certain range and depend on individual features and general metabolic situation.

The aforementioned data say nothing about the tolerance of tissues against a transient decrease in the O_2 supply. Diminished supply of O_2 is very critical to the brain as brain cells consume about 20%–25% of total inspired O_2 and energy stores are limited. Thus, neuronal cell death occurs within minutes after cessation of cerebral O_2 delivery [155].

4.6.2 Normoxia, Hyperoxia, and Hypoxia

Thus, normal or physiological tissue O_2 values are much lower than in inspired air. Moreover, there are tissue-specific normal values for pO_2 ranging from 3%—7% in relation to the standard atmospheric pressure. Normoxia or physioxia are terms commonly used to describe these physiological O_2 levels in peripheral tissues [154].

In vivo, hyperoxic conditions are a rare event. They can result from underwater diving, the application of hyperbaric oxygen therapy, or after second- and third-degree burning of the skin [156,157].

Otherwise, investigations on cell cultures are usually performed with 95% air and 5% CO_2, i. e. at a pO_2 value around 2O %. Similarly, many chemical and biochemical in vitro experiments are done in buffered aqueous solutions, in which dioxygen is dissolved corresponding to a pO_2 value of about 2O%. Thus, many experiments including those ones investigating O_2-dependent processes are executed under clearly hyperoxic conditions with respect to O_2 values in cells and tissues.

In tissues, diminished O_2 values below normal O_2 pressures are much more common. The reasons for hypoxia can be quite different, including, e.g., decreased perfusion of tissues and organs, stay at high altitude, breathing problems, intoxications, and altered physiological conditions in numerous diseases.

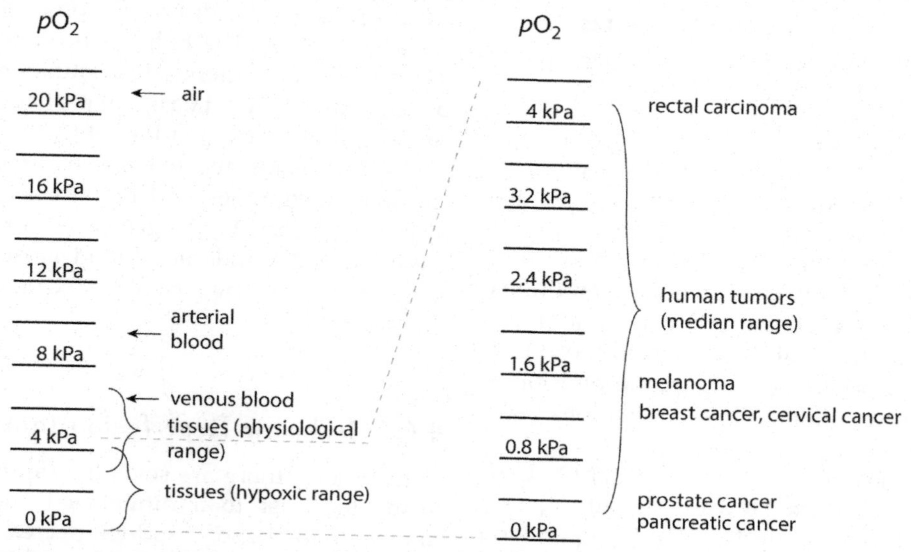

FIGURE 4.9 Selected ranges and values for partial pressure of dioxygen (pO_2) in human tissues under normoxic and hypoxic conditions. *Data are taken from McKeown SR. Defined normoxia, physoxia and hypoxia in tumours – implications for treatment response, Br. J. Radiol. 2014;87:20130676, 1–12. https://doi.org/10.1259/brj.20130676.*

In many tumors, the O_2 levels are clearly below the levels in adjacent healthy tissue [153,154,158]. A median range from 0.3%—4.2% was determined for pO_2 values in human tumors [154]. This issue is addressed in more detail in Sections 9.8.1 and 9.8.2.

4.6.3 Adaptations to Hypoxia

A mild hypoxia induces several protective mechanisms balancing the deficient O_2 supply. These adaptations are promoted by hypoxia-inducible factors (HIFs) that affect transcription of many genes. Under hypoxic conditions, these factors are responsible for vascularization in differentiating tissues, during wound healing, and in growing tumors. They control the production of red blood cells and the iron metabolism and increase glycolysis and cellular uptake of glucose [159—161]. In humans, to date, three HIFs are known: the ubiquitously expressed HIF-1α, HIF-2α, which is predominantly expressed in the lung, endothelium, and carotid body, and HIF-3α, a less-characterized species [159].

Activity of HIF-1α is predominantly induced in response to acute hypoxic events. This activity disappears under prolonged hypoxia. In contrast, HIF-2α becomes more important during long-lasting O_2 deficiency [153,162].

In cells, HIF-1α forms a heterodimer with the constitutively expressed HIF-1β. While HIF-1α is rapidly degraded under normoxic conditions, the level of HIF-1β is maintained unchanged independent of the O_2 supply. Hydroxylation of HIF-1α by an O_2-dependent HIF prolyl hydroxylase induces ubiquitination and proteasome degradation. This regulatory mechanism deactivates HIFs at normal O_2 values in tissues, but not under hypoxia [163,164]. As the O_2-dependent HIF prolyl hydroxylase requires ascorbate as a cofactor, ascorbate deficiency prolongs HIF activity [165,166].

4.6.4 Role of Superoxide Dismutase, Cytochrome c Oxidase, and Catalase in Recovery of Dioxygen

Among reactions, which are known to form dioxygen in biological material, are the catalase-mediated destruction of H_2O_2, dismutation of $O_2^{\bullet-}$, oxidation of $O_2^{\bullet-}$ by cytochrome *c*, recombination of peroxyl radicals, and some others. Under normal metabolic conditions, most of these reactions occur at low level. Here, the attention will be drawn to those processes of O_2 formation having (patho)physiological relevance under stress situations and hypoxic conditions.

Dismutation of $O_2^{\bullet-}$ yields O_2 and H_2O_2. Usually the focus is directed toward detoxification of $O_2^{\bullet-}$ and deleterious effects of H_2O_2. Superoxide anion radicals are formed as a result of one-electron reduction of O_2, a process that contributes to further decrease of the already low O_2 concentration. Hence, dismutation of $O_2^{\bullet-}$ can partially replace the O_2 deficit.

Another reaction of $O_2^{\bullet-}$ concerns the oxidation of $O_2^{\bullet-}$ to O_2 by oxidized cytochrome *c* [137,138]. This reaction takes mainly place in the intermembrane space of mitochondria where about 0.7 mM cytochrome *c* is present [167]. Moreover, cytochrome *c* is usually kept in its oxidized state by interaction with cytochrome *c* oxidase [168]. This reaction is favored under conditions of enhanced formation of $O_2^{\bullet-}$ in mitochondria in association with redox processes in the intermediary Q-cycle. Thus, this recovery of O_2 is important to adapt mitochondria and cells to a deficient supply of O_2.

In human organism, although being widely distributed, catalase is mainly found in red blood cells, kidney, and liver cells and intracellularly in peroxisomes. Enzymatic processes in peroxisomes are driven by the reduction of O_2 to H_2O_2. In these organelles, the role of catalase is not only directed to detoxify H_2O_2 but also to recover a part of O_2 diminishing, thus, the O_2 burden in these processes. Apparently, similar conclusions can be drawn for other main locations of catalase. The kidney and liver are very susceptible to different stress situations. Degradation of H_2O_2 by catalase can partially compensate the stress-mediated loss of O_2.

As H_2O_2 is less reactive and freely permeable through membranes, red blood cell catalase is also a sink for H_2O_2 generated in the near environment of inflamed blood vessels. Dioxygen liberated by catalase can bind to hemoglobin.

4.6.5 Ischemia-Reperfusion Injury

In tissues, there are several adaptive mechanisms to resist to a diminished supply with dioxygen. In case of a sudden decrease in tissue oxygenation, more drastic metabolic changes can occur. For example, this might happen after local bleeding, rupture of aneurysm, or closure of an afferent vessel by a thrombus. Formidable life-threatening conditions are stroke and cardiac arrest [169,170].

During ischemic periods, tissues are less supplied by dioxygen and nutrients, and waste products are insufficiently removed. After the onset of ischemia, numerous biochemical and physiological events are initiated most of all due to deficits in energy metabolism and reduced removal of metabolic products. There is an accumulation of metabolic products derived from ATP such as ADP, AMP, adenosine, inosine, and hypoxanthine [171]. Adenosine can act as vasodilator and help to restore tissue oxygenation [172]. Necrotic cell death in affected area, in particular death of neurons, is a further consequence of ischemia [173].

Sometimes, after ischemia a reperfusion period follows. Hence, tissues are better supplied with dioxygen and nutrients. However, dioxygen can react with products accumulated during the preceding ischemic phase and reinforce, thus, already existing tissue damage [174,175]. A likely reaction is the O_2-driven oxidation of hypoxanthine to xanthine and urate by xanthine oxidase under formation of $O_2^{\bullet-}$ and H_2O_2 [176]. Moreover, sequestration of transition metal ions can be disturbed in necrotic areas during ischemia. These now available free metal ions and metal complexes contribute to formation of highly reactive species and substrate damage. Similarly, several proteases and other destructive enzymes complete this injury as they are released during ischemia and from invading immune cells and necrotic tissue.

The spectrum of tissue damage varies considerably in ischemia-reperfusion injury [174,175]. This depends mainly on duration and strength of ischemic period, tissue location of ischemic area, and general health condition of the patient.

4.6.6 Critical Role of O_2 in Processes of Tissue Destruction

Processes of cell and tissue destruction are promoted under pathological conditions by deficient supply of O_2 and nutrients, the presence of catalytic active transition metal centers, and by the release of degrading enzymes. Moreover, exhaustion of naturally protective mechanisms additionally favors these degradations.

Dioxygen participates in these events as a source of $O_2^{\bullet-}$, H_2O_2, and highly reactive species and in its reaction with different substrate radicals under the formation of organic peroxyl radicals and peroxides. Otherwise, the tissue pO_2 value can considerably be reduced under hypoxic conditions. In defining hypoxia, it should be differentiated between physiological hypoxia, a condition where numerous adaptive responses are induced to balance the diminished O_2 supply, and pathological hypoxia, which is beyond the regulatory potential in tissues and is accompanied by increasing necrosis. As a rough guide for pO_2 values under these conditions, it has been proposed that 5% corresponds to a normal value, 2% to physiological hypoxia, and 1% and lower to pathological hypoxia [154]. According to Henry's law, which relates to each other partial pressure and molar concentration of a dissolved gas [177], a pO_2 value of 1% corresponds to 10 μM O_2 dissolved in an aqueous solution at 37 °C. Thus, only very low micromolar concentrations of O_2 are present in strong hypoxic tissues.

It remains unknown, is O_2 really able to participate in formation of reactive species, peroxyl radicals, and peroxides in cells and tissues under strong and very strong hypoxic conditions. Otherwise, numerous protective mechanisms depend also on the presence of O_2 such as the formation of reactive species by activated neutrophils, degradation of free heme by heme oxygenase, and formation of hydroperoxides by lipoxygenases and cyclooxygenases. The question arises, are these mechanisms really active at a highly diminished O_2 content?

4.7 GLUCOSE AS AN ENERGY SUBSTRATE

4.7.1 The Danger of Hypo- and Hyperglycemia

In healthy persons, the serum level of glucose is well-regulated. The fasting level should be between 3.9–5.5 mmol L^{-1} that means 70–100 mg dL^{-1}. Of course, this level increases after food intake. Under normal circumstances, this transient increase is over about 2 hours after the meal. There are several factors responsible for the maintenance of the glucose levels in humans. Glucose and other dietary monosaccharides can be stored in form of glycogen and triglycerides. Otherwise, glucose can be mobilized from these internal stores by enhanced exercise, metabolic activities, and also by starvation.

Lower values of blood glucose lead to hypoglycemia [178,179]. Among causes of hypoglycemia are medications for diabetes treatment, alcohol abuses, malnutrition, kidney failure, severe infections, certain tumors, hypothyroidism, and others. As the brain metabolism highly depends on continuous glucose supply, mild hypoglycemia starts with neurological symptoms such as fatigue, weakness, lethargy, confusion, memory loss, and feeling of dizziness. Others symptoms result from counterregulation of hypoglycemia by epinephrine and glucagon. Anxiety, nervousness, palpitations, tachycardia, sweating, and others belong to these symptoms. A strong decline of blood glucose is very dangerous to the organism and results in stupor, coma, abnormal breathing, and finally lethal outcome.

Hyperglycemia is very common in industrialized countries. It is associated with diabetes mellitus and various other disease complications [180,181]. Most of all, hyperglycemia is a consequence of our modern lifestyle and results especially from the combination of overeating and physical inactivity. Increased values of glucose are mainly associated with increasing insulin resistance in muscle and fat tissues. Enhanced blood glucose levels lead to an accumulation of glucose in insulin-independent cells such as red blood cells, endothelial cells, and neurons, as well as to increased values of glycated products [182,183]. Finally, numerous fatal pathological consequences are associated with long-term hyperglycemia.

4.7.2 Absorption of Glucose and Other Monosaccharides From Foodstuffs in the Intestine

D-Glucose is the most important source of energy for cellular respiration. It is widely distributed in nature and regularly taken up with our foodstuffs. Two other dietary monosaccharides, namely fructose and galactose, are also absorbed in the intestine and transported via the portal vein to the liver.

Glucose is mainly incorporated in poly- and disaccharides in our food. In the digestive tract, polysaccharides such as starch are degraded by amylase into smaller units. The final digestion of disaccharides to monosaccharides is conducted by so-called "brush border hydrolases" that are tethered to enterocytes at the luminal plasma membrane [184]. These hydrolases cleave maltose into two molecules of glucose, lactose into glucose and galactose, and sucrose into glucose and fructose.

In the intestinal epithelium, enterocytes separate the intestinal lumen from blood capillaries. For absorption, foodstuffs have to permeate through enterocytes or the space between these cells. Glucose and galactose are taken up by enterocytes at the luminal site by a cotransport with sodium ions. The sodium–glucose transporter is driven by the influx of sodium ions. Glucose or galactose is taken up by enterocytes against its concentration gradient by this cotransporter [185].

The absorption of fructose to enterocytes is conducted by the hexose transporter GLUT5 at the luminal site and GLUT2 at the basolateral pool. Maybe, a not well-characterized active transport mechanism is also involved [186]. Inside enterocytes, the fructose concentration is low. Hence, a facilitated diffusion is sufficient for this monosaccharide unlike the situation for glucose and galactose. Otherwise, absorption of pure fructose is limited. This absorption increases when fructose and glucose are present together in the intestine [187,188].

4.7.3 Metabolism of Dietary Monosaccharides in the Liver

Glucose, galactose, and fructose are transported into hepatocytes via the GLUT2 transporter. GLUT2 is a bidirectional transporter that means the flow direction is determined by concentration differences besides the membrane. Other locations of GLUT2 are the intestine, kidney, β cells of pancreatic islets, neurons, and astrocytes [189].

An important function of the liver is to generate glycogen from the dietary monosaccharides glucose, fructose, and galactose and to store glycogen for demand during enhanced activity or fasting periods [190]. Glycogen pool in liver can be used as fuel for other tissues via the bloodstream. During glycogenolysis, glycogen is stepwise broken down to glucose-1-phosphate and further transformed into glucose-6-phosphate. Glucose-6-phosphatase, which is present in hepatocytes but not in myocytes, removes the phosphate group. The resulting glucose leaves hepatocytes via the GLUT2 transporter and is distributed by circulating blood to other tissues [191].

Besides the liver, glycogen is also synthesized and stored in skeletal muscle and in small amounts in the kidneys, red blood cells, leukocytes, and glial cells [192–195].

In liver, excess glucose is also utilized to produce pyruvate, acetyl-CoA, and fatty acids. The synthesis of fatty acids from acetyl-CoA via de novo lipogenesis occurs mainly not only in the liver but also in adipose tissue and lactating mammary glands [196].

Besides glycogenolysis, glucose can also be produced in the liver and mobilized to the circulating blood from de novo synthesis from precursors such as lactate, alanine, glycerol, and fructose via gluconeogenesis [197].

In hepatocytes, fructose is rapidly phosphorylated. Fructokinase always converts fructose completely into fructose-1-phosphate as this enzyme is characterized by a very low K_m and a high V_{max} value [198]. Thus, the majority of absorbed fructose is utilized in the liver. Only a small portion of fructose is metabolized in muscle and adipose tissue [199]. Moreover, the hepatic fructose pathway is independent of phosphofructokinase, a rate-determining enzyme in glucose metabolism [200,201]. As the corresponding K_m is much higher for glucokinase, much of the glucose passes through the liver and is metabolized in peripheral tissues [202]. In contrast to fructose, only 15%–30% of ingested glucose is consumed in liver [203].

In liver, metabolisms of fructose (fructolysis) and glucose are closely interlinked to each other [204]. Fructose is finally converted into glucose, lactate, glycogen, and to a minor percentage into triglycerides. Similar to fructose, galactose can also be transformed into glucose.

4.7.4 Deleterious Effects of Excess Fructose on Liver

Massive uptake of fructose by the liver depletes the ATP pool and intracellular phosphate due to enhanced activity of fructokinase C [205]. In turn, low phosphate favors the formation of uric acid [206,207] and low ATP blocks protein synthesis and mediates oxidative stress and mitochondrial dysfunction [205,208,209]. High fructose also impairs the hepatic copper metabolism, causes a copper deficiency, and induces an iron overload [210–212].

Several studies demonstrate that fructose causes fatty liver. Via fructokinase C, this monosaccharide stimulates de novo lipogenesis, suppresses β-fatty acid oxidation, and contributes, thus, to enhanced triglyceride synthesis [199,208,209,213,214]. Fructose promotes overexpression of fructokinase C and fatty acid synthase [199,215,216]. Fructokinase levels are elevated in obese humans with nonalcoholic steatohepatitis [216]. Fructose-mediated mechanisms also contribute to the development of an insulin resistance [217].

Increased intestinal permeability, endotoxemia, and enhanced inflammatory response are further consequences of excessive intake of fructose [209].

4.7.5 Insulin-dependent Uptake of Glucose by Skeletal Muscle Cells and Adipocytes

Cells take up their essential energy substrate glucose by special glucose receptors all belonging to the GLUT family. In humans, 14 different GLUT proteins are known [218,219]. The transporters GLUT1–GLUT4, which comprise the subclass I of this family, and GLUT5 are well-characterized. Details about the functional response and the specific role in glucose metabolism of the other transporters need to be further evaluated.

The peptide hormone insulin promotes the regulated uptake of glucose via GLUT4 into fat cells and skeletal muscle cells [220,221]. On binding of insulin to its receptor, the intracellularly prestored glucose transporter GLUT4 is translocated into the plasma membrane, and numerous metabolic pathways are activated. Increased blood level of glucose is usually accompanied by increase of insulin that is released from pancreatic β-cells. The insulin-dependent glucose transporter GLUT4 provides a rapid uptake of glucose into cells.

In muscle cells, glucose is mainly stored as glycogen [195], while glucose is predominantly converted into triglycerides in fat tissue [222]. These mechanisms ensure a continuous supply of glucose for cellular processes.

Binding of insulin to the insulin receptor generates signals that are involved in regulation of glucose, lipid, and energy metabolism. A critical pathway of insulin signaling is the activation of phosphoinositide 3-kinase (PI3-kinase) followed by enzymes of the Akt (protein kinase B) pathway [223]. Downstream from Akt, the activated mTORC1 complex mediates regulation of numerous genes controlling metabolic routes, protein synthesis, and growth processes [224].

4.7.6 Insulin Resistance

Under normal conditions, the pancreatic hormone insulin triggers the uptake of glucose by muscle and fat cells. This glucose uptake is, however, reduced during insulin resistance. As a result enhanced values of glucose in blood might be found, which provide the basis for developing of diabetes.

Under the condition of insulin resistance, skeletal muscle and fat cells, which are equipped with insulin receptors, do not adequately respond to increased insulin. In consequence, the uptake of glucose by these cells is delayed, and the β-cells of pancreas release more insulin [225]. Prolonged stress of β-cells due to insulin overproduction finally impairs these cells. This leads to diabetes mellitus type 2 characterized by hyperglycemia and diminished insulin production.

Insulin resistance per se is a regulatory mechanism of the organism to adapt to altered metabolic conditions, for example, during pregnancy or during starvation periods [226]. This resistance ensures a better supply of brain tissues with glucose by reducing the sensitivity of muscle cells to insulin [227]. Neuronal tissue has a high demand for glucose. It has been assumed that the insulin response is determined by certain threshold values. According to the adjustable threshold hypothesis of insulin resistance, the translocation of GLUT4 into the plasma membrane depends on factors affecting the onset of insulin sensitivity [227].

Several dietary factors are under discussion for their promoting effect on insulin resistance [228,229]. Elevated blood and tissue levels of free fatty acids and triglycerides are important for insulin resistance. Excess body weight is a risk factor for insulin resistance and diabetes mellitus type 2. High dietary levels of fructose or sucrose lead to enhanced triglyceride production in the liver. Long-lasting physical inactivity also promotes the development of insulin resistance. In skeletal muscle, insulin sensitivity is a dynamic parameter that adapts the glucose uptake to the real demand [230,231].

4.7.7 Insulin-independent Uptake of Glucose by Cells

In other cells and tissues, glucose uptake is mainly provided by the transporters GLUT1 and GLUT3 or by other special, not yet specified, transporters of the GLUT family. These transporters function independent of insulin. GLUT1 facilitates the glucose uptake in red blood cells, endothelial cells, and through barriers such as the blood–brain barrier [232,233]. This transporter allows many cells to uptake glucose for basal level of respiration. The expression of GLUT1 is known to increase with reduced glucose level in blood and vice versa. Nevertheless, prolonged hyperglycemia increases the intracellular glucose level in a number of cells such as red blood cells, endothelial cells, retinal cells, and others. The accumulation of glucose in these cells leads to serious metabolic deviations (Sections 4.7.8 and 4.7.9) and provides the basis for the development of pathologies associated with diabetes mellitus (Sections 9.3.1 to 9.3.7).

The uptake of glucose by neurons is provided by the glucose transporter GLUT3. This high-affinity transporter allows the transport even at diminished glucose concentrations. In neurons, this transporter is predominantly expressed in axons and dendrites but less prominently in the cell body [234].

4.7.8 Formation of Glycated Products

Chemical side reactions of monosaccharides are caused by the presence of the carbonyl moiety in the open-chain form. Although having the lowest carbonyl yield among monosaccharides [235,236], D-glucose is known to cause different glycation reactions of biological substrates (see Section 3.3.4). The probability of chemical side reactions of glucose rises considerably under hyperglycemic conditions and in cells accumulating a surplus of glucose.

The glycation of proteins and other constituents is especially evident for long-living biological material. The object of choice for glycation is hemoglobin. Red blood cells have a lifetime of about 120 days. As red blood cells are not equipped with any mechanisms to control the uptake of glucose, their intracellular content of glucose corresponds to the serum concentration of this monosaccharide. Thus, glycation of hemoglobin rises under hyperglycemia. The detection of hemoglobin A1c, where glucose is coupled to the *N*-terminal valine of the β-chain of hemoglobin A1, is an important diagnostic parameter reflecting the long-term burden of glucose [237,238]. This parameter accounts for the blood glucose levels of the last 8–12 weeks.

On hyperglycemia, increased glycated products were also detected in serum proteins, for example, in human serum albumin (HSA). Because of its high abundance in serum and a lifetime of about 18 d, this protein is a preferred subject for studying glycation reactions. In patients with diabetes mellitus, the level of glycated HSA is about 2–3 times higher in comparison with healthy persons [239,240]. Glycation occurs mainly on ε-amino groups of lysine residues and also on arginine residues [241,242]. Glycated HSA exhibits lower ligand binding capacity [243]. The binding of glucose to Lys-195 in HSA favors a ring opening of the cyclic form [244].

In general, side reactions of glucose with proteins, lipids, nucleic acids, and others contribute to formation of advanced glycation end products, a very heterogenous group of molecules and compounds [245–247].

4.7.9 Reduction of Glucose to Sorbitol

Under hyperglycemic conditions, enhanced intracellular glucose levels in insulin-independent cells also favor the conversion of glucose into sorbitol via aldose reductase [248,249]. During this reaction, which represents the first reaction in the polyol glucose pathway, NADPH is oxidized to $NADP^+$. In pathogenesis of diabetes, this reaction is important in capillary endothelial cells in retina, mesangial cells in the renal glomerulus, and neurons and Schwann cells in peripheral nerves, as the glucose entry into these cells is insulin-independent. Elevated cellular glucose levels favor NADPH depletion and cause reduced values of NADPH-dependent metabolites such as glutathione, nitric monoxide, myo-inositol, and taurine. Dreadful effects of enhanced glucose levels on the retina, kidney glomeruli, and nerve cells are explained by the intensification of the polyol pathway of glucose in hyperglycemic patients. Moreover, the resulting sorbitol does not freely permeate through cell membranes. Its accumulation causes an osmotic stress that contributes to the development of these pathologies.

Aldose reductase is also contained in red blood cells. The activity of this enzyme is regulated via a redox-sensitive cysteine residue, Cys-298, by nitric monoxide [250]. Normal levels of nitric monoxide depress its activity. Thus, less than 3% of total glucose is metabolized into sorbitol in nondiabetic patients. With decreased bioavailability of nitric monoxide, the activity of aldose reductase raises considerably. In hyperglycemia patients, about 30% of glucose is converted by aldose reductase. There is a general association between enhanced glucose concentrations and decreased bioavailability of nitrogen monoxide in diabetes [251,252].

4.8 SUMMARY

As open systems, cells, tissues, and organisms are highly dependent on the regular uptake of energy substrates. The permanent cellular generation of chemical energy equivalents is of paramount importance for the long-term survival of organisms. In humans and higher animals, energy metabolism is mainly based on the supply and utilization of dioxygen, glucose, and triglycerides. Specialized biomolecules, cell organelles, and cells are involved in transport, storage, and exploitation of energy substrates as well as in formation of ATP as universal energy source. As all living structures, these complex systems are also prone to numerous destructions and side reactions, which can finally disturb the capability of energy metabolism.

Transport and storage of dioxygen can primarily be affected by enhanced formation of methemoglobin and increased intravascular hemolysis and/or rhabdomyolysis. These processes are associated with a diminished supply of dioxygen to tissues, reduced bioavailability of nitric monoxide, and the formation of the highly cytotoxic free heme. Protective mechanisms such as methemoglobin reductase, haptoglobin, and hemopexin are designed to act efficiently against weak and medium impacts. Their capacity is, however, limited to provide an adequate protection against severe interferences.

Under conditions of hyperglycemia, glucose promotes different glycation reactions of biological substrates, an uncontrolled uptake of glucose, and accumulation of osmotically sensitive metabolites in erythrocytes and other insulin-independent cells. Moreover, high long-term uptake of fructose together with glucose in our food has several negative impacts on regulation of main metabolic routes and favors condition of insulin resistance. These processes strongly impair metabolism of affected cells and provide the basis for serious disease complications in hyperglycemic patients.

In mitochondria, electron transfer processes, which are highly necessary for oxidative phosphorylation, can be a source for reactive species under certain stress situations. Enhanced formation of superoxide anion radicals and hydrogen peroxide by mitochondria induces numerous adaptations of cell metabolism but contributes to induction of cell death mechanisms as well.

An efficiently functioning energy metabolism depends not only on the regular supply of energy substrates but also on the intactness of protective mechanisms to counterregulate low-level and medium disturbances. Limited capacity or exhaustion of defense reactions can convert usually harmless disturbances into dangerous impacts to the organism. These destructive mechanisms play a substantial role in the pathogenesis of different diseases, which are often associated with the development of chronic inflammatory states in tissues and organs.

References

[1] L. Laffel, Ketone bodies: a review of physiology, pathophysiology and application of monitoring to diabetes, Diabet. Metabol. Res. Rev. 15 (1999) 412–426.

[2] M.F. Perutz, M.G. Rossmann, A.F. Cullis, H. Muirhead, G. Will, A.C.T. North, Structure of haemoglobin: a three-dimensional Fourier synthesis at 5.5 Å resolution, obtained by X-ray analysis, Nature 185 (1960) 416–422.

[3] L.R. Davis, Changing blood picture in sickle cell anemia from shortly after birth to adolescence, J. Clin. Pathol. 29 (1976) 898–901.

[4] J. Rochette, J.E. Craig, S.L. Thein, Fetal hemoglobin in adults, Blood Rev. 8 (1994) 213–224.

[5] C. Thomas, A.B. Lumb, Physiology of hemoglobin, Cont. Educ. Anaesth. Crit. Care Pain 12 (2012) 251–256.

[6] M.-H. Mihailescu, I.M. Russu, A signature of the T → R transition in human hemoglobin, Proc. Natl. Acad. Sci. U.S.A. 98 (2001) 3773–3777.

[7] R.I. Weed, C.F. Reed, G. Berg, Is hemoglobin an essential structural component of human erythrocyte membrane? J. Clin. Investig. 42 (1963) 581–588.

[8] E. Dominguez de Villota, M.T. Garcia Carmona, J.J. Rubio, S. Ruiz de Andrés, Equality of the in vivo and in vitro oxygen-binding capacity of haemoglobin in patients with severe respiratory disease, Br. J. Anaesth. 53 (1981) 1325–1328.

[9] M. Tsuruga, A. Matsuoka, A. Hachimori, Y. Sugawara, K. Shikama, The molecular mechanism of autoxidation for human oxyhemoglobin, J. Biol. Chem. 273 (1998) 8607–8615.

[10] Y. Sugawara, M. Sakoda, N. Shibata, H. Sakamoto, Autoxidation of human hemoglobin: kinetic analysis of the pH profile, Jpn. J. Physiol. 43 (1993) 21–34.

[11] K. Shikama, The molecular mechanism of autoxidation for myoglobin and hemoglobin: a venerable puzzle, Chem. Rev. 98 (1998) 1357–1373.

[12] L. Zhang, A. Levy, J.M. Rifkind, Autoxidation of hemoglobin enhanced by dissociation into dimers, J. Biol. Chem. 266 (1991) 24698–24701.

[13] J. Umbreit, Methemoglobin – it's not just blue: a concise review, Am. J. Hematol. 82 (2007) 134–144.

[14] M.J. Percy, T.R. Lappin, Recessive congenital methaemoglobinemia: cytochrome b5 reductase deficiency, Br. J. Haematol. 141 (2008) 298–308.

[15] R.F. Grace, A. Zanella, E.J. Neufeld, D.H. Morton, S. Eber, H. Yaish, B. Glader, Erythrocyte pyruvate kinase deficiency: 2015 status report, Am. J. Hematol. 90 (2015) 825–830.

[16] E. Beutler, Glucose-6-phosphate dehydrogenase deficiency: a historical perspective, Blood 111 (2008) 16–24.

[17] A.G. Stucke, M.L. Riess, L.A. Connolly, Hemoglobin M (Milwaukee) affects arterial oxygen saturation and makes pulse oximetry unreliable, Anesthesiology 104 (2006) 887–888.

[18] S. Fucharoen, V. Viprakasit, Hb H disease: clinical course and disease modifiers, Hematol. Am. Soc. Hematol. Ed. Program (2009) 26–34.

[19] J.A. Cortazzo, A.D. Lichtman, Methemoglobinemia: a review and recommendations for management, J. Cardiothorac. Vasc. Anesth. 28 (2014) 1043–1047.

[20] A. Mansouri, A.A. Lurie, Concise review: methemoglobinemia, Am. J. Hematol. 42 (1993) 7–12.

[21] J.R. Lancester Jr., Simulation of the diffusion and reaction of endogenously produced nitric oxide, Proc. Natl. Acad. Sci. U.S.A. 91 (1994) 8137–8141.

[22] T.H. Han, D.R. Hyduke, M.W. Vaughn, J.M. Fukuto, J.C. Liao, Nitric oxide reaction with red blood cells and hemoglobin under heterogeneous conditions, Proc. Natl. Acad. Sci. U.S.A. 99 (2002) 7763–7768.

[23] L. Jia, C. Bonaventura, J. Bonaventura, J.S. Stamler, S-Nitrosohemoglobin: a dynamic activity of blood involved in vascular control, Nature 380 (1996) 221–226.

[24] A.J. Gow, Nitric oxide, hemoglobin, and hypoxic vasodilation, Am. J. Respir. Cell Mol. Biol. 32 (2005) 479–482.

[25] J.M. Robinson, J.R. Lancester, Hemoglobin-mediated, hypoxia-induced vasodilation via nitric oxide: mechanism(s) and physiologic versus pathophysiologic

relevance, Am. J. Respir. Cell Mol. Biol. 32 (2005) 257–261.

[26] C. Helms, D.B. Kim-Shapiro, Hemoglobin-mediated nitric oxide signaling, Free Rad. Biol. Med. 61 (2013) 464–472.

[27] K.D. Vandegriff, L. Benazzi, M. Ripamonti, M. Parrella, Y.C. Le Tellier, A. Zegna, R.M. Winslow, Carbon dioxide binding to human hemoglobin cross-linked between the α chains, J. Biol. Chem. 266 (1991) 2697–2700.

[28] F.L. Rodkey, J.D. O'Neal, H.A. Collison, Oxygen and carbon monoxide equilibria of human adult hemoglobin at atmospheric and elevated pressure, Blood 33 (1969) 57–65.

[29] M. Goldstein, Carbon monoxide poisoning, J. Emerg. Nurs. 34 (2008) 538–542.

[30] J.A. Raub, M. Mathieu-Nolf, N.B. Hampson, S.R. Thom, Carbon monoxide poisoning – a public health perspective, Toxicology 145 (2000) 1–14.

[31] S.N. Vogel, T.R. Sultan, Cyanide poisoning, Clin. Toxicol. 18 (1981) 367–383.

[32] J.L. Way, Cyanide intoxication and its mechanism of antagonism, Annu. Rev. Pharmacol. Toxicol. 24 (1984) 451–481.

[33] M.A. Holland, L.M. Kozlowski, Clinical features and management of cyanide poisoning, Clin. Pharm. 5 (1986) 737–741.

[34] D.M.G. Beasley, W.I. Glass, Cyanide poisoning: pathophysiology and treatment recommendations, Occup. Med. 48 (1998) 427–431.

[35] C.D. Reiter, X. Wang, J.E. Tanus-Santos, N. Hogg, R.O. Cannon III, A.N. Schechter, M.T. Gladwin, Cell-free hemoglobin limits nitric oxide bioavailability in sickle-cell disease, Nat. Med. 8 (2002) 1383–1389.

[36] S. Sarkar, D. Prakash, R.K. Marwaha, G. Garewal, L. Kumar, S. Singhi, B.N. Walia, Acute intravascular haemolysis in glucose-6-phosphate dehydrogenase deficiency, Ann. Trop. Paediatr. 13 (1993) 391–394.

[37] R.P. Rother, L. Bell, P. Hillman, M.T. Gladwin, The clinical sequelae of intravascular hemolysis and extracellular plasma hemoglobin, J. Am. Med. Assoc. 293 (2005) 1653–1662.

[38] C. Parker, Paroxysmal nocturnal hemoglobinuria, Curr. Opin. Hematol. 19 (2012) 141–148.

[39] G.J. Kato, M.H. Steinberg, M.T. Gladwin, Intravascular hemolysis and the pathophysiology of sickle cell disease, J. Clin. Investig. 127 (2017) 750–760.

[40] R. Gozzelino, V. Jeney, M.P. Soares, Mechanisms of cell protection by heme oxygenase-1, Annu. Rev. Pharmacol. Toxicol. 50 (2010) 323–354.

[41] D. Chiabrando, F. Vinchi, V. Fiorito, E. Tolosano, Haptoglobin and hemopexin in heme detoxification and iron recycling, in: F. Veas (Ed.), Acute Phase Proteins – Regulation and Functions of Acute Phase Proteins, Intech, Rijeka, Croatia, 2011, pp. 261–288.

[42] D.J. Schaer, P.W. Buehler, A.I. Alayash, J.D. Belcher, G.M. Vercelotti, Hemolysis and free hemoglobin revisited: exploring hemoglobin and hemin scavengers as a novel class of therapeutic proteins, Blood 121 (2013) 1276–1284.

[43] W.G. Land, Transfusion-related acute lung injury: the work of DAMPs, Transfus. Med. Hemotherapy 40 (2013) 3–13.

[44] A. Ferreira, J. Balla, V. Jeney, G. Balla, M.P. Soares, A central role for free heme in the pathogenesis of severe malaria: the missing link? J. Mol. Med. 86 (2008) 1097–1111.

[45] R.F. Eich, T. Li, D.D. Lemon, D.H. Doherty, S.R. Curry, J.F. Aitken, A.J. Mathews, K.A. Johnson, R.D. Smith, G.N. Phillips Jr., J.S. Olson, Mechanism of NO-induced oxidation of myoglobin and hemoglobin, Biochemistry 35 (1996) 6976–6983.

[46] J.S. Olson, E.W. Foley, C. Rogge, A.L. Tsai, M.P. Doyle, D.D. Lemon, NO scavenging and the hypertensive effect of hemoglobin-based blood substitutes, Free Rad. Biol. Med. 36 (2004) 685–697.

[47] M.S. Hargrove, J.S. Olson, The stability of holomyoglobin is determined by heme affinity, Biochemistry 35 (1996) 11310–11318.

[48] H.F. Bunn, J.H. Jandl, Exchange of heme among hemoglobins and between hemoglobin and albumin, J. Biol. Chem. 243 (1968) 465–475.

[49] J. Balla, G.M. Vercellotti, V. Jeney, A. Yachie, Z. Varga, J.W. Eaton, G. Balla, Heme, heme oxygenase and ferritin in vascular endothelial cell injury, Mol. Nutr. Food Res. 49 (2005) 1030–1043.

[50] M. Kristiansen, J.H. Graversen, C. Jacobsen, O. Sonne, H.J. Hoffmann, S.K. Law, S.K. Moestrup, Identification of the haemoglobin scavenger receptor, Nature 409 (2001) 198–201.

[51] J.H. Thomsen, A. Etzerodt, P. Svendsen, S.K. Moestrup, The haptoglobin-CD163-heme oxygenase-1 pathway for hemoglobin scavenging, Oxid. Med. Cell. Longev. (2013), article ID 523652, 1–11, https://doi.org/10.1155/2013/523652.

[52] K. Theilgaard-Mönch, L.C. jacobsen, M.J. Nielsen, T. Rasmussen, L. Udby, M. Gharib, P.D. Arkwright, A.F. Gombart, J. Calafat, S.K. Moestrup, B.T. Porse, N. Borregaard, Haptoglobin is synthesized during granulocyte differentiation, stored in specific granules, and released by neutrophils in response to activation, Blood 108 (2006) 353–361.

[53] M. Paoli, B.F. Anderson, H.M. Baker, W.T. Morgan, A. Smith, E.N. Baker, Crystal structure of hemopexin reveals a novel high-affinity heme site formed

between two beta-propeller domains, Nat. Struct. Biol. 6 (1999) 926–931.
[54] V. Hvidberg, M.B. Maniecki, C. Jacobson, P. Hojrup, H.J. Moller, S.K. Moestrup, Identification of the receptor scavenging hemopexin-heme complexes, Blood 106 (1999) 2572–2579.
[55] T. Lin, D. Maita, S.R. Thundivalappil, F.E. Riley, J. Hambsch, L.J. van Marter, H.A. Christou, L. Berra, S. Fagan, D.C. Christiani, H.S. Warren, Hemopexin in severe inflammation and infection: mouse models and human diseases, Crit. Care 19 (166) (2015) 1–8, https://doi.org/10.1186/s13054-015-0885-x.
[56] G.A. Ordway, D.J. Garry, Myoglobin: an essential hemoprotein in striated muscle, J. Exp. Biol. 207 (2004) 3441–3446.
[57] G. Liu, F.M. Gabhann, A.S. Popel, Effects of fiber type and size on the heterogeneity of oxygen distribution in exercising skeletal muscle, PLoS One 7 (2012), e44375, 1–13, https://doi.org/10.1371/journal.pone.0044375.
[58] E. Antonini, Interrelationship between structure and function in hemoglobin and myoglobin, Physiol. Rev. 45 (1965) 123–170.
[59] A. Rossi-Fanelli, E. Antonini, Studies on the oxygen and carbon monoxide equilibria of human myoglobin, Arch. Biochem. Biophys. 77 (1958) 478–492.
[60] U. Flögel, A. Fago, T. Rassaf, Keeping the heart in balance: the functional interactions of myoglobin with nitrogen oxides, J. Exp. Biol. 213 (2010) 2726–2733.
[61] L. Hagler, R.I. Coppes Jr., R.H. Herman, Metmyoglobin reductase. Identification and purification of a reduced nicotinamide adenine dinucleotide-dependent enzyme from bovine heart which reduces metmyoglobin, J. Biol. Chem. 254 (1979) 6505–6514.
[62] D.J. Livingston, S.J. McLachlan, G.N. La Mar, W.D. Brown, Myoglobin: cytochrome b5 interactions and the kinetic mechanism of metmyoglobin reductase, J. Biol. Chem. 260 (1985) 15699–15707.
[63] G. Gros, B.A. Wittenberg, T. Jue, Myoglobin's old and new clothes: from molecular structure to function in living cells, J. Exp. Biol. 213 (2010) 2713–2725.
[64] G.B. Postnikova, E.A. Shekhovtsova, Myoglobin and mitochondria: how does the "oxygen store" work? J. Phys. Chem. Biophys. 3 (2013) 1–7, https://doi.org/10.4172/2161-0398.1002126.
[65] Y. Chung, D. Xu, T. Jue, Nitrite oxidation of myoglobin in perfused myocardium: implications for energy coupling in respiration, Am. J. Physiol. 271 (1996) H1166–H1173.
[66] J.M. Sauret, G. Marinides, G.K. Wang, Rhabdomyolysis, Am. Fam. Phys. 65 (2002) 907–912.
[67] J.D. Hunter, K. Gregg, Z. Damani, Rhabdomyolysis, Cont. Ed. Anaesth. Crit. Care Pain 6 (2006) 141–143.
[68] N. Petejova, A. Martinek, Acute renal failure due to rhabdomyolysis and renal replacement therapy: a critical review, Crit. Care 18 (2014), 224, 1–8, https://doi.org/10.1188/cc13897.
[69] P.A. Torres, J.A. Helmstetter, A.M. Kaye, A.D. Kaye, Rhabdomyolysis: pathogenesis, diagnosis, and treatment, Ochsner J. 15 (2015) 58–69.
[70] T. Burmeister, B. Weich, S. Reinhard, T. Hankeln, A vertebrate globin expressed in the brain, Nature 407 (2000) 520–523.
[71] T. Burmeister, B. Ebner, B. Weich, T. Hankeln, Cytoglobin: a novel globin type ubiquitously expressed in vertebrate tissues, Mol. Biol. Evol. 19 (2002) 416–421.
[72] J.T. Trent III, M.S. Hargrove, A ubiquitously expressed human hexacoordinate hemoglobin, J. Biol. Chem. 277 (2002) 19538–19545.
[73] A. Pesce, M. Bolognesi, A. Bocedi, P. Ascenzi, S. Dewilde, L. Moens, T. Hankeln, T. Burmeister, Neuroglobin and cytoglobin. Fresh blood for the vertebrate globin family, EMBO Rep. 3 (2002) 1146–1151.
[74] B. Casado, L.K. Pannell, G. Whalen, D.J. Clauw, J.N. Baraniuk, Human neuroglobin protein in cerebrospinal fluid, Proteome Sci. 3 (2) (2005) 1–8, https://doi.org/10.1186/1477-5956-3-2.
[75] Y. Sun, K. Jin, X.O. Mao, Y. Zhu, D.A. Greenberg, Neuroglobin is up-regulated by and protects neurons from hypoxic-ischemic injury, Proc. Natl. Acad. Sci. U.S.A. 98 (2001) 15306–15311.
[76] L. Moens, S. Dewilde, Globins in the brain, Nature 407 (2000) 461–462.
[77] S. Reuss, S. Saaler-Reinhardt, B. Weich, S. Wystub, M. Reuss, T. Burmeister, T. Hankeln, Expression analysis of neuroglobin mRNA in rodent tissues, Neuroscience 115 (2002) 645–656.
[78] N. Kawada, D.B. Kristensen, K. Asahina, K. Nakatani, Y. Minamiyama, S. Seki, K. Yoshizato, Characterization of a stellate cell activation-associated protein (STAP) with peroxidase activity found in rat hepatic stellate cells, J. Biol. Chem. 276 (2001) 25318–25323.
[79] I.C. Vermeulen Windsant, N.C.J. de Wit, J.T.C. Sertorio, A.A. van Bijnen, Y.M. Ganushchak, J.H. Heijmans, J.E. Tanus-Santos, M.J. Jacobs, J.G. Maessen, W.A. Buurman, Hemolysis during cardiac surgery is associated with increased intravascular nitric oxide consumption and perioperative kidney and intestinal tissue damage, Front. Physiol. 5 (2014), 340, 1–9, https://doi.org/10.3389/fphys.2014.00340.eCollection 2014.
[80] N. Tabibzadeh, C. Estournet, S. Placier, J. Perez, H. Bilbault, A. Girshovich, A. Vandermeersch,

C. Jouanneau, E. Letavenier, N. Hammoudi, F. Lionnet, J.P. Haymann, Plasma heme-induced renal toxicity is related to capillary rarefaction, Sci. Rep. 7 (2017), 40156, 1–8, https://doi.org/10.1038/srep40156.

[81] S. Kumar, U. Bandyopadhyay, Free heme toxicity and its detoxification systems in human, Toxicol. Lett. 157 (2005) 175–188.

[82] J. Flemmig, D. Schlorke, F.-W. Kühne, J. Arnhold, Inhibition of the heme-induced hemolysis of red blood cells by the chlorite-based drug WF10, Free Rad. Res. 50 (2016) 1386–1395.

[83] A.C. Chou, C.D. Fitch, Mechanism of hemolysis induced by FP (IX), J. Clin. Investig. 68 (1981) 672–677.

[84] S.C. Liu, S. Zhai, J. Lawler, J. Palek, Hemin mediated dissociation of erythrocyte skeletal proteins, J. Biol. Chem. 260 (1985) 12234–12239.

[85] G. Balla, H.S. Jacob, J.W. Eaton, J.D. Belcher, G.M. Vercelotti, Hemin: a possible physiological mediator of low density lipoprotein oxidation and endothelial injury, Arterioscler. Thromb. 11 (1991) 1700–1711.

[86] V. Jeney, J. Balla, A. Yachie, Z. Varga, G.M. Vercellotti, J.W. Eaton, G. Balla, Pro-oxidant and cytotoxic effects of circulating heme, Blood 100 (2002) 879–887.

[87] T. Lin, F. Sammy, H. Yang, S. Thundivalappil, J. Hellman, K.C. Tracey, H.S. Warren, Identification of hemopexin as an anti-inflammatory factor that inhibits synergy of hemoglobin with HMGB1 in sterile and infectious inflammation, J. Immunol. 189 (2012) 2017–2022.

[88] J.W. Deuel, C.A. Schaer, F.S. Boretti, L. Opitz, I. Garcia-Rubio, J.H. Baek, D.R. Spahn, P.W. Buehler, D.J. Schaer, Hemoglobinuria-related acute kidney injury is driven by intrarenal oxidative reactions triggering a heme toxicity response, Cell Death Dis. 7 (2016), e2064, 1–12, https://doi.org/10.1038/cddis.2015.392.

[89] R.T. Figueiredo, P.L. Fernandez, D.S. Mourao-Sa, B.N. Porto, F.F. Dutra, L.S. Alves, M.F. Oliviera, P.L. Oliviera, A.V. Graca-Souza, M.T. Bozza, Characterization of heme as activator of toll-like receptor 4, J. Biol. Chem. 282 (2007) 20221–20229.

[90] J.D. Belcher, C. Chen, J. Nguyen, L. Milbauer, F. Abdulla, A.I. Alayash, A. Smith, K.A. Nath, R.P. Hebbel, G.M. Vercelotti, Heme triggers TLR4 signaling leading to endothelial cell activation and vaso-occlusion in murine sickle cell disease, Blood 123 (2014) 377–390.

[91] K. Elbrit, H. Bonkovsky, Heme oxygenase: recent advances in understanding its regulation and role, Proc. Assoc. Am. Phys. 111 (1999) 438–447.

[92] J.P. Evans, F. Niemevz, G. Buldain, P.O. de Montellano, Isoporphyrin intermediate in heme oxygenase catalysis: oxidation of alpha-meso-phenylheme, J. Biol. Chem. 283 (2008) 19530–19539.

[93] S. Iwahara, H. Satoh, D.-X. Song, J. Webb, A.L. Burlingame, Y. Nagae, U. Muller-Eberhard, Purification, characterization and cloning of a heme-binding protein (23 kDa) in rat liver cytosol, Biochemistry 34 (1995) 13398–13406.

[94] A.M. Santoro, M.C. Lo Giudice, A. D'Urso, R. Lauceri, R. Purello, D. Milardi, Cationic porphyrins are reversible proteasome inhibitors, J. Am. Chem. Soc. 134 (2012) 10451–10457.

[95] F. Vallelian, J.W. Deuel, L. Opitz, C.A. Schaer, M. Puglia, M. Lönn, W. Engelsberger, S. Schauer, E. Karnaukhova, D.R. Spahn, R. Stocker, P.W. Buehler, D.J. Schaer, Proteasome inhibition and oxidative reactions disrupt cellular homeostasis during heme stress, Cell Death Differ. 22 (2015) 597–611.

[96] P. Peterson, M.F. Cornacchia, Anemia: pathophysiology, clinical features, and laboratory evaluation, Lab. Med. 30 (1999) 463–467.

[97] V.M. Hodges, S. Rainey, T.R. Lappin, A.P. Maxwell, Pathophysiology of anemia and erythrocytosis, Crit. Rev. Oncol. Hematol. 64 (2007) 139–158.

[98] J. O'Neil, Diagnosis and classifying anemia in adult primary care, Clin. Rev. (2017) 28–35.

[99] D.M. Ridley, F. Dawkins, E. Perlin, Erythropoietin: a review, J. Natl. Med. Assoc. 86 (1994) 129–135.

[100] W. Junge, N. Nelson, ATP synthase, Annu. Rev. Biochem. 84 (2015) 631–657.

[101] P.R. Rich, The molecular machinery of Keilin's respiratory chain, Biochem. Soc. Trans. 31 (2003) 1095–1105.

[102] K. Bartlett, S. Eaton, Mitochondrial beta-oxidation, Eur. J. Biochem. 271 (2004) 462–469.

[103] S.M. Houten, R.J.A. Wanders, A general introduction to the biochemistry of mitochondrial fatty acid β-oxidation, J. Inherit. Metab. Dis. 33 (2010) 469–477.

[104] P. Mitchell, Chemiosmotic coupling in oxidative and photosynthetic phosphorylation, Biol. Rev. Camb. Philos. Soc. 41 (1966) 445–502.

[105] Y. Yaniv, M. Jushazova, H.B. Nuss, S. Wang, D.B. Zorov, E.G. Lakatta, S.J. Sollott, Matching ATP supply and demand in mammalian heart: in vivo, in vitro, and in silico perspectives, Ann. N. Y. Acad. Sci. 1188 (2010) 133–142.

[106] L.D. Zorova, V.A. Popkov, E.Y. Plotnikov, D.N. Silachev, I.B. Pevzner, S.S. Jankauskas, V.A. Babenko, S.D. Zorov, A.V. Balakireva, M. Juhaszova, S.J. Sollott, D.B. Zorov, Mitochondrial membrane potential, Anal. Biochem. 552 (2018) 50–59.

[107] A. Eastman, Deoxyribonuclease II in apoptosis and the significance of intracellular acidification, Cell Death Differ. 1 (1994) 7–9.

[108] S.J. Morana, C.M. Wolf, J. Li, J.E. Reynolds, K.M. Brown, A. Eastman, The involvement of protein phosphatases in the activation of ICE/CED-3 protease, intracellular acidification, DNA digestion, and apoptosis, J. Biol. Chem. 271 (1996) 18263–18271.

[109] M. Khacho, M. Tarabay, D. Patten, P. Khacho, J.G. MacLaurin, J. Guadagno, R. Bergeron, S.P. Cregan, M.-E. Harper, D.S. Park, R.S. Slack, Acidosis overrides oxygen deprivation to maintain mitochondrial function and cell survival, Nat. Commun. 5 (2014), 3550, 1–15, https://doi.org/10.1038/ncomms4550.

[110] J.D. Ly, D.R. Grubb, A. Lawen, The mitochondrial membrane potential ($\Delta\Psi_m$) in apoptosis: an update, Apoptosis 8 (2003) 115–128.

[111] R.L. Goncalves, C.L. Quinlan, I.V. Perevoshchikova, M. Hey-Mogensen, M.D. Brand, Site of superoxide and hydrogen peroxide production by muscle mitochondria assessed ex vivo under conditions mimicking rest and exercise, J. Biol. Chem. 290 (2015) 209–227.

[112] M.D. Brand, Mitochondrial generation of superoxide and hydrogen peroxide as the source of mitochondrial redox signaling, Free Rad. Biol. Med. 100 (2016) 14–31.

[113] K.R. Pryde, J. Hirst, Superoxide is produced by the reduced flavin in mitochondrial complex I: a single, unified mechanism that applies during both forward and reverse electron transfer, J. Biol. Chem. 286 (2011) 18056–18065.

[114] E.L. Robb, A.R. Hall, T.A. Prime, S. Eaton, M. Szibor, C. Viscomi, A.M. James, M.P. Murphy, Control of mitochondrial superoxide production by reverse electron transport at complex I, J. Biol. Chem. 293 (2018) 9869–9879.

[115] J.F. Turrens, A. Alexandre, A.L. Lehninger, Ubisemiquinone is the electron donor for superoxide formation by complex III of heart mitochondria, Arch. Biochem. Biophys. 237 (1985) 408–414.

[116] R.G. Hansford, B.A. Hogue, V. Mildaziene, Dependence of H_2O_2 formation by rat heart mitochondria on substrate availability and donor age, J. Bioenerg. Biomembr. 29 (1997) 89–95.

[117] M.D. Brand, The sites and topology of mitochondrial superoxide production, Exp. Gerontol. 45 (2010) 466–472.

[118] J.F. Turrens, Mitochondrial formation of reactive oxygen species, J. Physiol. 552 (2003) 335–344.

[119] A.J. Swallow, Physical chemistry of quinones, in: P.L. Trumpower (Ed.), Function of Quinones in Energy Conserving Systems, Academic Press, New York, 1982, pp. 59–72.

[120] Y.A. Ilan, G. Czapski, D. Meisel, The one-electron transfer redox potentials of free radicals. I. The oxygen/superoxide system, Biochim. Biophys. Acta 430 (1976) 209–224.

[121] O.V. Demin, B.N. Khodolenko, V.P. Skulachev, A model of $O_2^{\bullet -}$ generation in the complex III of the electron transport chain, Mol. Cell. Biochem. 184 (1998) 21–33.

[122] P.R. Gardner, Superoxide-driven aconitase Fe-S cycling, Biosci. Rep. 17 (1997) 33–42.

[123] P.R. Gardner, Aconitase: sensitive target and measure of superoxide, Methods Enzymol. 349 (2002) 9–23.

[124] D.H. Flint, J.F. Tominello, M.H. Emptage, The investigation of Fe-S cluster containing hydro-lyases by superoxide, J. Biol. Chem. 268 (1993) 22369–22376.

[125] A. Hausladen, I. Fridovich, Superoxide and peroxynitrite inactivate aconitases, but nitric oxide does not, J. Biol. Chem. 269 (1994) 29405–29408.

[126] R.S. Eisenstein, K.P. Blemings, Iron regulatory proteins, iron responsive elements and iron homeostasis, J. Nutr. 128 (1998) 2295–2298.

[127] P. Ponka, C. Beaumont, D.R. Richardson, Function and regulation of transferrin and ferritin, Semin. Hematol. 35 (1998) 35–54.

[128] Z. Lacza, E. Pankotai, D.W. Busija, Mitochondrial nitric oxide synthase: current concepts and controversies, Front. Biosci. 14 (2009) 4436–4443.

[129] Z. Lacza, M. Puskar, J.P. Figueroa, J. Zhang, N. Rajapakse, D.W. Busija, Mitochondrial nitric oxide synthase is constitutively active and is functionally upregulated in hypoxia, Free Radic. Biol. Med. 31 (2001) 1609–1615.

[130] Z. Lacza, J.A. Snipes, J. Zhang, E.M. Horváth, J.P. Figueroa, C. Szabó, D.W. Busija, Mitochondrial nitric oxide synthase is not eNOS, nNOS or iNOS, Free Radic. Biol. Med. 35 (2003) 1217–1228.

[131] I. Fridovich, Superoxide radical and superoxide dismutases, Annu. Rev. Biochem. 64 (1995) 97–112.

[132] A. Okado-Matsumoto, I. Fridovich, Subcellular distribution of superoxide dismutases (SOD) in rat liver: Cu, Zn-SOD in mitochondria, J. Biol. Chem. 276 (2001) 38388–38393.

[133] Y. Li, T.T. Huang, E.J. Carlson, S. Melov, P.C. Ursell, J.L. Olson, L.J. Noble, M.P. Yoshimura, C. Berger, P.H. Chan, D.C. Wallace, C.J. Epstein, Dilated cardiomyopathy and neonatal lethality in mutant mice lacking manganese superoxide dismutase, Nat. Genet. 11 (1995) 376–381.

[134] R.M. Lebovitz, H. Zhang, H. Vogel, J. Cartwright Jr., L. Dionne, N. Lu, S. Huang, M.M. Matzuk, Neurodegeneration, myocardial injury, and peripheral death

in mitochondrial superoxide dismutase-deficient mice, Proc. Natl. Acad. Sci. U.S.A. 93 (1996) 9782–9787.

[135] Y.S. Ho, M. Gargano, J. Cao, R.T. Bronson, I. Heimler, R.J. Hutz, Reduced fertility in female mice lacking copper-zinc superoxide dismutase, J. Biol. Chem. 273 (1998) 7765–7769.

[136] B.H.J. Bielski, D.E. Cabelli, R.L. Arudi, Reactivity of $HO_2/O_2^{\bullet-}$ radicals in aqueous solution, J. Phys. Chem. Ref. Data 14 (1985) 1041–1100.

[137] A.A. Starkov, The role of mitochondria in reactive oxygen species metabolism and signaling, Ann. N. Y. Acad. Sci. 1147 (2008) 37–52.

[138] P. Pasdois, J.E. Parker, E.J. Griffiths, A.P. Halestrap, The role of oxidized cytochrome c in regulating mitochondrial reactive species production and its perturbation in ischemia, Biochem. J. 436 (2011) 493–505.

[139] J. Butler, G.G. Jayson, A.J. Swallow, The reaction between the superoxide anion radical and cytochrome c, Biochim. Biophys. Acta 408 (1975) 215–222.

[140] J. Butler, W.H. Koppenol, E. Margoliash, Kinetics and mechanism of the reduction of ferricytochrome c by superoxide anion, J. Biol. Chem. 257 (1982) 10747–10750.

[141] L. Bao, M.V. Avshalumov, J.C. Patel, C.R. Lee, E.W. Miller, C.J. Chang, M.E. Rice, Mitochondria are the source of hydrogen peroxide for dynamic brain-cell signaling, J. Neurosci. 29 (2009) 9002–9010.

[142] M.E. Rice, H_2O_2: a dynamic neuromodulator, Neuroscientist 17 (2011) 389–406.

[143] S.G. Rhee, H_2O_2, a necessary evil for cell signaling, Science 312 (2006) 1882–1883.

[144] E.A. Veal, A.M. Day, B.A. Morgan, Hydrogen peroxide sensing and signaling, Mol. Cell 26 (2007) 1–14.

[145] E.W. Miller, O. Tulyathan, E.Y. Isacoff, C.J. Chang, Molecular imaging of hydrogen peroxide produced for cell signaling, Nat. Chem. Biol. 3 (2007) 263–267.

[146] V. Bernard, J.P. Bolam, Subcellular and synaptic distribution of the NR1 subunit of the NMDA receptor in the neostriatum and globus pallidus of the rat: colocalization at synapses with the GluR2/3 subunit of the AMPA receptor, Eur. J. Neurosci. 10 (1998) 3721–3728.

[147] E. Wehage, J. Eisfeld, I. Heiner, E. Jüngling, C. Zitt, A. Lückhoff, Activation of the cation channel long transient receptor potential channel 2 (LTRPC2) by hydrogen peroxide. A splice variant reveals a mode of activation independent of ADP-ribose, J. Biol. Chem. 277 (2002) 23150–23156.

[148] A.L. Perraud, C.L. Takanishi, B. Shen, S. Kang, M.K. Smith, C. Schmitz, H.M. Knowles, D. Ferraris, W. Li, J. Zhang, B.L. Stoddard, A.M. Scharenberg, Accumulation of free ADP-ribose from mitochondria mediates oxidative-stress-induced gating of TRPM2 cation channels, J. Biol. Chem. 280 (2005) 6138–6148.

[149] C.M. Hecquet, A.B. Malik, Role of H_2O_2-activated TRPM2 calcium channel in oxidant-induced endothelial injury, Thromb. Haemostasis 101 (2009) 619–625.

[150] C.M. Atkins, J.D. Sweatt, Reactive oxygen species mediate activity-dependent neuron-glia signaling in output fibers of the hippocampus, J. Neurosci. 19 (1999) 7241–7248.

[151] A. Kamsler, A. Avital, V. Greenberger, M. Segal, Aged SOD overexpressing mice exhibit enhanced spatial memory while lacking hippocampus neurogenesis, Antioxidants Redox Signal. 9 (2007) 181–189.

[152] J.T. Greenamyre, T.B. Sherer, R. Betarbet, A.V. Panov, Complex I and Parkinson's disease, IUMBM Life 52 (2001) 135–141.

[153] A. Carreau, B. El Hafny-Rahbi, A. Matejuk, C. Grillon, C. Kieda, Why is the partial oxygen pressure of human tissues a crucial parameter? Small molecules and hypoxia, J. Cell Mol. Med. 15 (2011) 1239–1253.

[154] S.R. McKeown, Defined normoxia, physoxia and hypoxia in tumours – implications for treatment response, Br. J. Radiol. 87 (2014), 20130676, 1–12, https://doi.org/10.1259/brj.20130676.

[155] S.R. Wagner, W.L. Lanier, Metabolism of glucose, glycogen, and high-energy phosphates during complete cerebral ischemia: a comparison of normoglycemic, chronically hyperglycemic diabetic, and acutely hyperglycemic nondiabetic rats, Anesthesiology 81 (1994) 1516–1526.

[156] P. Ebbesen, E.O. Pettersen, J. Denekamp, B. Littbrand, J. Keski-Oja, A. Schousboe, U. Sonnewald, Ø. Åmellam, V. Zachar, Hypoxia, normoxia and hyperoxia, Acta Oncol. 39 (2000) 247–248.

[157] N. Bitterman, CNS oxygen toxicity, Undersea Hyperb. Med. 31 (2004) 63–72.

[158] P. Vaupel, A. Mayer, M. Höckel, Impact of hemoglobin levels on tumor oxygenation: the higher, the better? Strahlenther. Onkol. 182 (2006) 63–71.

[159] Q. Ke, M. Costa, Hypoxia-inducible factor-1 (HIF-1), Mol. Pharmacol. 70 (2006) 1469–1480.

[160] J.E. Ziello, I.S. Jovin, Y. Huang, Hypoxia-inducible factor (HIF)-1 regulatory pathway and its potential for therapeutic intervention in malignancy and ischemia, Yale J. Biol. Med. 80 (2007) 51–60.

[161] T.G. Smith, P.A. Robbins, P.J. Ratcliffe, The human side of hypoxia-inducible factor, Br. J. Haematol. 141 (2008) 325–334.

[162] T. Löfstedt, E. Fredlund, L. Holmquist-Mengelbier, A. Pietras, M. Overnberger, L. Pollinger, S. Påhlman, Hypoxia inducible factor-2α in cancer, Cell Cycle 6 (2007) 919–926.

[163] P.H. Maxwell, M.S. Wiesener, G.W. Chang, S.C. Clifford, E.C. Vaux, M.E. Cockman, C.C. Wykoff, C.W. Pugh, E.R. Maher, P.J. Ratcliffe, The tumor suppressor protein VHL targets hypoxia-inducible factors for oxygen-dependent proteolysis, Nature 399 (1999) 271–275.

[164] G.L. Semenza, Hydroxylation of HIF-1: oxygen sensing at the molecular level, Physiology 19 (2004) 176–182.

[165] R.K. Bruick, S.L. McKnight, A conserved family of prolyl-4-hydroxylases that modify HIF, Science 294 (2001) 1337–1340.

[166] A. Grano, M.C. de Tullio, Ascorbic acid as a sensor of oxidative stress and a regulator of gene expression: the ying and yang of vitamin C, Med, Hypotheses 69 (2007) 953–954.

[167] C.R. Hagenbrock, B. Chazotte, S.S. Gupte, The random collision model and a critical assessment of diffusion and collision in mitochondrial electron transport, J. Bioenerg. Biomembr. 18 (1986) 331–368.

[168] A.Y. Andreyev, Y.E. Kushnareva, A.A. Starkov, Mitochondrial metabolism of reactive oxygen species, Biochemistry (Mosc.) 70 (2005) 200–214.

[169] R.B. Jennings, K.A. Reimer, The cell biology of acute myocardial ischemia, Annu. Rev. Med. 42 (1991) 225–246.

[170] C. Xing, K. Arai, E.H. Lo, M. Hommel, Pathophysiologic cascades in ischemic stroke, Int. J. Stroke 7 (2012) 378–385.

[171] D.E. Farthing, C.A. Farthing, L. Xi, Inosine and hypoxanthine as novel biomarkers for cardiac ischemia: from bench to point-of-care, Exp. Biol. Med. 240 (2015) 821–831.

[172] H.V. Sparks, Mechanism of vasodilation during and after ischemic exercise, Fed. Proc. 39 (1980) 1487–1490.

[173] O. Miyamoto, R.N. Auer, Hypoxia, hyperoxia, ischemia, and brain necrosis, Neurology 54 (2000) 362–371.

[174] T. Kalogeris, C.P. Baines, M. Krenz, R.J. Korthuis, Cell biology of ischemia/reperfusion injury, Int. Rev. Cell Mol. Biol. 298 (2012) 299–317.

[175] M.-Y. Wu, G.-T. Yiang, W.-T. Liao, A.P.-Y. Tsai, Y.-L. Cheng, P.-W. Cheng, C.-Y. Li, C.-J. Li, Current mechanistic concepts in ischemia and reperfusion injury, Cell. Physiol. Biochem. 46 (2018) 1650–1657.

[176] M.G. Battelli, S. Musiani, M. Valgimigli, L. Gramantieri, F. Tomassoni, L. Bolondi, F. Stirpe, Serum xanthine oxidase in human liver disease, Am. J. Gastroenterol. 96 (2001) 1194–1199.

[177] R. Sander, Compilation of Henry's law constants (version 4.0) for water as solvent, Atmos. Chem. Phys. 15 (2015) 4399–4981.

[178] P.E. Cryer, L. Axelrod, A.B. Grossman, S.R. Heller, V.M. Montori, E.R. Seaquist, F.J. Service, Evaluation and management of adult hypoglycemic disorder: an endocrine society clinical practice guideline, J. Clin. Enodcrinol. Metab. 94 (2009) 709–728.

[179] J. Morales, D. Schneider, Hypoglycemia, Am. J. Med. 127 (2014) S17–S24.

[180] D.M. Nathan, R.R. Holman, J.B. Buse, R. Sherwin, M.B. Davidson, B. Zinman, E. Ferrannini, Medical management of hyperglycemia in type 2 diabetes: a consensus algorithm for the initiation and adjustment of therapy, Diabetes Care 32 (2009) 193–203.

[181] D. Brealey, M. Singer, Hyperglycemia in critical illness: a review, J. Diab. Sci. Technol. 3 (2009) 1250–1260.

[182] F.G. Njoroge, V.M. Monnier, The chemistry of the Maillard reaction under physiological conditions: a review, Prog. Clin. Biol. Res. 304 (1989) 85–91.

[183] A.L. Olson, J.E. Pessin, Structure, function, and regulation of the mammalian facilitative glucose transporter gene family, Annu. Rev. Nutr. 16 (1996) 235–256.

[184] D. Miller, R.K. Crane, The digestive function of the epithelium of the small intestine: II. Localization of disaccharide hydrolysis in the isolated brush border portion of intestinal epithelial cells, Biochim. Biophys. Acta 52 (1961) 293–298.

[185] M.A. Hediger, D.B. Rhoads, Molecular physiology of sodium-glucose cotransporters, Physiol. Rev. 74 (1994) 993–1026.

[186] H.F. Jones, R.N. Butler, D.A. Brooks, Intestinal fructose transport and malabsorption in humans, Am. J. Physiol. Gastrointest. Liver Physiol. 300 (2011) G202–G206.

[187] C.M.F. Kneepkens, R.J. Vonk, J. Fernandes, Incomplete intestinal absorption of fructose, Arch. Dis. Child. 59 (1984) 735–738.

[188] A.S. Truswell, J.M. Seach, A.W. Thorburn, Incomplete absorption of pure fructose in healthy subjects and the facilitating effect of glucose, Am. J. Clin. Nutr. 48 (1988) 1424–1430.

[189] B. Thorens, GLUT2, glucose sensing and glucose homeostasis, Diabetologia 58 (2015) 221–232.

[190] M.M. Adeva-Andany, N. Pérez-Felpete, C. Fernández-Fernández, C. Donapetry-Garcia, C. Pazos-Garcia, Liver glucose metabolism in humans, Biosci, For. Rep. 29 (2016), 00416, 1–15, https://doi.org/10.1042/BSR20160385.

[191] M. Bollen, S. Keppens, W. Stalmans, Specific features of glycogen metabolism in the liver, Biochem. J. 336 (1998) 19–31.

[192] I. Miwa, S. Suzuki, An improved quantitative assay of glycogen in erythrocytes, Ann. Clin. Biochem. 39 (2002) 612–613.
[193] A.M. Brown, B.R. Ransom, Astrocyte glycogen and brain energy metabolism, Glia 55 (2007) 1263–1271.
[194] A. Mitrakou, Kidney: its impact on glucose homeostasis and hormonal regulation, Diabetes Res. Clin. Pract. 98 (2011) S66–S72.
[195] J. Jensen, P.I. Rustad, A.J. Kolnes, Y.-C. Lai, The role of skeletal muscle glycogen breakdown for regulation of insulin sensitivity by exercise, Front. Physiol. 2 (2011), 112, 1–11, https://doi.org/10.3389/fphys.2011.00112.
[196] A. Kolderup, B. Svihus, Fructose metabolism and relation to atherosclerosis, type 2 diabetes, and obesity, J. Nutr. Metab. (2015), 823081, 1–12, https://doi.org/10.1155/2015/823081.
[197] F.Q. Nutfall, A. Ngo, M.C. Cannon, Regulation of hepatic glucose production and the role of gluconeogenesis in humans: is the rate of gluconeogenesis constant? Diab. Metab. Res. Rev. 24 (2008) 438–458.
[198] F. Heinz, W. Lamprecht, J. Kirsch, Enzymes of fructose metabolism in human liver, J. Clin. Investig. 47 (1968) 1826–1832.
[199] J.S. Lim, M. Mietus-Snyder, A. Valente, J.M. Schwarz, R.H. Lustig, The role of fructose in the pathogenesis of NAFLD and the metabolic syndrome, Nat. Rev. Gastroenterol. Hepatol. 7 (2010) 251–264.
[200] S.S. Elliott, N.L. Keim, J.S. Stern, K. Teff, P.J. Havel, Fructose, weight gain, and the insulin resistance syndrome, Am. J. Clin. Nutr. 76 (2002) 911–922.
[201] K.L. Teff, S.S. Elliott, M. Tschöp, T.J. Kieffer, D. Rader, M. Heiman, R.R. Townsend, N.L. Keim, D. D'Alessio, P.J. Havel, Dietary fructose reduces circulating insulin and leptin, attenuates postprandial suppression of ghrelin, and increases triglycerides in women, J. Clin. Endocrinol. Metab. 89 (2004) 1907–1913.
[202] P.B. Iynedijan, Mammalian glucokinase and its gene, Biochem. J. 293 (1993) 1–13.
[203] L. Tappy, K.-A. Lê, Does fructose consumption contribute to non-alcoholic fatty liver disease? Clin. Res. Hepatol. Gastroenterol. 36 (2012) 554–560.
[204] L. Tappy, K.-A. Lê, Metabolic effects of fructose and the worldwide increase in obesity, Physiol. Rev. 90 (2010) 23–46.
[205] P.H. Maenpaa, K.O. Raivio, M.P. Kekomaki, Liver adenine nucleotides: fructose-induced depletion and its effects on protein synthesis, Science 161 (1968) 1253–1254.
[206] B.T. Emmerson, Effect of oral fructose on urate production, Ann. Rheum. Dis. 33 (1974) 276–280.
[207] G. van den Berghe, Fructose: metabolism and short-term effects on carbohydrate and purine metabolic pathways, Prog. Biochem. Pharmacol. 21 (1986) 1–32.
[208] M.A. Lanaspa, L.G. Sanchez-Lozada, Y.J. Choi, C. Cicerchi, M. Kanbay, C.A. Roncal-Jiminez, T. Ishimoto, N. Li, G. Marek, M. Duranay, G. Schreiner, B. Rodriguez-Iturbe, T. Nakagawa, D.-H. Kang, Y.Y. Sautin, R.J. Johnson, Uric acid induces hepatic steatosis by generation of mitochondrial oxidative stress: potential role in fructose-dependent and -independent fatty liver, J. Biol. Chem. 287 (2012) 40732–40744.
[209] T. Jensen, M.F. Abdelmalek, S. Sullivan, K.J. Nadeau, M. Green, C. Roncal, T. Nakagawa, M. Kuwabara, Y. Sato, D.-H. Kang, D.R. Tolan, L.G. Sanchez-Lozada, H.R. Rosen, M.A. Lanaspa, A.M. Diehl, R.J. Johnson, Fructose and sugar: a major mediator of non-alcoholic liver disease, J. Hepatol. 68 (2018) 1063–1075.
[210] E. Aigner, I. Theurl, H. Haufe, M. Seifert, F. Hohla, L. Scharinger, F. Stickel, F. Mourlane, G. Weiss, C. Datz, Copper availability contributes to iron perturbations in human nonalcoholic fatty liver disease, Gastroenterology 135 (2008) 680–688.
[211] M. Song, D.A. Schuschke, Z. Zhou, T. Chen, W.M. Pierce Jr., R. Wang, W.T. Johnson, C.J. McClain, High fructose feeding induces copper deficiency in Sprague-Dawley rats: a novel mechanism for obesity related fatty liver, J. Hepatol. 56 (2012) 433–440.
[212] Y. Yilmaz, Review article: fructose in non-alcoholic fatty liver disease, Aliment. Pharmacol. Ther. 35 (2012) 1135–1142.
[213] M.B. Vos, J.E. Lavine, Dietary fructose in nonalcoholic fatty liver disease, Hepatology 57 (2013) 2525–2531.
[214] S. Softic, D.E. Cohen, C.R. Kahn, Role of dietary fructose and hepatic de novo lipogenesis in fatty liver disease, Dig. Dis. Sci. 61 (2016) 1282–1293.
[215] S.S. Chirala, S.J. Wakil, Structure and function of animal fatty acid synthase, Lipids 39 (2004) 1045–1053.
[216] S. Softic, M.K. Gupta, G.X. Wang, S. Fujisaka, B.T. O'Neill, T.N. Rao, J. Willoughby, C. Harbison, K. Fitzgerald, O. Ilkeyeva, C.B. Newgard, D.E. Cohen, C.R. Kahn, Divergent effects of glucose and fructose on hepatic lipogenesis and insulin signaling, J. Clin. Investig. 127 (2017) 4059–4074.
[217] T. Ishimoto, M.A. Lanaspa, M.T. Le, G.E. Garcia, C.P. Diggle, P.S. Maclean, M.R. Jackman, A. Asipu, C.A. Roncal-Jiminez, T. Kosugi, C.J. Rivard, S. Maruyama, B. Rodriguez-Iturbe, L.G. Sánchez-Lozada, D.T. Bonthron, Y.Y. Sautin, R.J. Johnson, Opposing effects of fructokinase C and A isoforms on fructose-induced metabolic syndrome in mice, Proc. Natl. Acad. Sci. U.S.A. 109 (2012) 4320–4325.

[218] B. Thorens, M. Mueckler, Glucose transporters in the 21st century, Am. J. Physiol. Endocrinol. Metab. 298 (2010) E141−E145.
[219] J.S. Bogan, Regulation of glucose transporter translocation in health and diabetes, Annu. Rev. Biochem. 81 (2012) 507−532.
[220] S. Huang, M.P. Czech, The GLUT4 glucose transporter, Cell Metabol. 5 (2007) 237−252.
[221] P.D. Brewer, E.N. Habtemichael, I. Romenskaia, C.C. Mastick, A.C.F. Coster, Insulin-regulated Glut4 translocation. Membrane protein trafficking with six distinctive steps, J. Biol. Chem. 289 (2014) 17280−17298.
[222] E.D. Rosen, B.M. Spiegelman, Adipocytes as regulators of energy balance and glucose homeostasis, Nature 444 (2006) 847−853.
[223] J. Boucher, A. Kelinridders, R. Kahn, Insulin receptor signaling in normal and insulin-resistant states, Cold Spring Harb. Perspect. Biol. 6 (2014), a009191, 1−23, https://doi.org/10.1101/cshperspect.a009191.
[224] K. Duvel, J.L. Yecies, S. Menon, P. Raman, A.I. Lipovsky, A.L. Souza, E. Triantafellow, Q. Ma, R. Gorski, S. Cleaver, M.G. van der Heiden, J.P. MacKeigan, P.M. Finan, C.B. Clish, L.O. Murphy, B.D. Manning, Activation of a metabolic gene network downstream of mTOR complex 1, Mol. Cell 39 (2010) 171−183.
[225] M.E. Cerf, Beta cell dysfunction and insulin resistance, Front. Endocrinol. 4 (37) (2013) 1−12, https://doi.org/10.3389/fendo.2013.00037.
[226] S.R. Stannard, N.A. Johnson, Insulin resistance and elevated triglyceride in muscle: more important for survival than 'thrifty' genes? J. Physiol. 554 (2004) 595−607.
[227] G. Wang, Raison d'être of insulin resistance: the adjustable threshold hypothesis, J. R. Soc. Interface 11 (2014), 0892, 1−11, https://doi.org/10.1098/rsif.2014.0892.
[228] D.M. Muoio, C.B. Newgard, Molecular and metabolic mechanisms of insulin resistance and β-cell failure in type 2 diabetes, Nat. Rev. Mol. Cell Biol. 9 (2008) 193−205.
[229] V.T. Samuel, G.I. Shulman, Mechanisms for insulin resistance: common threads and missing links, Cell 148 (2012) 852−871.
[230] D.S. King, G.P. Dalsky, W.E. Clutter, D.A. Young, M.A. Staten, P.E. Cryer, J.O. Holloszy, Effects of exercise and lack of exercise on insulin sensitivity and responsiveness, J. Appl. Physiol. 64 (1988) 1942−1946.
[231] R.S. Biensø, S. Ringholm, K. Kiilerich, N.-J. Aachmann-Andersen, R. Krogh-Madsen, B. Guerra, P. Plomgaard, G. van Hall, J.T. Treebak, B. Saltin, C. Lundby, J.A.L. Calbet, H. Pilegaard, J.F.P. Wojtaszewski, GLUT4 and glycogen synthase are key players in bed rest-induced insulin resistance, Diabetes 61 (2012) 1090−1099.
[232] M. Mueckler, C. Caruso, S.A. Baldwin, M. Panico, I. Blench, H.R. Morris, W.J. ASllard, G.E. Lienhard, H.F. Lodish, Sequence and structure of human glucose transporter, Science 229 (1985) 941−945.
[233] F. Maher, S.J. Vannucci, I.A. Simpson, Glucose transport proteins in brain, FASEB J. 8 (1994) 1003−1011.
[234] S.J. Vannucci, F. Maher, I.A. Simpson, Glucose transporter proteins in brain: delivery of glucose to neurons and glia, Glia 21 (1997) 2−21.
[235] L.D. Hayward, S.J. Angyal, A symmetry rule for the circular dichroism of reducing sugars, and the proportion of carbonyl forms in aqueous solutions thereof, Carbohydr. Res. 53 (1977) 13−20.
[236] J.P. Dworkin, S.L. Miller, A kinetic estimate of the free aldehyde content of aldoses, Carbohydr. Res. 329 (2000) 359−365.
[237] R.J. Koenig, C.M. Peterson, R.L. Jones, C. Saudek, M. Lehrmann, A. Cerami, Correlation of glucose regulation and hemoglobin A1c in diabetes mellitus, N. Engl. J. Med. 295 (1976) 417−420.
[238] H.F. Bunn, K.H. Gabbay, P.M. Gallop, The glycosylation of hemoglobin: relevance to diabetes mellitus, Science 200 (1978) 21−27.
[239] N. Shaklai, R.L. Garlick, H.F. Bunn, Nonenzymatic glycosylation of human serum albumin alters its confirmation and function, J. Biol. Chem. 259 (1984) 3812−3817.
[240] E. Bourdon, N. Loreau, D. Blache, Glucose and free radicals impair the antioxidant properties of serum albumin, FASEB J. 13 (1999) 233−244.
[241] C. Wa, R.L. Cerny, W.A. Clarke, D.S. Hage, Characterization of glycation adducts on human serum albumin by matrix-assisted laser desorption/ionization time-of-flight mass spectrometry, Clin. Chim. Acta 385 (2007) 48−60.
[242] O.S. Barnaby, R.L. Cerny, W. Clarke, D.S. Hage, Comparison of modification sites formed on human serum albumin at various stages of glycation, Clin. Chim. Acta 412 (2011) 277−285.
[243] P.S. Lim, Y.M. Cheng, S.M. Yang, Impairments of the biological properties of serum albumin in patients on haemodialysis, Nephrology 12 (2007) 18−24.
[244] Y. Wang, H. Yu, X. Shi, Z. Luo, D. Lin, M. Huang, Structural mechanism of ring-opening reaction of glucose by human serum albumin, J. Biol. Chem. 288 (2013) 15980−15987.
[245] S.Y. Goh, M.E. Cooper, Clinical review: the role of advanced glycation end products in progression and complications of diabetes, J. Clin. Endocrinol. Metab. 93 (2008) 1143−1152.

[246] P. Gkogkolou, M. Böhm, Advanced glycation end products. Key players in skin aging? Derm. Endocrinol. 4 (2012) 259–270.

[247] C. Ott, K. Jacobs, E. Haucke, A.N. Santos, T. Grune, A. Simm, Role of advanced glycation end products in cellular signaling, Redox Biol 2 (2014) 411–429.

[248] A. Gugliucci, Glycation as the glucose link to diabetic complications, JAOA 100 (2000) 621–634.

[249] M. Brownlee, The pathophysiology of diabetic complications. A unifying mechanism, Diabetes 54 (2005) 1615–1625.

[250] D. Chandra, E.B. Jackson, K.V. Ramana, R. Kelley, S.K. Srivastava, A. Bhatnagar, Nitric oxide prevents aldose reductase activation and sorbitol accumulation during diabetes, Diabetes 51 (2002) 3095–3101.

[251] G.M. Pieper, Review of alterations in endothelial nitric oxide production in diabetes, Hypertension 31 (1998) 1047–1060.

[252] U. Hink, H. Li, H. Mollnau, M. Oelze, E. Matheis, M. Hartmann, M. Skatchov, F. Thaiss, R.A. Stahl, A. Warnholtz, T. Meinertz, K. Grinedling, D.G. Harrison, U. Forstermann, T. Munzel, Mechanisms underlying endothelial dysfunction in diabetes mellitus, Circ. Res. 88 (2001) E14–E22.

CHAPTER

5

Mechanisms of Cell Death

5.1 OVERVIEW

In higher animals, the permanent formation of novel unperturbed cells and the removal of unwanted or deteriorated cells ensure the long-term stability and functioning of the organism. Cell death is an important feature of life. During the last decades, investigations about the two basic cell death mechanisms, apoptosis and necrosis, were in the focus of the scientific community. Cell death by apoptosis, a process that is strictly controlled by the genetic program, leads to a safe removal of affected cells without damaging surrounding cells and tissues. Contrary, the nonphysiological necrotic cell death can cause substantial deteriorations in their neighborhood.

Cell death is, however, more complex than originally thought. Novel mechanisms how a cell can die have been described for the last years. These processes considerably enlarge our knowledge about tissue homeostasis under pathological conditions, where the pure description of cell death by apoptosis or uncontrolled necrosis is insufficient. Among these additional mechanisms of programmed cell death are pyroptosis, ferroptosis, necroptosis, eryptosis, cell death in neutrophils and other immune cells, and some other variations.

In this chapter, main attention will be directed on basic mechanisms of detection and elimination of unwanted, damaged, or falsely programmed host's cells. Of course, the programmed cell death by apoptosis will play a substantial role. There are two main routes of apoptosis induction, once via defective mitochondria by the so-called intrinsic pathway or via signaling of death receptors that represents the extrinsic pathway of apoptosis induction. Other aforementioned mechanisms of programmed cell death and the totally uncontrolled necrosis will also be highlighted. Relationship between apoptosis and mitosis will be discussed.

The removal of damaged cell organelles and injured proteins as well as the accumulation of waste products is additionally addressed in this chapter as these processes are tightly linked to cell death.

A thorough knowledge of cell death mechanisms is highly necessary for a better understanding of the involvement of these processes in pathogenesis of different diseases.

5.2 APOPTOSIS: MITOCHONDRIAL PATHWAY

5.2.1 Sensing of Mitochondria

The intrinsic pathway of apoptosis induction starts with alterations in mitochondria. As main suppliers of chemical energy equivalents in form of adenosine triphosphate (ATP), these organelles reduce dioxygen to water. The correct functionality of mitochondria highly depends on

Cell and Tissue Destruction
https://doi.org/10.1016/B978-0-12-816388-7.00005-X

sufficient supply of energy substrates. Under stress situations, mitochondria act as primary sensor in cells detecting and responding to altered conditions.

5.2.2 Release of Mitochondrial Proteins

In intact mitochondria, the heme protein cytochrome *c* is mainly localized in the intermembrane space between the outer and inner mitochondrial membranes [1]. It forms a complex with cardiolipin, a unique phospholipid in the inner membrane of mitochondria [2]. On the onset of apoptosis, enhanced formation of reactive species in mitochondria promotes cardiolipin oxidation and detachment of cytochrome *c* from cardiolipin [3]. The release of cytochrome *c* from disturbed mitochondria is regulated by the mitochondrial apoptosis–induced channel, an ion channel formed in the outer mitochondrial membrane. The formation and opening of this channel is controlled by members of the B-cell lymphoma 2 (Bcl-2) protein family [4]. This protein family includes proapoptotic factors such as Bax and Bak and antiapoptotic components such as Bcl-2 and Bcl-xL [5].

Small amounts of cytochrome *c* released from mitochondria interact with the inositol trisphosphate receptor on endoplasmic reticulum (ER) and promote a calcium release. Calcium and cytochrome *c* form a positive feedback loop that drastically enhances further release of these components from their stores [6].

Damaged mitochondria release cytochrome *c* and some other proteins into the cytosol. While cytochrome *c* release initiates apoptosome formation and is, thus, a key factor for caspase activation, the second mitochondria-derived activator of caspase (Smac) binds and inactivates inhibitors of apoptosis (IAP) induction. IAP binding to caspases protects cells from degradation. The concerted action of cytochrome *c* and Smac release is apparently a protection against incidental release of cytochrome *c* from mitochondria. Cytochrome *c* is a smaller protein than Smac. If mitochondria are severely damaged, both proteins are released [7].

Other proteins released from damaged mitochondria are apoptosis-inducing factor (AIF) and endonuclease G (endoG). These proteins act independently of caspase activation. AIF translocates to the nucleus and causes chromatin condensation and large-scale DNA fragmentation [8]. EndoG is also involved in nucleosomal DNA fragmentation [9].

5.2.3 Caspase Activation

Caspases are a family of proteolytic enzymes responsible for the degradation of cellular components during programmed cell death. Caspase 1 is an important mediator of innate immune response (see Section 6.2.4). Caspase stands for cysteine-dependent aspartate-directed protease. In living cells, these proteins exist in a nonactive form.

On initiation of the suicidal cell death program, monomeric procaspases are activated by dimerization and oligomerization, often followed by cleavage into a small and large subunit and formation of active heterodimers [10]. In apoptosis, caspases are divided into initiator caspases and executioner caspases [11]. The latter enzymes are responsible for the controlled cleavage of cellular components in apoptotic cells avoiding inflammation and damage to surrounding cells and tissues.

Initiator caspases transmit signals from altered mitochondria or clustered death receptors to executioner caspases. In the mitochondrial or intrinsic pathway, the signal for initiator caspase activation is the release of cytochrome *c* into the cytosol [12]. Cytochrome *c* binds an adaptor protein, the apoptotic protease activating factor 1 (APAF1). This binding facilitates the formation of an apoptosome, a wheel-like particle where seven activated APAF1 monomers, cytochrome *c*, and deoxyadenosine triphosphate are complexed together, that allows the binding of procaspase 9 [13,14]. The apoptosome generates active caspase 9, the initiator caspase of the mitochondrial pathway [15].

Active caspase 9 further mobilizes executioner caspases such as caspase 3, caspase 6, and caspase 7 [16].

5.2.4 Degradation of Cellular Components

During apoptosis executioner caspases are responsible for characteristic alterations in cell morphology as they degrade numerous macromolecules. The main executioner caspase is caspase 3 that is able to cleave different components of the cytoskeleton and the nucleus [17].

Caspases 6 and 7 play apparently a more specialized role during the execution phase of apoptosis. Caspase 6 is responsible for degradation of lamin A, whereas caspase 7 is involved in poly (ADP-ribose) polymerase degradation [18].

Caspase-induced cytoskeleton alterations include actin reorganization, disruption of link proteins for intermediate filaments, and depolymerization of microtubules [19].

Actin alterations start with the caspase-mediated cleavage of focal adhesion kinase pp125FAK and structural proteins that link actin to focal adhesions [20]. This contributes to reorganization of actin into a peripheral ring and cell rounding. Membrane blebbing results from increased contractility of actin–myosin fibers followed by myosin light chain phosphorylation. Bleb formation is an important requirement for subsequent phagocytosis by macrophages [19].

Caspases cleave different intermediate filaments such as keratins, vimentin, and lamins as well as cross-linking proteins that link intermediate filaments with actin-binding domains [21,22]. Lamins are important for the maintenance of nuclear structure. Lamin A is degraded by caspase 6, whereas lamin B is possessed by caspase 3. Lamin degradation facilitates chromatin condensation [18,23].

In the nucleus, about 150 base pairs of DNA are wrapped in about one and half turns around a histone octamer forming a nucleosome. Repeating units of nucleosomes provide the structural basis of chromatin, where neighbored nucleosomes are linked by DNA segments of about 20–80 base pairs. Caspase 3 mediates fragmentation of DNA by activated endogenous endonucleases [24]. They cleave DNA in the internucleosomal linker region with formation of fragments of about 180 base pairs and multiples thereof. In electrophoretic investigations, typical DNA ladders were observed [25].

Caspase 3 activates Mst1 kinase, promotes its nuclear translocation, and initiates chromatin condensation via a specific phosphorylation on histone H2B [26]. Another effect of Mst1 activation is the recruitment of the regulator of chromosome condensation 1 protein that leads to the inactivation of the nuclear transport machinery and in consequence to inactivation of nuclear survival factors [27].

5.2.5 Release of Signaling Molecules

The execution of the apoptotic program made only sense if the formed apoptotic vesicles are rapidly recognized and phagocytosed by professional phagocytes [28]. Otherwise, the apoptotic vesicles become leaky and release their content uncontrolled. This nonessential disruption of vesicles is called secondary necrosis [29,30].

Recruitment of macrophages occurs by special molecules released from cells undergoing apoptosis. These molecules are collectively called "find-me signals." Among them are fractalkine and nucleotides, which have been proved to function also in vivo [31,32]. Other "find-me signals" are lysophosphatidylcholine [33] and sphingosine-1-phosphate [34].

In addition, apoptotic vesicles also express special surface patterns enabling a better contact to phagocytes. These patterns are known as "eat-me signals." Phosphatidylserine epitopes are well-known surface marker on apoptotic cells. These epitopes appear on the surface of apoptotic blebs [35,36].

5.3 APOPTOSIS: DEATH RECEPTOR PATHWAY

5.3.1 Activation of Death Receptors

The extrinsic pathway of apoptosis induction starts with the binding of ligands to so-called death receptors. In humans, there are six death receptors (Table 5.1) all belonging to the tumor necrosis factor (TNF) receptor superfamily [44]. These integral proteins contain 2–4 cysteine-rich repeats in their extracellular domain for ligand binding. A death domain is localized on the cytoplasmic site of these receptors. This domain is capable of recruiting specific adaptor molecules necessary for downstream signaling.

Death receptors involved in the extrinsic pathway of apoptosis induction are CD95, TRAIL-R1, and TRAIL-R2. The corresponding ligands are CD95L and TRAIL. Death receptor–mediated apoptosis induction is also known as activation-induced cell death (AICD). In activated T cells, death receptor clustering can be induced as a result of repeated cell stimulation [45]. This helps to maintain peripheral immune tolerance. Autoimmune processes can be linked to defective AICD [45].

5.3.2 Activation of the Initiator Caspase 8

Clustered death receptors form a multiprotein death-inducing signaling complex that attracts procaspase 8 and activates this protein. Caspase 8 is the main initiator caspase of the death receptor signaling pathway [46]. There are two principle modes of action of caspase 8 [47–49]. In type I cells, caspase 8 is able, like caspase 9, to activate different executioner caspases such as caspase 3, caspase 6, and caspase 7. In type II cells, caspase 8 induces alterations in mitochondria that lead to activation of the intrinsic pathway of apoptosis.

All further metabolic processes of degradation of cellular components and formation of signaling molecules are as described in Sections 5.2.4 and 5.2.5.

Caspase 8 is also involved in suppression of necrosis [50], T-cell homeostasis, and leukocyte differentiation [11].

5.4 NECROSIS

5.4.1 Unprogrammed Form of Necrosis

In contrast to apoptosis, the necrotic cell death is a response to cell injury and severe stress associated with the loss of plasma membrane

TABLE 5.1 Human Death Receptors and Their Ligands.

Death Receptor	Other Names	Ligands	Functional Response	References
TNF-R1		TNF	Gene activation	[37]
DR3	TRAMP	TL1A	Gene activation	[38]
CD95	Fas, APO-1	CD95L (FasL)	Apoptosis	[39,40]
TRAIL-R1	DR4	TRAIL (Apo2L)	Apoptosis	[41]
TRAIL-R2	DR5	TRAIL (Apo2L)	Apoptosis	[42]
DR6		N-APP	Unknown	[43]

integrity [30]. Cytoplasmic products are released from necrotic cells in the extracellular space, where they might damage other cells and tissues. Necrosis of cells is mostly initiated by external factors such as traumata, toxins, infections, osmotic stress, mechanic stress, and others. In some cases, internal factors are discussed for causing necrosis [51].

Morphologically, necrosis represents an ischemic cell death (also called oncosis) characterized by a rapid depletion of ATP followed by swelling of the cells, formation of blebs, changes in the nucleus, and plasma membrane disruption. Sequence of nuclear changes includes nucleus shrinkage and condensation of chromatin, followed by disappearance of chromatin due to the loss of DNA [52–54].

Another pathway concerns secondary necrosis of apoptotic vesicles [29,30].

5.4.2 Necroptosis

Necrosis can also occur in a programmed fashion, known as programmed necrosis or necroptosis [55,56]. This mechanism is observed in some viral infections, where virus components act inhibitory on caspases. Necroptosis is also known in association with inflammatory diseases such as Crohn's disease, pancreatitis, and myocardial infarction. This kind of cell death can mediate the release of proinflammatory cytokines and damage-associated molecular pattern and thus promote inflammation [57].

For initiation of necroptosis signals mediated by the proinflammatory cytokine, tumor necrosis factor α can induce kinase activities of receptor-interacting protein kinase (RIPK) 1 and 3 and mixed lineage kinase domain-like pseudokinase (MLKL). These kinases are key elements of signaling pathways of cells undergoing necroptosis [56,58,59].

Necroptosis plays apparently an important role in aborting defective embryos [57]. These mechanisms can also be activated in response to acute and chronic insults in adults and elder persons. Targeting of RIPK1 is under investigation to block cell death and inflammation in certain human pathologies [60–63].

5.5 SPECIAL FORMS OF PROGRAMMED CELL DEATH

5.5.1 Pyroptosis

On infection with intracellular pathogens, macrophages can undergo a programmed cell death known as pyroptosis [64]. This process involves the formation of an inflammasome and activation of caspase 1 [65]. Inflammasome formation does not lead to recruitment of initiator and executioner caspases as seen in apoptosis. This kind of cell death is also known as caspase 1–dependent programmed cell death.

Cell death by pyroptosis results in the disruption of plasma membrane of the infected cell and release of numerous molecules and cytokines that further perpetuate the inflammatory cascade. More details about pyroptosis and inflammasome formation are given in Section 6.2.4.

5.5.2 Ferroptosis

This form of programmed cell death depends on the presence of free iron ions and the accumulation of lipid peroxides. It is associated with a diminished activity of glutathione peroxidase 4 (GPx4) [66,67]. Lipophilic antioxidants and iron chelators suppress ferroptosis supporting the role of iron-mediated oxidation in lipid phases [68].

Different molecular mechanisms, which are known to cause intracellular oxidative stress, are involved in ferroptosis induction. Erastin and sulfasalazine act as a competitive inhibitor of the glutamate/cystine antiporter [69,70].

This increases cytotoxic effects of glutamate and leads to intracellular cysteine and glutathione depletion. The drug RSL3 directly targets GPx4 [71]. As a result, lipid hydroperoxides are not efficiently reduced within biological membranes.

Phosphatidylethanolamines (PEs) containing doubly- and triply oxygenated arachidonic or adrenic acid residues are identified as a key factor to drive cells into ferroptosis. Oxidation of these lipid species occurs in the ER and is driven by 15-lipoxygenase (15-LOX) [72]. The PE-binding scaffold protein PEBP1 binds to 15-LOX and changes the substrate specificity of this enzyme [73]. Another study reassures the role of lipoxygenases in formation of lipid hydroperoxides in ferroptosis initiation but reinforces lipid autoxidation as driving mechanism in this cell death process [74].

5.5.3 Mitotic Catastrophe

This term described a form of cell death associated with a premature or inappropriate entry of cells into mitosis. Chemical and physical stress can induce a mitotic catastrophe [75–77]. This kind of cell death can also be triggered by agents affecting the stability of microtubules and various anticancer drugs acting on cell cycle checkpoints. Mitotic catastrophe is also observed in ionized cancer cells.

5.5.4 Cell Death of Nerve Cells

Excessive stimulation of nerve cells by glutamate and related neurotransmitters can damage or kill nerve cells. This process is known as excitotoxicity [78–80]. Excessive activation of glutaminergic receptors increases Ca^{2+} entry into cells and induces the activation of Ca^{2+}-dependent enzymes that substantially damage cell components [81,82]. This condition is found in numerous neurological destructive processes such as spinal cord injury, stroke, traumatic brain injury, and hearing loss. Neurodegenerative diseases are also characterized by a progressive loss of neurons.

In general, the lack of dioxygen and glucose during ischemic phases promotes an extracellular accumulation of glutamate and aspartate. Calcium overload, collapse of the membrane potential in mitochondria, and enhanced formation of reactive species are consequences of glutamate-mediated overexcitation in neurons [83,84]. These conditions can induce either apoptosis via the mitochondrial pathway or necrosis in case of severe energy failure [85,86]. Often, ER is also affected during excitotoxicity, resulting in accumulation of misfolded proteins [87].

Retinal ganglion cells, glial cells, in particular oligodendrocytes, and astrocytes, can also undergo excitotoxicity [88,89].

5.5.5 Anoikis

This form of programmed cell death is found in cells after their detachment from the extracellular matrix. The close cell matrix interaction provides essential signals for growth and survival. The lack of these signals may induce anoikis [90].

Invading tumor cells may prevent the anoikis process. However, the mechanism of how invading tumor cells can survive remains largely unknown [91].

5.5.6 Cornification

This form of cell death is unique to epidermal keratinocytes. During cornification, an epidermal barrier is formed in stratified squamous epithelial tissue [92].

5.6 DEGRADATION OF DYSFUNCTIONAL COMPONENTS AND WASTE PRODUCT FORMATION

5.6.1 Recycling of Cell Components

Cells are equipped with two main systems for degradation and removal of dysfunctional cellular components and especially unwanted proteins. These systems are autophagy and the ubiquitin–proteasome system. They contribute to the homeostasis between synthesis of novel cellular components and removal of damaged and dysfunctional parts.

In contrast to the aforementioned mechanisms of cell death, the affected cells survive. Only selected components of these cells or defective proteins are disposed and, in turn, replaced by novel ones. Degradation processes are highly regulated and controlled by several mechanisms. This allows an efficient regeneration of cellular components.

5.6.2 Autophagy

During autophagy, damaged cell organelles, deteriorated cytoplasmic components, aggregated proteins, and sometimes invaded pathogenic microbes are cleared by means of lysosomes [93,94]. The latter organelles are rich in acidic hydrolases and are able to degrade a variety of substrates. Among lysosomal enzymes are sulfatases, lipases, nucleases, phosphatases, glycosidases, and different proteinases such as aspartate-, cysteine-, and serine-dependent proteinases [95]. Autophagy can be induced as a nonselective process, e.g., by starvation or as selective mechanism under activation of p62 and others that label damaged organelles, microbes, and dysfunctional proteins [96,97].

There are three main mechanisms of autophagy: macroautophagy, microautophagy, and chaperone-mediated autophagy. In macroautophagy, cell organelles, other cell constituents, or sometimes invading microorganisms are surrounded by a double membrane forming an autophagosome [98,99]. Then lysosomes fuse with the autophagosome [100]. During microautophagy, smaller cellular debris and defective proteins are directly taken up by lysosomes [101]. In the third process, unwanted proteins are delivered as a protein–chaperon complex to lysosomes [102].

These processes are controlled by autophagy-related genes and associated enzymes [100,103]. An important negative regulator of autophagy is the serine-threonine protein kinase mTORC1 [104,105]. This abbreviation stands for mechanistic target of rapamycin complex 1. This protein kinase senses the intracellular energy status. If sufficient nutrients are available, the activated mTORC1 promotes net protein synthesis and cell growth and inhibits autophagy via phosphorylation of the ULK1 protein [106,107].

A key regulator of the cellular energy status is also the AMP-activated protein kinase (AMPK) that is activated by increasing AMP levels [108]. This enzyme activates ATP production, inhibits energy consuming pathways such as protein synthesis, and inhibits mTORC1 [109]. Thus, autophagy is favored by activation of AMPK. Conditions leading to an enhanced AMP/ATP ratio are nutrient deprivation, hypoxia, stress situations, infections, and others.

Inhibition of autophagy has profound effects on aging [110] and is associated with the pathogenetic process in diabetes, cardiovascular disease, neurodegenerative diseases, and cancer [111].

5.6.3 The Ubiquitin–Proteasome System

A second intracellular degradation mechanism of defective proteins is realized by the ubiquitin–proteasome system [112,113]. Inside cells, defective or misfolded proteins are tagged with multiple ubiquitin residues under catalyzation of ubiquitin ligases. Subsequently proteins with polyubiquitin chains are subjected to intracellular proteolysis by the proteasome. The whole cascade of responsive events includes recognition of injured proteins by chaperones, coordinated reactions of ubiquitin-activating (E1), ubiquitin-conjugating (E2), and ubiquitin-ligating (E3) enzymes, and finally a concerted action of the proteasome on the protein target consisting of partially unfolding, deubiquitylation, and proteolytic degradation [114,115]. The resulting short polypeptides are used for antigen presentation via MHC I molecules or further degraded to single amino acids [116]. Ubiquitylation is a reversible process [117]. About 85 different enzymes are known for catalyzing the reverse reaction.

The 26S proteasome is an elongated complex with a central hole consisting of 33 different subunits. Proteolytic active sites are inside the core particle of this complex. Only ubiquitin-labeled and unfolded proteins can enter the narrow translocation channel of the proteasome complex, where these proteins are degraded and deubiquitylated. Protein unfolding and translocation into the core particle is driven by ATP hydrolysis [118,119].

The physiological significance of these mechanisms is to keep cells clean from any damaged, unwanted, and potentially harmful structural elements [120]. The ubiquitin–proteasome system is also important for regulatory processes of the cell cycle, cell growth and differentiation, gene transcription, signal transduction, and apoptosis [121,122]. Disturbances of this system contribute to the pathogenesis of many diseases [123,124].

5.6.4 Quality Control Over Protein Folding and Unfolded Protein Response

In ER, a set of molecular chaperones and enzymes contributes to correct folding of newly synthesized proteins. *N*-Glycosylation and formation of disulfide linkages are key events in the folding process. Important requirements for successful protein folding are the presence of glucose and calcium ions for functioning of chaperones and the existence of an oxidized environment [125].

Quality control over folding is also performed in these organelles. Properly folded proteins are exocytozed from ER and transferred to the Golgi apparatus for final modification [126,127]. Misfolded proteins are recognized by the heat shock protein glucose regulate protein 78 (Grp78) and guided through ER-associated degradation [128]. Misfolded proteins, which are additionally labeled by chaperones, enter the ubiquitin–proteasome pathways for final degradation.

Under stress conditions, the unfolded protein response (UPR) pathway is activated in ER to increase the ER protein folding capacity and to decrease the ER protein folding load [129–131]. This is achieved via activation of the three stress sensors activating transcription factor 6 (ATF6), inositol requiring enzyme 1 (IRE1), and double-stranded RNA-activating protein kinase-like ER kinase (PERK). By these three mechanisms, UPR is directed to reestablish cellular homeostasis. Cells are driven into apoptosis, when UPR acts too long and is insufficient to combat ER stress [132].

5.6.5 Accumulation of Intracellular Waste Products

Biological material is usually recycled during metabolic activity of cells. Accumulation of some waste products inside cells might occur in tissues containing postmitotic cells with a long lifetime.

Under stress situations, the number of misfolded and oxidized proteins raises, and in consequence, all subsequent processes of intracellular proteolysis increases too. Although most damaged material can be degraded without any traces, some insoluble aggregated material accumulates and resists intracellular degradation and removal. This kind of debris is known as lipofuscin and ceroid pigments [133–135]. They accumulate most of all in postmitotic cells, whereby higher depositions are found in tissues of aged individuals. Hence, lipofuscin is also called age pigment.

Lipofuscin and ceroid pigments originate from lysosomal catabolism of defective cell components. Apparently, they are end products of macroautophagy [136]. Otherwise, these waste products can also originate from cytosolic sources [137]. They contain different oxidized lipids, advanced glycated end products, and metal ions such as copper, zinc, and iron [134].

5.6.6 Formation of Amyloid Fibrils

Other forms of waste products are amyloid protein fibrils that are mainly found in the extracellular space of tissues and organs. On amyloid formation, sequential changes in protein folding occur resulting in deposition of soluble proteins in form of insoluble fibrils [138]. In other words, the involved proteins change their physical state without any change in the primary sequence. It is expected that a large energy barrier exists between the "normal" conformation of a soluble protein and its amyloid form.

In amyloid fibrils, the β-sheet conformation largely increases. These β-sheets are arranged perpendicular to the fiber axis [139,140], forming stacks that are stabilized by interstrand hydrogen bonds. In X-ray diffraction investigations, amyloid fibers display a so-called cross-β fiber diffraction pattern [141]. Amyloid fibers can also be arranged into fibrous deposits in plaques.

Mechanisms of amyloid formation are only poorly understood. On the basis of in vitro experiments with serum amyloid A (SAA), it has been hypothesized that acidic pH values around 3.5–4.5 promote the α-helix to β-sheet conversion of SAA and other proteins [142–144]. This favors a lysosomal origin of amyloid fibrils [144] (see also Section 7.2.2).

Amyloidosis is a heterogeneous group of about 50 disorders associated with the formation of amyloid fibrils [145,146]. Among them are neurodegenerative diseases such as Morbus Alzheimer, Parkinson's disease, Huntington's disease, and others. Diabetes mellitus, atherosclerosis, and rheumatoid arthritis are further examples of diseases associated with amyloid formation.

In humans, 27 extracellular fibril proteins are known [147]. Important amyloid proteins and their appearance in diseases are listed in Table 5.2. It is matter of intensive debate whether amyloid fibrils are the cause or the consequence of the associated disease process.

Mechanistic details of amyloid formation are largely unknown as well as factors contributing to the structural rearrangement of soluble proteins into insoluble fibril fibers [156]. Anyway several factors are discussed favoring amyloid formation. Among them are denaturation of normally folded proteins, overexpression of proteins that shifts the balance between proteins and chaperones, cleavage of peptides from a folded protein, overproduction of disordered proteins, and diminished degradation of a given protein [157]. According to the nucleation hypothesis, fibril formation starts with the very slow nucleation phase, a rare process of arrangement of 3–4 proteins into fibril spine [141]. This phase is followed by a more rapid growth phase [158]. Fibril growth kinetics is affected by breakage of fibrils [159].

Amyloidogenesis has striking similarities to the formation of prions [140]. Prions are a special form of amyloids. These infectious amyloids can act as a template for conversion of other noninfectious proteins into infectious form [145].

TABLE 5.2 Common Types of Amyloids in Human Disorders.

Amyloid Protein	Preferred Location	Disorder	Abbreviation (Disorder)	References
Serum amyloid A	Kidney, gastrointestinal tract, spleen, liver, autonomic nerves	Reactive amyloidosis, rheumatoid arthritis, atherosclerosis	AA	[148]
Monoclonal light chain	Heart, kidney, liver, gastrointestinal tract, nerves, soft tissue	Immunoglobulin light-chain amyloidosis	AL	[149]
Fibrinogen	Kidney	Fibrinogen amyloidosis	AFb	[150]
Apolipoprotein A1	Liver, kidney, brain	Atherosclerosis	AApoAI	[151]
α-Synuclein	Brain	Parkinson's disease		[152]
β-Amyloid precursor protein	Brain	Alzheimer's disease	Aβ	[152,153]
Huntingtin	Brain	Huntington's disease		[154]
Islet amyloid polypeptide (amylin)	Pancreatic islets	Diabetes mellitus type 2	AIAPP	[155]

5.7 RED BLOOD CELLS

5.7.1 Removal of Senescent Red Blood Cells

Human red blood cells have neither a nucleus nor mitochondria. Their normal life span approaches about 120 d. In senescent erythrocytes, a clustering of band 3 proteins occurs followed by attachment of complement C3 fragments and antiband 3 immunoglobulins. These labeled cells are removed from circulation by macrophages in spleen [160,161].

Investigation of fine mechanisms of band 3 clustering revealed that a moderate peroxidation of lipids in the plasma membrane of senescent red blood cells is a prerequisite for methemoglobin-mediated clustering of band 3 [162]. In intact red blood cells, the band 3 protein forms an immobile tetramer that is attached on the cytoplasmic domain via ankyrin to the spectrin network. Peroxidation detaches ankyrin from band 3 protein. Under the influence of methemoglobin, the more mobile band 3 proteins forms a cluster that is recognized by complement C3 and antiband 3 immunoglobulins [162].

Senescent red blood cells contain a higher amount of free heme in their membranes than younger cells [163].

5.7.2 Eryptosis

Red blood cells can also undergo a suicidal cell death, known as eryptosis [164]. This mechanism is induced on injury independent of the senescent status of these cells. Similar to apoptosis induction in nucleated cells, eryptosis is also a strictly regulated physiological mechanism designed not to induce any harm to the organism.

Eryptosis is characterized by increased cytosolic Ca^{2+}, cell shrinkage, membrane blebbing, and translocation of phosphatidylserine to the outer leaflet in the plasma membrane [164,165]. Among causes of eryptosis are osmotic stress, energy depletion, enhanced temperature, oxidative stress, and enhanced activities of several kinases. A wide variety of endogenous mediators and xenobiotics may elicit eryptosis [164,166,167]. Otherwise, eryptosis is inhibited by erythropoietin, nitric monoxide, catecholamines, adenosine, caffeine, high concentrations of urea, and some others [164].

On eryptosis induction, nonspecific cation channels are activated in the plasma membrane and an enhanced influx of Ca^{2+} takes place [168,169]. The Ca^{2+}-induced formation of prostaglandin E_2 by cyclooxygenase further potentiates the Ca^{2+} entry. Another consequence of Ca^{2+} entry is the activation of scramblases leading to disruption of phospholipid asymmetry in the plasma membrane and exposure of phosphatidylserine on the outer leaflet. This process is facilitated by ceramide that result from Ca^{2+}-dependent sphingomyelinase activity [166]. The activation of calpain followed by ankyrin degradation is also caused by enhanced cytosolic Ca^{2+} level.

Ceramide is known to trigger apoptosis in a wide variety of cells [170] as well as eryptosis. In red blood cells, it is formed from sphingosine by ceramide synthase. Sphingosine, but not sphingosine-1-phosphate, which is derived from sphingosine by sphingosine kinase 1, is a strong inducer of eryptosis [171].

In eryptotic cells, increased Ca^{2+} permeability affects also other ion fluxes. Activation of Ca^{2+}-sensitive K^+-channels causes cell membrane hyperpolarization, loss of K^+ and Cl^-, and cell shrinkage [172–174]. Under condition of oxidative stress, Cl^- is released via activated Cl^--channels. In turn with fluxes of K^+ and Na^+, a membrane depolarization and cell swelling result. Excessive swelling is associated with hemolysis [175,176].

5.7.3 Excessive Eryptosis and Hemolysis

As long as eryptosis occurs on a low level, no pathophysiological problems are expected. Defective red blood cells are usually removed from circulation before hemolysis by this useful mechanism. A number of diseases are characterized by an excessive eryptosis. Among them are iron deficiency, diabetes, renal insufficiency, sepsis, malaria, sickle cell disease, and many others [166,177]. In general, disease conditions leading to energy depletion and iron deficiency are known to trigger eryptosis. Under these conditions, an intravascular hemolysis is observed before the injured cells are cleared from circulation. Hemolysis is a strong complication of the pathological situation (see Section 4.2.6).

5.8 NEUTROPHILS

Polymorphonuclear leukocytes or neutrophils as the most abundant type of white blood cells can undergo apoptosis and necrosis in dependence on the conditions on infected/inflammatory sites. This issue is outlined in more detail in Sections 6.1.3 and 6.4.

A special form of cell death of neutrophils is associated with the formation and release of extracellular traps. This form of cell death is also known as NETosis [178]. NET stands for neutrophil extracellular trap, a network of extracellular DNA–derived fibers with attached neutrophil granule proteins such as myeloperoxidase, elastase, cathepsin G, gelatinase, lactoferrin, and the cytoplasmic protein complex calprotectin [179,180].

During suicidal NETosis, activated protein arginine deiminase 4 promotes citrullination of histones followed by decondensation of chromatin. This process is further promoted by myeloperoxidase and elastase entering the nucleus [181]. The release of calcium ions from internal stores amplifies additionally the NETosis pathway [182]. Intracellular formed traps are released into the extracellular space after rupture of the plasma membrane.

In another form of NETosis, the vital NETosis, the plasma membrane remains intact and neutrophils do not die. After blebbing of the nucleus, DNA-filled vesicles are exocytosed [181].

5.9 SUMMARY

Elimination of dysfunctional proteins, damaged cell organelles, and dead cells without disturbing neighbored intact structural elements is of paramount importance to maintain cell and tissue homeostasis. These well-regulated mechanisms prevent an excessive accumulation of waste products and eliminate potential sources for novel destructions. Moreover, proteins and other biological materials are efficiently recycled. Thus, the aforementioned processes are an indispensable part of survival strategies in higher multicellular organisms.

We do not forget that mechanisms of programmed cell death, autophagy, quality control over protein folding, and other mechanisms discussed in this chapter act efficiently as long as the disturbing impacts do not exceed certain limits. Under more severe stress conditions, there is a rising disbalance between disturbances in cell components and the capacity of protecting systems to combat them. This disbalance can cause the occurrence of nonregulated disturbance mechanisms such as the uncontrolled cell death by necrosis, the accumulation of misfolded proteins, intravascular hemolysis, and the release of proinflammatory agents. These processes can initiate additional damage to so far unperturbed structural elements and cells and finally lead to an inflammatory response in the affected tissues.

References

[1] W. Neupert, Protein import into mitochondria, Annu. Rev. Biochem. 66 (1997) 863–917.

[2] P. Ascenzi, M. Coletta, M.T. Wilson, L. Fiorucci, M. Marino, F. Polticelli, F. Sinibaldi, R. Santucci, Cardiolipin-cytochrome c complex: switching cytochrome c from an electron-transfer shuttle to a myoglobin- and a peroxidase-like heme-protein, IUMBM Life 67 (2015) 98–109.

[3] V.E. Kagan, V.A. Tyurin, J. Jiang, Y.Y. Tyurina, V.B. Ritov, A.A. Amoscato, A.N. Osipov, N.A. Belikova, A.A. Kapralov, V. Kini, I.I. Vlasova, Q. Zhao, M. Zou, P. Di, D.A. Svistunenko, I.V. Kurnikov, G.G. Borisenko, Cytochrome c acts as a cardiolipin oxygenase required for the release of proapoptotic factors, Nat. Chem. Biol. 1 (2005) 223–232.

[4] Y. Tsujimoto, L.R. Finger, J. Yunis, P.C. Nowell, C.M. Croce, Cloning of the chromosome breakpoint of neoplastic B cells with the t(14;18) chromosome translocation, Science 226 (1984) 1097–1099.

[5] R.J. Youle, A. Strasser, The Bcl-2 protein family: opposing activities that mediate cell death, Nat. Rev. Mol. Cell Biol. 9 (2008) 47–59.

[6] D. Boehning, R.L. Patteson, L. Sedaghat, N.O. Glebova, T. Kurosaki, S.H. Snyder, Cytochrome c binds to inositol (1,4,5) trisphosphate receptors, amplifying calcium-dependent apoptosis, Nat. Cell Biol. 5 (2003) 1051–1061.

[7] X. Wang, The expanding role of mitochondria in apoptosis, Genes Dev. 15 (2001) 2922–2933.

[8] S.A. Susin, H.K. Lorenzo, N. Zamzani, I. Marzo, B.E. Snow, G.M. Brothers, J. Mangion, E. Jacotot, P. Constantini, M. Loeffler, N. Larochette, D.R. Goodlett, R. Aedersold, D.P. Siderovski, J.M. Penninger, G. Kroemer, Molecular characterization of mitochondrial apoptosis-inducing factor, Nature 397 (1999) 441–446.

[9] X. Liu, P. Li, P. Widlak, H. Zou, X. Luo, W.T. Gaarrard, X. Wang, The 40-kDa subunit of DNA fragmentation factor induces DNA fragmentation and chromatin condensation during apoptosis, Proc. Natl. Acad. Sci. U.S.A. 95 (1998) 8461–8466.

[10] Y. Shi, Caspase activation, Cell 117 (2004) 855–858.

[11] D.R. McIlwain, T. Berger, T.W. Mak, Caspase functions in cell death and disease, Cold Spring Harbor Perspect. Biol. 5 (2013), a008656, 1–28, https://doi.org/10.1101/cshperspect.a008656.

[12] Z. Hao, G.S. Duncan, C.C. Chang, A. Elia, M. Fang, A. Wakeham, H. Okada, T. Calzascia, Y. Jang, A. You-Ten, W.C. Yeh, P. Ohashi, X. Wang, T.W. Mak, Specific ablation of the apoptotic functions of cytochrome c reveals a differential requirement for cytochrome c and Apaf-1 in apoptosis, Cell 121 (2005) 579–591.

[13] D. Acehan, X. Jiang, D.G. Morgan, J.E. Heuser, X. Wang, C.W. Akey, Three-dimensional structure of the apoptosome: implications for assembly, procaspase-9 binding, and activation, Mol. Cell 9 (2002) 423–432.

[14] K. Cain, S.B. Bratton, G.M. Cohen, The Apaf-1 apoptosome: a large caspase-activating complex, Biochimie 84 (2002) 203–214.

[15] M. D'Amelio, E. Tino, F. Cecconi, The apoptosome: emerging insights and new potential targets for drug design, Pharmaceut. Res. 25 (2008) 740–751.

[16] K. Kuida, Caspase 9, Int. J. Biochem. Cell Biol. 32 (2000) 121–124.

[17] A.G. Porter, R.U. Jänicke, Emerging roles of caspase-3 in apoptosis, Cell Death Differ 6 (1999) 99–104.

[18] E.A. Slee, C. Adrain, S.J. Martin, Executioner caspase-3. -6, and -7 perform distinct, non-reductant roles during the demolition phase of apoptosis, J. Biol. Chem. 276 (2001) 7320–7326.

[19] O. Ndozangue-Touriguine, J. Hamelin, J. Brèard, Cytoskeleton and apoptosis, Biochem. Pharmacol. 76 (2008) 11–18.

[20] B. Levkau, B. Herren, H. Koyama, R. Ross, E.W. Raines, Caspase-mediated cleavage of focal adhesion kinase pp125FAK and disassembly of focal adhesions in human endothelial cell apoptosis, J. Exp. Med. 187 (1998) 579–586.

[21] Y. Byun, F. Chen, R. Chang, M. Trivedi, K.J. Green, V.L. Cryns, Caspase cleavage of vimentin disrupts intermediate filaments and promotes apoptosis, Cell Death Differ. 8 (2001) 443–450.

[22] S. Aho, Plakin proteins are coordinately cleaved during apoptosis but preferentially through the action of different caspases, Exp. Dermatol. 13 (2004) 700–707.

[23] L. Rao, D. Perez, E. White, Lamin proteolysis facilitates nuclear events during apoptosis, J. Cell Biol. 135 (1996) 1441–1455.

[24] J. Füllgrabe, N. Hajji, B. Joseph, Cracking the death code: apoptosis-related histone modifications, Cell Death Differ. 17 (2010) 1238–1243.

[25] S. Elmore, Apoptosis: a review of programmed cell death, Toxicol. Pathol. 35 (2007) 495–516.

[26] W.L. Cheung, K. Ajiro, K. Samejima, M. Kloc, P. Cheung, C.A. Mizzen, A. Beeser, L.D. Etkin, J. Cherhoff, W.C. Earnshaw, C.D. Allis, Apoptotic phosphorylation of histone H2B is mediated by mammalian sterile twenty kinase, Cell 113 (2003) 507–517.

[27] C.H. Wong, H. Chan, C.Y. Ho, S.K. Lai, K.S. Chan, C.G. Koh, H.Y. Li, Apoptotic histone modification inhibits nuclear transport by regulating RCC1, Nat. Cell Biol. 11 (2009) 36–45.

[28] K.S. Ravichandran, Find-me and eat-me signals in apoptotic cell clearance: progress and conundrums, J. Exp. Med. 207 (2010) 1807–1817.

[29] M.T. Silva, Secondary necrosis: the natural outcome of the complete apoptotic program, FEBS Lett. 584 (2010) 4491–4499.

[30] T. Van den Berghe, N. Vanlangenakker, E. Parthoens, W. Deckers, M. Devos, N. Festjens, C.J. Guerin, U.T. Brunk, W. Declecq, P. Vandenabeele, Necroptosis, necrosis and secondary necrosis converge on similar cellular disintegration features, Cell Death Differ. 17 (2010) 922–930.

[31] L.A. Truman, C.A. Ford, M. Pasikowska, J.D. Pound, S.J. Wilkinson, I.E. Dumitriu, L. Melville, L.A. Melrose, C.A. Ogden, R. Nibbs, G. Graham, C. Combadiere, C.D. Gregory, CX3CL1/fractalkine is released from apoptotic lymphocytes to simulate macrophage chemotaxis, Blood 112 (2008) 5026–5036.

[32] M.R. Elliott, F.B. Chekani, P.C. Trampont, E.R. Lazarowski, A. Kadl, S.F. Walk, D. Park, R.I. Woodson, M. Ostankovich, P. Sharma, J.J. Lysiak, T.K. Harden, N. Leitlinger, K.S. Ravichandran, Nucleotides released by apoptotic cells act as a find-me signal to promote phagocytic clearance, Nature 461 (2009) 282–286.

[33] K. Lauber, E. Bohn, S.M. Kröber, Y.J. Xiao, S.G. Blumenthal, R.K. Lindemann, P. Marini, C. Wiedig, A. Zobywalski, S. Baksh, Y. Xu, I.B. Authenrieth, K. Schulze-Osthoff, C. Belka, G. Stuhler, S. Wesselborg, Apoptotic cells induce migration of phagocytes via caspase-3-mediated release of a lipid attraction signal, Cell 113 (2003) 717–730.

[34] D.R. Gude, S.E. Alvarez, S.W. Paugh, P. Mitra, J. Yu, R. Griffiths, S.E. Barbour, S. Milstien, S. Spiegel, Apoptosis induces expression of sphingosine kinase 1 to release sphingosine-1-phosphate as "come-and-get-me" signal, FASEB J. 22 (2008) 2629–2638.

[35] V.A. Fadok, D.R. Voelker, P.A. Campbell, J.J. Cohen, D.L. Bratton, P.M. Henson, The role of phosphatidylserine on the surface of apoptotic lymphocytes triggers specific recognition and removal of macrophages, J. Immunol. 148 (1992) 2207–2216.

[36] C. Segundo, F. Medina, C. Rodriguez, R. Martinez-Palencia, F. Leyva-Cobian, J.A. Brieva, Surface molecule loss and bleb formation by human germinal center B cells undergoing apoptosis: role of apoptotic blebs in monocyte chemotaxis, Blood 6 (1999) 6–12.

[37] H. Loetscher, Y.C. Pan, H.W. Lahm, R. Gentz, M. Brockhaus, H. Tabuchi, W. Lesslauer, Molecular cloning and expression of the human 55 kd tumor necrosis factor receptor, Cell 61 (1990) 351–359.

[38] J.L. Bodmer, K. Burns, P. Schneider, K. Hofmann, V. Steiner, M. Thome, T. Bornand, M. Hahne, M. Schröter, K. Becker, A. Wilson, L.E. French, J.L. Browning, H.R. MacDonald, J. Tschopp, TRAMP,

a novel apoptosis-mediating receptor with sequence homology to tumor necrosis factor receptor 1 and Fas(Apo-1/CD95), Immunity 6 (1997) 79–88.
[39] N. Itoh, S. Yonehara, A. Ishii, M. Yonehara, S. Mizushima, M. Sameshima, A. Hase, Y. Seto, S. Nagata, The polypeptide encoded by the cDNA for human cell surface antigen Fas can mediate apoptosis, Cell 66 (1991) 233–243.
[40] A. Oehm, I. Behrmann, W. Falk, M. Pawlita, G. Maier, C. Klas, M. Li-Weber, S. Richards, J. Dhein, B.C. Trauth, H. Ponstingl, P.H. Krammer, Purification and molecular cloning of the APO-1 cell surface antigen, a member of the tumor necrosis factor/nerve growth factor receptor superfamily. Sequence identity with the Fas antigen, J. Biol. Chem. 267 (1992) 10709–10715.
[41] G. Pan, K. O'Rourke, A.M. Chinnaiyan, R. Gentz, R. Ebner, J. Ni, V.M. Dixit, The receptor for the cytotoxic ligand TRAIL, Science 276 (1997) 111–113.
[42] H. Walczak, M.A. Degli-Esposti, R.S. Johnson, P.J. Smolak, J.Y. Waugh, N. Bolani, M.S. Timour, M.J. Gerhart, K.A. Schooley, C.A. Smith, R.G. Goodwin, C.T. Rauch, TRAIL-R2: a novel apoptosis-mediating receptor for TRAIL, EMBO J. 16 (1997) 5386–5397.
[43] G. Pan, J.H. Bauer, V. Haridas, S. Wang, D. Liu, G. Yu, C. Vincenz, B.B. Aggarwal, J. Ni, V.M. Dixit, Identification and functional characterization of DR6, a novel death domain-containing TNF receptor, FEBS Lett. 431 (1998) 351–356.
[44] H. Walczak, Death receptor-ligand systems in cancer, cell death, and inflammation, Cold Spring Harbor Perspect. Biol. 5 (2013), a008698, 1–18, https://doi.org/10.1101/cshperspect.a008698.
[45] J. Zhang, X. Xu, Y. Liu, Activation-induced cell death in T cells and autoimmunity, Cell. Mol. Immunol. 1 (2004) 186–192.
[46] M. Kruidering, G.I. Evan, Caspase-8 in apoptosis: the beginning of "the end"? IUBMB Life 50 (2000) 85–90.
[47] C. Scaffidi, S. Fulda, A. Srinivasan, C. Friesen, F. Li, K.J. Tomaselli, K.M. Debatin, P.H. Krammer, M.E. Peter, Two CD95 (APO-1/Fas) signaling pathways, EMBO J. 17 (1998) 1675–1687.
[48] A.K. Samraj, E. Keil, N. Ueffing, K. Schulze-Osthoff, I. Schmitz, Loss of caspase-9 provides genetic evidence for the type I/II concept of CD95-mediated apoptosis, J. Biol. Chem. 281 (2006) 29652–29659.
[49] P.J. Jost, S. Grabow, D. Gray, M.D. McKenzie, U. Nachbar, P. Bouillet, H.E. Thomas, C. Borner, J. Silke, A. Strasser, T. Kaufmann, XIAP discriminates between type I and type II FASD-induced apoptosis, Nature 460 (2009) 1035–1039.
[50] A. Oberst, D.R. Green, It cuts both ways: reconciling the dual roles of caspase 8 in cell death and survival, Nat. Rev. Mol. Cell Biol. 12 (2011) 757–763.
[51] M. Raffray, G.M. Cohen, Apoptosis and necrosis in toxicology: a continuum or distinct modes of cell death? Pharmacol. Ther. 75 (1997) 153–177.
[52] J.U. Schweichel, H.J. Merker, The morphology of various types of cell death in prenatal tissues, Teratology 7 (1973) 253–266.
[53] M. Leist, M. Jaattela, Four deaths and a funeral: from caspases to alternative mechanisms, Nat. Rev. Mol. Cell Biol. 2 (2001) 589–598.
[54] N. Festjens, T. Vanden Berghe, P. Vandenabeele, Necrosis, a well-orchestrated form of cell demise: signaling cascades, important mediators and concomitant immune response, Biochim. Biophys. Acta 1757 (2006) 1371–1387.
[55] D.E. Christofferson, J. Yuan, Necroptosis as an alternative form of programmed cell death, Curr. Opin. Cell Biol. 22 (2010) 263–268.
[56] R. Weinlich, A. Oberst, H.M. Beere, D.R. Green, Necroptosis in development, inflammation and disease, Nat. Rev. Mol. Cell Biol. 18 (2017) 127–136.
[57] B. Shan, H. Pan, A. Najafov, J. Yuan, Necroptosis in development and diseases, Genes Dev. 32 (2018) 327–340.
[58] D. Ofengeim, J. Yuan, Regulation of RIP1 kinase signaling at the crossroads of inflammation and cell death, Nat. Rev. Mol. Cell Biol. 14 (2013) 727–736.
[59] J.M. Murphy, P.E. Czabotar, J.M. Hildebrand, I.S. Lucet, J.G. Zhang, S. Alvarez-Diaz, R. Lewis, N. Lalaoui, D. Metcalf, A.I. Webb, S.N. Young, L.N. Varghese, G.M. Tannahill, E.C. Hatchell, I.J. Majewski, T. Okamoto, R.C. Dobson, D.J. Hilton, J.J. Babon, N.A. Nicola, A. Strasser, J. Silke, W.S. Alexander, The pseudokinase MLKL mediated necroptosis via a molecular switch mechanism, Immunity 39 (2013) 443–453.
[60] A. Degterev, J. Hitomi, M. Germscheid, I.L. Ch'en, O. Korkina, X. Teng, D. Abbott, G.D. Cuny, C. Yuan, G. Wagner, S.M. Hedrick, S.A. Gerber, A. Lugovskoy, J. Yuan, Identification of RIP1 kinase as specific cellular target of necrostatins, Nat. Chem. Biol. 4 (2008) 313–321.
[61] T. Xie, W. Peng, Y. Liu, C. Yan, J. Maki, A. Degterev, J. Yuan, Y. Shi, Structural basis of RIP1 inhibition by necrostatins, Structure 21 (2013) 493–499.
[62] P.A. Harris, S.B. Berger, J.U. Jeong, R. Nagilla, D. Bandyopadhyay, N. Campobasso, C.A. Capriotti, J.A. Cox, L. Date, X. Dong, P.M. Eidam, J.N. Finger, S.J. Hoffman, J. Kang, V. Kasparcova, B.W. King, R. Lehr, Y. Lan, L.K. Leister, J.D. Lich, T.T. MacDonald, N.A. Miller, M.T. Ouelette,

C.S. Pao, A. Rahman, M.A. Reilly, A.R. Rendina, E.J. Rivera, M.C. Schaeffer, C.A. Sehon, R.R. Singhaus, H.H. Sun, B.A. Swift, R.D. Totoritis, A. Vossenkämper, P. Ward, D.D. Wisnoski, D. Zhang, R.W. Marquis, P.J. Gough, J. Berlin, Discovery of a first-in-class receptor interacting protein 1 (RIP1) kinase specific clinical candidate (GSK2982772) for the treatment of inflammatory diseases, J. Med. Chem. 60 (2017) 1247–1261.

[63] D. Ofengeim, S. Mazzitelli, Y. Ito, J.P. DeWitt, L. Mifflin, C. Zou, S. Das, X. Adiconis, H. Chen, H. Zhu, M.A. Kelliher, J.Z. Levin, J. Yuan, RIPR1 mediates a disease-associated microgial response in Alzheimer's disease, Proc. Natl. Acad. Sci. U.S.A. 114 (2017) E8788–E8797.

[64] T. Bergsbaken, S.L. Fink, B.T. Cookson, Pyroptosis: host cell death and inflammation, Nat. Rev. Microbiol. 7 (2009) 99–109.

[65] A. Lu, Y. Li, F.I. Schmidt, Q. Yin, S. Chen, T.M. Fu, A.B. Tong, H.L. Ploegh, Y. Mao, H. Wu, Molecular basis of caspase-1 polymerization and its inhibition by a new capping mechanism, Nat. Struct. Mol. Biol. 23 (2016) 416–425.

[66] W.S. Yang, B.R. Stockwell, Ferroptosis: death by lipid peroxidation, Trends Cell Biol. 26 (2016) 165–176.

[67] Y. Xie, W. Hou, X. Song, Y. Yu, J. Huang, X. Sun, R. Kang, D. Tang, Ferroptosis: process and function, Cell Death Differ. 23 (2016) 369–379.

[68] S.J. Dixon, K.M. Lemberg, M.R. Lamprecht, R. Skouta, E.M. Zaitsev, C.E. Cleason, D.N. Patel, A.J. Bauer, A.M. Cantley, W.S. Yang, B. Morrison III, B.R. Stockwell, Ferroptosis: an iron-dependent form of nonapoptotic cell death, Cell 149 (2012) 1060–1072.

[69] S. Bannai, E. Kitamura, Transport interaction of L-cystine and L-glutamate in human diploid fibroblasts in culture, J. Biol. Chem. 255 (1980) 2372–2376.

[70] R.J. Bridges, N.R. Natale, S.A. Patel, System xc(-) cystine/glutamate antiporter: an update on molecular pharmacology and roles within the CNS, Br. J. Pharmacol. 165 (2012) 20–34.

[71] W.S. Yang, B.R. Stockwell, Synthetic lethal screening identifies compounds activating iron-dependent, nonapoptotic cell death in oncogenic-RAS-harboring cancer cells, Chem. Biol. 15 (2008) 234–245.

[72] V.E. Kagan, G. Mao, F. Qu, J.P.F. Angeli, S. Doll, C. St Croix, H.H. Dar, B. Liu, V.A. Tyurin, V.B. Ritov, A.A. Kapralov, A.A. Amoscato, J. Jiang, T. Anthonymuthu, D. Mohammadyani, Q. Yang, B. Proneth, J. Klein-Seetharaman, S. Watkins, I. Bahar, J. Greenberger, R.K. Mallampalli, B.R. Stockwell, Y.Y. Tyurina, M. Conrad, H. Bayir, Oxidized arachidonic/adrenic phosphatidylethanolamines navigate cells to ferroptosis, Nat. Chem. Biol. 13 (2017) 81–90.

[73] S.E. Wenzel, Y.Y. Tyurina, J. Zhao, C.M. St Croix, H.H. Darr, G. Mao, V.A. Tyurin, T.S. Anthonymuthu, A.A. Kapralov, A.A. Amoscato, K. Mikulska-Ruminska, I.H. Shrivastava, E.M. Kenny, Q. Yang, J.C. Rosenbaum, L.J. Sparvero, D.R. Emlet, X. Wen, Y. Minami, F. Qu, S.C. Watkins, T.R. Holman, A.P. Van Demark, J.A. Kellum, I. Bahar, H. Bayir, V.E. Kagan, PEBP1 wardens ferroptosis by enabling lipoxygenase generation of lipid death signals, Cell 171 (2017) 628–641.

[74] R. Shah, M.S. Shchepinov, D.A. Pratt, Resolving the role of lipoxygenases in the initiation and execution of ferroptosis, ACS Cent. Sci. 4 (2018) 387–396.

[75] F. Ianzini, M.A. Mackey, Spontaneous premature chromosome condensation and mitotic catastrophe following irradiation of HeLa S3 cell, Int. J. Radiat. Biol. 72 (1997) 409–421.

[76] M. Castedo, J.L. Perfettini, T. Roumier, K. Andreau, R. Medema, G. Kroemer, Cell death by mitotic catastrophe: a molecular definition, Oncogene 23 (2004) 2825–2837.

[77] H. Vakifahmetoglu, M. Olsson, B. Zhivotovsky, Death through a tragedy: mitotic catastrophe, Cell Death Differ. 15 (2008) 1153–1162.

[78] J.W. Olney, Role of excitotoxins in developmental neuropathology, APMIS Suppl. 40 (1993) 103–112.

[79] L.P. Mark, R.W. Prost, J.L. Ulmer, M.M. Smith, D.L. Daniels, J.M. Strottmann, W.D. Brown, L. Hacein-Bey, Pictorial review of glutamate excitotoxicity: fundamental concepts for neuroimaging, Am. J. Neuroradiol. 22 (2001) 1813–1824.

[80] X.-X. Dong, Y. Wang, Z.-H. Qin, Molecular mechanisms of excitotoxicity and their relevance to pathogenesis of neurodegenerative diseases, Acta Pharmacol. Sin. 30 (2009) 379–387.

[81] H. Manev, M. Favaron, A. Guidotti, E. Costa, Delayed increase of Ca^{2+} influx elicited by glutamate: role in neuronal death, Mol. Pharmacol. 36 (1989) 106–112.

[82] M.K. Jaiswal, W.D. Zech, M. Goos, C. Leutbecher, A. Ferri, A. Zippelius, M.T. Carri, R. Nau, B.U. Keller, Impairment of mitochondrial calcium handling in a mtSOD1 cell culture model of motoneuron disease, BMC Neurosci. 10 (64) (2009) 1–16, https://doi.org/10.1186/1471-2202-10-64.

[83] A.Y. Abramov, M.R. Duchen, Mechanisms underlying the loss of mitochondrial membrane potential in glutamate excitotoxicity, Biochim. Biophys. Acta 1777 (2007) 953–964.

[84] A.Y. Abramov, A. Scorziello, M.R. Duchen, Three distinct mechanisms generate oxygen free radicals in neurons and contribute to cell death during anoxia and reoxygenation, J. Neurosci. 27 (2007) 1129–1138.

[85] V.L. Dawson, T.M. Dawson, Deadly conversions: nuclear-mitochondrial cross-talk, J. Bioenerg. Biomembr. 36 (2004) 287–294.

[86] H. Prentice, J.P. Modi, J.-Y. Wu, Mechanisms of neuronal protection against excitotoxicity, endoplasmic reticulum stress, and mitochondrial dysfunction in stroke and neurodegenerative diseases, Oxid. Med. Cell. Longev. (2015), article 964518, 1–7, https://doi.org/10.1155/2015/964518.

[87] W. Paschen, T. Mengesdorf, Endoplasmic reticulum stress response and neurodegeneration, Cell Calcium 38 (2005) 409–415.

[88] D.R. Lucas, J.P. Newhouse, The toxic effect of sodium L-glutamate on the inner layers of the retina, Arch. Ophthalmol. 58 (1957) 193–201.

[89] C. Matute, E. Alberdi, G. Ibarretxe, M.V. Sánchez-Gómez, Excitotoxicity in glial cells, Eur. J. Pharmacol. 447 (2002) 239–246.

[90] S.M. Frisch, R.A. Screaton, Anoikis mechanisms, Curr. Opin. Cell Biol. 13 (2001) 555–562.

[91] P. Paoli, E. Giannoni, P. Chiarugi, Anoikis molecular pathways and its role in cancer progression, Biochim. Biophys. Acta 1833 (2013) 3481–3498.

[92] L. Eckhart, S. Lippens, E. Tschachler, W. Declercq, Cell death by cornification, Biochim. Biophys. Acta 1833 (2013) 3471–3480.

[93] N. Mizushima, M. Komatsu, Autophagy: renovation of cells and tissues, Cell 147 (2011) 728–741.

[94] S. Kobayashi, Choose delicately and reuse adequately: the newly revealed process of autophagy, Biol. Pharmaceut. Bull. 38 (2015) 1098–1103.

[95] H. Xu, D. Ren, Lysosomal physiology, Annu. Rev. Physiol. 77 (2015) 57–80.

[96] L. Shang, S. Chen, F. De, S. Li, L. Zhao, X. Wang, Nutrient starvation elicits an acute autophagic response mediated by Ulk1 dephosphorylation and its subsequent dissociation from AMPK, Proc. Natl. Acad. Sci. U.S.A. 108 (2011) 4788–4793.

[97] W.J. Liu, L. Ye, W.F. Huang, L.J. Guo, Z.G. XDu, H.L. Wu, C. Yang, H.F. Liu, p62 links the autophagy pathway and the ubiquitin-proteasome system upon ubiquitinated protein degradation, Cell. Mol. Biol. Lett. 21 (29) (2016) 1–14, https://doi.org/10.1186/s11658-016-0031-z.

[98] Z. Xie, D.J. Klionsky, Autophagosome formation: core machinery and adaptations, Nat. Cell Biol. 27 (2011) 107–132.

[99] B. Levine, N. Mizushima, H.W. Virgin, Autophagy in immunity and inflammation, Nature 469 (2011) 323–335.

[100] N. Mizushima, Y. Ohsumi, T. Yoshimori, Autophagosome formation in mammalian cells, Cell Struct. Funct. 27 (2002) 421–429.

[101] W.W. Li, J. Li, J.K. Bao, Microautophagy: lesser-known self-eating, Cell. Mol. Life Sci. 69 (2012) 1125–1136.

[102] S. Kaushik, A.M. Cuervo, Chaperone-mediated autophagy: a unique way to enter the lysosome world, Trends Cell Biol. 22 (2012) 407–417.

[103] R.C. Russell, H.-X. Yuan, K.-L. Guan, Autophagy regulation by nutrient signaling, Cell Res. 24 (2014) 42–57.

[104] E.F. Blommaart, J.J. Luiken, P.J. Blommaartm, G.M. van Woerkom, A.J. Meijer, Phosphorylation of ribosomal protein S6 is inhibitory for autophagy in isolated rat hepatocytes, J. Biol. Chem. 270 (1995) 2320–2326.

[105] A. Iwamaru, Y. Kondo, E. Iwado, H. Aoki, K. Fujiwara, T. Yokoyama, G.B. Mills, S. Kondo, Silencing mammalian target of rapamycin signaling by small interfering RNA enhances rapamycin-induced autophagy in malignant glioma cells, Oncogene 26 (2007) 1840–1851.

[106] T. Nobukuni, M. Joaquin, M. Roccio, S.G. Dann, S.Y. Kim, P. Gulati, M.P. Byfield, J.M. Backer, F. Natt, J.L. Bos, F.J. Zwartkruis, G. Thomas, Amino acids mediate mTOR/raptor signaling through activation of class 3 phosphatidylinositol 3OH-kinase, Proc. Natl. Acad. Sci. U.S.A. 102 (2005) 14238–14243.

[107] M. Laplante, D.M. Sabatini, mTOR signaling in growth control and disease, Cell 149 (2012) 274–293.

[108] S. Wang, P. Song, M.-H. Zou, AMP-activated protein kinase, stress responses and cardiovascular diseases, Clin. Sci. 122 (2012) 555–573.

[109] D.M. Gwinn, D.B. Shackelford, D.F. Egan, M.M. Mihaylova, A. Mery, D.S. Vasquez, B.E. Turk, R.J. Shaw, AMPK phosphorylation of raptor mediates a metabolic checkpoint, Mol. Cell 30 (2008) 214–226.

[110] A.M. Cuervo, J.F. Dice, Age-related decline in chaperone-mediated autophagy, J. Biol. Chem. 275 (2000) 31505–31513.

[111] P. Codogno, A.J. Meijer, Autophagy and signaling: their role in cell survival and cell death, Cell Death Differ. 12 (2005) 1509–1518.

[112] M.H. Glickman, A. Ciechanover, The ubiquitin-proteasome proteolytic pathway: destruction for the sake of construction, Physiol. Rev. 82 (2002) 373–428.

[113] C.M. Pickart, M.J. Eddins, Ubiquitin: structures, functions, mechanisms, Biochim. Biophys. Acta 1695 (2004) 55–72.

[114] M.J. Clague, C. Heride, S. Urbé, The demographics of the ubiquitin system, Trends Cell Biol. 25 (2015) 417–426.

[115] M. Akutsu, I. Dikic, A. Bremm, Ubiquitin chain diversity at a glance, J. Cell Sci. 129 (2016) 875–880.

[116] K. Tanaka, T. Mizushima, Y. Saeki, The proteasome: molecular machinery and pathophysiological roles, Biol. Chem. 79 (2012) 217–234.

[117] M.J. Clague, I. Barsukov, J.M. Coulson, H. Liu, D.J. Rigden, S. Urbé, Deubiquitylases from genes to organism, Physiol. Rev. 93 (2013) 1289–1315.

[118] E. Kish-Trier, C.P. Hill, Structural biology of the proteasome, Annu. Rev. Biophys. 42 (2013) 29–49.

[119] Y. Saeki, Ubiquitin recognition by the proteasome, J. Biochem. 161 (2017) 113–124.

[120] I. Amm, T. Sommer, D.H. Wolf, Protein quality control and elimination of protein waste: the role of the ubiquitin-proteasome system, Biochim. Biophys. Acta 1843 (2014) 182–196.

[121] I.A. Voutsadakis, The ubiquitin-proteasome system and signal transduction pathways regulating epithelial mesenchymal transition of cancer, J. Biomed. Sci. 19 (67) (2012) 1–12, https://doi.org/10.1186/1423-0127-19-67.

[122] F. Bassermann, R. Eichner, M. Pagano, The ubiquitin proteasome system – implications for cell cycle control and the targeted treatment of cancer, Biochim. Biophys. Acta 1843 (2014) 150–162.

[123] C. McKinnon, S.J. Tabrizi, The ubiquitin-proteasome system in neurodegeneration, Antioxidants Redox Signal. 21 (2014) 2302–2321.

[124] A. Tramutola, F. DiDomenico, E. Barone, M. Perluigi, D.A. Butterfield, It is all about (u)biquitin: role of altered ubiquitin-proteasome system and UCHL1 in Alzheimer disease, Oxid. Med. Cell. Longev. (2016), article 2756068, 1–12, https://doi.org/10.1155/2016/2758068.

[125] C. Hammond, I. Braakman, A. Helenius, Role of N-linked oligosaccharide recognition, glucose trimming, and calnexin in glycoprotein folding and quality control, Proc. Natl. Acad. Sci. U.S.A. 91 (1994) 913–917.

[126] X. Chen, J. Shen, R. Prywes, The luminal domain of ATF6 senses endoplasmic reticulum (ER) stress and causes translocation of ATF6 from the ER to the Golgi, J. Biol. Chem. 277 (2002) 13045–13052.

[127] L. Eilgaard, A. Helenius, Quality control in the endoplasmic reticulum, Nat. Rev. Mol. Cell Biol. 4 (2003) 181–191.

[128] J.W. Brewer, J.A. Diehl, PERK mediates cell-cycle exit during the mammalian unfolded protein response, Proc. Natl. Acad. Sci. U.S.A. 97 (2000) 12625–12630.

[129] Y.C. Tsai, A.M. Weissman, The unfolded protein response, degradation from the endoplasmic reticulum, and cancer, Genes Cancer 1 (2010) 764–778.

[130] P. Walter, D. Ron, The unfolded protein response; from stress pathway to homeostatic regulation, Science 334 (2011) 1081–1086.

[131] R. Bravo, V. Parra, D. Gatica, A.E. Rodriguez, N. Torrealba, F. Paredes, Z.V. Wang, A. Zorzano, J.A. Hill, E. Jaimovich, A.F.G. Quest, S. Lavandero, Endoplasmic reticulum and the unfolded protein response: dynamics and metabolic integration, Int. Rev. Cell Mol. Biol. 301 (2013) 215–290.

[132] I. Tabas, T. Ron, Integrating the mechanisms of apoptosis induced by endoplasmic reticulum stress, Nat. Cell Biol. 13 (2011) 184–190.

[133] S.H. Benavides, A.J. Montserrat, S. Farina, E.A. Porta, Sequential histochemical studies of neuronal lipofuscin in human cerebral cortex from the first to the ninth decade of life, Arch. Gerontol. Geriatr. 34 (2002) 219–231.

[134] K.L. Double, V.N. Dedov, H. Fedorow, E. Kettle, G.M. Halliday, B. Garner, U.T. Brunk, The comparative biology of neuromelanin and lipofuscin in the human brain, Cell. Mol. Life Sci. 65 (2008) 1669–1682.

[135] A. Höhn, T. Grune, Lipofuscin: formation, effects and role of macroautophagy, Redox Biol. 1 (2013) 140–144.

[136] U.T. Brunk, A. Terman, Lipofuscin: mechanisms of age-related accumulation and influence on cell function, Free Radic. Biol. Med. 33 (2002) 611–619.

[137] A. Höhn, A. Sittig, T. Jung, S. Grimm, T. Grune, Lipofuscin is formed independently of macroautophagy and lysosomal activity in stress-prematurely senescent human fibroblasts, Free Radic. Biol. Med. 53 (2012) 1760–1769.

[138] G. Merlini, V. Bellotti, Molecular mechanisms of amyloidosis, N. Engl. J. Med. 349 (2003) 583–596.

[139] M. Sunde, L.C. Serpell, M. Bartlam, P.E. Fraser, M.B. Pepys, C.C. Blake, Common core structure of amyloid fibrils by synchrotron X-ray diffraction, J. Mol. Biol. 273 (1997) 729–739.

[140] B.H. Toyama, J.S. Weissman, Amyloid structure conformational diversity and consequences, Annu. Rev. Biochem. 80 (2011) 557–585.

[141] R. Nelson, M.R. Sawaya, M. Balbirnie, A.O. Madsen, C. Riekel, R. Grothe, D. Eisenberg, Structure of the cross-beta spine of amyloid-like fibrils, Nature 435 (2005) 773–778.

[142] Y. Fezoui, D.B. Teplow, Kinetic studies of amyloid-beta protein fibril assembly. Differential effects of alpha-helix stabilization, J. Biol. Chem. 277 (2002) 36948–36954.

[143] V.N. Uversky, Protein misfolding in lipid-mimetic environments, Adv. Exp. Med. Biol. 855 (2015) 33–66.

[144] S. Jayaraman, D.L. Gantz, C. Haupt, O. Gursky, Serum amyloid A forms stable oligomers that disrupt vesicles at lysosomal pH and contribute to the pathogenesis of reactive amyloidosis, Proc. Natl. Acad. Sci. U.S.A. 114 (2017) E6507–E6515.

[145] C. Soto, L. Estrada, J. Castilla, Amyloids, prions and the inherent infectious nature of misfolded protein aggregates, Trends Biochem. Sci. 31 (2006) 150–155.

[146] W. Pulawski, U. Ghoshdastider, V. Andrisano, S. Filipek, Ubiquitous amyloids, Appl. Biochem. Biotechnol. 166 (2012) 1626–1643.

[147] J.D. Sipe, M.D. Benson, J.N. Buxbaum, S.-I. Ikeda, G. Merlini, M.J.M. Saraiva, P. Westermark, Amyloid fibril protein nomenclature: 2010 recommendations from the nomenclature committee of the International Society of Amyloidosis, Amyloid 17 (2010) 101–104.

[148] H.J. Lachmann, H.J. Goodman, J.A. Gilbertson, J.R. Gallimore, C.A. Sabin, J.D. Gillmore, P.N. Hawkins, Natural history and outcome in systemic AA amyloidosis, N. Engl. J. Med. 356 (2007) 2361–2371.

[149] R.A. Kyle, A. Linos, C.M. Beard, R.P. Linke, M.A. Gertz, W.M. O'Fallon, L.T. Kurland, Incidence and natural history of primary systemic amyloidosis in Olmsted County, Minnesota, 1950 through 1989, Blood 79 (1992) 1817–1822.

[150] M.M. Picken, Fibrinogen amyloidosis: the clot thickens!, Blood 115 (2010) 2985–2986.

[151] J. Genschel, R. Haas, M.J. Pröpsting, H.H.-J. Schmidt, Apolipoprotein A-I induced amyloidosis, FEBS Lett. 430 (1998) 145–149.

[152] G.B. Irvine, O.M. El-Agnaf, G.M. Shankar, D.M. Walsh, Protein aggregation in the brain: the molecular basis for Alzheimer's and Parkinson's diseases, Mol. Med. 14 (2008) 451–464.

[153] S.T. Ferreira, M.N. Vieira, F.G. de Felipe, Soluble protein oligomers as emerging toxins in Alzheimer's and other amyloid diseases, IUBMB Life 59 (2007) 332–345.

[154] R. Truant, R.S. Atwal, C. Desmond, L. Munsie, T. Tran, Huntington's disease: revisiting the aggregation hypothesis in polyglutamine neurodegenerative diseases, FEBS J. 275 (2008) 4252–4562.

[155] J.W. Höppener, B. Ahrén, C.J. Lips, Islet amyloid and type 2 diabetes mellitus, N. Engl. J. Med. 343 (2000) 411–419.

[156] G. Merlini, D.C. Seldin, M.A. Gertz, Amyloidosis: pathogenesis and new therapeutic options, J. Clin. Oncol. 29 (2011) 1924–1933.

[157] D. Eisenberg, M. Jucker, The amyloid state of prions in human diseases, Cell 148 (2012) 1188–1203.

[158] J.T. Jarrett, P.T. Lansbury Jr., Seeding "one-dimensional crystallization of amyloid: a pathogenic mechanism in Alzheimer's disease and scrapie? Cell 73 (1993) 1055–1058.

[159] T.P. Knowles, C.A. Waudby, G.L. Devlin, S.I. Cohen, A. Aguzzi, M. Vemdruscolo, E.M. Terentjev, M.E. Welland, C.M. Dobson, An analytical solution to the kinetics of breakable filament assembly, Science 326 (2009) 1533–1537.

[160] H.U. Lutz, F. Bussolino, R. Flepp, S. Falser, P. Stammler, M.D. Kazatchkine, P. Arese, Naturally occurring anti-band-3 antibodies and complement together mediate phagocytosis of oxidatively stressed human erythrocytes, Proc. Natl. Acad. Sci. U.S.A. 84 (1987) 7368–7372.

[161] F. Turrini, P. Arese, J. Yuan, P.S. Low, Clustering of integral membrane proteins of the human erythrocytes membrane stimulates autologous IgG binding, complement depositions, and phagocytosis, J. Biol. Chem. 266 (1991) 23611–23617.

[162] N. Arashiki, N. Kimata, S. Manno, N. Mohandas, Y. Takakuwa, Membrane peroxidation and methemoglobin formation are both necessary for band 3 clustering: mechanistic insights into human erythrocyte senescence, Biochemistry 52 (2013) 5760–5769.

[163] S. Kumar, U. Bandyopadhyay, Free heme toxicity and its detoxification systems in human, Toxicol. Lett. 157 (2005) 175–188.

[164] F. Lang, E. Lang, M. Föller, Physiology and pathophysiology of eryptosis, Transfus. Med. Hemother. 39 (2012) 308–314.

[165] L. Repsold, A.M. Joubert, Eryptosis: an erythrocyte's suicidal type of cell death, BioMed Res. Int. (2018), article 9405617, 1–10, https://doi.org/10.1155/2018/9405617.

[166] F. Lang, E. Gulbins, P.A. Lang, D. Zappulla, M. Föller, Ceramide in suicidal death of erythrocytes, Cell. Physiol. Biochem. 26 (2010) 21–28.

[167] E. Pretorius, J.N. du Plooy, J. Bester, A comprehensive review on eryptosis, Cell. Physiol. Biochem. 39 (2016) 1977–2000.

[168] A.D. Maher, P.W. Kuchel, The Gárdos channel: a review of the Ca^{2+}-activated K^+ channel in human erythrocytes, Int. J. Biochem. Cell Biol. 35 (2003) 1182–1197.

[169] J. Schneider, J.P. Nicolay, M. Föller, T. Wieder, F. Lang, Suicidal erythrocyte death following cellular K^+ loss, Cell. Physiol. Biochem. 20 (2007) 25–44.

[170] E. Gulbins, Regulation of death receptor signaling and apoptosis by ceramide, Pharmacol. Res. 47 (2003) 393–399.

[171] S.M. Qadri, J. Bauer, C. Zelenak, H. Mahmoud, Y. Kucherenko, S.H. Lee, K. Ferlinz, F. Lang, Sphingosine but not sphingosine-1-phosphate stimulates erythrocyte death, Cell. Physiol. Biochem. 28 (2011) 339–346.

[172] C. Brugnara, I. de Franceschi, S.I. Alper, Inhibition of Ca^{2+}-dependent K^+ transport and cell dehydration in sickle erythrocytes by clotrimazole and other imidazole derivatives, J. Clin. Investig. 93 (1993) 520–526.

[173] P.A. Lang, S. Kaiser, S. Myssina, T. Wieder, F. Lang, S.M. Huber, Role of Ca^{2+}-activated K^+ channels in human erythrocyte apoptosis, Am. J. Physiol. Cell Physiol. 285 (2003) C1553–C1560.
[174] E. Lang, S.M. Qadri, F. Lang, Killing me softly – suicidal erythrocyte death, Int. J. Biochem. Cell Biol. 44 (2012) 1236–1243.
[175] F. Lang, G.L. Busch, M. Ritter, H. Volkl, S. Waldegger, E. Gulbins, D. Häussinger, Functional significance of cell volume regulatory mechanisms, Physiol. Rev. 78 (1998) 247–306.
[176] C. Duranton, S.M. Huber, F. Lang, Oxidation induces Cl^--dependent cation conductance in human red blood cells, J. Physiol. 539 (2002) 847–855.
[177] E. Lang, F. Lang, Mechanisms and pathophysiological significance of eryptosis, the suicidal erythrocyte death, Semin. Cell Dev. Biol. 39 (2015) 35–42.
[178] T.A. Fuchs, U. Abed, C. Goosmann, R. Hurwitz, I. Schulze, V. Wahn, Y. Weinrauch, V. Brinkmann, A. Zychlinsky, Novel cell death program leads to neutrophil extracellular traps, J. Cell Biol. 176 (2007) 231–241.
[179] V. Brinkmann, U. Reichard, C. Goosmann, B. Fauler, Y. Uhlemann, D.S. Weiss, Y. Weinrauch, A. Zychlinsky, Neutrophil extracellular traps kill bacteria, Science 303 (2004) 1532–1535.
[180] C.F. Urban, D. Ermert, M. Schmid, U. Abu-Abed, C. Goosmann, W. Nacken, V. Brinkmann, P.R. Jungblut, A. Zychlinsky, Neutrophil extracellular traps calprotectin, a cytosolic protein complex involved in host defense against Candida albicans, PLoS Pathog. 5 (2009), 1–18, e1000639, https://doi.org/10.1371/journal.ppat.1000639.
[181] S.K. Jorch, P. Kubes, An emerging role for neutrophil extracellular traps in noninfectious disease, Nat. Med. 23 (2017) 279–287.
[182] H. Yang, M.H. Biermann, J.M. Brauner, Y. Liu, Y. Zhao, M. Herrmann, New insights into neutrophil extracellular traps: mechanisms of formation and role in inflammation, Front. Immunol. 7 (2016), article 302, 1–8, https://doi.org/10.3389/fimmu.2016.00302.

CHAPTER

6

Immune Response and Tissue Damage

6.1 SHORT CHARACTERIZATION OF IMMUNE CELLS AS KEY PLAYERS OF IMMUNITY

6.1.1 Innate and Acquired Immunity

In higher animals, immune defense is ensured by the innate and acquired immune system. While innate immunity represents an evolutionary old mechanism that is found in nearly all animals, the acquired or adaptive immune response is only found in most vertebrates. Main properties and differences between both systems are summarized in Table 6.1.

Normally, an immune response starts with the recruitment and activation of cells of the innate immune system. If this system is unable to manage with the invading pathogen, cells of the acquired immune system are additionally activated to facilitate antigen elimination by cytotoxic lymphocytes and specific antibodies.

6.1.2 The Dual Role of the Immune System in Cell and Tissue Destruction

Structure and functions of the human immune system are very complex. It is not the aim of this chapter to review all aspects of immune activation. The main focus will be directed here on major biochemical and physiological phenomena of immune response associated with cell and tissue destruction.

The immune system is primarily employed to eliminate any threat from the organism, to help to replace damaged material, and to resist invasion of pathogenic microorganisms. In this respect, activated immune cells clearly diminish the degree of tissue damage in combination with the response of several additional systems (see Chapter 7). This is a very useful function for surveillance of our organism.

Otherwise activated immune cells can release numerous proteins and generate reactive species that are able to deteriorate host's own cells and tissues. Although these agents are predominantly directed to kill and destroy unwanted microbes, they can also affect the integrity of biological material in the organism. Undergoing host's cells can additionally release agents that enhance immune response. The severity of cell and tissue damage depends on numerous factors such as the strength of the initial immune activation, the recognition of foreign and own components as antigen, the general state of individual immune response, and the presence of any agents suppressing or modulating immune activation.

Thus, concerning cell and tissue damage, the immune system plays a dual role. On the one hand, it is highly efficient to remove any harm that can disturb the integrity of biological material. On the other hand, as a result of their activity, immune cells can be very harmful to host's cells and tissues if their activation is not

Cell and Tissue Destruction
https://doi.org/10.1016/B978-0-12-816388-7.00006-1

TABLE 6.1 Key Features of the Innate and Acquired Immune System.

Feature	Innate Immune System	Acquired Immune System
Specificity	Nonspecific response	Specific response to antigens
Onset	Nearly immediate activation	Lag time between exposure and response
Formation of specific antibodies	No	Yes
Memory function	No memory	Immunological memory
Key cell types	Neutrophils, monocytes/macrophages, natural killer cells, mast cells, eosinophils	T- and B lymphocytes

fine-tuned or affected by target cells. It would be too easy to categorize these both diametrical activities as good and evil. These activities are closely linked to each other. Their interplay is determined by multiple factors being in part outside of the immune system.

In this chapter, immune cells will be at first shortly characterized concerning their role in immune defense and potential participation in tissue damage. Main principles of regulation of immune and inflammatory response will be highlighted afterward. Important features of polymorphonuclear leukocytes (PMNs) and macrophages, the activation of T cells, the formation of antibodies, and the significance of immune suppression are further topics of this section.

6.1.3 Neutrophils

Like all other blood cells, PMNs (also called neutrophils) are produced in the bone marrow from the corresponding stem cells [1]. After formation, mature neutrophils are stored in the bone marrow for further 4–6 days before they are released into blood [2]. In response to infection or inflammation, additional neutrophils can rapidly be mobilized from this storage pool [3,4].

In 1 L blood of a healthy person, there are about $2.5–7.5 \times 10^9$ neutrophils. Only the number of red blood cells is higher. Neutrophils are the most abundant type of white blood cells. A daily release into blood of approximately 10^{11} neutrophils is assumed [5]. In blood, a mean lifetime of circulating neutrophils is given with about 7 h [6]. Neutrophils can leave the blood and penetrate into tissues, where they reside for several days before they undergo apoptotic cell death [6,7]. Under inflammatory conditions, tissue invasion of neutrophils is considerably enhanced [8,9]. The reverse process, the mobilization of these cells from tissue into blood, is also observed, for example, under the action of steroids [10]. Thus, the mean lifetime of human neutrophils can be longer, up to 90 h was reported, than the above mentioned value [11].

Neutrophils belong to the first cells that are rapidly recruited from circulating blood to inflammatory sites, followed later by monocytes. Their infiltration into infected or injured tissue is highly regulated by adhesion molecules, cytokines, chemotactic agents, and components of the extracellular matrix in both blood vessel wall and adjacent tissue [12,13]. On paving their way through endothelium and contiguous connective tissue to inflammatory loci, PMNs are stepwise activated, being normally fully activated at destination site [14]. These cells are able to recognize, phagocytose, kill, and destroy foreign microorganisms. Special functions of neutrophils such as chemotaxis, phagocytosis, degranulation, formation of reactive species, and formation of extracellular traps are described in more detail in Section 6.4.

During their short lifetime in circulating blood, neutrophils undergo structural and functional alterations. On release into blood, the CXCR2 receptor is well expressed on neutrophils. This receptor is important for migration of neutrophils toward CXC chemokines and thus for recruitment of these cells to inflammatory sites in tissues [15]. Within several hours, the surface CXCR2 receptor expression declines, whereas the previously not expressed CXCR4 receptor is upregulated. On aging, L-selectin is progressively downregulated in neutrophils [16]. Hence, senescent neutrophils have reduced ability to migrate into tissues. Several hours later, neutrophils become apoptotic. In apoptotic neutrophils, the CCR5 receptor is additionally expressed. Apoptotic cells are unable to migrate. Under inflammatory conditions, aged neutrophils show a proinflammatory activity and prone to extracellular trap formation [16].

Senescent neutrophils are cleared in the spleen, liver, and bone marrow, where they are phagocytosed by macrophages [4,11,17,18]. Clearance of neutrophils by the spleen and liver is independent of the expression of CXCR4, while senescent neutrophils expressing this receptor are predominantly cleared by the bone marrow [4]. The latter pathway provides a feedback mechanism stimulating granulopoiesis.

6.1.4 Eosinophils

These immune cells are important to combat helminthic and parasitic infections as well as viral infections [19]. Eosinophils can migrate into digestive tract (without esophagus), the female reproductive tract, and the mammary gland. These cells are also involved in a variety of allergic/atopic diseases such as allergic rhinitis, asthma, eczema formation, and in the pathology of some kinds of cancer [20–22].

They are larger than neutrophils and account for less than 5% of circulating leukocytes. Similar to neutrophils, eosinophils invade into tissues where they can survive up to 12 days in the absence of stimulation. On immune activation, these cells become closely attached to parasites and others and discharge their granule's content into target cells and organisms. The released granule proteins can deactivate and damage target cells and host cells [19,23]. Among these proteins are eosinophil peroxidase, major basic protein, eosinophil cationic protein, and eosinophil-derived neurotoxin. Some of these proteins exhibit an antiviral activity as they act as ribonuclease [24].

6.1.5 Basophils and Mast Cells

These two types of immune cells are very similar to each other. However, they are derived from different lineages of hematopoiesis [25]. Basophils are the largest circulating granulocytes. In unaffected blood, they represent less than 1% of white blood cells. Like other granulocytes, they reside in tissues if needed. Mast cells are not found in circulating blood. They leave the bone marrow in an immature form and mature only after residing in tissue [26]. Preferred resident loci of mast cells are connective tissues surrounding blood vessels and nerves as well as boundaries to the external milieu such as the skin and mucous surfaces in the lungs, gut, nose, mouth, and eye [25].

Mast cells resemble in their properties basophils in the blood. They contain the vasodilatory agent histamine, the anticoagulant heparin, chondroitin sulfate, neutral proteases such as chymase, tryptase, and carboxypeptidases, which they release on activation by immunoglobulin E [27,28]. Both cell types are involved in allergic reactions, cardiovascular diseases, anaphylaxis, asthma, gastrointestinal diseases, and others [29,30].

Similar to neutrophils, mast cells are able to release extracellular traps that provide a phagocytosis-independent antimicrobial activity [31,32].

6.1.6 Macrophages

Nearly all tissues contain macrophages. These tissue-resident cells are derived from embryonically established precursors in the yolk sac and fetal liver [33–35]. In tissues, resident macrophages maintain themselves through local proliferative processes [36,37]. In addition, tissue-resident macrophages can also be derived from circulating monocytes in the postnatal period. In the gut, macrophages are permanently recruited from blood monocytes [38].

In tissues, homeostatic functions of macrophages are quite different and depend on tissue specificity [39]. Some organs such as the liver, spleen, and lung contain a higher number of macrophages. Here, the main focus is directed on macrophage functions during innate and adaptive immune response. In inflamed tissues, additional monocytes/macrophages are recruited from circulating blood, which normally accumulate time delayed after neutrophils.

Although many cell types are able to phagocytose other cells and cell debris, the macrophages are the intrinsic scavenger cells in our organism. They are able to eliminate apoptotic and necrotic cells, cell and tissue debris, and pathogens. With this activity, they clear tissues from any waste. These cells are also highly important for regulation of immune response. More details about macrophage functions in relation of inflammatory response are given in Sections 6.3 and 6.5.

6.1.7 Dendritic Cells

Dendritic cells are a subset of macrophages. These cells present after phagocytosis of pathogenic and host material antigens on their surface to T cells. Other professional antigen-presenting cells are macrophages and B cells. Preferred location sites of dendritic cells are the skin, lung, nose, stomach, and intestinal tract [40]. At these locations attacks of external antigens are more likely. In blood, dendritic cells are in a premature state.

On activation, dendritic cells migrate into the lymph nodes, where they interact with lymphocytes and initiate the acquired immune response [41].

6.1.8 Natural Killer Cells

Natural killer cells, a special subtype of lymphocytes, recognize tumor- and virus-infected cells in host organism due to a diminished number of specific surface markers, the class I major histocompatibility complex (MHC) molecules [42]. They have the ability to detect stressed cells in the absence of antibodies and MHC molecules. After recognizing affected cells with altered MHC profile, natural killer cells destroy them by releasing cytotoxic contents from secretory lysosomes such as granzymes and perforin [43].

On receptor-mediated binding of natural killer cell to a target cell, apoptosis is induced in the latter cell via Fas ligands [44]. There is an interplay between activating and inhibitory receptors in natural killer cell action against target cells [45]. Natural killer cells also recognize infected cells opsonized with antibodies via CD16 receptors [46].

6.1.9 T and B Cells

Together with natural killer cells, T cells and B cells comprise the group of lymphocytes. T cells and B cells are the main cell types of the acquired immune system, whereas natural killer cells are attributed to the innate immune system.

Cytotoxic $CD8^{+}$ T cells are equipped with a similar cell-killing machinery like natural killer cells. They recognize their target cells via the T-cell receptor complex that is sensitive against antigen fragments bound to class I MHC molecules on the cell surface. This killing mechanism differs from natural killing cells as it acts at normal expression of class I MHC molecules.

T-helper cells are able to recognize specific antigens fragments bound to class II MHC molecules on the surface of professional antigen-presenting cells such as dendritic cells, B cells, and macrophages. Following this antigen presentation, these cells respond with proliferation, differentiation, and release of cytokines. Important subsets of T-helper cells are T_h1, T_h2, T_h17, and T-regulatory cells [47].

In lymphoid organs, B cells are responsible in interaction with T-helper cells for the production of high-affine antibodies toward target antigens.

Activities of T and B cells are directed to better eliminate pathogens and infected cells. Moreover, a subset of these lymphocytes is involved in developing an immunological memory that facilitates future pathogen recognition. Involvement of T cells and B cells in adaptive immune reactions is outlined in Section 6.6.

6.2 REGULATION OF IMMUNE PROCESSES

6.2.1 Initiation of Immune Reactions

A huge number of physical factors, chemical compounds, and foreign microorganisms can disturb the continuous supplementation of energy and nutritional equivalents to host's cells and tissues and causes thus, any threat. In consequence, cells are notably restricted in their functional responses and many of them die. Our immune system recognizes these alterations in cell properties and tissue homeostasis by help of specialized cells that are spread over all tissues. Sensing by these resident immune cells might activate a whole cascade of immune reactions that are directed to remove any threat from the organism and to ensure normal functioning of cells and organs. This complex answer is known as inflammation [48].

For example, neutrophils are known to permeate through vessel walls and reside in tissues for several days. In the absence of any substantial activation, they undergo apoptotic cell death [7,49]. Similarly, subsets of monocytes patrol extravascular tissues and intravascular space to clear any cell debris and antigens [50–52]. Tissue monocytes migrate further to the lymph vessels, where they are finally destroyed in lymph knots [50]. In addition, there are populations of tissue-specific macrophages, which reside independent of blood monocytes and fulfill specific homeostatic functions within these tissues [52,53]. Thus, an important function of immune cells in nonaffected tissue is the monitoring of their nearby environment for any deviations from the resting state. Other cell types such as endothelial cells, fibroblasts, keratinocytes, and mast cells contribute to this supervision as well [54,55]. Finally, sensitive synaptic end plates of the nociceptive nervous system are also involved in control of the functional state in tissues [56].

On ongoing inflammation, the functional response of tissue-residing immune cells and other surrounding cells changes considerably [57,58]. Activated cells generate specific mediators that alter properties of adjacent cells, induce vasodilatation, increase vessel wall permeability, attract numerous other immune cells to inflammatory sites, and promote differentiation of monocytes to macrophages. Communication between cells at inflammatory loci, immune cells, and the bone marrow, where all white blood cells are originated from, is realized by a set of cytokines, chemokines, growth factors, and others.

6.2.2 Pattern Recognition Receptors

Immune, endothelial, and other cells are equipped with special receptors recognizing structural elements from pathogens and damaged host's cells. These receptors are collectively called pattern recognition receptors (PRRs) [59]. There are five main types of evolutionary old PRRs (Table 6.2) having a broad specificity [61]. Further atypical forms of PRRs are receptors for advanced glycation end products [63].

Of the PRRs, the toll-like receptors (TLRs) are most intensively characterized. In humans, 10 TLRs are known to date. As transmembrane receptors, they are found both in the plasma and endosomal membranes. They sense the extracellular environment for specific structural elements from microorganisms or undergoing host's cells. In cytosol, main ligands for PRRs are viral RNA and DNA [61].

Activation of PRR signaling initiates two major events in immune cells [64]. First, various cytokines and antimicrobial compounds are released, which are directed to attract further immune cells, to combat against invading microorganisms, and to decrease tissue damage. The second activity mediated by PRRs consists in induction of dendritic cell maturation for presentation of antigens to T cells (Section 6.6.1).

6.2.3 DAMPs and PAMPs as Ligands to Pattern Recognition Receptors

Ligands to PRRs are known as pathogen-associated molecular patterns (PAMPs) [65] and damage-associated molecular patterns (DAMPs) [66]. The concept of immune activation by PAMPs and DAMPs is very attractive as it combines a rather broad spectrum of initiating events ranging from pathogens (viruses, bacteria, fungi, and others) to various kinds of host's cellular constituents released by traumata and other disturbances during silent inflammation. Thus, PRRs do not differentiate between PAMPs and DAMPs. They are able to recognize both kinds of ligands. For example, the TLR4 receptor is targeted by the bacterial product lipopolysaccharide (LPS) and free heme.

Common ligands to TLRs are summarized in Table 6.3. In plasma membrane, PRRs respond to specific bacterial and fungal components such as LPSs, flagellin, lipoteichoic acid, peptidoglycan, viral envelope proteins, oxidized phospholipids, and β-glucans [59,81,82]. Endosomal and cytoplasmic PRRs ligate viral DNA and RNA as well as bacterial molecules containing

TABLE 6.2 Main Types of Pattern Recognition Receptors (PRRs).

PRR Type	Abbreviation	Location	Ligands and Specificity	References
Toll-like receptors	TLRs	Plasma and endosomal membranes	Bacterial, viral, fungal, and protozoal products and host's components	[59]
Nucleotide oligomerization domain–like receptors	NLRs	Cytoplasm	Bacterial peptidoglycans; inflammasome formation	[60]
Retinoic acid–inducible gene I–like receptors	RLRs	Cytoplasm	Viral RNA	[59,61]
C-type lectin receptors	CLRs	Plasma membrane	Microbial carbohydrates	[62]
Absent in melanoma 2-like (AIM2-like) receptors	ALRs	Cytoplasm	Foreign and host's cytosolic dsDNA; inflammasome formation	[61]
Advanced glycation end product receptors	RAGE	Plasma membrane	Advanced glycation end products, high-mobility group protein B1, calcium-binding proteins	[63]

TABLE 6.3 Pathogen-Associated Molecular Patterns (PAMPs) and Damage-Associated Molecular Patterns (DAMPs) as Ligands to Human Toll-Like Receptors (TLRs).

Receptor	Expression	PAMPs	DAMPs	References
TLR2	Neutrophils, monocytes, macrophages, B cells, T cells, dendritic cells, epithelia	Different bacterial lipids	Heat shock protein 60	[67,68]
TLR1/TLR2 heterodimer	Gut epithelium, neutrophils, monocytes	Mycobacterial lipopeptides, lipoarabinomannan	Heat shock protein 70	[69,70]
TLR3	Fibroblasts, dendritic cells, macrophages	Double-stranded microbe DNA, polyinosinic:polycytidylic acid	Double-stranded DNA from necrotic cells	[71–73]
TLR4	Neutrophils, monocytes, macrophages, endothelial cells, gut epithelium	Lipopolysaccharide, teichuronic acid, mannuronic acid polymers, F protein	Heat shock proteins, free heme, cholesteryl hydroperoxides	[74]
TLR5	Monocytes, macrophages, gut epithelium	Flagellin, profilin		[75]
TLR6	Monocytes, macrophages, mast cells, B cells	Diacyl lipopeptides		[76]
TLR7	Monocytes, macrophages, dendritic cells, B cells	Single-stranded viral RNA		[77,78]
TLR8	Monocytes, macrophages, mast cells, gut epithelium	Single-stranded viral RNA, bacterial RNA		[77,78]
TLR9	Monocytes, macrophages, dendritic cells, killer cells	CpG oligodeoxynucleotide DNA		[79]
TLR10	Monocytes, macrophages, gut epithelium, B cells	Triacylated lipopeptides		[80]

D-glutamyl-meso-diaminopimelic acid or muramyl dipeptide [61,83].

DAMPs are also known as danger-associated molecular pattern, danger signals, or alarmins. On tissue injury, a number of nuclear and cytosolic proteins can be released from cells. During the release, these proteins denature as they move from a reducing to an oxidizing milieu [84]. Among them are heat shock proteins and high-mobility group box 1 [85]. Nonprotein DAMPs are hyaluronan fragments, ATP, uric acid, heparan sulfate, and free heme [86–89]. Another DAMP is tumor DNA that is released from undergoing cells [90].

6.2.4 Some Regulatory Aspects in Pattern Recognition Receptors Activation

Inflammatory response is a strong answer of the organism to any threat. This energy-consuming activation makes only sense when other protecting mechanisms are limited or exhausted. Many potential molecular patterns can efficiently be inactivated before they interact with PRRs. For example, free heme is normally complexed by hemopexin and cleared as heme–hemopexin complex from circulation (see Section 4.2.7). Noncomplexed free heme can ligate to TLR4 on endothelial and other cells and promote inflammatory reactions [89,91].

There are also different species-dependent threshold concentrations for some PRR ligands to cause activation. While humans respond already to low doses of LPS with TLR4 activation, much higher doses of LPS are necessary in mice [92–96]. Mouse serum contains a yet unknown factor inactivating LPS [95].

Any differentiation between PAMPs and DAMPs is only useful when the inflammatory process is newly initiated. Under chronic inflammatory conditions, death of host's cells and infection by pathogens occur often concomitantly. Thus, multiple different molecular patterns might be present at inflammatory sites. The continuous appearance of DAMPs and alterations in deactivating systems contribute to repeated change of acute and nonacute phases during the pathological process in long-lasting chronic inflammations.

Immune cells are equipped with hazardous components primarily designed to inactivate and destroy foreign microorganisms and to degrade tissue debris. These components can also participate in tissue destruction. Moreover, coactivation of PRRs by several ligands might lead to a stronger cytotoxic reaction to cells and tissues. Immune cell–induced tissue destruction seriously aggravates an already existing pathological process. At severe inflammations, it is highly necessary to combat any concurrent infection by immune cells even at the price of additional cell damage.

6.2.5 Inflammasome Formation

In myeloid cells, on ligand binding to some PRRs (receptors of the NLR and ALR subfamily), an intracellular cascade of events is induced, leading to the formation of an inflammasome [97]. This multiprotein oligomer is comparable with the apoptosome formed during the intrinsic pathway of apoptosis induction. The inflammasome is composed of assembled cytoplasmic receptors linked by the adaptor protein ASC that recruits procaspase 1 to the complex [98]. Fig. 6.1 shows the principal pathway of inflammasome formation.

The inflammasome generates active caspase 1 that maturates proinflammatory cytokines IL-1β and IL-18 from their inactive forms [99,100]. Other effects of caspase 1 activation are induction of interferon-γ (IFN-γ) secretion, inactivation of IL-33, activation of natural killer cells, DNA fragmentation and cell pore formation, inhibition of glycolytic enzymes, activation of lipid biosynthesis, and secretion of tissue repair

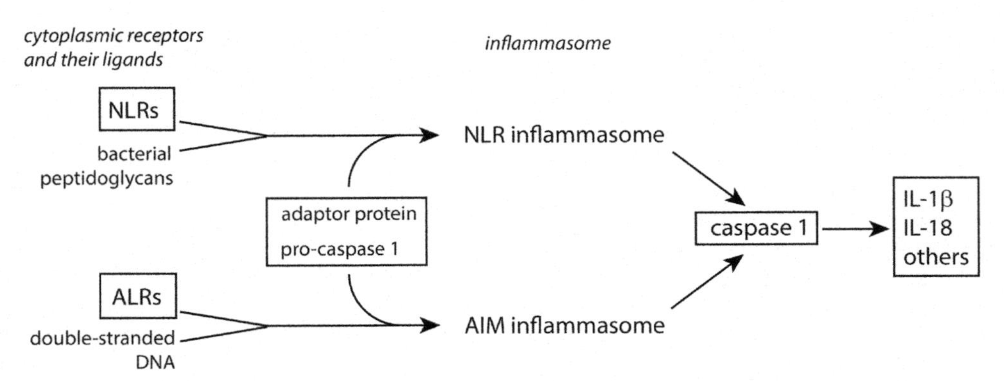

FIGURE 6.1 Schematic view of the inflammasome formation and caspase 1 activation on interaction of suitable ligands with cytoplasmic pattern recognition receptors. *AIM*, absent in melanoma-2; *ALRs*, AIM-like receptors; *IL*, interleukin; *NLRs*, nucleotide oligomerization-like receptors. Further explanations are given in the text.

factors [101–103]. The associated cell death mechanism, the pyroptosis, is different from apoptosis (Section 5.5.1) [104].

6.2.6 Energy Metabolism in Activated Immune Cells

Activation of macrophages by LPS/IFN-γ or by ligands of TLRs, namely TLR2, -3, -4, and -9 ligands, is accompanied by a switch from oxidative phosphorylation to glycolysis [105]. This switch does not occur on incubation of macrophages with IL-4, IL-13, glucocorticoids, immune complexes, or interleukin-10 (IL-10) [105]. A shift to glycolysis occurs also on maturation of dendritic cells initiated by TLR2, -4, and -9 ligands [106]. In LPS-treated macrophages, the switch to glycolysis increases succinylation of several proteins and enhances IL-1β production [107]. IL-10 inhibits LPS-mediated glycolysis induction, promotes oxidative phosphorylation, and suppresses mammalian target of rapamycin (mTOR) activity [108].

A polarization of metabolic pathways is observed in differentiating T-helper cells as well. T_h1, T_h2, and T_h17 cells follow the glycolysis pathway via mTOR signaling in contrast to T-regulatory cells, where mTOR signaling is suppressed [109–111].

6.2.7 Circadian Rhythm of Immune Responses

It is without any doubt that the intensity of immune response depends on the activity status of the organism. In humans, activity of the immune system is triggered by several hormones, particularly by cortisol and melatonin [112–114]. While cortisol suppresses immune activity during the day, the sleeping hormone melatonin promotes immune response at night times. However, many details about the circadian rhythm in immunity are not well understood.

Melatonin is known to promote early phase of inflammatory response [115]. It activates cytoplasmic phospholipase A_2 and thus provides arachidonic acid for production of prostaglandins and leukotrienes. Otherwise, this action of melatonin extinguishes within 2–3 h [116,117]. Hence, this hormone contributes also to resolution of a time-limited inflammation.

6.3 INFLAMMATORY RESPONSE

6.3.1 Acute versus Chronic Inflammation

An acute inflammation is a nonrecurring cascade of events initiated by defined physical, chemical, pathogenic, or other factors. Under assumption that the initiating agent was not too strong, the organism answers by a sequence of biochemical and physiological events that can be roughly divided into three main phases. These parts are exudation, infiltration, and proliferation phases (Fig. 6.2). There is some overlapping between these phases. It is impossible to separate these phases clearly. The total duration of an acute inflammation depends on many factors suh as the strength of the initiator, the degree of damage, and the individual immune status. Thus, in some cases, an acute inflammatory process can last one week or even longer. Normally, this kind of inflammation is well terminated without any visible traces or under formation of replacement tissue. In textbooks, heat, pain, redness, swelling, and loss of function are usually listed as the five classical signs of an acute inflammation.

On the cellular levels, the transition from inflammatory state to normal tissue homeostasis includes a transient immunosuppression as well as apoptosis and removal of unneeded immune cells (Fig. 6.2).

Another picture is found for chronic inflammations. Many diseases are often associated with chronic inflammatory states. Here, it

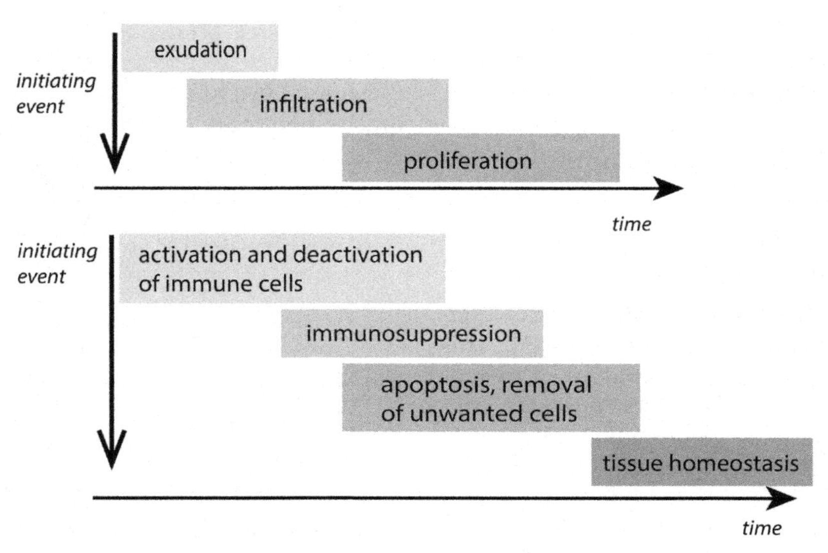

FIGURE 6.2 Differentiation of an acute inflammatory response into dominating physiological processes. In the upper panel, the focus is directed on main general events affecting inflammatory areas. Sequential changes in the state of immune and tissues cells occurring during inflammation are presented below. There is an overlap between different phases.

remains often unclear what was the initiating event. In chronically affected tissues, damaged areas can be colocalized with unperturbed regions. At the same time, attempts to repair defects are observed. Thus, the molecular image of chronic inflammatory response is very heterogeneous. Moreover, acute and silent phases of inflammation can alter to each other. A serious problem is the insufficient termination of the inflammatory process. This is associated with the recurrent action of molecular patterns, the weakness of individual immune defense, and the low capacity or exhaustion of other protecting systems.

Despite these complications in understanding the real nature of chronic inflammations, main features of physiological phases of an acute inflammation will be shortly characterized. These features are also analogously valid for the chronic process.

6.3.2 Exudation Phase

In affected tissues, an inflammatory response starts with alterations in blood vessels. Several mediators are released by immune and other cells that locally increase the blood flow and vessel wall permeability. Among these mediators are histamine, bradykinin, prostaglandin E_2, leukotriene B_4, and nitrogen monoxide.

Histamine released from basophils and mast cells increases the blood flow and endothelial permeability for about 10–20 min [118]. It exerts its action via G protein–coupled receptors that are highly expressed on venular endothelial cells in comparison with the arteriolar and capillary endothelium [119]. Uncontrolled release of histamine and other mediators from basophils and mast cells causes anaphylactic reactions [120]. Bradykinin is another short-time acting dilator of blood vessels. Its release is linked to factor

XII of the coagulation system (see Sections 7.4.4 and 7.5.1). However, fine mechanisms of its production in the human body are unknown [121].

During inflammatory response, long-lasting exudation followed by leukocyte infiltration takes place on activation of venular endothelial cells by proinflammatory cytokines such as IL-1α, IL-1β, and TNF-α [122]. The activation of the corresponding cytokine receptors on endothelial cells leads to signaling events and transcriptional processes. In addition, PAMPs and DAMPs induce also activation responses via their receptors [123]. As a result, properties of the adjacent blood endothelium are altered to allow recruitment and extravasation of circulating neutrophils and other immune cells.

Mast cell–derived histamine and bradykinin, a product of the contact system (Section 7.5.1), enhance the vascular permeability [124]. At inflammatory sites, blood plasma penetrates also into tissue spaces, where plasma components such as complement factors, lysozyme, and antibodies are involved in opsonization and damage to microbes. Within the tissue, extravasated plasma proteins help to promote migration of neutrophils and other immune cells.

Moreover, several cell types such as fibroblasts and endothelial cells start to synthesize and release the long-chain carbohydrate hyaluronan at inflammatory sites. Hyaluronan is thereby fixed via CD44 to cell surfaces [125]. As very hygroscopic macromolecule, it attracts enormous amounts of water and influences physiochemical properties in their microenvironment [126]. Molecular crowding by high molecular weight hyaluronan excludes other macromolecules from their territory, concentrates other molecules in the residual water, and potentiates osmotic effects [127].

The aforementioned actions facilitate the subsequent infiltration of immune cells and prepare foreign and damaged host's material to be better recognized and phagocytosed by these cells.

In addition to the action of mediators from immune and other cells, several protecting systems mediated by plasma components can be activated during inflammatory response. These additional systems are the complement system, the kinin system, the coagulation system, and the fibrinolysis system. Contribution of these systems to prevention of tissue damage will be summed up in Chapter 7.

6.3.3 Infiltration Phase

Numerous immune cells are attracted to inflammatory sites. These cells inactivate and kill any pathogens and phagocytose damaged material. Usually neutrophils invade first into inflamed loci followed by monocytes/macrophages. At strong inflammatory events, several waves of invasion of these cells can be observed. Recruitment of further cell types such as eosinophils and lymphocytes depends on the local cytokine milieu. Important cytokines involved in inflammatory immune response and their functions are listed in Table 6.4.

The extravasation of neutrophils from blood to tissues can be roughly divided into the following steps:

i) margination of cells from circulating blood and endothelial adhesion,
ii) diapedesis, t. e. migration across the endothelium, and
iii) chemotactic movement of cells within the tissue.

All processes of recruitment, infiltration, and activation of neutrophils and other immune cells are triggered by cytokines and supported by interaction of adhesion molecules with their receptors [149–152]. At inflammatory sites, neutrophils are normally fully activated. Here, they phagocytose microbes, death cells, and cell debris and contribute thus to elimination of tissue-damaging agents.

TABLE 6.4 Important Cytokines Involved in Inflammatory Immune Response.

Cytokine	Short Name	Released Cells	Target Cells	Remarks	References
Interleukin-1α/1β	IL-1α/IL-1β	Macrophages, many other cells	Lymphocytes, natural killer cells, macrophages, endothelial cells	Regulation of immune and inflammatory responses	[128]
Tumor necrosis factor-α	TNF-α	Macrophages, many other cells	Immune cells, tissue cells	Regulation of immune and inflammatory response	[129,130]
Interleukin-18	IL-18	Macrophages	T_h1 cells, natural killer cells	Increases IFNγ production and activity of natural killer cells	[131]
Interferon-γ	IFN-γ	T cells, natural killer cells, mucosal epithelial cells	T_h1 cells, T_h2 cells, macrophages	Promotes T_h1 cell response, suppresses T_h2 cells	[132,133]
Interleukin-33	IL-33	Endothelial cells, epithelial cells, fibroblasts	T cells, natural killer cells	Nuclear cytokine, promotes T-cell responses	[134]
Interleukin-6	IL-6	Monocytes, macrophages, and others	Hepatocytes, B cells, T cells	Effects on acute-phase proteins, involved in antibody production	[135]
Interleukin-8	IL-8	Endothelial cells and others	Neutrophils	Chemotactic agent to neutrophils	[136]
Interleukin-12	IL-12	Lymphocytes, dendritic cells, macrophages	T_h1 cells	Formation of cytotoxic T cells (with IL-2)	[137]
Interleukin-10	IL-10	T_h2 cells, T_{reg} cells, and others	Lymphocytes, macrophages, mast cells	Antiinflammatory activity	[138,139]
Transforming growth factor-β	TGF-β	Macrophages, many other cells	Macrophages and other cells	Antiinflammatory activity	[140,141]
Interleukin-2	IL-2	T-helper cells	T-helper cells, B cells, natural killer cells	Growth factor	[142]
Interleukin-4	IL-4	T_h2 cells and others	Many tissue cells	Regulation of inflammatory response, induction of alternative pathways	[143]
Interleukin-5	IL-5	T_h2 cells, mast cells	Eosinophils, B cells	Regulates eosinophil activation, IgA production	[144]
Interleukin-17	IL-17	T_h17 cells	Epithelial and endothelial cells	Stimulates angiogenesis, osteoclastogenesis	[145]

TABLE 6.4 Important Cytokines Involved in Inflammatory Immune Response.—cont'd

Cytokine	Short Name	Released Cells	Target Cells	Remarks	References
Granulocyte macrophage colony-stimulating factor	GM-CSF	Immune cells, endothelial cells	Monocytes, macrophages	Activation and differentiation of macrophages	[146]
Interleukin-21	IL-21	T_h17 cells, natural killer cells	Lymphocytes, macrophages, dendritic cells	Different proinflammatory effects	[147]
Interleukin-23	IL-23	Macrophages	T cells, natural killer cells, monocytes, dendritic cells	Effects on T_h17 cells	[145]
Interleukin-35	IL-35	T_{reg} cells	Lymphocytes	Suppression of T-helper cells	[148]

Recruitment of neutrophils is mainly supported by the expression of E-selectin on inflamed endothelial cells. This adhesion molecule mediates tethering and rolling of neutrophils on vascular endothelium. Its expression starts about 2 h after receptor activation, is maximal after 6–10 h, and declines afterward to baseline levels within 24 h [153]. The expression of the adhesion molecules VCAM-1 occurs more time delayed on activated endothelium. VCAM-1 is involved in recruitment of monocytes, eosinophils, and lymphocytes [154].

At inflammatory sites, neutrophils contribute by several mechanisms to recruitment and invasion of monocytes/macrophages. Some neutrophil's granule proteins such as azurocidin and proteinase 3 promote extravasation of monocytes via receptor activation on neighboring endothelial cells and macrophages [155,156]. Release of IL-6 and IL-6 receptors shed from neutrophils favor expression of monocyte receptors on endothelial cells. Mediators released from activated neutrophils such as CCL2, CCL19, CCL20, and IL-8 are involved in monocyte recruitment [157].

TNF-α and IFN-γ are two important signaling molecules for proinflammatory activation of macrophages [158]. Activation of macrophages must be tightly controlled as the produced cytokines and mediators can contribute to cell and tissue damage [159].

The duration of the infiltration and activity of immune cells depends on the presence of DAMPs and PAMPs. As long as these molecular patterns exist, the corresponding proinflammatory cytokines and chemokines are generated and induce signals that promote recruitment of further leukocytes to inflammatory site. Only after the decline of the corresponding DAMPs and PAMPs, antiinflammatory mediators are released, causing a stop of further invasion of immune cells and providing the condition for proliferation of novel tissue cells or cells of replacement tissues. This signaling is also important for efficient termination of inflammation.

6.3.4 Proliferation Phase and Resolution of Inflammation

During exudation and infiltration phases of inflammation, activities of the immune system are mainly directed to recognize and combat against foreign pathogens and altered host's components (the PAMP and DAMP concept), to limit any sites of tissue destruction, to eliminate damaged material, and to create better conditions for the ongoing immune reactions. Major objectives of the following proliferation phase are regeneration of destroyed cells and tissues, downregulation of all proinflammatory activities, and restoration of former tissue homeostasis.

This terminal phase of inflammation is also strongly regulated. Although many biochemical and regulatory aspects are only scarcely known, there is a progressive decline in proinflammatory triggers and agents in later phases of inflammatory response. In addition, a switch in main signaling routes of macrophages and other cells leads to release of antiinflammatory cytokines and inflammation-resolving mediators [138,160,161]. Apoptosis of proinflammatory cells, especially apoptosis of neutrophils, downregulation of proinflammatory signal cascades, and desensitization and downregulations of receptors for proinflammatory signals are further actions playing a substantial part in resolution of inflammation. In addition, a transient immunosuppression supports these activities.

On clearance of apoptotic neutrophils at inflammatory sites, macrophages change their phenotype [162]. For this phenotype switch, two distinct signals are required [161]. One signal alone, e.g., the presence of immune complexes, adenosine, or apoptotic cells, is usually without effect. In combination with a second stimulus like TLR ligands, macrophages are reprogrammed to produce IL-10 and to downregulate IL-12 [138,139].

Macrophage signaling is crucial for initiating of tissue regeneration, or at least repair with scar formation, after acute injury and removal of proinflammatory stimuli. As tissues differ in their regenerative capacities, functional responses of macrophages may also differ among tissue types. In skeletal muscle and the liver, tissues with high regeneration rate, normally complete reconstitution is achieved after injury. Otherwise, tissues with limited regeneration capacity such as the heart and brain tend to heal wounds by fibrotic scarring with loss of tissue structure and function [163–165]. Macrophages contribute to tissue regeneration not only by the switch from pro- to antiinflammatory signaling but also by controlling the expansion and differentiation of the stem and progenitor cells, by supporting survival of parenchymal and stromal cells via growth factors, and by regulating processes of angiogenesis and extracellular matrix production [166].

6.3.5 Mediators of Resolution of Inflammation

Important inflammation-resolving agents are transforming growth factor β (TGF-β), IL-10, and specialized proresolving lipid mediators such as lipoxins, resolvins, neuroprotectins, and maresins.

The TGF-β is secreted from activated macrophages on ingestion of apoptotic neutrophils [140]. It inhibits proliferation and activity of cytotoxic T cells [167,168], promotes the generation of T-regulatory cells [169], and represses cytokine production and cytotoxicity of natural killer cells [170,171]. In activated dendritic cells, the expression of class II MHC molecules is blocked by TGF-β [172]. These activities of TGF-β contribute to suppression of immune reaction during resolution of inflammation [173–176]. Otherwise, the production of TGF-β by many types of tumors dampens antitumor attacks of the host's immune system and contributes to tumor surveillance and progression [173,176].

In immunosuppression activity, TGF-β and IL-10 collaborate together [177,178]. In proinflammatory macrophages, IL-10 inhibits glycolysis-dependent pathways and expression of glycolytic genes as well as promotes inhibition of mTOR activity and drives dysfunctional mitochondria into mitophagy [108]. This cytokine can be released by many immune cells at inflammatory sites. As a master regulator, IL-10 suppresses proinflammatory activities of macrophages, dendritic cells, T_h1, and T_h2 cells [179]. In infections, response of IL-10 should be delayed in time to ensure that the proinflammatory activity is only downregulated after clearance of pathogens [179]. There again, IL-10 can contribute to an uncontrolled growth of pathogens and persistent infections [180–182].

During resolution of inflammation, proinflammatory metabolites of arachidonic acid like prostaglandins and leukotrienes are replaced by terminating lipoxins [183,184]. Latter mediators are also derived from arachidonic acid, but two different oxidizing enzymes presumably from distinct cell types are necessary for their biosynthesis. There are two major routes for lipoxin formation. For example, lipoxin A_4 is formed by sequential action of neutrophil 5-lipoxygenase and platelet 12-lipoxygenase on arachidonic acid [185]. Transcellular biosynthetic pathways are possible due to close contact of distinct cell types at inflammatory loci.

Lipoxins and related epi-lipoxins exert their antiinflammatory activity via high-affinity G protein–coupled receptors. They redirect neutrophils to apoptosis and impede lifetime-prolonging effect on neutrophils by myeloperoxidase and the acute-phase protein serum amyloid A [186,187]. These mediators delay apoptosis in macrophages and inhibit the secretion of cytokines from T cells [188].

Arachidonic acid is an ω-6 polyunsaturated fatty acid. Similar terminating lipid mediators such as lipoxins are derived from ω-3 polyunsaturated fatty acids. Resolvins, neuroprotectins, and maresins promote resolution of inflammation by enhancing the uptake of apoptotic neutrophils by macrophages and diminishing the recruitment of neutrophils [189–191].

Important inflammation-resolving lipid mediators are depicted in Fig. 6.3. In many cases, resolution of inflammation occurs without any defects or persistent damages.

6.3.6 The Interplay Between Neutrophils and Macrophages During Inflammation

In most inflammations, neutrophils and macrophages are the key players at inflammatory sites. The interplay between these cells is thought to determine the further outcome of an inflammation. Monocytes/macrophages are usually recruited after neutrophils to inflammatory sites. These cells phagocytose any cell debris including dying neutrophils. As long as necrotic cell material predominates at inflammatory loci, macrophages secrete proinflammatory mediators such as TNF-α, IL-6, and IL-12 that further promote the inflammation [192]. For example, elastase, cathepsin G, and heat shock proteins released from defective neutrophils can modify surface molecules of macrophages and induce a proinflammatory activation of these cells [193–195]. Thereby, it does not play a role whether necrotic neutrophils were formed by direct necrosis or secondary necrosis from previously apoptotic vesicles.

In contrast, macrophages are triggered toward an antiinflammatory pathway on rapid clearance of apoptotic neutrophils [7,140,196–198]. Removal of mainly intact apoptotic vesicles by macrophages initiated signals that stop the further recruitment of novel unperturbed immune cells, deactivated proinflammatory cytokines, and initiated tissue repair and other inflammation-resolving activities.

This very simplified concept provides indeed a good working basis to understand how

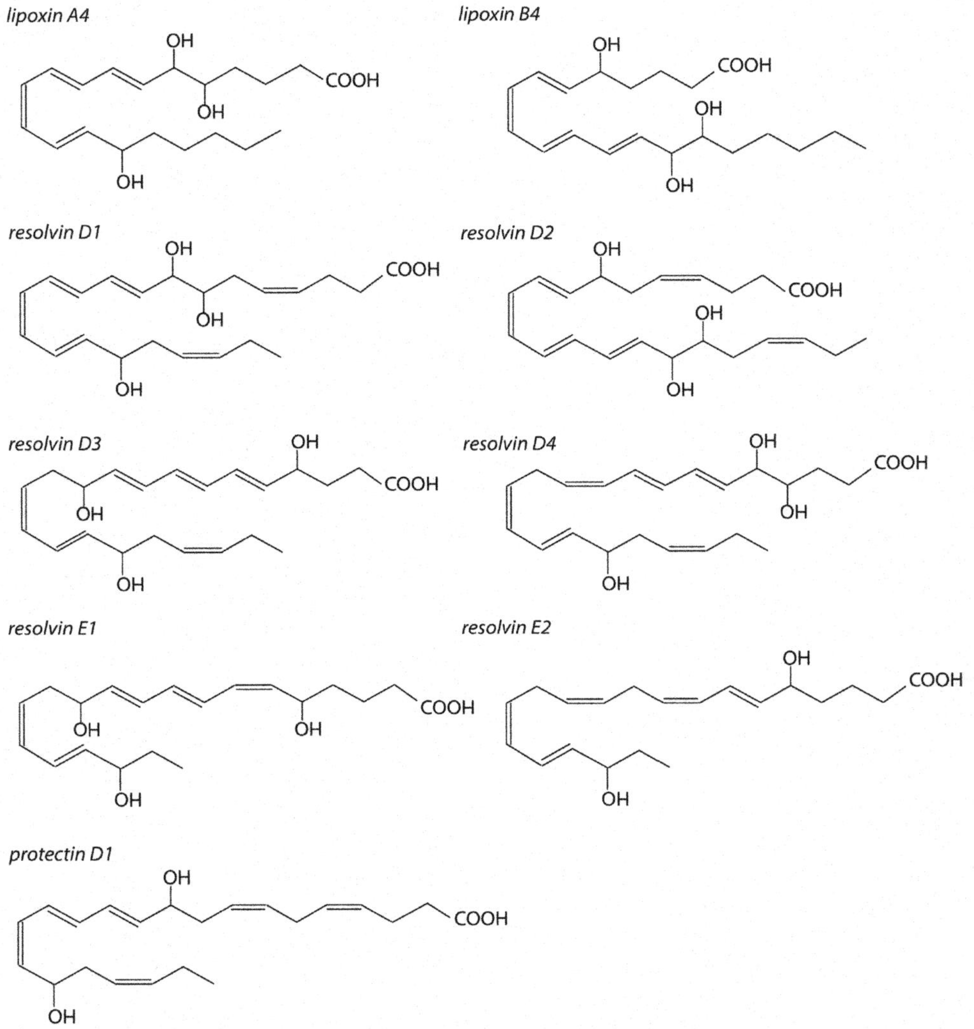

FIGURE 6.3 Selected lipid-based mediators of resolution of inflammation. Lipoxins are derived from arachidonic acid. Resolvins E1 and E2 are products of eicosapentaenoic acid. All other mediators shown are generated from docosahexaenoic acid.

properties of macrophages regulate the fate of inflammatory processes and contribute, thus, to maintenance of tissue homeostasis. However, this concept does not reflect the huge variety of damaged material, undergoing cells, and cell debris, as well as the concurrent existence of opposite, parallel, and redundant pathways at inflammatory loci. The real in vivo situation is rather multifaceted even within an inflamed area of a given patient. To date, it is impossible to give a clear answer on how this switch occurs from a proinflammatory state to inflammation-resolving phase. It seems to be that this switch also highly depends on the general health status, energy supply, state of blood circulation, age, and other factors.

6.3.7 Persistence of Chronic Inflammation

Insufficient termination can lead to long-lasting chronic inflammation and associated diseases. It is challenging to give a clear answer to which factors contribute to this development. A critical checkpoint in regulation of inflammation is the switch from proinflammatory state to resolution. Deficiencies in this switch are often considered as a major source of chronic inflammation such as defective apoptosis of neutrophils. This subject is outlined later (Section 6.7.2).

I will stress here on another very crucial aspect. A serious reason for chronicity of inflammation is the potential of activated immune cells to damage host's cells and tissues by unbalanced release of aggressive components (Fig. 6.4). Then, novel DAMPs and antigens are generated from affected unperturbed cells that again aggravate the inflammation. How strong the real cell and tissue damage depends not only on the strength of the inflammation-initiating agents but also on the individual status of protecting systems. The latter status can vary considerably from one patient to the next.

For instance, some serum proteins can scavenge and inactivate aggressive components of invading defective neutrophils. The serpin α_1-antitrypsin inactivates elastase and cathepsin G, whereas α1-antichymotrypsin is active against cathepsin G and some other proteases. Ceruloplasmin antagonizes myeloperoxidase. During inflammatory response, although being acute-phase proteins, the activity of these serum proteins can be affected by reactive species (see Sections 7.2.3 and 7.2.5). Moreover, they can locally be exhausted under conditions of strong inflammation and limited supply.

Besides release of proteases and other hydrolyzing enzymes, reactive species generated by neutrophils and other cells also contribute to damaging reactions. Scavenging and inactivation of reactive species depends on the presence and efficiency of detoxifying proteins. Free transition metal iron and copper ions as well as their complexes are very dangerous against the integrity of cells and tissues. Diminished control over these mechanisms in combination with the

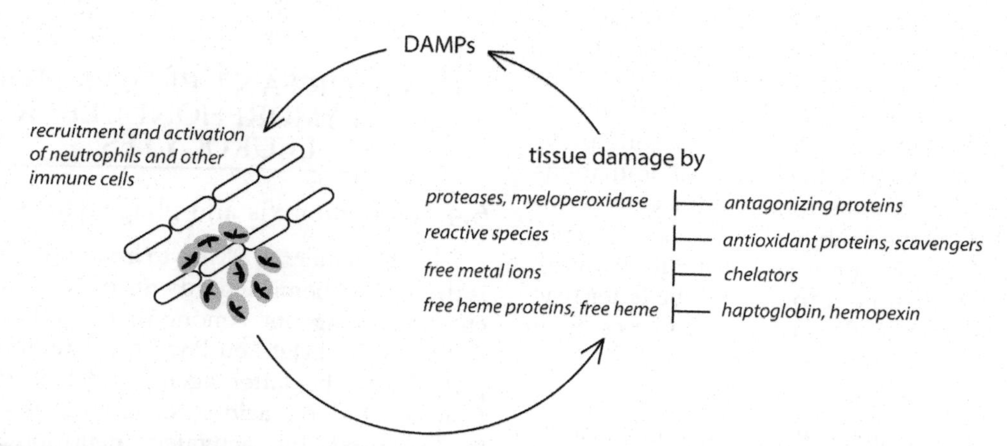

FIGURE 6.4 Simplified scheme for the persistence of chronic inflammatory states. This vicious cycle results when tissue-damaging agents, which are released from activated immune cells and dysfunctional blood and tissue cells, are insufficiently inactivated due to limitation or exhaustion of protecting mechanisms. Major cell and tissue-damaging agents and their inhibitors/scavengers are indicated. *DAMPs*, damage-associated molecular patterns.

presence of reactive species favors cell and tissue destruction. Release of hemoglobin from red blood cells and myoglobin from damaged skeletal muscles are very hazardous when the protecting haptoglobin and hemopexin are exhausted (see Sections 4.4.1 and 4.4.2).

Taken together, the activated immune cells release and generate numerous cell and tissue-damaging agents. However, the impact of these agents on destruction of adjacent biological material is fundamentally determined by the status of protecting systems, the inactivation of proteases, the scavenging of free transition metal ions, the degradation and neutralization of reactive species, and binding and removal of free hemoglobin/myoglobin and free heme (Fig. 6.4). In other words, during inflammatory response, the degree of inevitable tissue damage depends on the general patient's health status, on its life habit, on the presence of risk factors, on patient's age, on genetic predisposition, and on numerous other factors. A large quantity of parameters affects cytotoxicity to host's cells and tissues at inflammatory sites and determines whether and to what extent destruction of intact biological material occurs.

Thus, cell and tissue damage is mainly favored by already exhausted or low-capacity protecting systems. This explains the general rise of chronic inflammatory diseases with increasing age as well as the augmentation of disease complications by already existing previous illness conditions. In chronic inflammations, proinflammatory phases are insufficiently terminated due to the inability of host's protecting systems to prevent or at least to limit unwanted destructive reactions. Products of the latter reactions foment proinflammatory processes again and again.

6.3.8 Fibrosis and Other Outcomes of Inflammation

Fibrosis, as an outcome of chronic inflammatory processes, is characterized by excess deposition of fibrous connective tissue including collagen. In fibrotic disorders, normal tissue architecture is progressively remodeled and destroyed in favor of connective tissue elements [199]. Known diseases with fibrotic etiology are pulmonary fibrosis, liver cirrhosis, cardiovascular fibrosis, systemic sclerosis, and nephritis.

Various matrix metalloproteinases and their inhibitors regulate collagen turnover and extracellular matrix remodeling [200]. In fibrotic tissues, imbalance between formation of new collagen by myofibroblasts and collagen degradation can be caused by several mechanisms such as conserved PAMPs derived from bacteria, viruses, and other pathogens [201,202], autocrine TGF-β signals [203], T-helper cell type 2 cytokines [204], autoantibodies [205], and angiotensin II [206,207]. In affected areas, scars can be formed, which are primarily composed of collagen. Hence, functional impairment may occur at these loci.

Formation of abscesses filled with pus is another kind of inflammatory outcome. Pus consists of remains of neutrophils, dead cells, cell debris, and fluid. Purulent inflammation is typical for infection with pyogenic bacteria such as staphylococci [208,209].

6.4 IMPORTANT PROPERTIES OF POLYMORPHONUCLEAR LEUKOCYTES

6.4.1 Chemotaxis and phagocytosis

The movement of PMNs to inflammatory sites is driven by chemical gradients of cytokines and chemotactic agents. Among them are IL-8, IFN-γ, C3a, C5a, fMet-Leu-Phe, and leukotriene B_4 [210–214]. The latter agent, which is derived from arachidonic acid by 5-lipoxygenase [215], is produced by activated neutrophils and ensures via contraction of microfilaments a directed movement of the phagocytes to their targets [216].

Activated neutrophils recognize opsonized material and phagocytose them. They can internalize many microbes and deactivate and kill them by secretion of antimicrobial and hydrolytic proteins and formation of reactive species. This activity increases the consumption of dioxygen. Hence, the term "respiratory burst" is sometimes used for this phase of neutrophil activation.

In the newly formed phagosomes, a special internal milieu is created by fusion with preformed neutrophil granules, ion fluxes, and activation of the fused components [217]. During the first 3 minutes, the phagosomal pH raises to about 7.8–8.0 driven by NADPH oxidase. Then, the pH value diminishes gradually to about 7.0 after 10–15 min and to about 6.0 during the next hour [218,219].

6.4.2 Release of Granule Constituents

Neutrophils are equipped with numerous small vesicular granules that contain preassembled hydrolytic and bactericidal proteins. On cell activation, granules either fuse with the plasma membrane and release their content in the surrounding milieu or discharge their constituents into the formed phagosomes. In PMNs, there are four types of granules differing in their internal and membranous components [220], threshold cytoplasmic Ca^{2+} concentration for degranulation [221,222], and thus in their functional role during cell activation. These granule types are secretory vesicles, tertiary granules, specific (or secondary) granules, and azurophilic (or primary) granules. Important granule constituents released on activation of neutrophils and their main functions are listed in Table 6.5.

At low levels of cell activation, secretory vesicles and tertiary granules discharge their components. These proteins help to degrade components of the endothelium and adjacent extracellular matrix and enable penetration of neutrophils through these layers [222]. At destination sites, where bacteria or fungi are phagocytosed, specific and azurophilic granules release their contents. These granules contain numerous antimicrobial and hydrolytic proteins that contribute to inactivation and degradation of microbial components and to activation and regulation of granule components. The most aggressive proteins are found in azurophilic granules that usually detach their content into the phagosomes. At inflammatory sites, release of azurophilic components into the surrounding medium might occur due to premature activation of neutrophils or incomplete, also so-called frustrated, phagocytosis. This release contributes to host's cell and tissue damage.

Among components of azurophilic granules are myeloperoxidase (MPO), the serine proteases elastase, cathepsin G, proteinase 3, bactericidal/permeability-increasing protein, numerous defensins, and others. Elastase and cathepsin G are able to degrade bacterial proteins. Bactericidal/permeability increasing protein, defensins, and lysozyme act by disrupting anionic bacterial surfaces and increase thus the membrane permeability.

In azurophilic granules of resting neutrophils, myeloperoxidase and other cationic proteins are sequestered and inactivated by chondroitin 4-sulfate and other proteoglycans [240,241].

6.4.3 Formation of Reactive Species

Besides degranulation, PMNs generate reactive species on cell activation. The NADPH oxidase reduces dioxygen to superoxide anion radicals that further supply hydrogen peroxide by dismutation. During cell activation, this enzyme is rapidly assembled from membranous and cytoplasmic components [242,243]. The heme protein myeloperoxidase (Section 2.5.1) that is unique to PMNs utilizes hydrogen peroxides to produce HOCl, HOBr, and hypothiocyanite and to oxidize numerous small molecules [239].

TABLE 6.5 Main Secreted Granule Constituents of Human Neutrophils and Their Functions.

Constituent	Granule Type	Functions	References
Albumin and other plasma components	Secretory	Release from neutrophils rolling along the activated endothelium	[223,224]
Type IV collagenases, gelatinases	Specific, tertiary	Degradation of collagen IV in the basement membrane	[225]
Lysozyme	Specific	Degradation of bacterial wall peptidoglycans	[226]
Lactoferrin	Specific	Sequestration of free iron, interaction with outer bacterial membranes	[227,228]
Histaminase	Specific	Inactivation of histamine	[229]
Bactericidal/permeability-increasing protein	Azurophilic	Killing activity against Gram-negative bacteria, binds to lipopolysaccharides	[230]
Defensins	Azurophilic	Antimicrobial and antiviral activities	[231]
Azurocidin	Azurophilic, specific	Initiation of immune response, activation of monocytes and macrophages	[232]
Elastase	Azurophilic	Shedding of surface molecules on macrophages, killing of Gram-negative bacteria	[233,234]
Cathepsin G	Azurophilic, specific, tertiary	Degradation of extracellular matrix components	[234]
Proteinase 3	Azurophilic, specific, tertiary	On the surface of resting and dying neutrophils	[235,236]
Neutrophil serine protease 4	Azurophilic	Apparently involved in regulation of neutrophil response	[237,238]
Myeloperoxidase	Azurophilic	Generation of HOCl, HOBr, and $^{-}OSCN$, oxidation of small molecules	[239]

The involvement of the MPO–H_2O_2–halide system in killing of microbes has been demonstrated in numerous reports and verified by MPO inhibitors or applying neutrophils from MPO-deficient individuals [244–248]. However, another concept support the view that under the high load of granule proteins and a pH value around 7.8–8.0 in the newly formed phagosome [217], MPO largely halogenates other granule proteins rather than the phagocytosed material [249,250]. By this activity, other granule proteins may be either activated or depressed [251]. For example, HOCl is known to inactivate elastase, cathepsin G, proteinase 3, and matrix metallopeptidase 9 [251].

Bacteria and fungi are phagocytosed by neutrophils and affected by a concerted action of hydrolyzing and microbicidal proteins and reactive species. In most cases, these microbes are successfully deactivated. However, some pathogens developed strategies to survive these attacks. In general, there is a redundancy in deactivating pathways in PMNs. Furthermore, complex interaction between granule constituents also takes place. Although, there exist many investigations on how neutrophils kill pathogens, molecular details are not yet understood in each detail.

As already noticed, the role of MPO remains questionable during phagocytosis of microbes by neutrophils. Total and partial MPO deficiency occurs in humans with an abundance of 1 case on 2000–4000 individuals [252,253]. These persons live normally. Only in some cases problems with *Candida albicans* infections were reported [254–256]. Another picture is found in MPO-knockout mice, where more serious complications were observed [247,251,257–259]. It seems that human neutrophils developed additional strategies to combat with microbes in comparison with murine cells. For example, azurophilic granules of human neutrophils contain four types of defensins, a group of small cysteine-rich cationic proteins that induce defects in microbial membranes [260]. Mouse neutrophils do not express defensins [261].

During the first phase of phagocytosis, the formation of sufficient HOCl by MPO is unlikely due to the slightly alkaline pH. In vitro investigations demonstrated no or very low HOCl formation above pH 7 by the MPO–H_2O_2–Cl^- system [262,263]. Under neutral conditions, HOCl formation by MPO is mostly detected by taurine. This agent binds near active site to the enzyme and is halogenated without formation of free HOCl [264,265]. Otherwise, neutrophils produce concomitantly with phagosome formation increasing amounts of superoxide anion radicals that can support the transient formation of Compound III of MPO [266]. This MPO compound is unable to oxidize (pseudo)halides. Another situation is found at later phases of phagocytosis of microbes, when the pH within phagosomes drops down to acidic values up to pH 5. Under these conditions, the halogenating activity of MPO increases considerably and sufficient free HOCl is generated [267]. These thoughts support more likely a role of MPO products in control and termination of the phagocytic activities and protecting of surrounding tissues from uncontrolled proteolysis rather than an active contribution of this enzyme in microbe killing.

6.4.4 Formation of Neutrophil Extracellular Traps

Under inflammatory conditions, dying neutrophils can extrude so-called extracellular traps, where myeloperoxidase, proteases, and other granule components form a tight complex with DNA and citrullinated histones [268]. This special kind of cell death is known as NETosis (see Section 5.8). The extruded webs of fibers can trap and kill external microbes independent of phagocytic uptake [269,270]. Some strains of microbes developed strategies to escape scavenging by traps such as expression of chromatin degrading nucleases or modification of outer polysaccharide capsules to reduce binding to trap fibers [271,272].

In contrast to apoptotic neutrophils, cells undergoing NETosis do not express "eat-me" signals and are not cleared by macrophages [273]. The fine-tuning of signals to drive neutrophils into apoptosis or necrosis or trap formation remains unknown.

6.4.5 Tissue Damage by Polymorphonuclear Leukocytes

Components of azurophilic granules can also be released from PMNs into surroundings of the cells during the so-called frustrated phagocytosis or by premature activation of these cells [274,275]. Then, these components contribute to tissue degradation despite deactivating mechanisms by some plasma proteins. Another source for the release of these components is the necrotic cell death of neutrophil. As outlined in more detail in Chapter 5, primary and secondary necrotic events of neutrophils are detrimental for adjacent tissues. Excessive NETosis also contributes to disease progression including autoimmune disorders and sepsis [276–279].

Some disease scenarios such as bacterial meningitis, acute glomerulonephritis, ischemia-reperfusion injury, adult respiratory distress syndrome, or hypersensitivity reactions are accompanied by neutrophil-mediated enhanced damage of host material [57,280,281]. Several lines of evidence indicate the crucial role of neutrophil apoptosis and clearance of apoptotic neutrophils by macrophages in termination of inflammation. Disruption of these mechanisms promotes inflammation and associated pathologies [194,282,283].

6.5 MACROPHAGES AS MAIN PHAGOCYTES

6.5.1 Tissue-Resident and Monocyte-Derived Macrophages

Although many different cells are able to phagocytose other cell material, most intense phagocytosis is realized by macrophages. In addition to tissue-resident macrophages, monocytes are recruited from circulating blood to inflammatory sites, where they maturate to macrophages.

Under inflammatory conditions, macrophages fulfill several important functions. They remove any undergoing material from hosts or foreign origin. They release anti- and proinflammatory mediators as well in dependence on the dominance of apoptotic or necrotic material in their near environment. Furthermore they are able, in particular dendritic cells, to present antigens to T cells. With these activities macrophages are key regulators of inflammation [161].

6.5.2 Phagocytosis

Macrophages are enabled to engulf and digest large amounts of undergoing cells and waste material. Interaction between macrophages and phagocytosing material is driven by different "find-me" and "eat-me" signals. The latter ones correspond to PAMPs and DAMPs. Apoptotic cells express "eat-me" signals as apoptotic cell–associated molecular patterns (ACAMPs) [284,285].

Otherwise, cell membrane–based and loosely membrane-attached "do not eat-me" signals protect intact host cells from engulfment, hindering macrophages to phagocytose these cells. These signals are also known as self-associated molecular patterns (SAMPs) [285]. Among them are CD31, CD47, CD55, OX2/CD200, sialic acid, complement C1 inhibitor, factor H, factor I, and some others. These signals are expressed by host cells, but they are absent on microorganisms. It has been hypothesized that "do not eat-me" signals control the humoral and/or innate immune responses. They can interact with inhibitory PRRs expressed on the surface of macrophages. This interaction protects host cells from engulfment. On apoptosis of cells, the expression of "do not eat-me" signals decreases [286]. Both the presence of ACAMPs and SAMPs on apoptotic cells favors the phagocytosis of apoptotic cells by macrophages, maintains a noninflammatory response, and promotes tissue repair [285,287].

Among "find-me" signals of undergoing cells are the nucleotides ATP and uridine-5′-triphosphate (UTP), the soluble fragment CX3CL1, lysophosphatidylcholine, and sphingosine 1-phosphate [288]. These soluble agents are formed and released during apoptosis of cells. Following interaction with their counterpart receptors on the surface of macrophages, the phagocytes start to move in direction to the signal source.

A variety of "eat-me" signals on dying cells and receptors to these signals on macrophages provide a close contact between these players. Among these signals are phosphatidylserine epitopes, binding sites for complement proteins C1q or C3b/bi, binding sites for the mannose binding lectin, lung surfactant proteins, thrombospondin-1, and others [289]. In most living cells, phosphatidylserine is normally found on the inner leaflet of the plasma membrane [290]. Phosphatidylserine accumulates at the outer leaflet in dying cells, where these epitopes serve as a matrix for a number of proteins such as annexin 1, thrombospondin, and β2-glycoprotein 1 [7,289]. In apoptotic cells, myeloperoxidase binds also to phosphatidylserine epitopes [291,292]. Exposure of phosphatidylserine on the outer leaflet is associated with activation of scramblases and inhibition of translocases [293]. In apoptotic neutrophils, both caspase-dependent and -independent mechanisms lead to appearance of phosphatidylserine epitopes [294,295]. Otherwise, a transfer of phosphatidylserine from internal granule stores to the cell surface cannot be excluded during apoptosis [296].

The significance of phosphatidylserine epitopes as "eat-me" signal is under discussion as some investigations state that the sole presence of these epitopes is insufficient for engulfment of material by macrophages [288,297]. The presence of additional "eat-me" signals such as calreticulin or others have been proposed [298]. Another possibility is the presence of yet existing "do not eat-me" signals such as CD47 and CD31 together with appearance of phosphatidylserine epitopes [298,299]. These survival signals need to be reduced to allow phagocytosis by macrophages. Other explanations are based on the existence of a threshold value for phosphatidylserine exposure in the outer leaflet [297] or the assumption of the presence of oxidized phosphatidylserine [300].

Macrophages are highly flexible and rapidly engulf the phagocytosing material that is further processed by phagolysosomal degradation. These cells have a great arsenal of different receptors for "eat-me" signals. Originally, a single phosphatidylserine receptor has been proposed [301]. The present knowledge supports the existence of multiple distinct receptors for phosphatidylserine that recognize phosphatidylserine either directly or via soluble bridging molecules [288,302].

6.5.3 Classification of Macrophages

At inflammatory sites, activated macrophages are known to secrete various mediators that either promote or depress the inflammatory process. Their release depends on the material phagocytosed by macrophages. As long as PAMPs and DAMPs and necrotic cell material dominates at inflammatory loci, macrophages will secrete proinflammatory mediators signaling that further immune cells are needed to be attracted to inflammatory sites. Among molecular patterns, DAMPs are important arising from necrotic neutrophils as well as tissue damage caused by activated neutrophils.

Only the dominance of apoptotic cells, particularly apoptotic neutrophils, and their uptake by macrophages provide the basis for internal switches to secrete antiinflammatory mediators by macrophages. This switch initiates tissue repair and downregulates and terminates the inflammatory process. With these activities, macrophages play a key role in regulation of inflammation.

Often macrophages are classified into M1 (classically activated) and M2 (alternatively activated) macrophages. This polarization is based on gene expression profiles and identification of surface markers on murine macrophages. For each subset, a panel of markers exist [303]. In macrophages of human origin, the identification of M1 and M2 subsets is more challenging due to species-related differences of macrophage properties. Some M2 mouse markers are absent in men which arginase 1, Fizz1, matrix metalloproteinase-1, and Ym1, whereas Ykl39, fibrinoligase, and platelet-derived growth factor C are lacking in mice [304]. Otherwise, this classification does not reflect the enormous functional plasticity of macrophages. A more functional-based subdivision into classically activated, wound healing, and regulatory macrophages was proposed [161]. A number of recommendations were currently addressed to find a consensus in confusing terminology of macrophage activation [305].

Enormous differences exist between murine and human macrophages in the arginine and tryptophan metabolism. In mice, activated inducible nitric monoxide synthase (iNOS) can produce large amounts of nitric monoxide and L-citrulline. This enzyme is upregulated by LPS and IFN-γ in murine macrophages by several orders of magnitude but not in human cells [303,306,307]. However, other reports state an expression of iNOS and enhanced NO production under disease conditions [308–310]. Maybe, other cytokine combinations are responsible for these observations.

6.6 SPECIAL ASPECTS OF ACQUIRED IMMUNE RESPONSE

6.6.1 Presentation of Antigens

Fragments of both exogenous and endogenous proteins and other molecules can be presented by all cell types but red blood cells on the cell surface as a complex with MHC molecules. These surface complexes activate either cytotoxic lymphocytes or initiate the formation of antibodies against antigens. Human MHC molecules are also known as human leukocyte antigens. Presentation of antigens provides the link between innate and adaptive immune response.

There are two main principal strategies for antigen presentation. In our organism, all nucleated cells and platelets display class I MHC molecules that can form a complex with small peptides of intracellularly digested antigens [311]. The presentation of these complexes on the plasma membrane enables patrolling cytotoxic CD8-positive T cells, also known as killer T cells, to recognize and kill transformed or infected cells [312].

The second more specific pathway involves the expression of class II MHC molecules by the professional antigen-presenting cell types such as dendritic cells, B cells, or macrophages [313,314]. These cells internalize exogenous and sometimes also endogenous antigens by endocytosis, autophagy, and related pathways and process them by endosomal and lysosomal degradation. In endoplasmic reticulum, class II MHC molecules form a complex with antigen peptide fragments. These complexes are finally transported to the plasma membrane. In addition to the stepwise maturation during antigen presentation, these cells migrate to lymph nodes, where they communicate with CD4-positive T-helper cells [42].

The presentation of extracellular antigens by antigen-presenting cells via class I MHC molecules to cytotoxic T cells is known as cross-presentation [315,316]. By this mechanism, exogenous antigens can also be processed by the class I MHC pathway. This pathway is of particular importance in immune response against viruses, bacteria, and tumors [316–318] and in promotion of central and peripheral immune tolerance [316,319].

During inflammatory response, dendritic cells as main antigen-presenting cells undergo maturation that is highly necessary to activate T cells. Appropriate triggers for dendritic cell maturation are ligands docking on TLRs and tumor necrosis factor receptors [320]. Otherwise, immature dendritic cells present antigens from peripheral tissues to naïve T cells in lymph nodes resulting in deletion of the T cell or conversion into T-regulatory cell, a mechanism important for peripheral immune tolerance [320].

Presentation of specific antigens on the surface of dysfunctional or damaged cells or by professional antigen-presenting cells activates cytotoxic T cells or T-helper cells in dependence on the involved MHC molecules.

6.6.2 Activation of Cytotoxic T Cells ($CD8^+$ T Cells)

In response to the class I MHC pathway, $CD8^+$ T cells form a tight complex with affected cells and induce pores in the plasma membrane of their targets [321,322]. Affected cells are finally lysed or driven into apoptosis. Numerous factors contribute to regulate and control differentiation, migration, effector functions, and exhaustion of $CD8^+$ T cells [323]. Important regulatory factors for immune functions of $CD8^+$ T cells are interleukin-2 (IL-2) and IFN-γ that are released by $CD4^+$ T cells [324,325]. Natural killer T cells are also involved in regulation of $CD8^+$ T-cell functions [326].

6.6.3 Activation of T-Helper Cells ($CD4^+$ T Cells)

Presentation of antigen fragments bound to class II MHC molecules on the surface of dendritic cells and other professional antigen-presenting cells provokes binding, proliferation, and differentiation of naïve T-helper cells, as well as cytokine release. Differentiation of T-helper cells depends on the cytokine milieu, strength of T-cell receptor interaction, and other factors in the microenvironment [327,328]. There are several subsets of T-helper cells such as T_h1, T_h2, T_h9, T_h17, T_h22, T follicular helper cells, and T-regulatory cells [47]. Moreover, these subsets are heterogeneous in cytokine production due to epigenetic modification, variable expression of some transcription factors, and stochastic events during cell signaling [328]. Another aspect is the plasticity of different $CD4^+$ T-cell lineages, in particular in T_h17 and T-regulatory cells [329,330].

On interaction of a naïve T-helper cell with an antigen-presenting cell, a tight complex is initially formed between the T-cell receptor and the antigen-MHC II molecule. For T-cell activation, a second independent signal is mandatory, indicating that really a foreign antigen is presented. For instance, this co-stimulatory signal is provided through interaction of CD28 expressed on T cells and CD86 found on dendritic cells exposed to proinflammatory agents [331,332]. In the absence of co-stimulation, adherent T cells are not activated and become anergic.

6.6.4 Differentiation of Activated T-Helper Cells

Activated T-helper cells secrete IL-2 that initiates as an autocrine and paracrine signal T cell proliferation and differentiation [333]. About three decades ago, activated murine T-helper cells were first subdivided into two major groups, T_h1 and T_h2 cells, on the basis of their functions and secreted cytokines [334,335]. This simplified model is principally valid for humans with the difference that human IL-10 does not promote T_h2 response.

At present, three main T-effector cell pathways (Fig. 6.5) are known according to differentiation of activated T-helper cells [336]. T_h1 cells mediate host defense against intracellular bacteria, protozoa, and viruses. On activation, these cells secrete IFN-γ, lymphotoxin α, and IL-2. IFN-γ is responsible for monocyte/macrophage recruitment and activation and triggers B cells to produce IgG immunoglobulins [337]. In humans, T_h1 cells are thought to be involved in resistance to mycobacterial infections and induction of some autoimmune diseases such as autoimmune type 1 diabetes and multiple sclerosis [338,339].

The T_h2 cell pathway is involved in immune response against extracellular parasites including helminths predominantly at mucosal surfaces. The interleukins IL-4, IL-5, IL-9, IL-10, IL-13, and IL-25, and amphiregulin are released by T_h2 cells [340]. Other effector cells of T_h2 response are basophils, mast cells, and eosinophils [341,342]. T_h2 cytokines drive B cells to produce IgE immunoglobulins [343]. T_h2 cells

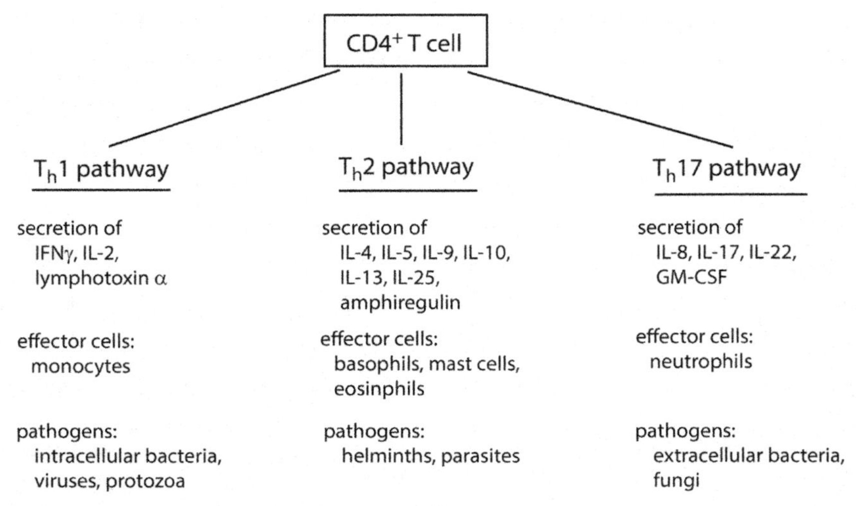

FIGURE 6.5 Major pathways resulting from activation of naïve $CD4^+$ T cells. Typical properties of these pathways are indicated. Further explanations are given in the text. *GM-CSF*, granulocyte macrophage colony-stimulating factor; *IFN*, interferon; *IL*, interleukin.

also play a substantial role in induction of allergic diseases such as asthma and others [344,345].

IL-17 secreting T_h17 cells are differentiated from activated T-helper cells under exposure of TGF-β, IL-6, IL-21, and IL-23 [145]. Granulocyte macrophage colony-stimulating factor, IL-8, and IL-22 are key mediators released from T_h17 cells. This pathway is directed against extracellular bacteria and fungi. Cytokines released from T_h17 cells are closely linked to neutrophil production and recruitment. Dysregulation of T_h17 cells is involved in autoimmune disorders such as multiple sclerosis, rheumatoid arthritis, and psoriasis [346].

6.6.5 Activation of B Cells and Production of Specific Antibodies

Functions of B cells include the binding and presentation of antigens, secretion of cytokines, and the production of antibodies. Activation of B cells occurs in secondary lymphoid organs such as the lymph nodes and spleen [347]. These cells are activated on recognition of antigens via B-cell receptors [348]. Complexes of antigen fragments bound to class II MHC molecules are presented on the cell surface and recognized by T follicular helper cells that were activated by the same antigen [349,350]. This co-stimulation initiates B-cell proliferation and differentiation and leads to formation of short-lived plasmablasts and long-lived plasma cells and memory B cells. Plasmablasts and plasma cells release antibodies toward their target antigens. Plasma cell–derived antibodies have a higher affinity compared with antibodies from plasmablasts. They are released later in an infection than plasmablast-derived antibodies [351]. Formation of plasma cells is associated with immunoglobulin class switching and affinity maturation directed by somatic hypermutation [352].

This efficient mechanism of formation of highly specific antibodies and memory B cells is known as T cell–dependent B-cell activation. It normally takes place in the presence of foreign proteins. Otherwise, some antigens such as foreign polysaccharides and unmethylated

CpG DNA can activate B cells independent of T cells. Then, the resulting antibodies have lower affinity [351].

The formation of memory B cells is crucial for recurring exposure to antigens and initiates an accelerated and robust generation of antibodies during immune response [353].

6.6.6 Regulatory T Cells

A subset of T cells differentiates into T-regulatory cells. These cells regulate primarily induction and proliferation of effector T cells and also innate immune response and maintain tolerance to self-antigens. Their former name, T suppressor cells, indicates that downregulation of immune responses is an essential part of their functions. This tight regulation limits pathogen-induced and sterile inflammatory responses, prevents autoimmune and allergic diseases, and maintains the protective memory. Otherwise, immunosuppression by excessive control reactivates infectious diseases, favors secondary infections, and contributes to tumor development [354].

T-regulatory cells formed in the thymus are known as natural T-regulatory cells (nT_{reg} cells). The so-called induced T-regulatory cells (iT_{reg} cells) are derived from mature T cells outside of the thymus [355,356]. Although sharing similar properties with nT_{reg} cells, different iT_{reg} cell populations have complementary functions in vivo in dependence on milieu conditions and degree of inflammation [355,357]. Induced T_{reg} cells play a pivotal role in maintenance of mucosal tolerance including tolerance to intestinal commensal bacteria [358,359].

Expression and release of inhibitory cytokines, such as IL-10 [360], TGF-β [361], and IL-35 [362], is a common function of T_{reg} cells. Further, activated human nT_{reg} cells express cytotoxic granzyme A and perforin [363]. T_{reg} cells are able to disrupt several metabolic pathways. They cause depletion of IL-2 [364] and generate suppressive adenosine [365]. Modulation of functions of dendritic cells, T-effector cells, monocytes, and macrophages is another inhibitory mechanism of T_{reg} cells [366,367]. Expression of indoleamine 2,3-dioxygenase by T_{reg} cells depletes tryptophan necessary for functional responses of activated T cells [368,369].

6.6.7 Immune Tolerance

Under normal conditions, the human adaptive immune system does not produce antibodies against own components. Immune tolerance is classified into central tolerance, which is taking place in the thymus and bone marrow, and peripheral tolerance allocated to other tissues and lymph nodes. Already in the thymus, developing lymphocytes are selected to avoid reactivity against host's peptides. A second selection occurs in the periphery, where self-reactive T cells are deleted. Additionally mature B cells are also checked for peripheral tolerance [370].

Several mechanisms contribute to remove self-reactive T cells during central and peripheral tolerance. In secondary lymphoid organs, the presentation of self-antigens by immature dendritic cells to naïve T cells leads either to deletion of the T cell or its conversion into T-regulatory cell [320,371]. Fibroblastic reticular cells and lymph node stromal cells are also able to induce $CD8^+$ and $CD4^+$ T-cell tolerance [372]. T-regulatory cells contribute to suppression of autoimmune reactions by several mechanisms including IL-2 depletion and secretion of IL-10 and TGF-β [373]. In the absence of co-stimulatory signals from proinflammatory cytokines, anergy is induced in T cells on their interaction with dendritic cells [374,375].

A small number of B- and T cells, which can escape these tolerance mechanisms, does not normally induce autoimmune processes despite the presence of natural autoantibodies unless these processes are triggered by genetic predisposition or environmental factors [376].

Although central tolerance is necessary to discriminate self from non-self, the peripheral tolerance is directed to prevent overreactions of the immune system to harmless entities. Deficits and disturbances of tolerance mechanisms result in production of autoantibodies and development of autoimmune disorders.

6.7 DYSREGULATION OF IMMUNE RESPONSES

6.7.1 Autoantibodies and Autoimmune Diseases

As already proposed, the general development of chronic inflammatory or immune-related diseases is favored by an imbalance between immune stimulation and prevention of damage of unperturbed biological material (see Section 6.3.7). This mismatch is also valid in development of fibrotic disorders (Section 6.3.8). Further examples for imbalance between immune functions and protection against cell and tissue damage represent numerous autoimmune diseases, where antibodies against host's own components play a sadden role. Autoimmune diseases are closely related to disorders classified as hypersensitivity reactions. Autoantibodies are involved in the pathogenesis of types I, II, and III of hypersensitivity responses [377].

Many peripheral antigens are insufficiently processed and not presented by dendritic cells. These ignored antigens have the potential to initiate autoimmune reactions. An exposure of autoantigens to immune cells can occur during infections and is associated with disruption of peripheral tolerance and breakdown of vascular and cellular barriers [378]. Other mechanisms of autoimmunity include processing of host's components by mold proteases and exposure to sunlight, some chemicals, drugs, or foodstuffs. This processing can lead to presentation of intracellular molecules to T cells, when an imbalance exists between antigen presentation and self-antigen processing and intracellular removal of processed fragments [378]. A molecular mimicry between host molecules and nearly identical components of external antigens is also under discussion as potential source for autoimmune reactions [379]. These mechanisms of autoimmunity are favored by genetic predisposition, epigenetics, and the status of protective reactions against antigen processing.

There are a number of very rare monogenic autoimmune diseases supporting the concept of genetic predisposition [380]. In most cases, multiple genes seem to be involved. Despites numerous genome analyses in the last years, most results do not have additional predictive relevance expect data about the MHC, whose significance for autoimmune diseases was already shown in early studies [381,382]. The involvement of epigenetic mechanisms, including DNA methylation, is also under discussion in development of autoimmune disease [383,384].

Among environmental factors associated with the appearance of human autoimmune diseases are effects of nutrition, infections, xenobiotics, hormones, vaccines, smoking, and others [370,385,386]. Autoantibodies are directed against host's own components. Diseases associated with increased formation of autoantibodies are known as autoimmune diseases. About 100 disease scenarios are known for these pathologies [370].

It remains unclear how this kind of immune response develops. A delay in phagocytosis of damaged cells and tissues by macrophages promotes autoimmune processes. Defects in lysosomal degradation of ingested apoptotic and necrotic cell material can contribute to autoimmunity as well. In most cases, autoantibodies are developed against components that are normally not exposed to cell surface.

6.7.2 Defective Apoptosis of Neutrophils

Concerning defective apoptosis of neutrophils, two main scenarios are under discussion as source of the appearance of autoantibodies [387]. An increased accumulation of neutrophils at some loci can be associated with the existence of survival factors inhibiting apoptosis. For example, antiapoptotic effects are provoked by monosodium urate or calcium pyrophosphate crystals in crystalline arthritis [388,389], by the presence of adenosine in synovial fluid of rheumatoid arthritis patients [390], or by elevated values of granulocyte colony-stimulating factor (G-CSF) in inflamed mucosa in inflammatory bowel disease [391].

On contrary, other diseases are characterized by an enhanced apoptosis rate of neutrophils. Examples are the antineutrophil cytoplasmic antibody–related vasculitis [392,393] and systemic lupus erythematosus [394–396]. In these diseases, neutrophil apoptosis is caused by circulating autoantibodies. Defective clearance of apoptotic cells by macrophages delays the removal of apoptotic cells and favors secondary necrosis. In defective apoptosis, neutrophils are prone to necrotic cell death that promotes a proinflammatory response.

6.7.3 Defects in Lysosomal DNA Degradation

Engulfed apoptotic bodies and other ingested material are further degraded by the lysosomal apparatus of macrophages. Under acidic conditions with a pH value around 4.5–5.0, lysosomal enzymes hydrolyze remaining macromolecules into smaller units suitable for reuse in biosynthetic processes. This acidic luminal pH results from the action of a vacuolar-type H^+-ATPase combined with a counterion flux, whereby the chloride antiporter ClC-7 plays a crucial role in regulation and maintenance of lysosomal ion homeostasis [397,398].

In apoptotic cells, chromosomal DNA is cleaved by caspase-activated deoxyribonuclease into nucleosomal fragments of 180 base pairs. Complete digestion into nucleotides occurs by lysosomal deoxyribonuclease II (DNase II) in macrophages [399]. In mice models, deficient DNase II is associated with development of polyarthritis and anemia. These animals have swollen joints with severe synovitis and aggressive pannus formation. In addition, levels of proinflammatory cytokines IL-1β, IL-6, and TNF-α are enhanced in the affected joints [400,401]. Insufficiently degraded DNA acts as DAMP and causes formation of TNF-α and IFN-β similar to bacterial or viral DNA [401].

The significance of DNA fragments as immunogenic agents is further demonstrated by investigations about degradation of nuclei expelled from precursors cells of erythrocytes [401,402].

Another DNase, the three prime repair exonuclease (Trex1), is ubiquitously expressed in the cytosol of many cells. Its improper function is also associated with cytokine production leading to chronic inflammatory myocarditis [403].

In systemic lupus erythematosus, an autoimmune disease with manifestations in many parts of the body, autoantibodies against ribonucleoprotein, and DNA are found in patient's sera [404,405].

These data support the potential involvement of lysosomal defects in pathogenesis of chronic inflammatory diseases [401,406]. Concerning inflammatory diseases such as arthritis, myocarditis, and many others, it is necessary to prove in future research whether a disturbed maintenance of lysosomal parameters can contribute to disease development.

6.7.4 Disturbed Lysosomal Acidification

Lysosomes degrade not only endocytosed extracellular material but also material delivered from autophagic processes. Thus, association of defective lysosomes with chronic inflammatory diseases is more far reaching than discussed before in interaction between neutrophils and macrophages. There are very seldom-occurring diseases with gene defects of certain lysosomal enzymes, the so-called lysosomal storage diseases [407], which underscore the significance of intact lysosomes for cellular homeostasis. Here, another aspect will be highlighted, namely disturbances in lysosomal homeostasis.

In digestion by lysosomes, critical parameters are the high pH, ion, and electrical gradients across the lysosomal membranes that can be altered under conditions of long-lasting energy and nutrient deficits. Lysosomal hydrolases differ in their pH optima. On fusion of endosomes with lysosomes, the internal pH is stepwise lowered. This favors a sequential action of various hydrolases according to their pH profiles. Already a moderate increase of lysosomal pH can affect this concerted activity of lysosomal enzymes and promote atypical cleavages and accumulation of partially ingested material and maybe toxic products [408].

Several lines of evidence indicate an association between defective lysosomal acidification and neurological dysfunctions [408]. Lysosomal digestion is critically in neurons, as lysosomes are concentrated in the perikaryon, and any digestible material must be delivered over long distances. Waste products accumulate easily in neurons when lysosomal proteolysis is disturbed [409]. The acidic lysosomal pH value affects also dissociation of cholesterol and other lipids, metal ions, and receptor ligands from their counterparts [410–412]. Lysosomes serve additionally as sensor of nutritional and cellular stress [413]. Altered lysosomal pH may promote oxidative processes, impair substrate digestion, and weaken the integrity of lysosomal membrane [408].

6.7.5 Delayed-Type Hypersensitivity Reactions

Of the four types of originally proposed hypersensitivity responses [414], types II, III, and IV are summarized as delayed-type reactions, as symptom onset starts 1 h or later after antigen challenge.

In type II hypersensitivity reactions, IgM or IgG antibodies bind to antigens on the surface of a target cell followed by elimination of the affected cells. Examples are drug-induced or transfusion-mediated labeling of red blood cells by antibodies resulting in immune hemolytic anemia [415,416]. Nicotine acetylcholine receptors can be blocked by circulating antibodies. The associated disorder is known as myasthenia gravis [417]. Other consequences of hypersensitivity reactions of type II are neutropenia and thrombocytopenia.

The presence of small, blood-soluble antigen–antibody complexes is characteristic of type II hypersensitivity reactions. Soluble antigens, which are not bound to cell surfaces, form small complexes with IgG antibodies. Although large antigen–antibody complexes are cleared by macrophages of the reticuloendothelial system, small circulating immune complexes are often deposited in vessel walls of joint, kidney, and lung, where they induce local inflammatory reactions. Complement components such as C3a and C5a, which are bound to the antigen–antibody complexes at deposition site, activate mast cells and recruit neutrophils and macrophages [418].

Vasculitis, glomerulonephritis, arthritis, and serum sickness are typical clinical manifestations associated with type III hypersensitivity reactions [419]. In these pathologies, complement components are often diminished, for example, in renal complications of systemic lupus erythematosis [420]. Apparently, the complement factor C3a acts protective, favoring disintegration of antigen–antibody complexes [420,421].

Type IV hypersensitivity reactions are related to a modified interplay between T cells and presentation of antigens by MHC molecules. In contrast to the other types, type IV hypersensitivity response is not triggered by antibodies. Several hypotheses try to explain how small molecules interact with MHC molecules and T cells in inducing an immune reaction [422]. Four subtypes of type IV responses are distinguished according to the involvement of different T cells and the dominating cells in following inflammation [423]. Among diseases caused by type IV hypersensitivity mechanisms are allergic contact dermatitis, type 1 diabetes, multiple sclerosis, inflammatory bowel disease, autoimmune myocarditis, and rheumatoid arthritis [424–428].

6.7.6 Immediate Hypersensitivity

Type I of hypersensitivity reactions, also known as allergic hypersensitivity, is related to activation of mast cells and basophils and accompanied often by recruitment and stimulation of eosinophils via T_h2 cytokines. The most prominent activation mechanism involves the recognition of normally harmless antigens by IgE, binding of antigen-IgE complexes to FcεR1 receptor on mast cells and basophils, and subsequent activation of these cells [429]. These reactions are fasten and amplified by exposure of the antigen to memory IgE antibodies. Allergic hypersensitivity reactions can occur within few minutes after antigen recognition. Mast cell activation can also occur independent of IgE via TLRs and other mechanisms [28].

Mast cells are predominantly located in connective tissue adjacent to skin and internal mucous surfaces. On degranulation of activated mast cells, potent preformed mediators are secreted, which act on neighbored vasculature, smooth muscle cells, connective tissue, mucous glands, and immune cells [429]. Physiological effects are vasodilation, outpouring of exudative fluid, goblet cell hyperplasia, mucin synthesis, and increased peristaltic movement [377]. These mechanisms are designed to combat against metazoan parasites.

Allergic reactions are favored by functional disturbances of epidermal and epithelial barriers as well as the association between microbiota and altered barrier properties [430]. For example, decreased expression of tight junction proteins in airway epithelium promotes allergic rhinitis [431]. Fine mechanisms of pathogenesis of allergic diseases are under intensive investigation. Different extracellular vesicles including those released from eosinophils [432,433] and effects of the alarmin IL-33 [434,435] play a crucial role in triggering inflammatory allergic responses. The nuclear factor IL-33 is secreted from platelets and other cells under inflammatory conditions and biomechanical stress [434,435].

The list of potential allergens is long. Common allergens are pollen, certain food, medications, insect stings, dust mite excretion, some metals, and some fungal spores. Allergic diseases most of all concern external and internal surfaces of the organism. Common allergic diseases are atopic dermatitis, asthma, hay fever, allergic rhinitis, allergic conjunctivitis, urticaria, and anaphylaxis [436].

A very dangerous, sometimes lethal, complication of allergic reactions is anaphylaxis with typical signs and symptoms on skin, respiratory tract, gastrointestinal system, cardiovascular system, and central nervous system [437].

6.8 SUMMARY

Similar to other protection systems, the immune system contributes to maintain structural and functional homeostasis in cell and tissues. It is mainly responsible to detect and eliminate foreign antigens and damaged host's material. To fulfill these missions, the functional state of cells and tissues is regularly checked, and in case of deviations, components of the immune system are activated and recruited to infection and inflammatory sites.

The immune system is crucial to combat with foreign bacteria, viruses, fungi, and other pathogens, which are recognized as PAMPs. It is also activated by traumata, surgery, and other impacts when host's own molecular patterns (DAMPs) initiate immune response. As a complex answer, an inflammation is induced and novel immune cells are recruited to affected sites to concentrate the whole power of the organism to eliminate any external and internal devastating agents and to restore the disturbed homeostasis. Immune response starts usually with activation of innate immune mechanisms and is sometimes followed by antigen presentation and activation of cells of the acquired immune system.

Tissue damage is on the one hand the reason for the activation of immune system. On the other hand, immune cells are equipped with an arsenal of toxic agents that can contribute to severe tissue destruction. Especially neutrophils are under discussion as a source of damaging reactions as these cells release numerous aggressive hydrolytic enzymes and generate massive amount of reactive species. At inflammatory sites, leakage of blood and tissue cells is another source for destructive molecules. A scheme about the role of cell and tissue damage during immune response is shown in Fig. 6.6.

The degree of immune cell–mediated tissue damage depends on the strength of the initiating events, the state of host's protecting systems, and the general supply of affected areas with dioxygen and nutrients. An immunocompromised and weak person will suffer from a more intense cell and tissue destruction than a healthy individual. The prosperousness of immune response is not only determined by intactness and functional state of immune cells but also by the general energy and health status of the affected organism. Cell and tissue damage by immune cells generates novel pattern molecules and can activate hidden antigens that can participate in further immune reactions. These repetitive reactions provide the basis for chronic inflammatory states. Main factors responsible for persistence of chronic inflammatory states are included in Fig. 6.6.

Many diseases are associated with long-lasting chronic inflammations. Diminished or exhausted protective systems and disturbances in homeostatic physiological parameters mainly contribute to chronicity of inflammatory states. The impossibility to terminate successfully an

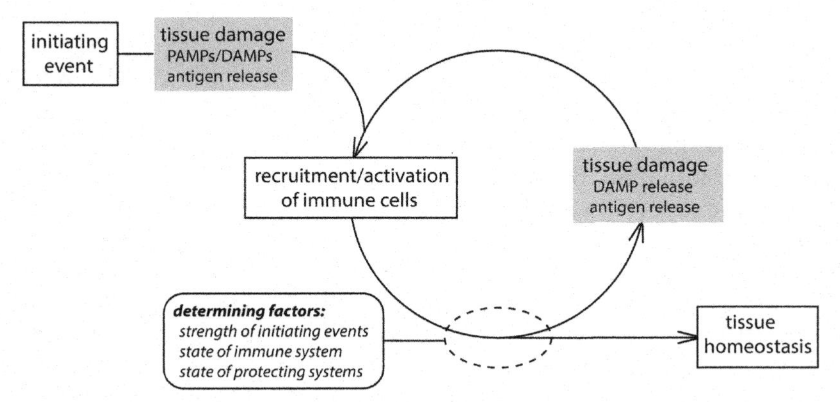

FIGURE 6.6 Role of tissue damage during inflammation. Tissue destruction results from the action of the initiating event and is additionally caused by agents released from activated immune cells (most of all neutrophils) and dysfunctional blood and tissue cells. Branching of pathways into restoration of tissue homeostasis or persistence of chronic inflammation is mainly determined by the severity of the initiating events as well as the individual state of immune response and protecting mechanisms. Long-lasting chronic inflammations and associated diseases prevail in immunocompromised persons. *DAMPs*, damage-associated molecular patterns; *PAMPs*, pathogen-associated molecular patterns.

inflammation due to the prevalence of cell and tissue destruction provides the basis for development of long-persisting chronic inflammatory processes, chronic disease states, and fibrosis. It is a great challenge to physicians and other scientists to evaluate individual factors impeding resolution of inflammatory states for a particular patient.

References

[1] N. Borregaard, Neutrophils, from marrow to microbes, Immunity 33 (2010) 657–670.

[2] J.T. Dancey, K.A. Deubelbeiss, L.A. Harker, C.A. Finch, Neutrophil kinetics in man, J. Clin. Investig. 58 (1976) 705–715.

[3] Y. Sato, S.F. van Eeden, D. English, J.C. Hogg, Pulmonary sequestration of polymorphonuclear leukocytes from bone marrow in bacteremic infection, Am. J. Physiol. 275 (1998) L255–L261.

[4] S.M. Rankin, The bone marrow: a site of neutrophil clearance, J. Leukoc. Biol. 88 (2010) 241–251.

[5] R.C. Furze, S.M. Rankin, Neutrophil mobilization and clearance in the bone marrow, Immunology 125 (2008) 281–288.

[6] C. Summers, S.M. Rankin, A.M. Condliffe, N. Singh, A.M. Peters, E.R. Chilvers, Neutrophil kinetics in health and disease, Trends Immunol. 31 (2010) 318–324.

[7] A. Walker, C. Ward, E.I. Taylor, I. Dransfield, S.P. Hart, C. Haslett, A.G. Rossi, Regulation of neutrophil apoptosis and removal of apoptotic cells, Curr. Drug Targets - Inflamm. Allergy 4 (2005) 447–454.

[8] S.H. Savermuttu, A.M. Peters, H. Reavy, J.P. Lavender, Measurement of granulocyte migration and accumulation in inflammation in man, Clin. Exp. Immunol. 52 (1983) 607–612.

[9] J. Pillay, V.M. Kamp, E. van Hoffen, T. Visser, T. Tak, J.W. Lammers, L.H. Ulfman, L.P. Leenen, P. Pickkers, L. Koenderman, A subset of neutrophils in human systemic inflammation inhibits T cell responses through Mac-1, J. Clin. Investig. 122 (2012) 327–336.

[10] P.C. Vincent, A.D. Chanana, E.P. Cronkite, D.D. Joel, The intravascular survival of neutrophils labeled in vivo, Blood 43 (1974) 371–377.

[11] T. Tak, K. Tesselaar, J. Pillay, J.A.M. Borghans, L. Koenderman, What's your age again? Determination of human neutrophil half-lives revisited, J. Leukoc. Biol. 94 (2013) 595–601.

[12] W.A. Muller, Leukocyte-endothelial interactions in the inflammatory response, Lab. Invest. 82 (2002) 521–533.

[13] K.R. Taylor, R.I. Gallo, Glycosaminoglycans and their proteoglycans: host-associated molecular pattern for initiation and modulation of inflammation, FASEB J. 20 (2006) 9–22.

[14] R.C. Allen, D.L. Stevens, The circulating phagocyte reflects the in vivo state of immune response, Curr. Opin. Infect. Dis. 5 (1992) 389–398.

[15] C. Martin, P.C. Burdon, G. Bridger, J.C. Gutierrez-Ramos, T.J. Williams, S.M. Rankin, Chemokines acting via CXCR2 and CXCR4 control the release of neutrophils from the bone marrow and their return following senescence, Immunity 19 (2003) 583–593.

[16] D. Zhang, G. Chen, D. Manwani, A. Mortha, C. Xu, J.J. Faith, R.D. Burk, Y. Kunisaki, J.-E. Jang, C. Scheiermann, M. Merad, P.S. Frenette, Neutrophil ageing is regulated by the microbiome, Nature 525 (2015) 528–532.

[17] B.T. Suratt, S.K. Young, J. Lieber, J.A. Nick, P.M. Henson, G.S. Worthen, Neutrophil maturation and activation determine anatomic site of clearance from circulation, Am. J. Physiol. 281 (2001) L913–L921.

[18] M. Casanova-Acebes, C. Pitaval, L.A. Weiss, C. Nombela-Arrieta, R. Chèvre, N.A. González, Y. Kunisaki, D. Zhang, N. van Rooijen, L.E. Silberstein, C. Weber, T. Nagasawa, P.S. Frenette, A. Castrillo, A. Hidalgo, Rhythmic modulation of the hematopoietic niche through neutrophil clearance, Cell 153 (2013) 1025–1035.

[19] M.E. Rothenberg, S.P. Hogan, The eosinophil, Annu. Rev. Immunol. 24 (2006) 147–174.

[20] D. Simon, H.U. Simon, Eosinophilic disorders, J. Allergy Clin. Immunol. 119 (2007) 1291–1300.

[21] H.F. Rosenberg, K.D. Dyer, P.S. Foster, Eosinophils: changing perspectives in health and disease, Nat. Rev. Immunol. 13 (2013) 9–22.

[22] B.N. Lambrecht, H. Hammad, The immunology of asthma, Nat. Immunol. 16 (2015) 45–56.

[23] H.Z. Shi, Eosinophils function as antigen-presenting cells, J. Leukoc. Biol. 76 (2004) 520–527.

[24] N.R. Slifman, D.A. Loegering, D.J. McKean, G.J. Gleich, Ribonuclease activity associated with human eosinophil-derived neurotoxin and eosinophil cationic protein, J. Immunol. 137 (1986) 2913–2917.

[25] C. Prussin, D.D. Metcalfe, IgE, mast cells, basophils, and eosinophils, J. Allergy Clin. Immunol. 111 (2003) S486–S494.

[26] Y. Kitamura, T. Kasugai, N. Arizono, H. Matsuda, Development of mast cells and basophils: processes

and regulation mechanisms, Am. J. Med. Sci. 306 (1993) 185–191.
[27] S.C. Bischoff, S. Krämer, Human mast cells, bacteria, and intestinal immunity, Immunol. Rev. 217 (2007) 329–337.
[28] E.Z. da Silva, M.C. Jamur, C. Oliver, Mast cell function: a new vision of an old cell, J. Histochem. Cytochem. 62 (2014) 698–738.
[29] M. Krystel-Whittemore, K.N. Dileepan, J.G. Wood, Mast cell: a multi-functional master cell, Front. Immunol. 6 (2016), article 620, 1–12, https://doi.org/10.3389/fimmu.2015.00620.
[30] C. Schwartz, J.U. Eberle, D. Voehringer, Basophils in inflammation, Eur. J. Pharmacol. 778 (2016) 90–95.
[31] O. Goldmann, E. Medina, The expanding world of extracellular traps: not only neutrophils but much more, Front. Immunol. 3 (2013), article 420, 1–10, https://doi.org/10.3389/fimmu.2012.00420.
[32] H. Möllerherm, M. von Köckritz-Blickwede, K. Branitzki-Heinemann, Antimicrobial activity of mast cells: role and relevance of extracellular DNA traps, Front. Immunol. 7 (2016), article 265, 1–7, https://doi.org/10.3389/fimmu.2016.00265.
[33] C. Schulz, E. Gomez Perdiguero, L. Chorro, H. Szabo-Rogers, N. Cagnard, K. Kierdorf, M. Prinz, B. Wu, S.E. Jacobson, J.W. Pollard, J. Frampton, K.J. Liu, F.A. Geissmann, A lineage of myeloid cells independent of Myb and hematopoietic stem cells, Science 336 (2012) 86–90.
[34] S. Epelman, K.J. Lavine, A.E. Beaudin, D.K. Sojka, J.A. Carrero, B. Calderon, T. Brija, E.L. Gautier, S. Ivanov, A.T. Satpathy, J.D. Schilling, R. Schwendener, I. Sergin, B. Razani, E.C. Forsberg, W.M. Yokoyama, E.R. Unanue, M. Colonna, G.J. Randolph, D.L. Mann, Embryonic and adult-derived resident cardiac macrophages are maintained through distinct mechanisms at steady state and during inflammation, Immunity 40 (2014) 91–104.
[35] F. Ginhoux, M. Guilliams, Tissue-resident macrophage ontogeny and homeostasis, Immunity 44 (2016) 439–449.
[36] S. Yona, K.W. Kim, Y. Wolf, A. Mildner, D. Varol, M. Breker, D. Strauss-Ayali, S. Viukov, M. Guilliams, A. Misharin, D.A. Hume, H. Perlman, B. Malissen, E. Zelzer, S. Jung, Fate mapping reveals origins and dynamics of monocytes and tissue macrophages under homeostasis, Immunity 38 (2013) 79–91.
[37] D. Hashimoto, A. Chow, C. Noizat, P. Teo, M.B. Beasley, M. Leboeuf, C.D. Becker, P. See, J. Price, D. Lucas, M. Greter, A. Mortha, S.W. Boyer, E.C. Forsberg, M. Tanaka, N. van Rooijen, A. Garcia-Sastre, E.R. Stanley, F. Ginhoux, P.S. Frenette, M. Merad, Tissue-resident macrophages self-maintain locally throughout adult life with minimal contribution from circulating monocytes, Immunity 38 (2013) 792–804.
[38] C.C. Bain, A. Bravo-Blas, C.L. Scott, E. Gomez Perdiguero, F. Geissmann, S. Henri, B. Malissen, L.C. Osborne, D. Artis, A.W. Mowat, Constant replenishment from circulating monocytes maintains the macrophage pool in the intestine of adult mice, Nat. Immunol. 15 (2014) 929–937.
[39] J. Jantsch, K.J. Binger, D.N. Müller, J. Titze, Macrophages in homeostatic immune function, Front. Physiol. 5 (2014), article 146, 1–7, https://doi.org/10.3389/fphys.2014.00146.
[40] A. Castell-Rodriguez, G. Piñón-Zárate, M. Herrera-Enriquez, K. Jarquin-Yáñez, I. Medina-Solares, Dendritic cells: location, function, and clinical implications, in: A. Ghosh (Ed.), Biology of Myelomonocytic Cells, IntechOpen, 2017, pp. 21–50, https://doi.org/10.5572/intechopen.68352.
[41] L. Flores-Romo, In vivo maturation and migration of dendritic cells, Immunology 102 (2001) 255–262.
[42] A. Mandal, C. Viswanathan, Natural killer cells: in health and disease, Hematol. Oncol. Stem Cell Ther. 8 (2015) 47–55.
[43] N.J. Topham, E.W. Hewitt, Natural killer cell cytotoxicity: how do they pull the trigger? Immunology 128 (2009) 7–15.
[44] J.R. Ortaldo, R.T. Winkler-Pickett, S. Nagata, C.F. Ware, Fas involvement in human NK cell apoptosis: lack of a requirement for CD16-mediated events, J. Leukoc. Biol. 6 (1997) 209–215.
[45] H. Terunuma, X. Deng, Z. Dewan, S. Fujimoto, N. Yamamoto, Potential role of NK cells in the induction of immune responses: implications for NK cell-based immunotherapy for cancers and viral infections, Int. Rev. Immunol. 27 (2008) 93–100.
[46] O. Mandelboim, P. Malik, D.M. Davis, C.H. Jo, J.E. Boyson, J.L. Strominger, Human CD16 as a lysis receptor mediating direct natural killer cell cytotoxicity, Proc. Natl. Acad. Sci. U.S.A. 96 (1999) 5640–5644.
[47] K. Hirahara, T. Nakayama, $CD4^+$ T-cell subsets in inflammatory diseases: beyond the T_h1/T_h2 paradigm, Int. Immunol. 28 (2016) 163–171.
[48] R. Medzhitov, Origin and physiological roles of inflammation, Nature 454 (2008) 428–435.

[49] H.-U. Simon, Neutrophil apoptosis pathways and their modification in inflammation, Immunol. Rev. 193 (2003) 101–110.

[50] C. Jabubzick, E.L. Gautier, S.L. Gibbings, D.K. Sojka, A. Schlitzer, T.E. Johnson, S. Ivanov, Q. Duan, S. Bala, T. Condon, N. van Rooijen, J.R. Grainger, Y. Belkaid, A. Ma'ayan, D.W. Riches, W.M. Yokoyama, F. Ginhoux, P.M. Henson, G.J. Randolph, Minimal differentiation of classical monocytes as they survey steady-state tissues and transport antigen to lymph nodes, Immunity 39 (2013) 599–610.

[51] L.C. Davies, S.J. Jenkins, J.E. Allen, P.R. Taylor, Tissue-resident macrophages, Nat. Immunol. 14 (2013) 986–995.

[52] S. Epelman, K.J. Lavine, G.J. Randolph, Origin and function of tissue macrophages, Immunity 17 (2014) 21–35.

[53] Y. Lavin, A. Mortha, A. Rahman, M. Merad, Regulation of macrophage development and function in peripheral tissues, Nat. Rev. Immunol. 15 (2015) 731–744.

[54] A.T. O'Neill, N.A. Monteiro-Riviere, G.M. Walker, Characterization of microfluidic human epidermal keratinocyte culture, Cytotechnology 56 (2008) 197–207.

[55] E. Dejana, K.K. Hirschi, M. Simons, The molecular basis of endothelial cell plasticity, Nat. Commun. 8 (2017), 1–11 14361, https://doi.org/10.1038/ncomms14361.

[56] A.E. Dubin, A. Patapoutin, Nociceptors : the sensors of the pain pathway, J. Clin. Investig. 120 (2010) 3760–3772.

[57] C. Nathan, Points of control in inflammation, Nature 420 (2002) 846–852.

[58] G.M. Barton, A calculated response: control of inflammation by the innate immune system, J. Clin. Investig. 118 (2008) 413–420.

[59] O. Takeuchi, S. Akira, Pattern recognition receptors and inflammation, Cell 140 (2010) 805–820.

[60] R. Caruso, N. Warner, N. Inohara, G. Núñez, NOD1 and NOD2: signaling, host defense, and inflammatory disease, Immunity 41 (2014) 898–908.

[61] M.R. Thompson, J.J. Kaminski, E.A. Kurt-Jones, K.A. Fitzgerald, Pattern recognition receptors and the innate immune response to viral infection, Viruses 3 (2011) 920–940.

[62] T.B. Geijtenbeek, S.I. Gringhuis, Signalling through C-type lectin receptors: shaping immune responses, Nat. Rev. Immunol. 9 (2009) 465–479.

[63] T. Chevakis, A. Bierhaus, P.P. Nawroth, RAGE (receptor for advanced glycation end products): a central player in the inflammatory response, Microb. Infect. 6 (2004) 1219–1225.

[64] R. Suresh, D.M. Moser, Pattern recognition in innate immunity, host defense, and immunopathology, Adv. Physiol. Educ. 37 (2013) 284–291.

[65] C.A. Janeway, R. Medzhitov, Innate immune recognition, Annu. Rev. Immunol. 20 (2002) 197–216.

[66] P. Matzinger, Tolerance, danger, and the extended family, Annu. Rev. Immunol. 12 (1994) 991–1045.

[67] S. Borello, C. Nicolò, G. Delogu, F. Pandolfi, F. Ria, TLR2: a crossroad between infections and autoimmunity? Int. J. Immunopathol. Pharmacol. 24 (2011) 549–556.

[68] K.N. Do, L.N. Fink, T.E. Jensen, L. Gautier, A. Parlesak, TLR2 controls intestinal carcinogen detoxication by CYP1A1, PLoS One 7 (2012), e32309 1–9, https://doi.org/10.1371/journal.pone.0032309.

[69] W.R. Berrington, T.R. Hawn, *Mycobacterium tuberculosis*, macrophages, and the innate immune response: does common variation matter? Immunol. Rev. 219 (2007) 167–186.

[70] E.A. Misch, M. Macdonald, C. Ranjit, B.R. Sapkota, R.D. Wells, M.R. Siddiqui, G. Kaplan, T.R. Hawn, Human TLR1 deficiency is associated with impaired mycobacterial signaling and protection from leprosy reversal reaction, PLoS Neglected Trop. Dis. 2 (2008), e231, 1–9, https://doi.org/10.1371/journal.pntd.0000231.

[71] L. Alexopoulou, A.C. Holt, R. Medzhitov, R.A. Flavell, Recognition of double-stranded RNA and activation of NF-kappaB by toll-like receptor 3, Nature 413 (2001) 732–738.

[72] M. Matsumoto, K. Funami, M. Tanabe, H. Oshiumi, M. Shingai, Y. Seto, A. Yamamoto, T. Seya, Subcellular location of toll-like receptor 3 in human dendritic cells, J. Immunol. 171 (2003) 3154–3162.

[73] J. Sun, K.E. Duffy, C.T. Ranjith-Kumar, J. Xiong, R.J. Lamb, J. Santos, H. Masarapu, M. Cunnigham, A. Holzenburg, R.T. Sarsky, M.L. Mbow, C. Kao, Structural and functional analyses of the human toll-like receptor 3, J. Biol. Chem. 281 (2006) 11144–11151.

[74] C. Vaure, Y. Liu, A comparative review of toll-like receptor 4 expression and functionality in different animal species, Front. Immunol. 5 (2014) 1–15, https://doi.org/10.3389/fimmu.2014.00316.

[75] E.A. Miao, E. Andersen-Nissen, S.E. Warren, A. Aderem, TLR5 and Ipaf: dual sensors of bacterial flagellin in the innate immune system, Semin. Immunopathol. 29 (2007) 275–288.

[76] O. Takeuchi, T. Kawai, H. Sanjo, N.G. Copeland, D.J. Gilbert, N.A. Jenkins, K. Takeda, S. Akira, TLR6: a novel of an expanding toll-like receptor family, Gene 231 (1999) 59–65.

[77] F. Heil, H. Hemmi, H. Hochrein, F. Ampernberger, C. Kirschning, S. Akira, G. Lipford, H. Wagner, S. Bauer, Species-specific recognition of single-stranded RNA via toll-like receptor 7 and 8, Science 303 (2004) 1526–1529.

[78] Y. Zhang, M. El-Far, F.P. Dupuy, M.S. Abdel-Hakeem, Z. He, F.A. Procopio, Y. Shi, E.K. Haddad, P. Ancuta, R.P. Sekaly, E.A. Said, HCV RNA activates APCs via TLR7/TLR8 while virus selectively stimulates macrophages without inducing antiviral responses, Sci. Rep. 6 (2016), 29447, 1–13, https://doi.org/10.1038/srep29447.

[79] C. Martinez-Campos, A.I. Burgete-Garcia, V. Madrid-Marina, Role of TLR9 in oncogenic virus-produced cancer, Viral Immunol. 30 (2017) 98–105.

[80] T. Regan, K. Nally, R. Carmody, A. Houston, F. Shanahan, J. Macsharry, E. Brint, Identification of TLR10 as a key mediator of the inflammatory response to Listeria monocytogenes in intestinal epithelial cells and macrophages, J. Immunol. 1991 (2013) 6084–6092.

[81] T.J. Silhavy, D. Kahne, S. Walker, The bacterial cell envelope, Cold Spring Harb. Perspect. Biol. 2 (2010), e000414, 1–17, https://doi.org/10.1101/cshperspect.a000414.

[82] W. Dammermann, L. Wollenberg, F. Bentzien, A. Lohse, S. Lüth, Toll like receptor 2 agonists lipoteichoic acid and peptidoglycan are able to enhance antigen specific IFNγ release in whole blood during recall antigen responses, J. Immunol. Methods 396 (2013) 107–115.

[83] C. Leimkuhler Grimes, L. De Zoysa Ariyananda, J.E. Melnyk, E.K. O'Shea, The innate immune protein Nod2 binds directly to MDP, a bacterial cell wall fragment, J. Am. Chem. Soc. 134 (2012) 13535–13537.

[84] A. Rubartolli, M.T. Lotze, Inside, outside, upside down: damage-associated molecular pattern molecules (DAMPs) and redox, Trends Immunol. 28 (2007) 429–436.

[85] P. Scaffidi, T. Misteli, M.E. Bianchi, Release of chromatin protein HMGB1 by necrotic cells trigger inflammation, Nature 418 (2002) 191–195.

[86] Y. Shi, J.E. Evans, K.L. Rock, Molecular identification of a danger signal that alerts the immune system to dying cells, Nature 425 (2003) 516–521.

[87] K.A. Scheibner, M.A. Lutz, S. Boodoo, M.J. Fenton, J.D. Powell, M.R. Horton, Hyaluronan fragments act as an endogenous danger signal by engaging TLR2, J. Immunol. 177 (2006) 1272–1281.

[88] M.J. Bours, E.L. Swennen, F. Di Virgilio, B.N. Cronstein, P.C. Dagnelie, Adenosine 5′-triphosphate and adenosine as endogenous signaling molecules in immunity and inflammation, Pharmacol. Ther. 112 (2006) 358–404.

[89] R.T. Figueiredo, P.L. Fernandez, D.S. Mourao-Sa, B.N. Porto, F.F. Dutra, L.S. Alves, M.F. Oliviera, P.L. Oliviera, A.V. Graca-Souza, M.T. Bozza, Characterization of heme as activator of toll-like receptor 4, J. Biol. Chem. 282 (2007) 20221–20229.

[90] A.M. Farkas, T.M. Kilgore, M.T. Lotze, Detecting DNA: getting and begetting cancer, Curr. Opin. Investig. Drugs 8 (2007) 981–986.

[91] J.D. Belcher, C. Chen, J. Nguyen, L. Milbauer, F. Abdulla, A.I. Alayash, A. Smith, K.A. Nath, R.P. Hebbel, G.M. Vercelotti, Heme triggers TLR4 signaling leading to endothelial cell activation and vaso-occlusion in murine sickle cell disease, Blood 123 (2014) 377–390.

[92] L.M. Glode, S.E. Mergenhagen, D.L. Rosenstreich, Significant contribution of spleen cells in mediating the lethal effects of endotoxin in vivo, Infect. Immun. 14 (1976) 626–630.

[93] C. Sauter, C. Wolfenberger, Interferon in human serum after injection of endotoxin, Lancet 316 (1980) 852–853.

[94] G.D. Martich, R.L. Danner, M. Ceska, A.F. Suffredini, Detection of interleukin 8 and tumor necrosis factor in normal humans after intravenous endotoxin: the effect of antiinflammatory agents, J. Exp. Med. 173 (1991) 1021–1024.

[95] H.S. Warren, C. Fitting, E. Hoff, M. Adib-Conquy, L. Beasley-Topliffe, B. Tesini, X. Liang, C. Valentine, J. Hellman, D. Hayden, J.M. Cavaillon, Resilience to bacterial infection: difference between species could be due to proteins in serum, J. Infect. Dis. 201 (2010) 223–232.

[96] J. Zschaler, D. Schlorke, J. Arnhold, Differences in innate immune response between man and mouse, Crit. Rev. Immunol. 34 (2014) 433–454.

[97] S. Mariathasan, K. Newton, D. Monack, D. Vucic, D. French, W. Lee, M. Roose-Girma, S. Erickson, V. Dixit, Differential activation of the inflammasome by caspase-1 adaptors ASC and Ipaf, Nature 430 (2004) 213–218.

[98] T. Fernandes-Alnemri, J. Wu, J.W. Yu, P. Datta, B. Miller, W. Jankowski, S. Rosenberg, J. Zhang, E.S. Alnemri, The pyroptosome: a supramolecular assembly of ASC dimers mediating inflammatory cell death via caspase-1 activation, Cell Death Differ. 14 (2007) 1590–1604.

[99] F. Martinon, K. Burns, J. Tschopp, The inflammasome: a molecular platform triggering activation of

inflammatory caspases and processing of proIL-beta, Mol. Cell. 10 (2002) 417–426.

[100] M. Dagenais, A. Skeldon, M. Saleh, The inflammasome in memory of Dr. Jürg Tschopp, Cell Death Differ. 19 (2012) 5–12.

[101] M. Keller, A. Rüegg, S. Werner, H.D. Beer, Active caspase-1 is a regulator of unconventional protein secretion, Cell 132 (2008) 818–831.

[102] L. Franchi, T. Eigenbrod, R. Muñoz-Planillo, G. Nuñez, The inflammasome: a caspase-1 activation platform regulating immune responses and disease pathogenesis, Nat. Immunol. 10 (2009) 241–247.

[103] G. Sollberger, G.E. Strittmatter, M. Garstkiewicz, J. Sand, H.-D. Beer, Caspase-I: the inflammasome and beyond, Innate Immun. 20 (2014) 115–125.

[104] S.L. Fink, B.T. Cookson, Apoptosis, pyroptosis, and necrosis: mechanistic description of dead and dying eukaryotic cells, Infect. Immun. 73 (2005) 1907–1916.

[105] J.C. Rodriguez-Prados, P.C. Través, J. Cuenca, D. Rico, J. Aragonés, P. Martin-Sanz, M. Casante, L. Boscá, Substrate fate in activated macrophages: a comparison between innate, classic, and alternative activation, J. Immunol. 185 (2010) 605–614.

[106] C.M. Krawczyk, T. Holowka, J. Sun, J. Blagih, E. Amiel, R.J. DeBerardinis, J.R. Cross, E. Jung, C.B. Thompson, R.G. Jones, E.J. Pearce, Toll-like receptor-induced changes in glycolytic metabolism regulate dendritic cell activation, Blood 115 (2010) 4742–4749.

[107] G.M. Tannahill, A.M. Curtis, J. Adamik, E.M. Palsson-McDermott, A.F. McGettrick, G. Goel, C. Frezza, N.J. Bernard, B. Kelly, N.H. Foley, L. Zheng, A. Gardet, Z. Tong, S.S. Jany, S.C. Corr, M. Haneklaus, B.E. Caffrey, K. Pierce, S. Walmsley, F.C. Beasley, E. Cummins, V. Nizet, M. Whyte, C.T. Taylor, H. Lin, S.L. Masters, E. Gottlieb, V.P. Kelly, C. Clisjh, P.E. Auron, R.T.J. Xavier, L.A.J. O'Neill, Succinate is a danger signal that induces IL-1β via HIF-1α, Nature 496 (2013) 238–242.

[108] W.K.E. Ip, N. Hoshi, D.S. Shouval, S. Snapper, R. Medzhitov, Anti-inflammatory effect of IL-10 mediated by metabolic reprogramming of macrophages, Science 356 (2017) 513–519.

[109] H. Kopf, G.M. de la Rosa, O.M. Horward, X. Chen, Rapamycin inhibits differentiation of Th17 cells and promotes generation of FoxP3^{+} T regulatory cells, Int. Immunopharmacol. 7 (2007) 1819–1824.

[110] G.M. Delgoffe, K.N. Polizzi, A.T. Waickman, E. Heikamp, D.J. Meyers, M.R. Horton, B. Xiao, P.F. Worley, J.D. Powell, The kinase mTOR regulates the differentiation of helper T cells through the selective activation of signaling by mTORC1 and mTORC2, Nat. Immunol. 12 (2011) 295–303.

[111] M.D. Buck, D. O'Sullivan, E.L. Pearce, T cell mechanism drives immunity, J. Exp. Med. 212 (2015) 1345–1360.

[112] A. Carrillo-Vico, J.M. Guerrero, P.J. Lardone, R.J. Reiter, A review of the multiple actions of melatonin on the immune system, Endocrine 27 (2005) 189–200.

[113] P.D. Mavroudis, J.D. Scheff, S.E. Calvano, I.P. Androulakis, Systems biology of circadian-immune interactions, J. Innate Immun. 5 (2013) 153–162.

[114] T.W. Kim, J.-H. Jeong, S.-C. Hong, The impact of sleep and circadian disturbance on hormones and metabolism, Int. J. Endocrinol. (2015), article ID 591729, 1–9, https://doi.org/10.1155/2015/591729.

[115] F. Radogna, M. Diederich, L. Ghibelli, Melatonin: a pleiotropic molecule regulating inflammation, Biochem. Pharmacol. 80 (2010) 1844–1852.

[116] D. Steinhilber, M. Brungs, O. Werz, I. Wiesenberg, C. Danielsson, J.P. Kahlen, S. Nayeri, M. Schräder, C. Carlberg, The nuclear receptor for melatonin repress 5-lipoxygenase gene expression in human B lymphocytes, J. Biol. Chem. 270 (1975) 7037–7040.

[117] B. Li, H. Zhang, M. Akbar, H.Y. Kim, Negative regulation of cytosolic phospholipase A(2) by melatonin in the rat pineal gland, Biochem. J. 351 (2000) 709–716.

[118] J.S. Pober, W.C. Sessa, Evolving functions of endothelial cells in inflammation, Nat. Rev. Immunol. 7 (2007) 803–815.

[119] C. Helthianu, M. Simionescu, N. Simionescu, Histamine receptors of the microvascular endothelium revealed in situ with a histamine-ferritin conjugate: characteristic high-affinity binding sites in venules, J. Cell Biol. 93 (1982) 357–364.

[120] R.D. Peavy, D.D. Metcalfe, Understanding the mechanisms of anaphylaxis, Curr. Opin. Allergy Clin. Immunol. 8 (2008) 310–315.

[121] Z. Hofman, S. de Maat, C.E. Hack, C. Maas, Bradykinin: inflammatory product of the coagulation system, Clin. Rev. Allergy Immunol. 51 (2016) 152–161.

[122] J.S. Pober, W.C. Sessa, Inflammation and the blood microvascular system, Cold Spring Harb. Perspect. Biol. 7 (2015) 1–11, https://doi.org/10.1101/cshperspect.a016345.

[123] B. Opitz, S. Hippenstiel, J. Eitel, N. Suttorp, Extra- and intracellular innate immune recognition in endothelial cells, Thromb. Haemost. 98 (2007) 319–326.

[124] L. Claesson-Welsh, Vascular permeability – the essentials, Ups. J. Med. Sci. 120 (2015) 135–143.

[125] J. Lesley, R. Hyman, CD44 structure and function, Front. Biosci. 3 (1998) 616–630.

[126] W.Y.J. Chen, G. Abatangelo, Functions of hyaluronan in wound repair, Wound Rep. 7 (1999) 79–89.

[127] J. Joerges, T. Schulz, J. Wegner, U. Schumacher, P. Prehm, Regulation of cell volume by glycosaminoglycans, J. Cell. Biochem. 113 (2012) 340–348.
[128] C.A. Dinarello, Interleukin-1 in the pathogenesis and treatment of inflammatory diseases, Blood 117 (2011) 3720–3732.
[129] H. Wajant, K. Pfizenmaier, P. Scheurich, Tumor necrosis factor signaling, Cell Death Differ. 10 (2003) 45–65.
[130] M.B. Olszewski, A.J. Groot, J. Dastych, E.F. Knol, TNF trafficking to mast cell granules: mature chain-dependent endocytosis, J. Immunol. 178 (2007) 5701–5709.
[131] C.A. Dinarello, D. Novick, S. Kim, G. Kaplanski, Interleukin-18 and IL-18 binding protein, Front. Immunol. 4 (2013), article 289, 1–10, https://doi.org/10.3389/fimmu.2013.00289.
[132] J.A. Green, S.R. Cooperband, S. Kibrick, Immune specific induction of interferon production in cultures of human blood lymphocytes, Science 164 (1969) 1415–1417.
[133] P.W. Gray, D.V. Goeddel, Structure of the human interferon gene, Nature 298 (1982) 859–863.
[134] C. Cayrol, J.P. Girard, Interleukin-33 (IL-33): a nuclear cytokine from the IL-1 family, Immunol. Rev. 281 (2018) 154–168.
[135] T. Tanaka, M. Narazaki, T. Kishimoto, IL-6 in inflammation, immunity, and disease, Cold Spring Harb. Perspect. Biol. 6 (2014), article 016295, 1–16, https://doi.org/10.1101/cshperspect.a016295.
[136] R.C. Russo, C.C. Garcia, M.M. Teixeira, F.A. Amaral, The CXCL8/IL-8 chemokine family and its receptors in inflammatory diseases, Expert Rev. Clin. Immunol. 10 (2014) 593–619.
[137] G. Trinchieri, Interleukin-12 and the regulation of innate resistance and adaptive immunity, Nat. Rev. Immunol. 3 (2003) 133–146.
[138] J.S. Gerber, D.M. Mosser, Reversing lipopolysaccharide toxicity by ligating the macrophage Fcγ receptors, J. Immunol. 166 (2001) 6861–6868.
[139] J.P. Edwards, X. Zhang, K.A. Frauwirth, D.M. Mosser, Biochemical and functional characterization of three activated macrophage populations, J. Leukoc. Biol. 80 (2006) 1298–1307.
[140] V.A. Fadok, D.L. Bratton, A. Konowal, P.W. Freed, J.Y. Westcott, P.M. Henson, Macrophages that have ingested apoptotic cells in vitro inhibit proinflammatory cytokine production through autocrine/paracrine mechanisms involving TGF-beta, PGE2, and PAF, J. Clin. Investig. 101 (1998) 890–898.
[141] J.J. Letterio, A.B. Roberts, Regulation of immune responses by TGF-beta, Annu. Rev. Immunol. 16 (1998) 137–161.
[142] O. Boyman, J. Sprent, The role of interleukin-2 during homeostasis and activation of the immune system, Nat. Rev. Immunol. 12 (2012) 180–190.
[143] I.G. Luzina, A.D. Keegan, N.M. Heller, G.A. Rock, T. Shea-Donohue, S.P. Atamas, Regulation of inflammation by interleukin-4: a review of "alternatives", J. Leukoc. Biol. 92 (2012) 753–764.
[144] K. Takatsu, Interleukin-5 and IL-5 receptor in health and diseases, Proc. Jpn. Acad. Ser. B Phys. Biol. Sci. 87 (2011) 463–485.
[145] C.T. Weaver, R.D. Hatton, P.R. Mangan, L.E. Harrington, IL-17 family cytokines and the expanding diversity of effector T cell lineages, Annu. Rev. Immunol. 25 (2007) 821–852.
[146] I. Ushach, A. Zlotnik, Biological role of granulocyte macrophage colony-stimulating factor (GM-CSF) and macrophage colony-stimulating factor (M-CSF) on cells of the myeloid lineage, J. Leukoc. Biol. 100 (2016) 481–489.
[147] W.J. Leonard, C.-K. Wan, IL-21 signaling in immunity, F1000Research 5 (2016), 224, 1–10, https://doi.org/10.12688/f1000research.7634.1.
[148] J. Choi, P.S. Leung, C. Bowlus, M.E. Gershwin, IL-35 and autoimmunity: a comprehensive perspective, Clin. Rev. Allergy Immunol. 49 (2015) 327–332.
[149] A. Roy, U.H. van Adrian, Chemokines in innate and adaptive host defense: basic chemokine grammar for immune cells, Annu. Rev. Immunol. 22 (2004) 891–928.
[150] C. Nathan, Neutrophils and immunity: challenges and opportunities, Nat. Rev. Immunol. 6 (2006) 173–182.
[151] K. Ley, C. Laudanna, M.I. Cybulsky, S. Nourshargh, Getting to the site of inflammation: the leukocyte adhesion cascade updated, Nat. Rev. Immunol. 7 (2007) 678–689.
[152] E.Y. Choi, S. Santoso, T. Chavakis, Mechanisms of neutrophil transendothelial migration, Front. Biosci. 14 (2010) 1596–1605.
[153] J.F. Leeuwenberg, E.F. Smeets, J.J. Neefjes, M.A. Shaffer, T. Cinek, T.M. Jeunhomme, T.J. Ahern, W.A. Buurman, E-selectin and intracellular adhesion molecule-1 are released by activated human endothelial cells in vitro, Immunology 77 (1992) 543–549.
[154] J.M. Cock-Mills, M.E. Marchese, H. Abdala-Valencia, Vascular cell adhesion molecule-1 expression and signaling during disease: regulation by reactive species and antioxidants, Antioxidants Redox Signal. 15 (2011) 1607–1638.
[155] A. Di Gennaro, E. Kenne, M. Wan, O. Soehnlein, L. Lindbom, J.Z. Haeggström, Leukotriene B4-induced changes in vascular permeability are mediated by neutrophil release of heparin-binding protein

(HBP/CAP37/azurocidin), FASEB J. 23 (2009) 1750–1757.
[156] R. Kahn, T. Hellmark, L.M. Leeb-Lundberg, N. Akbari, M. Todiras, T. Olofsson, J. Wieslander, A. Christensson, K. Westman, M. Bader, W. Müller-Esterl, D. Karpman, Neutrophil-derived proteinase 3 induces kallikrein-independent release of a novel vasoactive kinin, J. Immunol. 182 (2009) 7906–7915.
[157] O. Soehnlein, L. Lindbom, C. Weber, Mechanisms underlying neutrophil-mediated monocyte recruitment, Blood 114 (2009) 4613–4623.
[158] J.J. O'Shea, P.J. Murray, Cytokine signaling modules in inflammatory responses, Immunity 28 (2008) 477–487.
[159] Z. Szekanecz, A.E. Koch, Macrophages and their products in rheumatoid arthritis, Curr. Opin. Rheumatol. 19 (2007) 289–295.
[160] D.M. Mosser, The many faces of macrophage activation, J. Leukoc. Biol. 73 (2003) 209–212.
[161] D.M. Mosser, J.P. Edwards, Exploring the full spectrum of macrophage activation, Nat. Rev. Immunol. 8 (2008) 958–969.
[162] A.A. Filardy, D.R. Pires, M.P. Nunes, C.M. Takiya, C.G. Freire-de-Lima, F.L. Ribeiro-Gomes, G.A. DosReis, Proinflammatory clearance of apoptotic neutrophils induces an IL-12^{low} IL-10^{high} regulatory phenotype in macrophages, J. Immunol. 185 (2010) 2044–2050.
[163] A.L. Mescher, A.W. Neff, Regenerative capacity and the developing immune system, Adv. Biochem. Eng. Biotechnol. 93 (2005) 39–66.
[164] G.C. Gurtner, S. Werner, Y. Barrandon, M.T. Longaker, Wound repair and regeneration, Nature 453 (2008) 314–321.
[165] A.B. Aurora, E.N. Olson, Immune regeneration of stem cells and regeneration, Cell Stem Cell 15 (2014) 14–25.
[166] Y. Oishi, I. Manabe, Macrophages in inflammation, repair and regeneration, Int. Immunol. 30 (2018) 511–528.
[167] J.H. Kehrl, L.M. Wakefield, A.B. Roberts, S. Jakowlew, M. Alvarez-Mon, R. Derynck, M.B. Sporn, A.S. Fauci, Production of transforming growth factor β by human T lymphocytes and its potential role in the regulation of T cell growth, J. Exp. Med. 163 (1986) 1037–1050.
[168] T.R. Mempel, M.J. Pittet, K. Khazaie, W. Weniger, R. Weissleder, H. von Boehmer, U.H. von Andrian, Regulatory T cells reversibly suppress cytotoxic T cell function independent of effector differentiation, Immunity 25 (2006) 129–141.
[169] M. Lopez, R. Aguilera, C. Perez, A. Mendoza-Naranjo, C. Pereda, M. Ramirez, C. Ferrada, J.C. Aquillón, F. Salazar-Onfray, The role of regulatory T lymphocytes in the induced immune response mediated by biological vaccines, Immunobiology 211 (2006) 127–136.
[170] A.H. Rook, J.H. Kehrl, L.M. Wakefield, A.B. Roberts, M.B. Sporn, D.B. Burlington, H.C. Lane, A.S. Fauci, Effects of transforming growth factor β on the functions of natural killer cells: depressed cytolytic activity and blunting of interferon responsiveness, J. Immunol. 136 (1986) 3916–3920.
[171] G. Bellone, M. Aste-Amezaga, G. Trinchieri, U. Rodeck, Regulation of NK cell functions by TGF-β1, J. Immunol. 155 (1995) 1066–1073.
[172] F. Geissmann, P. Revy, A. Regnault, Y. Lepelletier, M. Dy, N. Brousse, S. Amigorena, O. Hermine, A. Durandy, TGF-β1 prevents the noncognate maturation of human dendritic Langerhans cells, J. Immunol. 162 (1999) 4567–4575.
[173] S.H. Wrzesinski, Y.Y. Wan, R.A. Flavell, Transforming growth factor-β and the immune response: implications for anticancer therapy, Clin. Cancer Res. 13 (2007) 5262–5270.
[174] S. Sanjabi, L.A. Zenewicz, M. Kamanaka, R.A. Flavell, Anti- and pro-inflammatory roles of TGF-β, IL-10, and IL-22 in immunity and autoimmunity, Curr. Opin. Pharmacol. 9 (2009) 447–453.
[175] A. Yoshimura, Y. Wakabayashi, T. Mori, Cellular and molecular basis for the regulation of inflammation by TGF-β, J. Biochem. 147 (2010) 781–792.
[176] B. Bierie, H.L. Moses, Transforming growth factor beta (TGF-β) and inflammation in cancer, Cytokine Growth Factor Rev. 21 (2010) 49–59.
[177] M.O. Li, Y.Y. Wan, S. Sanjabi, A.K. Robertson, R.A. Flavell, Transforming growth factor-beta regulation of immune responses, Annu. Rev. Immunol. 24 (2006) 99–146.
[178] M.O. Li, R.A. Flavell, Contextual regulation of inflammation: a duet by transforming growth factor-beta and interleukin-10, Immunity 28 (2008) 468–476.
[179] K.N. Couper, D.G. Blount, E.M. Riley, IL-10: the master regulator of immunity to infection, J. Immunol. 180 (2008) 5771–5777.
[180] C.G. Feng, M.C. Kullberg, D. Jankovic, A.W. Cheever, P. Caspar, R.L. Coffman, A. Sher, Transgenic mice expressing human interleukin-10 in the antigen-presenting cell compartment show increased susceptibility to infection with *Mycobacterium avium* associated with decreased macrophage effector function and apoptosis, Infect. Immun. 70 (2002) 6672–6679.
[181] D.G. Brooks, M.J. Trifilo, K.H. Edelmann, L. Teyton, B. McGavern, M.B. Oldstone, Interleukin-10 determines viral clearance or persistence in vivo, Nat. Med. 12 (2006) 1301–1309.

[182] S. Roque, C. Nobrega, R. Appelberg, M. Correia-Neves, IL-10 underlies distinct susceptibility of BALB/c and C57BL/6 mice to *Mycobacterium avium* infection and influences efficacy of antibiotic therapy, J. Immunol. 178 (2007) 8028–8035.

[183] M. Hamberg, B. Samuelsson, Prostaglandin endoperoxides. Novel transformations of arachidonic acid in human platelets, Proc. Natl. Acad. Sci. U.S.A. 71 (1974) 3400–3404.

[184] B.D. Levy, C.B. Clish, B. Schmidt, K. Gronert, C.N. Serhan, Lipid mediator class switching during acute inflammation: signals in resolution, Nat. Immunol. 2 (2001) 612–619.

[185] C.N. Serhan, K.A. Sheppard, Lipoxin formation during human neutrophil-platelet interactions. Evidence for the transformation of leukotriene A4 by platelet 12-lipoxygenase in vitro, J. Clin. Investig. 85 (1990) 772–780.

[186] D. El-Kebir, L. Jozsef, T. Khreiss, W. Pan, N.A. Petasis, C.N. Serhan, J.G. Filep, Aspirin-triggered lipoxins override the apoptosis-delaying action of serum amyloid A in human neutrophils: a novel mechanisms for resolution of inflammation, J. Immunol. 179 (2007) 616–622.

[187] D. El-Kebir, L. Jozsef, W. Pan, L. Wang, N.A. Petasis, C.N. Serhan, J.G. Filep, 15-epi-lipoxin A4 inhibits myeloperoxidase signaling and enhances resolution of acute lung injury, Am. J. Respir. Crit. Care Med. 180 (2009) 311–319.

[188] J.A. Chandrasekharan, N. Sharma-Walia, Lipoxins: nature's way to resolve inflammation, J. Inflamm. Res. 8 (2015) 181–192.

[189] C.N. Serhan, J. Savill, Resolution of inflammation: the beginning programs the end, Nat. Immunol. 6 (2005) 1191–1197.

[190] J.M. Schwab, N. Chiang, M. Arita, C.N. Serhan, Resolvin E1 and protectin D1 activate inflammation-resolution programmes, Nature 447 (2007) 869–874.

[191] C.N. Serhan, J. Dalli, R.A. Colas, J.W. Winkler, N. Chiang, Protectins and marensins: new pro-resolving families of mediators in acute inflammation and resolution bioactive metabolome, Biochim. Biophys. Acta 1851 (2015) 397–413.

[192] R.W. Vandivier, P.M. Henson, I.S. Douglas, Burying the death: the impact of failed apoptotic cell removal (efferocytosis) on chronic inflammatory lung disease, Chest 129 (2006) 1673–1682.

[193] V.A. Fadok, D.L. Bratton, L. Guthrie, P.M. Henson, Differential effects of apoptotic versus lysed cells on macrophage production of cytokines: role of proteases, J. Immunol. 166 (2001) 8647–8654.

[194] R.W. Vandivier, V.A. Fadok, P.R. Hoffmann, D.L. Bratton, C. Penvari, K.K. Brown, J.D. Brain, F.J. Accurso, P.M. Henson, Elastase-mediated phosphatidylserine receptor cleavage impairs apoptotic cell clearance in cystic fibrosis and bronchiectasis, J. Clin. Investig. 109 (2002) 661–670.

[195] R. Wang, T. Town, V. Gokarn, R.A. Flavell, R.Y. Chandawarkar, HSP70 enhances macrophage phagocytosis by interaction with lipid raft-associated TLR-7 and upregulating p38 MAPK and PI3K pathways, J. Surg. Res. 136 (2006) 58–69.

[196] B. Fadeel, Programmed cell clearance, Cell. Mol. Life Sci. 60 (2003) 2575–2585.

[197] D.V. Krysko, K. D'Herde, P. Vandenabeele, Clearance of apoptotic and necrotic cells and its immunological consequences, Apoptosis 11 (2006) 1673–1682.

[198] L.-P. Erwig, P.M. Henson, Immunological consequences of apoptotic cell phagocytosis, Am. J. Pathol. 171 (2007) 2–8.

[199] T.A. Wynn, Common and unique mechanisms regulate fibrosis in various fibroproliferative diseases, J. Clin. Investig. 117 (2007) 524–529.

[200] J.L. Lauer-Fields, D. Juska, G.B. Fields, Matrix metalloproteinases and collagen catabolism, Biopolymers 66 (2002) 19–32.

[201] J.M. Otte, I.M. Rosenberg, D.K. Podolsky, Intestinal myofibroblasts in innate immune responses of the intestine, Gastroenterology 124 (2003) 1866–1878.

[202] M.D. Meneghin, C. Hogaboam, Infectious disease, the innate immune response, and fibrosis, J. Clin. Investig. 117 (2007) 530–538.

[203] J. Varga, D. Abraham, Systemic sclerosis: a prototypic multisystem fibrotic disorder, J. Clin. Investig. 117 (2007) 557–567.

[204] C.G. Lee, R.J. Homer, Z. Zhu, S. Lanone, X. Wang, V. Koteliansky, J.M. Shipley, P. Gotwals, P. Noble, Q. Chen, R.M. Senior, J.A. Elias, Interleukin-13 induces tissue fibrosis by selectively stimulating and activating transforming growth factor-β1, J. Exp. Med. 194 (2001) 809–821.

[205] M. Hasegawa, M. Fujimoto, K. Takehara, S. Sato, Pathogenesis of systemic sclerosis: altered B cell function is the key linking systemic autoimmunity and tissue fibrosis, J. Dermatol. Sci. 39 (2005) 1–17.

[206] S.A. Mezzano, M. Ruiz-Ortega, J. Egido, Angiotensin II and renal fibrosis, Hypertension 38 (2001) 635–638.

[207] T. Watanabe, T.A. Barker, B.C. Berk, Angiotensin II and the endothelium: diverse signals and effects, Hypertension 45 (2005) 163–169.

[208] S.D. Kobayashi, N. Malachowa, F.R. DeLeo, Pathogenesis of *Staphylococcus aureus* abscesses, Am. J. Pathol. 185 (2015) 1518–1527.

[209] R. Feuerstein, M. Seidl, M. Prinz, P. Henneke, MyD88 in macrophages is critical for abscess resolution in

staphylococcal skin infection, J. Immunol. 194 (2005) 2735–2745.
[210] H.N. Fernandez, P.M. Henson, A. Otani, T.E. Hugli, Chemotactic response to human C3a and C5a anaphylatoxins. I. Evaluation of C3a and C5a leukotaxis in vitro and under simulated in vivo conditions, J. Immunol. 120 (1978) 109–115.
[211] V. Witko-Sarsat, P. Rieu, B. Descamps-Latscha, P. Lesavre, L. Halbwachs-Mecarelli, Neutrophils: molecules, functions and pathophysiological aspects, Lab. Invest. 80 (2000) 617–653.
[212] N. Mukaida, Pathophysiological roles of interleukin-8/CXCL8 in pulmonary diseases, Am. J. Physiol. Lung Cell Mol. Physiol. 284 (2003) L566–L577.
[213] M. Phillipson, P. Kubes, The neutrophil in vascular inflammation, Nat. Med. 17 (2011) 1381–1390.
[214] D. Schlorke, L. Thomas, S.A. Samsonov, D. Huster, J. Arnhold, A. Pichert, The influence of glycosaminoglycans on IL-8-mediated functions of neutrophils, Carbohydr. Res. 356 (2012) 196–203.
[215] O. Radmark, O. Werz, D. Steinhilber, B. Samuelsson, 5-Lipoxygenase, a key enzyme for leukotriene biosynthesis in health and disease, Biochim. Biophys. Acta 1851 (2015) 331–339.
[216] P.P. McDonald, S.R. McColl, P. Braquet, P. Borgeat, Autocrine enhancement of leukotriene synthesis by endogenous leukotriene B4 and platelet-activating factor in human neutrophils, Br. J. Pharmacol. 111 (1994) 852–860.
[217] A.W. Segal, How neutrophils kill microbes, Annu. Rev. Immunol. 23 (2005) 197–223.
[218] A.W. Segal, M. Geisow, R. Garcia, A. Harper, R. Miller, The respiratory burst of phagocytic cells is associated with a rise in vacuolar pH, Nature 290 (1981) 406–409.
[219] P. Cech, R.I. Lehrer, Phagolysosomal pH of human neutrophils, Blood 63 (1984) 88–95.
[220] S.W. Edwards, Biochemistry and Physiology of the Neutrophil, Cambridge University Press, 1994.
[221] H. Sengeløv, L. Kjeldsen, N. Borregaard, Control of exocytosis in early neutrophil activation, J. Immunol. 150 (1993) 1535–1543.
[222] H. Sengeløv, P. Follin, L. Kjeldsen, K. Lollike, C. Dahlgren, N. Borregaard, Mobilization of granules and secretory vesicles during in vivo exudation of human neutrophils, J. Immunol. 154 (1995) 4157–4165.
[223] N. Borregaard, L. Kjeldsen, K. Rygaard, L. Bastholm, M.H. Nielsen, H. Sengeløv, O.W. Bjerrum, A.H. Johnsen, Stimulus-dependent secretion of plasma proteins from human neutrophils, J. Clin. Investig. 90 (1992) 86–96.
[224] N. Borregaard, J.B. Cowland, Granules of the human neutrophilic polymorphonuclear leukocyte, Blood 89 (1997) 3503–3521.
[225] B.S. Nielsen, S. Timshel, L. Kjeldsen, M. Sehested, C. Pyke, N. Borregaard, K. Danø, 92 kDa type IV collagenase (MMP-9) is expressed in neutrophils and macrophages but not in malignant epithelial cell in human colon cancer, Int. J. Cancer 65 (1996) 57–62.
[226] S.A. Ragland, A.K. Criss, From bacterial killing to immune modulation: recent insights into the functions of lysozyme, PLoS Pathog. 13 (2017), e1006512, 1–22, https://doi.org/10.1371/journal.ppat.1006512.
[227] E.W. Odell, R. Sarra, M. Foxworthy, D.S. Chapple, R.W. Evans, Antibacterial activities of peptides homologous to a loop region in human lactoferrin, FEBS Lett. 382 (1996) 175–178.
[228] S. Farnaud, R.W. Evans, Lactoferrin – a multifunctional protein with antimicrobial properties, Mol. Immunol. 40 (2003) 395–405.
[229] E. Ringel, Localization of histaminase to the specific granule of the human neutrophil, Immunology 52 (1984) 649–658.
[230] P. Elsbach, The bactericidal/permeability-increasing protein (BPI) in antibacterial host defense, J. Leukoc. Biol. 64 (1998) 14–18.
[231] R.I. Lehrer, W. Lu, α-Defensins in human innate immunity, Immunol. Rev. 245 (2012) 84–112.
[232] O. Soehnlein, L. Lindbom, Neutrophil-derived azurocidin alarms the immune system, J. Leukoc. Biol. 85 (2009) 344–351.
[233] A.A. Belaaouaj, K.S. Kim, S.D. Shapiro, Degradation of outer membrane protein A in *Escherichia coli* killing by neutrophil elastase, Science 289 (2000) 1185–1187.
[234] C.T.N. Pham, Neutrophil serine proteases: specific regulators of inflammation, Nat. Rev. Immunol. 6 (2006) 541–550.
[235] C. Kantari, M. Pederzoli-Ribeil, O. Amir-Moazami, V. Gausson-Dorey, I. Cruz Moura, M.-C. Lecomte, M. Benhoume, V. Witko-Sarsat, Proteinase 3, the Wegener autoantigen, is externalized during neutrophil apoptosis: evidence for a functional association with phospholipid scramblase 1 and interference with macrophage phagocytosis, Blood 110 (2007) 4086–4095.
[236] J. Fleddermann, A. Pichert, J. Arnhold, Interaction of serine proteases from polymorphonuclear leucocytes with the cell surface and heparin, Inflammation 35 (2012) 81–88.
[237] N.C. Perera, O. Schilling, H. Kittel, W. Back, E. Kremmer, D.E. Jenne, NSP4, an elastase-related protease in human neutrophils with arginine specificity, Proc. Natl. Acad. Sci. U.S.A. 109 (2012) 6229–6234.

[238] P. Kasperkiewicz, M. Poreba, S.J. Snipas, S.J. Lin, D. Kirchhofer, G.S. Salvesen, M. Drag, Design of a selective substrate and activity based probe for human neutrophil serine protease 4, PLoS One 10 (2015), e0132818, 1–12, https://doi.org/10.1371/journal.pone.0132818.

[239] J. Arnhold, J. Flemmig, Human myeloperoxidase in innate and acquired immunity, Arch. Biochem. Biophys. 500 (2010) 92–106.

[240] K. Murata, Acidic glycosaminoglycans in human platelets and leukocytes: the isolation and enzymatic characterization of chondroitin 4-sulfate, Clin. Chim. Acta 57 (1974) 115–124.

[241] S.O. Kolset, J.T. Gallagher, Proteoglycans in haemopoietic cells, Biochim. Biophys. Acta 1032 (1990) 191–211.

[242] B.M. Babior, NADPH oxidase, Curr. Opin. Immunol. 16 (2004) 42–47.

[243] A.R. Cross, A.W. Segal, The NADPH oxidase of professional phagocytes – prototype of the NOX electron transport chain systems, Biochim. Biophys. Acta 1657 (2004) 1–22.

[244] R.I. Lehrer, Inhibition by sulfonamides of the candidacidal activity of human neutrophils, J. Clin. Investig. 50 (1971) 2498–2505.

[245] C. Koch, Effect of sodium azide upon normal and pathological granulocyte function, Acta Pathol. Microbiol. Scand. B 82 (1974) 136–142.

[246] K. Yamamoto, T. Miyoshi-Koshio, Y. Utsuki, S. Mizuno, K. Suzuki, Virucidal activity and viral protein modification by myeloperoxidase: a candidate for defense factor of human polymorphonuclear leukocytes against influenza virus infection, J. Infect. Dis. 164 (1991) 8–14.

[247] Y. Aratani, H. Koyama, S.-I. Nyui, K. Suzuki, F. Kura, N. Maeda, Severe impairment in early host defense against Candida albicans in mice deficient in myeloperoxidase, Infect. Immun. 67 (1999) 1828–1836.

[248] S.J. Klebanoff, Myeloperoxidase: friend and foe, J. Leukoc. Biol. 77 (2005) 598–625.

[249] A.L. Chapman, M.B. Hampton, R. Senthilmohan, C.C. Winterbourn, A.J. Kettle, Chlorination of bacterial and neutrophil proteins during phagocytosis and killing of *Staphylococcus aureus*, J. Biol. Chem. 277 (2002) 9757–9762.

[250] E.P. Reeves, M. Nagl, J. Godavac-Zimmermann, A.W. Segal, Reassessment of microbicidal activity of reactive oxygen species and hypochlorous acid with reference to the phagocytic vacuole of the neutrophil granulocyte, J. Med. Microbiol. 52 (2003) 643–651.

[251] T.O. Hirche, J.P. Gaut, J.W. Heinecke, A. Belaaouaj, Myeloperoxidase plays critical roles in killing *Klebsiella pneumoniae* and inactivating elastase: effects of host defense, J. Immunol. 174 (2005) 1557–1565.

[252] M.F. Parry, R.K. Root, J.A. Metcalf, K.K. Delaney, L.S. Kaplow, W.J. Richar, Myeloperoxidase deficiency: prevalence and clinical significance, Ann. Intern. Med. 95 (1981) 293–301.

[253] D. Kutter, Prevalence of myeloperoxidase deficiency: population studies using Bayer-Technicon automated hematology, J. Mol. Med. 76 (1998) 669–675.

[254] R.I. Lehrer, M.J. Cline, Leukocyte myeloperoxidase deficiency and disseminated candidiasis: the role of myeloperoxidase in resistance to Candida infection, J. Clin. Investig. 48 (1969) 1478–1488.

[255] P. Cech, H. Stalder, J.J. Widmann, A. Rohner, P.A. Miescher, Leukocyte myeloperoxidase deficiency and diabetes mellitus associated with Candida albicans liver abscess, Am. J. Med. 66 (1979) 149–153.

[256] C. Nguyen, H.P. Katner, Myeloperoxidase deficiency manifesting as pustular candida dermatitis, Clin. Infect. Dis. 24 (1997) 258–260.

[257] J.P. Gaut, G.C. Yeh, H.D. Tran, J. Byun, J.P. Henderson, G.M. Richter, M.-L. Brennan, A.J. Lusis, A. Belaaouaj, R.S. Hotchkiss, J.W. Heinecke, Neutrophils employ the myeloperoxidase system to generate antimicrobial brominating and chlorinating oxidants during sepsis, Proc. Natl. Acad. Sci. U.S.A. 98 (2001) 11961–11966.

[258] M.-L. Brennan, A. Gaur, A. Pahuja, A.J. Lusis, W.F. Reynolds, Mice lacking myeloperoxidase are more susceptible to experimental encephalomyelitis, J. Neuroimmunol. 112 (2001) 97–105.

[259] Y. Aratani, F. Kura, H. Watanabe, H. Akagawa, Y. Takano, A. Ishida-Okawara, K. Suzuki, N. Maeda, H. Koyama, Contribution of the myeloperoxidase-dependent oxidative system to host defence against Cryptococcus neoformans, J. Med. Microbiol. 55 (2006) 1291–1299.

[260] R.I. Lehrer, A.K. Lichtenstein, T. Ganz, Defensins: antimicrobial and cytotoxic peptides of mammalian cells, Annu. Rev. Immunol. 11 (1993) 105–128.

[261] P.B. Eisenhauer, R.I. Lehrer, Mouse neutrophils lack defensins, Infect. Immun. 60 (1992) 3446–3447.

[262] O.M. Panasenko, H. Spalteholz, J. Schiller, J. Arnhold, Leukocytic myeloperoxidase-mediated formation of bromohydrins and lysophospholipids from unsaturated phosphatidylcholines, Biochemistry (Mosc.) 71 (2006) 571–580.

[263] O.M. Panasenko, T. Vakhrusheva, V. Tretyakov, H. Spalteholz, J. Arnhold, Influence of chloride on modification of unsaturated phosphatidylcholines by

the myeloperoxidase/hydrogen peroxide/bromide system, Chem. Phys. Lipids 149 (2007) 40–51.

[264] L.A. Marquez, H.B. Dunford, Chlorination of taurine by myeloperoxidase: kinetic evidence for an enzyme-bound intermediate, J. Biol. Chem. 269 (1994) 7950–7956.

[265] D.R. Ramos, M. Victoria Garcia, L.M. Canle, J. Arturo Santaballa, P.G. Furtmüller, C. Obinger, Myeloperoxidase-catalyzed taurine chlorination: initial versus equilibrium rate, Arch. Biochem. Biophys. 466 (2007) 221–233.

[266] T. Odajima, I. Yamazaki, Myeloperoxidase of the leukocyte of normal blood. 3. The reaction of ferric myeloperoxidase with superoxide anion, Biochim. Biophys. Acta 284 (1972) 355–359.

[267] J.M. Zgliczynski, R.J. Selvaraj, B.B. Paul, T. Stelmaszynska, P.K. Poskitt, A.J. Sbarra, Chlorination by the myeloperoxidase-H_2O_2-Cl^- antimicrobial system at acid and neutral pH, Proc. Soc. Exp. Biol. Med. 154 (1977) 418–422.

[268] I. Neeli, S.N. Khan, M. Radic, Histone deamination as a result to inflammatory stimuli in neutrophils, J. Immunol. 180 (2008) 1895–1902.

[269] V. Brinkmann, U. Reichard, C. Goosmann, B. Fauler, Y. Uhlemann, D.S. Weiss, Y. Weinrauch, A. Zychlinsky, Neutrophil extracellular traps kill bacteria, Science 303 (2004) 1532–1535.

[270] V. Papayannopoulos, A. Zychlinsky, NETs: a new strategy for using old weapons, Trends Immunol. 30 (2009) 513–521.

[271] K. Beiter, F. Wartha, B. Albiger, S. Normark, A. Zychlinsky, B. Henriques-Normark, An endonuclease allows Streptococcus pneumoniae to escape from neutrophil extracellular traps, Curr. Biol. 16 (2006) 401–407.

[272] F. Wartha, K. Beiter, B. Albiger, J. Fernebro, A. Zychlinsky, S. Normark, B. Henriques-Normark, Capsule and D-alanylated lipoteichoic acids protect Streptococcus pneumoniae against neutrophil extracellular traps, Cell Microbiol. 9 (2007) 1162–1171.

[273] T.A. Fuchs, U. Abed, C. Goosmann, R. Hurwitz, I. Schulze, V. Wahn, Y. Weinrauch, V. Brinkmann, A. Zychlinsky, Novel cell death program leads to neutrophil extracellular traps, J. Cell Biol. 176 (2007) 231–241.

[274] F.R. Sheppard, M.R. Kelher, E.E. Moore, N.J. McLaughlin, A. Banerjee, C.C. Silliman, Structural organization of the neutrophil NADPH oxidase: phosphorylation and translocation during priming and activation, J. Leukoc. Biol. 78 (2005) 1025–1042.

[275] J. Hirahashi, D. Mekala, J. Van Ziffie, L. Xiao, S. Saffaripour, D.D. Wagner, S.D. Shapiro, C. Lowell, T.N. Mayadas, Mac-1 signalling via SRc-family and Syk kinases results in elastase-dependent thrombohemorrhagic vasculopathy, Immunity 25 (2006) 271–283.

[276] S.R. Clark, A.C. Ma, S.A. Tavener, B. McDonald, Z. Goodarzi, M.M. Kelly, K.D. Patel, S. Chakrabarti, E. McAvoy, G.D. Sinclair, E.M. Keys, E. Allen-Vercoe, R. Devinney, C.J. Doig, F.H. Green, P. Kubes, Platelet TLR4 activates neutrophil extracellular traps to ensnare bacteria in septic blood, Nat. Med. 13 (2007) 463–469.

[277] K. Kessenbrock, M. Krumbholz, U. Schönermarck, W. Back, W.L. Gross, Z. Werb, H.J. Gröne, V. Brinkmann, D.E. Jenne, Netting neutrophils in autoimmune small-vessel vasculitis, Nat. Med. 15 (2009) 623–625.

[278] A. Hakkim, B.G. Furnrohr, K. Amann, B. Laube, U.A. Abed, V. Brinkmann, M. Herrmann, R.E. Voll, A. Zychlinsky, Impairment of neutrophil extracellular trap degradation is associated with lupus nephritis, Proc. Natl. Acad. Sci. U.S.A. 107 (2010) 9813–9818.

[279] J.S. Knight, M.J. Kaplan, Lupus neutrophils: "NET" gain in understanding lupus pathogenesis, Curr. Opin. Rheumatol. 24 (2012) 441–450.

[280] C.R. Welbourn, G. Goldman, I.S. Paterson, C.R. Valeri, D. Shepro, H.B. Hechtman, Pathophysiology of ischemia reperfusion injury: central role of the neutrophil, Br. J. Surg. 78 (1991) 651–655.

[281] A.C. Windsor, P.G. Mullen, A.A. Fowler, H.J. Sugerman, Role of the neutrophil in adult respiratory distress syndrome, Br. J. Surg. 80 (1993) 10–17.

[282] P. Teder, R.W. Vandivier, D. Jiang, J. Liang, L. Cohn, E. Puré, P.M. Henson, P.W. Noble, Resolution of lung inflammation by CD44, Science 296 (2005) 155–158.

[283] T.N. Mayadas, X. Cullier, Neutrophil β_2 integrins: moderators of life or death decisions, Trends Immunol. 26 (2005) 388–395.

[284] N.C. Franc, K. White, R.A. Ezekowitz, Phagocytosis and development: back to the future, Curr. Opin. Immunol. 11 (1999) 47–52.

[285] K. Elward, P. Gasque, "Eat me" and "don't eat me" signals govern the innate immune response and tissue repair in the CNS: emphasis on the critical role of the complement system, Mol. Immunol. 40 (2003) 85–94.

[286] R. Medzhitov, C.A. Janeway Jr., Decoding the pattern of self and nonself by the innate immune system, Science 296 (2002) 298–300.

[287] J. Savill, I. Dransfield, C. Gregory, C. Haslett, A blast from the past: clearance of apoptotic cells regulates immune responses, Nat. Rev. Immunol. 2 (2002) 965–975.

[288] K.S. Ravichandran, Beginnings of a good apoptotic meal: the find-me and eat-me signaling pathways, Immunity 35 (2011) 445–455.

[289] K. Lauber, S.G. Blumenthal, M. Waibel, S. Wesselborg, Clearance of apoptotic cells: getting rid of the corpses, Mol. Cell 14 (2004) 277–287.
[290] V.A. Fadok, D.R. Voelker, P.A. Campbell, J.J. Cohen, D.L. Bratton, P.M. Henson, Exposure of phosphatidylserine on the surface of apoptotic lymphocytes triggers specific recognition and removal by macrophages, J. Immunol. 148 (1992) 2207–2216.
[291] J. Leßig, H. Spalteholz, U. Reibetanz, P. Salavei, M. Fischlechner, H.-J. Glander, J. Arnhold, Myeloperoxidase binds to nonvital spermatozoa on phosphatidylserine epitopes, Apoptosis 12 (2007) 1803–1812.
[292] J. Flemmig, J. Leßig, U. Reibetanz, P. Dautel, J. Arnhold, Non-vital polymorphonuclear leukocytes express myeloperoxidase on their surface, Cell. Physiol. Biochem. 21 (2007) 287–296.
[293] H. Yoshida, K. Kawane, M. Koike, Y. Uchiyama, S. Nagata, Phosphatidylserine-dependent engulfment by macrophages of nuclei from erythroid precursor cells, Nature 437 (2005) 754–758.
[294] E. Blink, N.A. Maianski, E.S. Alnemri, A.S. Zervos, S. Roos, T.W. Kuijpers, Intramitochondrial serine protease activity of Omi/Htr2A is required for caspase-independent cell death of human neutrophils, Cell Death Differ. 11 (2004) 937–939.
[295] H.-C. Chen, C.-J. Wang, C.-L. Chou, S.-M. Lin, C.-D. Huang, T.-Y. Lin, C.-H. Wang, H.-C. Lin, C.-T. Yu, H.-P. Kuo, C.-Y. Liu, Tumor necrosis factor-α induces caspase-independent cell death in human neutrophils via reactive oxidants and associated with calpain activity, J. Biomed. Sci. 13 (2006) 261–273.
[296] B. Mirnikjoo, K. Balasubramanian, A.J. Schroit, Suicidal membrane repair regulates phosphatidylserine externalization during apoptosis, J. Biol. Chem. 284 (2009) 22512–22516.
[297] G.G. Borisenko, T. Matsura, S.X. Liu, V.A. Tyurin, V.A. Jianfei, F.B. Serinkan, V.E. Kagan, Macrophage recognition of externalized phosphatidylserine and phagocytosis of apoptotic Jurkat cells – existence of a threshold, Arch. Biochem. Biophys. 413 (2003) 41–52.
[298] S.J. Gardai, K.A. McPhillips, S.C. Frasch, W.J. Jenssen, A. Starefeldt, J.E. Murphy-Ullrich, D.L. Bratton, P.A. Oldenborg, M. Michalak, P.M. Henson, Cell-surface calreticulin initiates clearance of viable or apoptotic cells through trans-activation of LRP on the phagocyte, Cell 123 (2005) 321–334.
[299] S. Brown, I. Heinisch, E. Ross, K. Shaw, C.D. Buckley, J. Savill, Apoptosis disables CD31-mediated cell detachment from phagocytes promoting binding and engulfment, Nature 418 (2002) 200–203.
[300] V.A. Tyurin, Y.Y. Tyurina, P.M. Kochanek, R. Hamilton, S.T. DeKosky, J.S. Greenberger, H. Bayir, V.E. Kagan, Oxidative lipidomics of programmed cell death, Methods Enzymol. 442 (2008) 375–393.
[301] P.M. Henson, D.L. Bratton, V.A. Fadok, The phosphatidylserine receptor: a crucial molecular switch? Nat. Rev. Mol. Cell Biol. 2 (2001) 627–633.
[302] D.L. Bratton, P.M. Henson, Apoptotic cell recognition: will the real phosphatidylserine receptor(s) please stand up? Curr. Biol. 18 (2008) R76–R79.
[303] P.J. Murray, T.A. Wynn, Protective and pathogenic functions of macrophage subsets, Nat. Rev. Immunol. 11 (2011) 723–737.
[304] F.O. Martinez, S. Gordon, M. Locati, A. Montavani, Transcriptional profiling of the human monocyte-to-macrophage differentiation and polarization: new molecules and patterns of gene expression, J. Immunol. 177 (2006) 7303–7311.
[305] P.J. Murray, J.E. Allen, S.K. Biswas, E.A. Fisher, D.W. Gilroy, S. Goerdt, S. Gordon, J.A. Hamilton, L.B. Ivashkiv, T. Lawrence, M. Locati, A. Mantovani, F.O. Martinez, J.-L. Mege, D.M. Mosser, G. Natoli, J.P. Saeij, J.L. Schultze, K.A. Shirey, A. Sica, J. Suttles, I. Udalova, J.A. van Ginderachter, S.N. Vogel, T.A. Wynn, Macrophage activation and polarization: nomenclature and experimental guidelines, Immunity 41 (2014) 14–20.
[306] G.D. Barish, M. Downes, W.A. Alaynick, R.T. Yu, C.B. Ocampo, A.L. Bookout, D.J. Mangelsdorf, R.M. Evans, A nuclear receptor atlas: macrophage activation, Mol. Endocrinol. 19 (2005) 2466–2477.
[307] M. Schneemann, G. Schoeden, Macrophage biology and immunology: man is not a mouse, J. Leukoc. Biol. 81 (2007) 579.
[308] J. MacMicking, Q.W. Xie, C. Nathan, Nitric oxide and macrophage function, Annu. Rev. Immunol. 15 (1997) 323–350.
[309] J.B. Weinberg, Nitric oxide production and nitric oxide synthase type 2 expression by human mononuclear phagocytes: a review, Mol. Med. 4 (1998) 557–591.
[310] F.C. Fang, C.F. Nathan, Man is not a mouse: reply, J. Leukoc. Biol. 81 (2007) 580.
[311] K.L. Rock, C. Gramm, L. Rothstein, K. Clark, R. Stein, L. Dick, D. Hwang, A.L. Goldberg, Inhibitors of the proteasome block the degradation of most cell proteins and the generation of peptides presented on MHC class I molecules, Cell 78 (1994) 761–771.
[312] E.W. Hewitt, The MHC class I antigen presentation pathway: strategies for viral immune evasion, Immunology 110 (2003) 163–169.

[313] J.M.M. den Haan, R. Ares, M.C. van Zelm, The activation of the adaptive immune system: crosstalk between antigen-presenting cells, T cells and B cells, Immunol. Lett. 162 (2014) 103–112.
[314] E.R. Mann, X. Li, Intestinal antigen-presenting cells in mucosal immune homeostasis: crosstalk between dendritic cells, macrophages, and B-cells, World J. Gastroenterol. 20 (2014) 9653–9664.
[315] S. Jung, D. Unutmaz, P. Wong, G. Sano, K. de los Santos, T. Sparwasser, S. Wu, S.D. Vuthoori, K. Ko, F. Zavala, E.G. Pamer, D.R. Littman, R.A. Lang, In vivo depletion of $CD11c^+$ dendritic cells abrogates priming of $CD8^+$ T cells by exogenous cell-associated antigens, Immunity 17 (2002) 211–220.
[316] O.P. Joffre, E. Segura, A. Savina, S. Amigorena, Cross-presentation by dendritic cells, Nat. Rev. Immunol. 12 (2012) 557–569.
[317] A.Y. Huang, P. Golumbek, M. Ahmadzadeh, E. Jaffee, D. Pardoll, H. Levitsky, Role of bone marrow-derived cells in presenting MHC class I-restricted tumor antigens, Science 264 (1994) 961–965.
[318] K. Nopora, C.A. Bernhard, C. Ried, A.A. Castello, K.M. Murphy, P. Marconi, U. Koszinowski, T. Brocker, MHC class I cross-presentation by dendritic cells counteracts viral immune evasion, Front. Immunol. 3 (2012), article 348, 1–10, https://doi.org/10.3389/fimmu.2012.00348.
[319] M.B. Lutz, C. Kurts, Induction of peripheral $CD4^+$ T-cell tolerance and $CD8^+$ T-cell cross-tolerance by dendritic cells, Eur. J. Immunol. 39 (2009) 2325–2330.
[320] R.M. Steinman, D. Hawiger, M.C. Nussenzweig, Tolerogenic dendritic cells, Annu. Rev. Immunol. 21 (2003) 685–711.
[321] R. McCormack, L. de Armas, M. Shiratsuchi, E.R. Podack, Killing machines: three pore-forming proteins of the immune system, Immunol. Res. 57 (2013) 268–278.
[322] I. Osińska, K. Popko, U. Demkow, Perforin: an important player in immune response, Cent. Eur. J. Immunol. 39 (2014) 109–115.
[323] N. Zhang, M.J. Bevan, $CD8^+$ T cells: foot soldiers of the immune system, Immunity 35 (2011) 161–168.
[324] Y. Yang, Z. Xiang, H.C.J. Ertl, J.M. Wilson, Upregulation of class I major histocompatibility complex antigens by interferon γ is necessary for T-cell-mediated elimination of recombinant adenovirus-infected hepatocytes in vivo, Proc. Natl. Acad. Sci. U.S.A. 92 (1995) 7257–7261.
[325] X.Y. Mo, R.A. Tripp, M.Y. Sangster, P.C. Doherty, The cytotoxic T-lymphocyte response to sendai virus is unimpaired in the absence of gamma interferon, J. Virol. 71 (1997) 1906–1910.
[326] H. Ito, M. Seishima, Regulation of the induction and function of cytotoxic T lymphocytes by natural killer T cell, J. Biomed. Biotechnol. 2010 (2010), article 641757, 1–8, https://doi.org/10.1155/2010/641757.
[327] X. Tao, S. Constant, P. Jorritsma, K. Bottomly, Strength of TCR signal determines the costimulatory requirements for Th1 and Th2 $CD4^+$ T cell differentiation, J. Immunol. 159 (1997) 5956–5963.
[328] J. Zhu, H. Yamane, W.E. Paul, Differentiation of effector CD4 T cell populations, Annu. Rev. Immunol. 28 (2010) 445–489.
[329] L. Xu, I. Kitani, I. Fuss, W. Strober, Cutting edge: regulatory T cells induce CD4+CD25-Foxp3-T cells or are self-induced to become Th17 cells in the absence of TGF-β, J. Immunol. 178 (2007) 6725–6729.
[330] Y.K. Lee, H. Turner, C.L. Maynard, C.R. Oliver, D. Chen, C.O. Elson, C.T. Weaver, Late developmental plasticity in the T helper 17 lineage, Immunity 30 (2009) 92–107.
[331] F. Borriello, M.P. Sethna, S.D. Boyd, A.N. Schweitzer, E.A. Tivol, D. Jacoby, T.B. Storm, E.M. Simpson, G.J. Freeman, A.H. Sharpe, B7-1 and B7-2 have overlapping, critical role in immunoglobulin class switching and germinal enter formation, Immunity 6 (1997) 303–313.
[332] Y. Zheng, C.N. Manzotti, M. Liu, F. Burke, K.I. Mead, D.M. Sansom, CD86 and CD80 differentially modulate the suppressive function of human regulatory T cells, J. Immunol. 172 (2004) 2778–2784.
[333] M.F. Bachmann, A. Oxenius, Interleukin 2: from immunostimulation to immunoregulation and back again, EMBO Rep. 8 (2007) 1142–1148.
[334] T.R. Mosmann, H. Cherwinski, M.W. Bond, M.A. Giedlin, R.L. Coffman, Two types of murine helper T cell clone. I. Definition according to profiles of lymphokine activities and secreted proteins, J. Immunol. 136 (1986) 2348–2357.
[335] T.R. Mosmann, R.L. Coffman, TH1 and TH2 cells: different patterns of lymphokine secretion lead to different functional properties, Annu. Rev. Immunol. 7 (1989) 145–173.
[336] C.T. Weaver, C.O. Elson, L.A. Fouser, J.K. Kolls, The Th17 pathway and inflammatory diseases of the intestines, lungs, and skin, Annu. Rev. Pathol. 8 (2013) 477–512.
[337] J.R. Schoenborn, C.B. Wilson, Regulation of interferon-γ during innate and adaptive immune responses, Adv. Immunol. 96 (2007) 41–101.
[338] U. Christen, M.G. von Herrath, Manipulating the type 1 versus type 2 balance in type 1 diabetes, Immunol. Res. 30 (2004) 309–325.
[339] M. Sospedra, R. Martin, Immunology of multiple sclerosis, Annu. Rev. Immunol. 23 (2005) 683–747.

[340] J. Zhu, W.E. Paul, CD4 T cells: fates, function, and faults, Blood 112 (2008) 1557–1569.

[341] R.L. Coffmann, B.W. Seymour, S. Hudak, J. Jackson, D. Rennick, Antibody to interleukin-5 inhibits helminth-induced eosinophilia in mice, Science 245 (1989) 308–310.

[342] M. Longphre, D. Li, M. Gallup, E. Drori, C.L. Ordoñez, T. Redman, S. Wenzel, D.E. Bice, J.V. Fahy, C. Basbaum, Allergen-induced IL-9 directly stimulates mucin transcription in respiratory epithelial cells, J. Clin. Investig. 104 (1999) 1375–1382.

[343] M. Kopf, G. Le Gros, M. Bachmann, M.C. Lamers, H. Bluethmann, G. Kohler, Disruption of the murine IL-4 gene blocks Th2 cytokine responses, Nature 362 (1993) 245–248.

[344] L. Cohn, J.A. Elias, G.L. Chupp, Asthma: mechanisms of disease persistence and progression, Annu. Rev. Immunol. 11 (2004) 789–815.

[345] D.T. Umetsu, R.H. DeKruyff, The regulation of allergy and asthma, Immunol. Rev. 212 (2006) 238–255.

[346] J.F. Zambrano-Zaragoza, E.J. Romo-Martinez, M. Durán-Avelar, N. Garcia-Magallanes, N. Vibanco-Pérez, Th17 cells in autoimmune and infectious diseases, Int. J. Inflamm. (2014), article 651503, 1–12, https://doi.org/10.1155/2014/651503.

[347] N.E. Harwood, F.D. Batista, Early events in B cell activation, Annu. Rev. Immunol. 28 (2010) 185–210.

[348] M.-I. Yuseff, P. Pierobon, A. Reversat, A.-M. Lennon-Muménil, How B cells capture, process and present antigens: a crucial role for cell polarity, Nat. Rev. Immunol. 13 (2013) 475–486.

[349] J.S. Blum, P.A. Wearsch, P. Cresswell, Pathways of antigen processing, Annu. Rev. Immunol. 31 (2013) 443–473.

[350] S. Crotty, A brief history of T cell help to B cells, Nat. Rev. Immunol. 15 (2015) 185–189.

[351] S.L. Nutt, P.D. Hodgkin, D.M. Tarlinton, L.M. Corcoran, The generation of antibody-secreting plasma cells, Nat. Rev. Immunol. 15 (2015) 160–171.

[352] M.J. Shlomchik, F. Weisel, Germinal center selection and the development of memory B and plasma cells, Immunol. Rev. 247 (2012) 52–63.

[353] T. Kurosaki, K. Kometani, W. Ise, Memory B cells, Nat. Rev. Immunol. 15 (2015) 149–159.

[354] Y. Belkaid, Regulatory T cells and infection: a dangerous necessity, Nat. Rev. Immunol. 7 (2007) 875–888.

[355] A.M. Bilate, J.J. Lafaille, Induced $CD4^{+}Foxp3^{+}$ regulatory T cells in immune tolerance, Annu. Rev. Immunol. 30 (2012) 733–758.

[356] E.G. Schmitt, C.B. Williams, Generation and function of induced regulatory T cells, Front. Immunol. 4 (2013), article 152, 1–13, https://doi.org/10.3389/fimmu.2013.00152.

[357] D.J. Campbell, M.A. Koch, Phenotypical and functional specialization of $FOXP3^{+}$ regulatory T cells, Nat. Rev. Immunol. 11 (2011) 119–130.

[358] S.K. Lathrop, S.M. Bloom, S.M. Rao, K. Nutsch, C.W. Lio, N. Santacruz, D.A. Peterson, T.S. Stappenbeck, C.S. Hsieh, Peripheral education of the immune system by colonic commensal microbiota, Nature 478 (2011) 250–254.

[359] U. Hadis, B. Wahl, O. Schulz, M. Hardtke-Wolenski, A. Schippers, N. Wagner, W. Müller, T. Sparwasser, R. Förster, O. Pabst, Intestinal tolerance requires gut homing and expansion of $FoxP3^{+}$ regulatory T cells in the lamina propria, Immunity 34 (2011) 237–246.

[360] C.M. Hawrylowicz, A. O'Garra, Potential role of interleukin-10-secreting regulatory T cells in allergy and asthma, Nat. Rev. Immunol. 5 (2005) 271–283.

[361] K. Nakamura, A. Kitani, W. Strober, Cell contact-dependent immunosuppression by CD4(+)CD25(+) regulatory T cells is mediated by cell surface-bound transforming growth factor beta, J. Exp. Med. 194 (2001) 629–644.

[362] L.W. Collison, C.J. Workman, T.T. Kuo, K. Boyd, Y. Wang, K.M. Vignali, R. Cross, D. Sehy, R.S. Blumberg, D.A. Vignali, The inhibitory cytokine IL-35 contributes to regulatory T-cell function, Nature 450 (2007) 566–569.

[363] W.J. Grossman, J.W. Verbsky, B.L. Tollefsen, C. Kemper, J.P. Atkinson, T.J. Ley, Differential expression of granzymes A and B in human cytotoxic lymphocyte subsets and T regulatory cells, Blood 104 (2004) 2840–2848.

[364] N. Oberle, N. Eberhardt, C.S. Falk, P.H. Krammer, E. Suri-Payer, Rapid suppression of cytokine transcription in human $CD4^{+}CD25^{+}$ T cells by $CD4^{+}Foxp3^{+}$ regulatory Tcells: independence of IL-2 consumption, TGF-beta, and various inhibitors of TCR signaling, J. Immunol. 179 (2007) 3578–3587.

[365] S. Deaglio, K.M. Dwyer, W. Gao, D. Friedman, A. Usheva, A. Erat, J.F. Chen, K. Enjyoji, J. Linden, M. Oukka, V.K. Kuchroo, T.B. Storm, S.C. Robson, Adenosine generation catalyzed by CD39 and CD73 expressed on regulatory T cells mediates immune suppression, J. Exp. Med. 204 (2007) 1257–1265.

[366] L.S. Taams, J.M. von Amelsfort, M.M. Tiemessen, K.M. Jacobs, E.C. de Jong, A.N. Akbar, J.W. Bijlsma, F.P. Lafeber, Modulation of monocyte/macrophage function by human $CD4^{+}CD25^{+}$ regulatory T cells, Hum. Immunol. 66 (2005) 222–230.

[367] D.A.A. Vignali, L.W. Collison, C.J. Workman, How regulatory T cells work, Nat. Rev. Immunol. 8 (2008) 523–532.

[368] F. Fallarino, U. Grohmann, K.W. Hwang, C. Orabona, C. Vacca, R. Bianchi, M.L. Belladonna, M.C. Fioretti, M.L. Alegre, P. Puccetti, Modulation of tryptophan catabolism be regulatory T cells, Nat. Immunol. 4 (2003) 1206–1212.

[369] A.L. Mellor, D.H. Munn, Ido expression by dendritic cells: tolerance and tryptophan catabolism, Nat. Rev. Immunol. 4 (2004) 762–774.

[370] L. Wang, F.-S. Wang, M.E. Gershwin, Human autoimmune diseases: a comprehensive update, J. Intern. Med. 278 (2015) 369–395.

[371] A. Jones, J. Bourque, L. Kuehm, A. Opejin, R. Teague, C. Gross, D. Hawiger, Immunomodulatory functions of BTLA and HVEM govern induction of extrathymic regulatory T cells and tolerance by dendritic cells, Immunity 45 (2016) 1066–1077.

[372] J. Dobret, F.V. Duraes, L. Potin, F. Capotosti, D. Brighouse, T. Suter, S. LeibundGut-Landmann, N. Garbi, W. Reith, Lymph node stromal cells acquire peptide-MHC II complexes from dendritic cells and induce antigen-specific CD4$^+$ T cell tolerance, J. Exp. Med. 211 (2014) 1153–1166.

[373] E. Cretney, A. Kallies, S.L. Stephen, Differentiation and function of Foxp3+ effector regulatory cells, Trends Immunol. 34 (2012) 74–80.

[374] F.M. Marelli-Berg, R.I. Lechler, Antigen presentation by parenchymal cells: a route to peripheral tolerance? Immunol. Rev. 172 (1999) 297–314.

[375] D.L. Mueller, Mechanisms maintaining peripheral tolerance, Nat. Immunol. 11 (2010) 21–27.

[376] S. Panda, J.L. Ding, Natural antibodies bridge innate and adaptive immunity, J. Immunol. 194 (2015) 13–20.

[377] T.V. Rajan, The Gell-Coombs classification of hypersensitivity reactions: a re-interpretation, Trends Immunol. 24 (2003) 376–379.

[378] I.R. Mackay, Tolerance and autoimmunity, Br. Med. J. 321 (2000) 93–96.

[379] M.B.A. Oldstone, Molecular mimicry and immune-mediated diseases, FASEB J. 12 (1998) 1255–1265.

[380] I. Ulmanen, M. Halonen, T. Ilmarinen, L. Peltonen, Monogenic autoimmune diseases – lessons of self-tolerance, Curr. Opin. Immunol. 17 (2005) 609–615.

[381] Y. Cui, Y. Sheng, X. Zhang, Genetic susceptibility to SLE: recent progress from GWAS, J. Autoimmun. 41 (2013) 25–33.

[382] Q. Lu, The critical importance of epigenetics in autoimmunity, J. Autoimmun. 41 (2013) 1–5.

[383] P. Quintero-Ronderos, G. Montoya-Ortiz, Epigenetics and autoimmune diseases, Autoimmune Dis. 593720 (2012) 1–16, https://doi.org/10.1155/2012/593720.

[384] Z. Wang, Y. Zheng, C. Hou, L. Yang, X. Li, G. Huang, Q. Lu, C.Y. Wang, Z. Zhou, DNA methylation impairs TLR9 induced Foxp3 expression by attenuating IRF-7 binding activity in fulminant type 1 diabetes, J. Autoimmun. 41 (2013) 50–59.

[385] S. Kivity, M.T. Arango, M. Ehrenfeld, O. Tehori, Y. Shoenfeld, J.M. Anaya, N. Agmon-Levin, Infection and autoimmunity in Sjogren's syndrome: a clinical study and comprehensive review, J. Autoimmun. 51 (2014) 17–22.

[386] J. Peng, S. Narasimhan, J.R. Marchesi, A. Benson, F.S. Wong, L. Wen, Long term effect of gut microbiota transfer on diabetes development, J. Autoimmun. 53 (2014) 85–94.

[387] S.L. Peng, Neutrophil apoptosis in autoimmunity, J. Mol. Med. 84 (2006) 122–125.

[388] T. Akahoshi, T. Nagaoka, R. Namai, N. Sekiyama, H. Kondo, Prevention of neutrophil apoptosis by monosodium urate crystals, Rheumatol. Int. 16 (1997) 231–235.

[389] C. Tudan, D. Fong, V. Duronio, H.M. Burt, J.K. Jackson, The inhibition of spontaneous and tumor necrosis factor-alpha induced neutrophil apoptosis by crystals of calcium pyrophosphate dehydrate and monosodium urate monohydrate, J. Rheumatol. 27 (2000) 2463–2472.

[390] L. Ottonello, M. Cutolo, G. Frumento, N. Arduino, M. Bertolotto, M. Mancini, E. Sottofattori, F. Dallegri, Synovial fluids from patients with rheumatoid arthritis inhibits neutrophil apoptosis: role of adenosine and proinflammatory cytokines, Rheumatology 41 (2002) 1249–1260.

[391] K. Ina, K. Kusugami, T. Hosokawa, A. Imada, T. Shimizu, T. Yamaguchi, M. Ohsuga, K. Kyokane, T. Sakai, Y. Nishio, Y. Yokoyama, T. Ando, Increased mucosal production by granulocyte colony-stimulating factor is related to a delay in neutrophil apoptosis in inflammatory bowel disease, J. Gastroenterol. Hepatol. 14 (1999) 46–53.

[392] L. Harper, P. Cockwell, D. Adu, C.O. Savage, Neutrophil priming and apoptosis in anti-neutrophil cytoplasmic autoantibody-associated vasculitis, Kidney Int. 59 (2001) 1729–1738.

[393] L. Rauova, B. Gilburd, N. Zurgil, M. Blank, L.L. Guegas, C.M. Brickman, L. Cebecauer, M. Deutsch, A. Wiik, Y. Shoenfeld, Induction of biologically active antineutrophil cytoplasmic antibodies by immunization with human apoptotic polymorphonuclear leukocytes, Clin. Immunol. 103 (2002) 69–78.

[394] S.C. Hsieh, K.H. Sun, C.Y. Tsai, Y.Y. Tsai, S.T. Tsai, D.F. Huang, S.H. Han, H.S. Yu, C.L. Yu, Monoclonal anti-double stranded DNA antibody is a leucocoyte-binding protein to up-regulate interleukin-8 gene expression and elicit apoptosis of normal human

polymorphonuclear neutrophils, Rheumatology 40 (2001) 851–858.

[395] Y. Ren, J. Tang, M.Y. Mok, A.W. Chan, A. Wu, C.S. Lau, Increased apoptotic neutrophils and macrophages and impaired macrophage phagocytic clearance of apoptotic neutrophils in systemic lupus erythematosus, Arthritis Rheum. 48 (2003) 2888–2897.

[396] W. Matsuyama, M. Yamamoto, I. Higashimoto, K. Oonakahara, M. Watanabe, K. Machida, T. Yoshimura, N. Eiraku, M. Kawabata, M. Osame, K. Arimura, TNF-related apoptosis-inducing ligand is involved in neutropenia of systemic lupus erythematosus, Blood 104 (2004) 184–191.

[397] J.A. Mindell, Lysosomal acidification mechanisms, Annu. Rev. Physiol. 74 (2012) 69–86.

[398] Y. Ishida, S. Nayak, J.A. Mindell, M. Grabe, A model of lysosomal pH regulation, J. Gen. Physiol. 141 (2013) 705–720.

[399] K. Kawane, H. Fukuyama, H. Yoshida, H. Nagase, Y. Ohsawa, Y. Uchiyama, K. Okada, T. Lida, S. Nagata, Impaired thymic development of mouse embryos deficient in apoptotic DNA degradation, Nat. Immunol. 4 (2003) 138–144.

[400] K. Kawane, M. Ohtani, K. Miwa, T. Kizawa, Y. Kanbara, Y. Yoshioka, H. Yoshikawa, S. Nagata, Chronic polyarthritis caused by mammalian DNA that escapes from degradation in macrophages, Nature 443 (2006) 998–1002.

[401] S. Nagata, R. Hanayama, K. Kawane, Autoimmunity and clearance of dead cells, Cell 140 (2010) 19–30.

[402] K. Kawane, H. Fukuyama, G. Kondoh, J. Takeda, Y. Ohsawa, Y. Uchiyama, S. Nagata, Requirement of DNase II for definitive erythropoiesis in mouse fetal liver, Science 292 (2001) 1546–1549.

[403] M. Morita, G. Stamp, P. Robins, A. Dulic, I. Rosewell, G. Hrivnak, G. Daly, T. Lindahl, D.E. Barnes, Gene-targeted mice lacking the Trex1 (DNase III) 3′→5′ DNA exonuclease develop inflammatory myocarditis, Mol. Cell Biol. 24 (2004) 6719–6727.

[404] P.M. Rumore, C.R. Steinman, Endogenous circulating DNA in systemic lupus erythematosus. Occurrence as multimeric complexes bound to histone, J. Clin. Investig. 86 (1990) 69–74.

[405] U.S. Gaipl, A. Kuhn, A. Sheriff, L.E. Munoz, S. Franz, R.E. Voll, J.R. Kalden, M. Herrmann, Clearance of apoptotic cells in human SLE, Curr. Dir. Autoimmun. 9 (2006) 173–187.

[406] K. Kawane, K. Motani, S. Nagata, DNA degradation and its defects, Cold Spring Harb. Perspect. Biol. 6 (2014), a016394, 1–14, https://doi.org/10.1101/cshperspect.a016394.

[407] G. Parenti, G. Andria, A. Ballabio, Lysosomal storage diseases: from pathophysiology to therapy, Annu. Rev. Med. 66 (2015) 471–486.

[408] D.J. Colacurcio, R.A. Nixon, Disorders of lysosomal acidification: the emerging role of v-ATPase in aging and neurodegenerative disease, Ageing Res. Rev. 32 (2016) 75–88.

[409] R.A. Nixon, The role of autophagy in neurodegenerative disease, Nat. Med. 19 (2013) 983–997.

[410] D.J. Yamashiro, F.R. Maxfield, Acidification of endocytic compartments and the intracellular pathways of ligands and receptors, J. Cell. Biochem. 26 (1984) 231–246.

[411] R. Singh, S. Kaushik, Y. Wang, Y. Xiang, I. Novak, M. Komatsu, K. Tanaka, A.M. Cuervo, M.J. Czaja, Autophagy regulates lipid metabolism, Nature 458 (2009) 1131–1135.

[412] T. Asano, M. Komatsu, Y. Yamaguchi-Iwai, F. Ishikawa, N. Mizushima, K. Iwai, Distinct mechanisms of ferritin delivery to lysosomes in iron-depleted and iron-replete cells, Mol. Cell Biol. 31 (2011) 2040–2052.

[413] S. Wang, Z.Y. Tsun, R.L. Wolfson, K. Shen, G.A. Wyant, M.E. Plovanich, E.D. Yuan, T.D. Jones, L. Chantranupong, W. Comb, T. Wang, L. Bar-Peled, R. Zoncu, C. Straub, C. Kim, J. Park, B.L. Sabatini, D.M. Sabatini, Lysosomal amino acid transporter SLC38A9 signals arginine sufficiency to mTORC1, Science 347 (2015) 188–194.

[414] P.G.H. Gell, R.R.A. Combs, The classification of allergic reactions underlying disease, in: R.R.A. Combs, P.G.H. Gell (Eds.), Clinical Aspects of Immunology, Blackwell, London, 1963.

[415] M.S. Wright, Drug-induced hemolytic anemias: increasing complications to therapeutic interventions, Clin. Lab. Sci. 12 (1999) 115–118.

[416] B.C. Gehrs, R.C. Friedberg, Autoimmune hemolytic anemia, Am. J. Hematol. 69 (2002) 258–271.

[417] A.J. Trouth, A. Dabi, N. Solieman, M. Kurukumbi, J. Kalyanam, Myasthenia gravis: a review, Autoimmune Dis. (2012), artivel 874680, 1–10, https://doi.org/10.1155/2012/874680.

[418] W.G. Couser, Mechanisms of glomerular injury in immune-complex disease, Kidney Int. 28 (1985) 569–583.

[419] D.P. Legrende, C.A. Muzny, G.D. Marshall, E. Swiatlo, Antibiotic hypersensitivity reactions and approaches to desensitization, Clin. Infect. Dis. 58 (2014) 1140–1148.

[420] G.W. Miller, V. Nussenzweig, A new complement function: solubilization of antigen-antibody aggregates, Proc. Natl. Acad. Sci. U.S.A. 72 (1975) 418–422.

[421] M. Takahashi, J. Czop, A. Ferreira, V. Nussenzweig, Mechanism of solubilization of immune aggregates by complement. Implications for immunopathology, Transplant. Rev. 32 (1976) 121–139.

[422] R. Schrijvers, L. Gilissen, A.M. Chiriac, P. Demoly, Pathogenesis and diagnosis of delayed-type drug hypersensitivity reactions, from bedside to bench and back, Clin. Transl. Allergy 5 (2015), 31, 1–10, https://doi.org/10.1186/s13601-015-0073-8.

[423] W.J. Pichler, Delayed drug hypersensitivity reactions, Ann. Intern. Med. 139 (2003) 683–693.

[424] M. Feldmann, F.M. Brennan, R.N. Maini, Role of cytokines in rheumatoid arthritis, Annu. Rev. Immunol. 14 (1996) 397–440.

[425] D.C. Baumgart, S.R. Carding, Inflammatory bowel disease: cause and immunobiology, Lancet 369 (2007) 1627–1640.

[426] S.V. Pakala, M. Chivetta, C.B. Kelly, J.D. Katz, In autoimmune diabetes the transition from benign to pernicious insulitis requires an islet cell response to tumor necrosis factor alpha, J. Exp. Med. 189 (2009) 1053–1062.

[427] B.J. Nickoloff, Skin innate immune system in psoriasis: friend of foe? J. Clin. Investig. 104 (1999) 1161–1164.

[428] J. Nakahara, M. Maeda, S. Aiso, N. Suzuki, Current concepts in multiple sclerosis: autoimmunity versus oligodendrogliopathy, Clin. Rev. Allergy Immunol. 42 (2012) 26–34.

[429] S. Galli, S. Nakae, M. Tsai, Mast cells in the development of adaptive immune responses, Nat. Immunol. 6 (2005) 135–142.

[430] M.E. Rothenberg, H. Saito, R.S. Peebles, Advances in mechanisms of allergic disease in 2016, J. Allergy Clin. Immunol. 140 (2017) 1622–1631.

[431] B. Steelant, R. Farré, P. Wawrzyniak, J. Belmans, E. Dekimpe, H. Vanheel, L. Van Gerven, I. Kortekaas Krohn, D.M.A. Bullens, J.L. Ceuppens, C.A. Akdis, G. Boeckxstaens, S.F. Seys, P.W. Hellings, Impaired barrier function in patients with house-dust mite-induced allergic rhinitis is accompanied by decreased occludin and zonula occludens-1 expression, J. Allergy Clin. Immunol. 137 (2016) 1043–1053.

[432] C. Mazzeo, J.A. Cañas, M.P. Zafra, A. Rojas Marco, M. Fernández-Nieto, V. Sanz, M. Mittelbrunn, M. Izquierdo, F. Baixaulli, J. Sastre, V. Del Pozo, Exosome secretion by eosinophils: a possible role in asthma pathogenesis, J. Allergy Clin. Immunol. 135 (2015) 1603–1613.

[433] S. Ueki, Y. Konno, M. Takeda, Y. Moritoki, M. Hirokawa, Y. Matsuwaki, K. Honda, N. Ohta, S. Yamamoto, Y. Takagi, A. Wada, P.F. Weller, Eosinophil extracellular trap cell death-derived DNA traps: their presence in secretion and functional attributes, J. Allergy Clin. Immunol. 137 (2016) 258–267.

[434] R. Kakkar, H. Hei, S. Dobner, R.T. Lee, Interleukin 33 as a mechanically responsive cytokine secreted by living cells, J. Biol. Chem. 287 (2012) 6941–6948.

[435] T. Takeda, H. Unno, H. Motira, K. Futamura, M. Emi-Sugie, K. Arae, T. Shoda, N. Okada, A. Igarashi, E. Inoue, H. Kitazawa, S. Nakae, H. Saito, K. Matsumoto, A. Matsuda, Platelets constitutively express IL-33 protein and modulate eosinophilic airway inflammation, J. Allergy Clin. Immunol. 138 (2016) 1395–1403.

[436] P.H. Howarth, ABC of allergies. Pathogenic mechanisms: a rational basis for treatment, Br. Med. J. 316 (1998) 758–761.

[437] M.L. Oswald, S.F. Kemp, Anaphylaxis: office management and prevention, Immunol. Allergy Clin. N. Am. 27 (2007) 177–191.

CHAPTER

7

Acute-Phase Proteins and Additional Protective Systems

7.1 ACUTE-PHASE RESPONSE

7.1.1 General Features

In this chapter, contributions of acute phase, complement, coagulation, and contact systems to local and systemic immune response are highlighted.

Different stimuli such as infection, stress, trauma, tissue injury, inflammatory events, or neoplasia may induce a complex systemic response, the acute-phase response, which is directed to restore disturbed homeostasis and to promote healing processes. In this response, hepatocytes release acute-phase proteins that are designed to inhibit the growth of unwanted microorganisms, to control the inflammatory state, and to diminish tissue damage by host's own cytotoxic components. Several acute-phase proteins are also additionally released by immune and tissue cells.

Systemic acute-phase response is triggered by the proinflammatory cytokines IL-1, TNF-α, and IL-6, which are mainly released from activated immune cells, fibroblasts, and endothelial cells. This response includes characteristic features such as fever, increased numbers of neutrophils, generation of further mediators, changes in lipid, carbohydrate, and protein metabolism, mobilization of amino acids from muscles to the liver, activation of the complement and coagulation system, changes in the hormone status, and induction of acute-phase proteins [1–3].

Besides induction of proinflammatory cytokines, activation of complement, coagulation, and contact systems via blood-based factors is directed to diminish any threat resulting from damaging agents and to restore tissue homeostasis. A scheme about the interplay between basic mechanisms in systemic acute response is given in Fig. 7.1.

7.1.2 Role of Hepatocytes

Hepatocytes synthesize permanently a number of proteins and release them into the circulating blood. These proteins are transport vehicles for hormones, fatty acids, other lipophilic compounds, metal ions, and a number of drugs. They bind and inactivate cytotoxic components at injured and inflamed sites and are involved in maintaining oncotic pressure. Human serum albumin constitutes about half of serum proteins.

As part of the innate immune response hepatocytes are able to actively change the amount of released proteins. These, mostly liver-derived serum proteins with an altered release pattern, are collectively called acute-phase proteins. They can be divided into positive and negative acute-phase proteins. Proteins of the first group

Cell and Tissue Destruction
https://doi.org/10.1016/B978-0-12-816388-7.00007-3

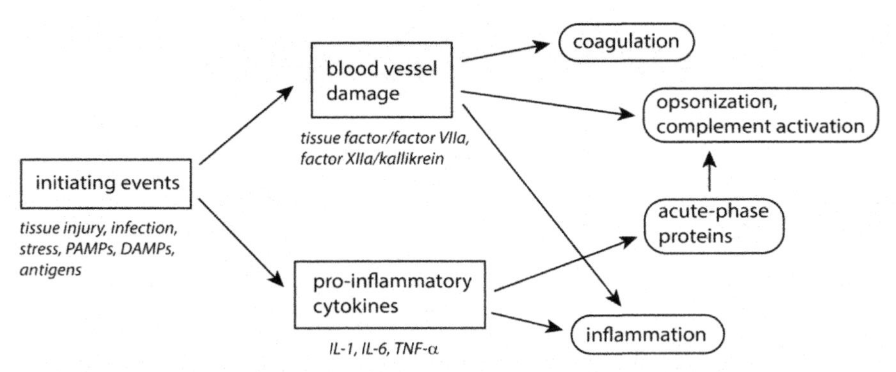

FIGURE 7.1 Schematic view of the systemic response to different events initiating tissue damage. Further explanations are given in the text.

are increasingly found in blood during acute-phase response. However, proteins of the second group are released at diminished rate into circulation [4]. Albumin is a typical example of a late-reacting negative acute-phase protein [5]. Other proteins of this group are transferrin, transthyretin, cortisol-binding protein, retinol-binding protein, antithrombin, and transcortin [6]. In activated hepatocytes, a switch in metabolic pathways occurs to allow the synthesis of positive acute-phase proteins instead of albumin and some others proteins. In addition, catabolic processes are induced in muscle tissue to provide amino acids for enhanced synthesis of positive acute-phase reactants.

Positive acute-phase proteins can further be divided into type I and type II proteins [2,7]. Differentiation of important acute-phase proteins into both groups is given in Fig. 7.2 for the human system. This differentiation is based on the activation of signaling pathways leading to their release. Hepatocytes are stimulated and regulated by four types of mediators including IL-6−related cytokines, IL-1−type cytokines, glucocorticoids, and growth factors [2]. The secretion of type I proteins is initiated by IL-1−like cytokines such as IL-1α, IL-1β, TNF-α, and TNF-β, while IL-6 and related cytokines induce the releases of type II proteins. Activation by IL-6 promotes a synergizing effect on acute-phase proteins of type I. Otherwise IL-1−like cytokines have no effect or inhibit the induction of type II proteins [2]. Glucocorticoids stimulate either the expression of some acute-phase proteins directly or enhance the effect of IL-1− and IL-6−like cytokines in a synergistic manner [2]. Modulating effects on cytokine-induced generation of acute-phase proteins are also reported for some growth factors [2].

Stimulation of the amino acid uptake by the liver belongs to the multiple functions of TNF-α [8,9]. IL-1 promotes the general amino acid flux in the whole organism [10]. There is also an interaction between hepatocytes and Kupffer cells in regulation of acute-phase response [11,12].

Main ligands and important functional effects of positive acute-phase proteins are listed in Table 7.1. The released proteins contribute by these mechanisms to eliminate cytotoxic components and to control and terminate inflammatory states.

7.1.3 Differences Between Men and Mice in Acute-Phase Response

Although many data about hepatocyte signaling were obtained on rodent models, there are some differences in activation of acute-phase proteins in the human organism. In mice,

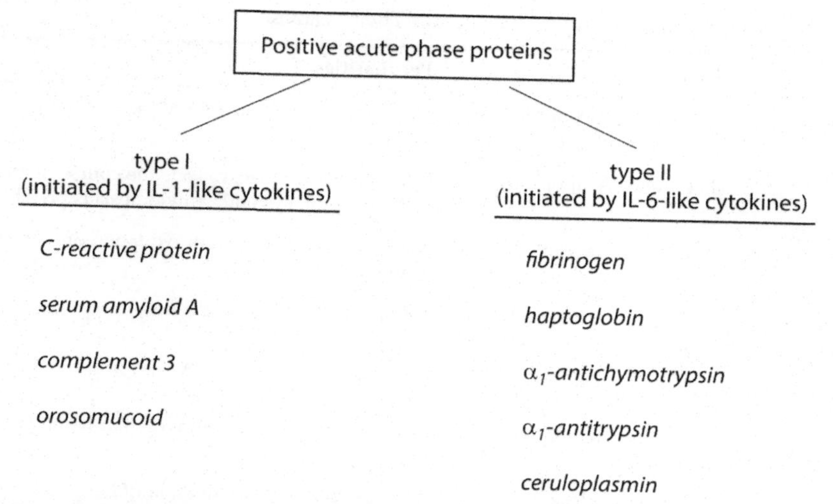

FIGURE 7.2 Differentiation of human positive acute-phase proteins according to the involved signaling pathways.

haptoglobin belongs to type I acute-phase protein, in contrast to men where it represents a type II protein [3]. In humans, haptoglobin is cleared from circulation by high-affinity binding to the macrophage receptor CD163. In contrast, mouse haptoglobin is rapidly taken up by the kidney and liver and does not promote high-affinity binding to CD163 [33]. Hemopexin is an acute-phase protein in mice, but not in humans [34].

Although the C-reactive protein (CRP) can increase by several orders of magnitude in human beings during acute-phase response, there is only a moderate increase of CRP in the mouse system [13]. In contrast, serum amyloid P, which shares a large homology with CRP, is an acute-phase reactant in mice but changes only slightly in humans [35].

7.1.4 Physiological Relevance of Acute-Phase Response

Acute-phase response is widely distributed throughout the animal kingdom. As an evolutionary old system, this response is very important in some invertebrates where the adaptive immune system is less evolved in comparison with higher vertebrates. Here, we will focus our attention most notably on the human system.

Blood levels of selected acute-phase proteins, most of all CRP, are widely used as important inflammatory markers. Their measurement provides useful data about diagnosis and prognosis of diseases such as myocardial infarction, infections, and autoimmune diseases as well as cancer treatment, general state after organ transplantations, traumata, and many others [13].

7.2 CONTROL OF INFLAMMATORY RESPONSE BY ACUTE-PHASE PROTEINS

7.2.1 C-Reactive Protein

Under normal conditions, the median concentration of CRP is 0.8 mg L^{-1} in blood of young adult volunteers [36]. Its concentration increases with age of individuals [37]. This protein is mainly produced and released by hepatocytes [13]. The production of CRP by adipocytes contributes apparently also to the plasma CRP value [38].

TABLE 7.1 Positive Acute-Phase Proteins in Humans and Their Ligands.

Acute-Phase Protein	Ligands	Peculiarities	References
C-reactive protein	Oxidized phosphocholines, lipopolysaccharides	Very strong increase, reflects the course of an inflammation	[13,14]
Serum amyloid A	High-density lipoprotein	Very strong increase, reflects the course of an inflammation, systemic amyloidosis is possible	[15–17]
α_1-Antitrypsin	Elastase, cathepsin G	Inactivation by heparin or oxidation of critical methionine	[18–20]
α_1-Antichymotrypsin	Cathepsin G, mast cell chymase, angiotensin-converting enzyme proteases	Can promote amyloidosis and hyperphosphorylation	[21,22]
Haptoglobin	Free hemoglobin, free myoglobin	Exhaustion by severe intravascular hemolysis	[23]
Ceruloplasmin	Myeloperoxidase	Mobilization and inactivation of Fe^{2+} and reactive species	[24,25]
Orosomucoid	E-selectin, P-selectin on inflamed endothelium	Very low p*I* value, high carbohydrate content, inhibits recruitment of neutrophils and immune cells	[26,27]
Mannose-binding protein	Mannose, glucose, fucose, *N*-acetylglucosamine	Initiates the lectin pathway of complement activation	[28,29]
Complement factors: C3, C4, C9, factor B, C4b-binding protein, C1 inhibitor	Other members of the complement cascade	Involved in activation and regulation of the complement cascade	[30]
Fibrinogen	Thrombin	Important for clot formation, fibrin fiber formation in the presence of heme iron	[31,32]

The concentration of CRP can considerably increase on IL-6 release by immune cells in response to inflammation. CRP is an important and easily detectable marker of inflammation. In mild inflammations and viral infections, about 10–40 mg L^{-1} CRP is measured. Higher values in the range of 40–200 mg L^{-1} are found during more strong inflammation and bacterial infection. In some severe bacterial infections and burns, values above 200 mg L^{-1} are detected [39,40]. The plasma half-life time of CRP is about 19 h both under healthy and disease conditions [13].

In most diseases, the rapid increase in CRP reflects very well the ongoing inflammation, the course, and resolution of this inflammation [40,41]. Moreover, this increase is independent of food uptake and day-time variation [13]. Hence, this parameter serves as a widely used nonspecific biochemical marker of inflammatory events. An only moderate or missing response of CRP is observed in some diseases such as systemic lupus erythematosus, scleroderma, dermatomyositis, ulcerative colitis, leukemia, and graft versus host disease [13]. Reasons for this failure are unknown.

CRP functions as soluble pattern recognition receptor [14]. In both oxidized low-density lipoprotein (LDL) and apoptotic cells, CRP binds to oxidized phosphatidylcholine in a calcium-dependent manner. In contrast, nonmodified phosphatidylcholine in native LDL or intact cells does not bind CRP [42]. The binding site for CRP is the phosphocholine moiety that is apparently more accessible under oxidized conditions. CRP is also known to bind to chromatin in the presence of histone H1 [43,44]. Other binding sites of CRP are laminin, fibronectin, and factor H [45].

Binding epitopes for CRP are also found on the surface of several microorganisms including the C-polysaccharide of *Streptococcus pneumoniae* [46,47], the repeating phosphorylated disaccharide on *Leishmania donovani* [48], and the lipopolysaccharide of *Haemophilus influenzae* [47].

Complexes of CRP with their ligands facilitate the activation of the complement system via binding of C1q [49,50].

7.2.2 Serum Amyloid A

Serum amyloid A (SAA) comprises a group of closely related small proteins with 104 amino acids [51]. In humans, two highly inducible forms, SAA1 and SAA2, and a constitutive form, SAA4, are known [17]. If we speak about SAA as acute-phase protein, the focus is directed on SAA1 and SAA2. Under inflammatory conditions, these isoforms are largely induced by hepatocytes, macrophages, and other cells both in the liver and tissues. In humans, SAA protein is the only positive acute-phase protein that displays a similar sensitivity, dynamic range, and response speed as CRP [15].

At least six different receptors are known for SAA [17]. In blood, SAA becomes tightly associated with high-density lipoprotein (HDL) under displacement of apo A-1 [52]. This binding favors lipid clearance from injured sites, promotes lipid recycling for tissue repair, stabilizes the α-helical structures of SAA, and protects it from proteolysis and misfolding [53,54].

There are numerous functional effects of SAA in blood and tissues. It facilitates the cholesterol efflux [52], acts as chemoattractant to neutrophils and induces the infiltration of these cells into inflammatory sites [55], prolongs the life span of neutrophils [56], induces the M2 phenotype in macrophages [57], potentiates the efferocytosis of apoptotic neutrophils by macrophages [57], and activates TLR2 [58] and TLR4 [59] signaling. SAA opsonizes also Gram-negative bacteria and promotes the clearance of these microbes [60].

As a pleotropic agent, SAA exhibits both pro- and antiinflammatory effects as well as homeostatic functions during the course of inflammation [17]. It contributes to resolution of inflammation [57].

Under inflammatory conditions, deposits of SAA can initiate the development of systemic AA amyloidosis [16]. The liver, spleen, and kidney are primarily affected by SAA deposition. Nephrotic syndrome and end-stage renal disease are formidable complications [61]. Acidic pH values around pH 3.5–4.5, which can be found in the stomach, in urine, or in lysosomes, favor the α-helix to β-sheet conversion of SAA, and formation of stable SAA fibrils [62]. Apparently, some SAA is filtered in the kidney, endocytosed, and finally localized in lysosomes. It is assumed that these SAA oligomers resist lysosomal degradation, damage cell membranes, and release intracellular amyloid [62].

7.2.3 Protease Inhibitors

Hepatocytes release some protease inhibitors. They are directed to inactivate active proteases from leukocytes and other cells. Among these protease inhibitors are the serpins α_1-antitrypsin and α1-antichymotrypsin. These inhibitors are several-fold upregulated during acute-phase response. α1-antichymotrypsin is also synthesized by astrocytes [63].

α_1-antitrypsin, also known as α_1-antiproteinase, is able to form a tight complex and to inactivate the human neutrophil proteases elastase, cathepsin G, and to a lesser degree also proteinase 3. In the presence of heparin, the inhibition of elastase and cathepsin G by α_1-antitrypsin fails [18,19], whereas the inactivation of proteinase 3 is enhanced [64]. Unlike elastase, cathepsin G is found on the surface of resting neutrophils. This protein is involved in degradation of specific components of the extracellular matrix on invasion of neutrophils. Elastase is released at inflammatory loci and involved in shedding of surface molecules of macrophages that further promotes the inflammatory response [65]. The expression of proteinase 3 on neutrophils is important for infiltration of unperturbed neutrophils and for delay of phagocytosis of these cells by macrophages [66].

Elastase has broad substrate specificity. This protease is able to cleave elastin degrading and thus elasticity in thus lung and other tissues [67]. This activity of elastase contributes to respiratory complications, lung emphysema, and development of chronic obstructive pulmonary disease under conditions of deficient or inadequate amount of α_1-antitrypsin in blood [68,69]. Moreover, myeloperoxidase-derived oxidants such as hypochlorous acid are known to inactivate α_1-antitrypsin by oxidation of critical methionine residues [20,70]. In the liver, defective α_1-antitrypsin induces stress response in endoplasmic reticulum and leads to development of cirrhosis [71].

Neutrophil elastase is crucial for killing of *Shigellae*, *Salmonella*, and *Yersinia* by degradation of virulence factors [72]. This protease is essential for dispatch of Gram-negative bacteria such as *Klebsiella pneumoniae* or *Escherichia coli* [73].

α_1-antichymotrypsin targets and inactivates cathepsin G [74], mast cell chymases [75], and angiotensin-converting enzyme proteases [76]. It promotes amyloid fibril formation [21,77] and tau protein hyperphosphorylation [63,78]. By these mechanisms, α_1-antichymotrypsin contributes to Alzheimer's disease [79]. Both α_1-antichymotrypsin and α_1-antitrypsin bind to DNA, a process that fails to inhibit cathepsin G [22].

7.2.4 Haptoglobin

Haptoglobin forms a complex with free hemoglobin and free myoglobin. These complexes are cleared by the spleen and liver macrophages. For more details, see Sections 4.2.7 and 4.3.2.

Otherwise, haptoglobin is taken up by human neutrophils and released from these cells on exposure to TNF-α or the chemotactic tripeptide fMLP. This secretion of haptoglobin modulates the acute inflammatory response at sites of infection or injury [80].

7.2.5 Ceruloplasmin

Ceruloplasmin is a late acute-phase protein that reached its maximum after most other acute-phase proteins. Under inflammatory conditions, its production in the liver cells can rise to about 50% [81]. This copper-containing proteins exhibits numerous antiinflammatory and antioxidant effects such as mobilization of iron from cellular stores and oxidation of Fe^{2+} to Fe^{3+} [24], binding and inhibition of myeloperoxidase released from neutrophils [25,82,83], and dismutation of superoxide anion radicals [84].

7.2.6 Orosomucoid

Orosomucoid or α_1-acid glycoprotein is characterized by a very low p*I* value of 2.8–3.8 and an extremely high carbohydrate content of 45% [26]. Its carbohydrate cover consists of di-, tri-, and tetraantennary *N*-linked glycans [85]. These peculiar properties determine the immunomodulatory actions of this protein. Orosomucoid exhibits antineutrophil and anticomplement activities [86]. Under inflammatory conditions, this protein expresses sialyl Lewis X elements that bind to E-selectin and P-selectin on inflamed endothelium [27]. This interaction inhibits the recruitment of neutrophils and other immune cell to inflammatory loci [85,87].

7.3 COMPLEMENT SYSTEM

7.3.1 Main Functions of the Complement System

The complement system, an important part of the innate immune system, consists of about 30 serum proteins and regulatory cell membrane components. Most of serum proteins of the complement are synthesized in the liver. Monocytes, tissue macrophages, and epithelial cells, especially in the genitourinary system and in the gastrointestinal tract, are also able to produce these proteins. In serum, proteins of the complement system comprise about 10% of the globulin fraction.

Components of the complement circulate as inactive precursor molecules in the blood. In the presence of target epitopes, some complement molecules bind and opsonize foreign and host's own damaged material. Subsequently, a cascade of complement activation events is initiated under involvement of complement-specific proteases. Finally, clusters of complement molecules form pores and lyse target cells. Thus, the complement system enables targeted attacks on membranes of microbes, attracts neutrophils and macrophages, and facilitates the recognition and engulfment of labeled material by professional phagocytes [88]. To prevent an unwanted lysis of intact host's cells, complement activation is tightly controlled and regulated [89–91].

7.3.2 Pathways of Activation of the Complement System

There are three main biochemical pathways of complement activation. These pathways are the classical complement pathway, the lectin pathway, and the alternative complement pathway [92,93]. They differ in the kind of primary target epitopes and the first steps of complement activation. All three pathways lead to formation of C3 and C5 convertases and finally to the formation of a membrane attack complex. An overview about these pathways and main functional responses and regulatory mechanisms is given in Fig. 7.3.

The classical complement pathway is initiated by binding of the complement protein C1q to antigen–antibody complexes containing IgG or IgM antibodies [91,93]. This pathway can also be activated by C1q binding to apoptotic cells, necrotic cell material, and CRP, if the latter is attached to target molecules. Binding of C1q induces a series of successive activation steps of further complement components such as C1r, C1s, C4, and C2.

The lectin pathway differs from the classical pathway in the kind of primary labeling. In this pathway, mannose-binding lectin and ficolin act as opsonins instead of C1q. Following binding of these opsonins to sugar residues on the microbe surface, specific mannose-binding lectin serine proteases, C4, and C2 activate this cascade [94,95]. While mannose-binding protein prefers as binding substrate carbohydrate residues with equatorial hydroxyl groups at C3 and C4 positions such as mannose, glucose, fucose, and *N*-acetylglucosamine [28], ficolins bind predominantly *N*-acetylated saccharides such as *N*-acetylglucosamine, *N*-acetylgalactosamine, and sialic acid [96]. No recognition by an antibody is involved in the lectin pathway.

In the classical complement and lectin pathways, the complexes C4b/C2b and C4b/C2b/C3b act as convertase 3 and 5, respectively, cleaving C3 and C5 into the fragments C3a, C3b, C5a, and C5b.

The alternative pathway of complement activation starts with the spontaneous formation of C3b on hydrolysis of C3. C3b is rapidly inactivated by factor H or factor I. Opsonization of microbes or damaged host tissues by C3b

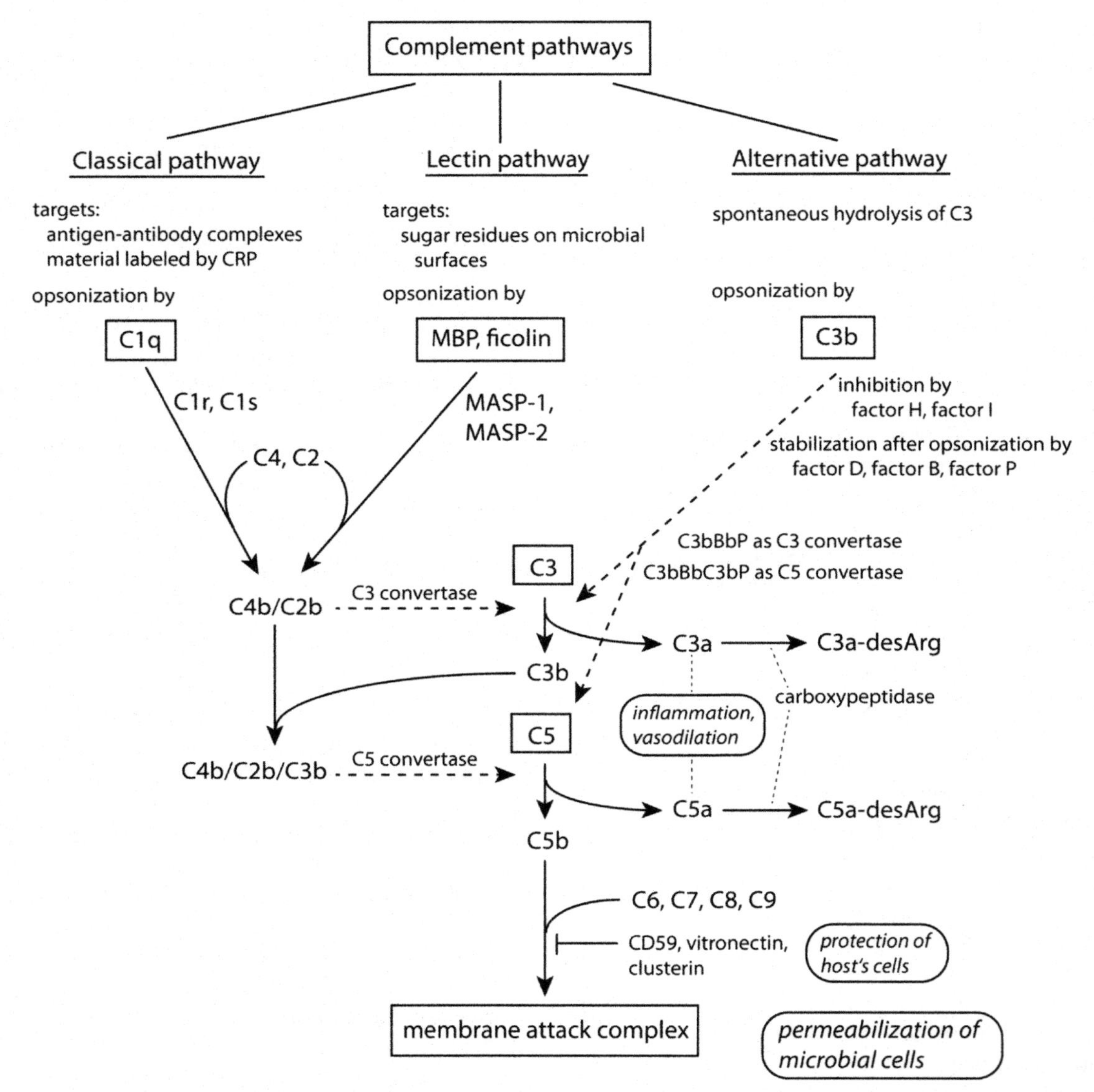

FIGURE 7.3 Overview of the three main pathways of complement activation. These pathways starts with the opsonization of microbes, antigen–antibody complexes, or deteriorated host's material labeled with C-reactive protein by components of the complement system. In these reactions, C3 and C5 convertases are activated. Their cleavage products, C3a and C5a, promote proinflammatory reactions by vasodilation, smooth muscle contraction, and enhanced blood vessel permeability. The complement fragment C5b forms with complement factors C6, C7, C8, and C9 an annular membrane attack complex on the surface of microbes that lyses these target cells. *CRP*, C-reactive protein; *MASP*, mannose-binding lectin serine protease; *MBP*, mannose-binding protein. Further explanations are given in the text.

protects this complement fragment against inhibition. Factors B, D, and P are involved in stabilization of this binding, and further recruitment of C3b. The complexes C3bBbP and C3bBbC3bP function as C3 and C5 convertases, respectively [97,98].

7.3.3 Formation of the Membrane Attack Complex

In all three pathways, a complex of the four different complement molecules Cb5, C6, C7, and C8 is formed on the outer surface of microbes or targeted host's cells. The complement protein C9 converts this complex into an annular structure forming a pore in the membrane [89]. If enough pores are formed in a cell, the lysed cell is unable to survive. Targeted cells dye by osmotic lysis.

There are several mechanisms to prevent an unwanted permeabilization of host's cells by the complement. The regulatory proteins clusterin and vitronectin as well as CD59 on the surface of host's cells can prevent pore formation [99]. For example, red blood cells lacking CD59, a condition found in paroxysmal nocturnal hemoglobinuria, are prone to enhanced rate of hemolysis [100,101].

7.3.4 Complement Components as Acute-Phase Proteins

Components of the complement system are always present in the organism. Under normal health conditions, they are mainly inactive or exhibit a low basal activity due to the presence of inhibitory elements. Tissue injury, invasion of microbes, and inflammatory state activate the complement system induced by the attachment of initiator molecules to foreign or host's own damaged surfaces. This response is further amplified by enhanced synthesis and release of some components of the complement system as acute-phase proteins.

Both CRP and mannose-binding proteins are upregulated during acute-phase response. As CRP is usually much stronger enhanced than mannose-binding protein, CRP is able to inhibit the lectin pathway by binding the regulatory protein H [29].

Besides mannose-binding protein, the proteins C3, C4, C9, factor B, C4b-binding protein, and C1 inhibitor have elevated plasma concentration during acute-phase response [30].

7.3.5 Complement Components as Anaphylatoxins

During complement activation, C3 and C5 are split by C3 convertase and C5 convertase, respectively, into the fragments C3a, C3b, C5a, and C5b. While C3b and C5b are involved in activation and formation of the membrane attack complex, the fragments C3a and C5a can act as anaphylatoxins by mediating vasodilation, contraction of smooth muscles, and enhancing permeability of small blood vessels [102]. Other proinflammatory effects of these complement fragments are stimulating activities on neutrophils, eosinophils, basophils, and macrophages as well as the release of mediators from mast cells. Inflammatory reactions mediated by C3a and C5a can be very severe as seen in type I hypersensitivity responses (see also Section 6.7.6) [102,103].

The complement fragment C4a is also listed as anaphylatoxin. However, it differs considerably in its functions and by absence of specific receptors from the two other members [104].

In addition to the formation of C3a and C5a as a result of complement activation, these fragments can also be generated by the so-called extrinsic pathway. This pathway is activated by microbial or parasite proteases [105–108] or by plasmin, thrombin, and some other components of the coagulation cascade [109]. Release of cathepsin D [110], mast cell β-tryptase [111], or granzyme B [112] also promotes C3a and C5a generation.

Activities of C3a and C5a are tightly regulated by carboxypeptidases that cleave off the C-terminal arginine residue [113]. The resulting C3a desArg affects the lipid metabolism in adipocytes [114]. Disturbances in the fine-tuning of the complement cascade, and in particular, in the control over anaphylatoxins, result in numerous disease scenarios [115]. Lack of C3a and C5a favor the development of autoimmune processes such as lupus erythematosus due to insufficient clearance of immune complexes. Otherwise, excessive generation of these fragments promotes inflammatory states and cell and tissue damages. High concentrations of C3a, C5a, and cleavage products C3a desArg and C5a desArg are found in septic patients [116].

7.4 COAGULATION SYSTEM

7.4.1 Main Functions of the Coagulation System

The liquid tissue blood delivers regularly dioxygen, glucose, and other essential nutrients to all regions of our body and transports carbon dioxide and waste products to the lung, spleen, liver, and kidney. The coagulation system is crucial to minimize blood loss as a result of vessel wall damage and any kinds of injury. In intact vessels, antithrombogenic factors inhibit efficiently any procoagulant activity and ensure thus an undisturbed blood flow. Among these anticoagulant components are heparin, thrombomodulin, tissue plasminogen activator, antithrombin, protein C, protein S, and plasminogen [117].

Heparin is widely used as an anticoagulant drug to prevent clot formation, thrombosis, and embolism in numerous diseases scenarios [118]. It enhances considerably the activity of antithrombin toward thrombin and factor Xa [119].

Owing to bleeding (also known as hemorrhaging), blood escapes from the circulatory system either internally or externally. This activates circulating thrombogenic factors, thrombocytes (also known as platelets), and several vessel wall–specific proteins. The complex interaction of these components is directed to stop bleeding, a process known as hemostasis. This process includes vasoconstriction, temporary formation of a platelet thrombus, and blood coagulation [120]. Vasoconstriction is promoted by vascular smooth muscle cells especially in smaller blood vessels.

7.4.2 Activation of Thrombocytes

Platelets adhere to collagen epitopes at injured vessel wall. Activated platelets release serotonin, ADP, and thromboxane A_2, which further potentiate vasoconstriction and assist in attraction of additional platelets and their aggregation [121]. Dense granules of human platelets contain the highly anionic polyphosphate that is secreted on cell activation [122]. Platelet polyphosphate consists of on average 60–100 phosphates. With this property, it differs from polyphosphates of microbes, which contain up to 1000 phosphate units [123]. Polyphosphates exert proinflammatory and prothrombotic effects and provide, thus, a link between coagulation and inflammation [124,125].

Adhesion of thrombocytes to subendothelial surface is mediated by the von Willebrand factor, a glycoprotein present in blood plasma and produced constitutively as large aggregate in endothelium, megakaryocytes, and connective tissue adjacent to endothelium [126]. Several positive feedback loops are involved in platelet activation and aggregation [127]. Formation of platelet aggregates is referred to as primary hemostasis. The platelet thrombus is further stabilized by formation of a dense fibrin mesh, a process that is called secondary hemostasis [128]. During this process, a stepwise activation of coagulation factors occurs.

7.4.3 Coagulation Cascade: Role of the Tissue Factor

Coagulation cascade was originally subdivided into an extrinsic, intrinsic, and common pathway. This classical theory is useful for understanding in vitro processes of blood coagulation but fails to incorporate interaction with cell wall components [129]. Current concepts of coagulation differentiate between initiation, amplification, propagation, and stabilization phase. Several amplifying loops are included in these pathways, for example, for robust formation of thrombin in interaction with activated platelets [117].

Numerous coagulation factors circulate in inactive form in blood. Most of these factors are serine proteases. Active forms are marked by the lowercase letter a after the Roman numeral, which identifies the corresponding factor. Coagulation can start with complex formation between the coagulation factor VII and tissue factor (CD142) [130], a surface glycoprotein present on subendothelial cells such as smooth muscle cells and fibroblasts. These cells are located near intact blood vessels, however, without immediate contact to circulating blood. On vessel injury, both proteins interact with each other, and this complex activates via factor X, the serine protease thrombin [131,132]. In the coagulation cascade (Fig. 7.4), factors IX and XI are involved in activation loops.

Under inflammatory conditions and on vascular injury, the tissue factor is upregulated and also expressed by endothelial cells and monocytes. This often results in hypercoagulability and increases the tendency of thrombosis development [133,134]. Numerous inflammatory diseases such as dysfunctions of the cardiovascular system are associated with enhanced risk of thrombosis [135]. CRP is known to upregulate tissue factor [136]. There is also a mutual interplay between inflammation and coagulation [134,137]. Different protease-activated receptors are targeted by factors Xa and IIa (thrombin) and are involved in proinflammatory activity [134].

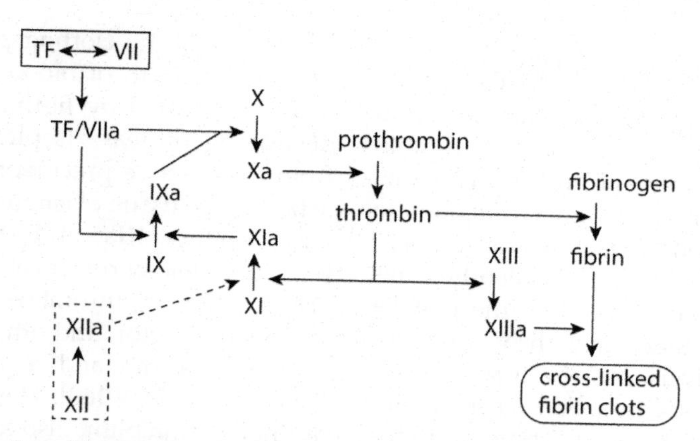

FIGURE 7.4 Schematic view of the coagulation cascade. Coagulation starts with the complex formation between tissue factor and factor VII. Activation of factor XII (dashed line) can also contribute to coagulation. *TF*, tissue factor. Further explanations are given in the text.

7.4.4 Coagulation Cascade: Role of the Coagulation Factor XII

The coagulation factor XII is activated on contact with negatively charged substances such as polyphosphates, nucleic acids, heparin, collagen, misfolded proteins, extracellular traps, and anionic bacterial surfaces [138–140]. This pathway is also called contact-dependent coagulation. Factor XII comprises together with factor XI, prekallikrein/kallikrein, kininogen, and C1 inhibitor in the contact system [141].

On activation, factor XIIa promotes coagulation via stepwise activation of further factors of the coagulation cascade that results as in the tissue factor pathway in formation of factor Xa and thrombin. In blood coagulation, this intrinsic pathway plays only a minor role in contrast to the extrinsic pathway (tissue factor–mediated pathway) as shown by analysis of severe deficiencies of single components and their consequences [142].

However, the activated factor XII is on the crossroad of several systems. Other outcomes of factor XIIa are effects on the complement, fibrinolysis, and the activation of the kallikrein–kinin pathway (see Section 7.5).

7.4.5 Blood Clotting and Fibrinogen

Thrombin is proteolytically cleaved from prothrombin in the clotting cascade. Thrombin converts fibrinogen into fibrin. In addition, thrombin exerts several regulatory functions in this cascade [31]. Thrombin is necessary for activation of factor XIII, which covalently links fibrin polymers and stabilizes thus the platelet thrombus [143]. Furthermore, thrombin activates an antifibrinolysis protein that protects the clot from fibrinolysis [144]. Fibrin contributes also in regulation of blood clotting. It binds thrombin and inhibits thus the further coagulation process [145].

Fibrinogen circulates as a soluble plasma glycoprotein with a half-life time of approximately 4 d [146]. It consists of three distinct polypeptide chains that are arranged to an elongated molecule, which binds to a second identical molecule by disulfide bonds forming a homodimer [147]. This protein is synthesized in the liver [148]. Blood clotting is a physiological important example of the formation of insoluble fibers. In this process, soluble fibrinogen is converted by thrombin to insoluble fibrin fibers that form a dense fibrin mesh and close any wounds. Fibrinogen plays also an important role in angiogenesis, revascularization, cell–matrix interaction, wound healing, tissue repair, and tumor progression [149–154].

Human fibrin is composed of 30% α-helices, 37% β-sheet, and 32% turns, loops, and random coils [155]. Conformational changes of fibrinogen/fibrin during normal blood clotting do not involve any changes in the α-helical to β-sheet ratio. On elongation and compression of fibrin clots, denser fibrin structures are formed, characterized by an α-helix to β-sheet transition [155].

7.4.6 Fibrinolysis

To control the clotting process and to avoid excessive growth of blood clots, coagulation is counterregulated by fibrinolysis. The main clot destroying protease is plasmin that is formed from the inactive precursor plasminogen under action of the tissue plasminogen activator, urokinase, factors XIa and XIIa, and kallikrein [156,157]. The serine protease plasmin cleaves also fibronectin, thrombospondin, laminin, von Willebrand factor, and other substrates [158].

α_2-antiplasmin and α_2-macroglobulin inactivate plasmin [159,160]. The thrombin-activatable fibrinolysis inhibitor, also known as carboxypeptidase B_2, protects fibrin against plasmin activity [144]. Plasminogen activator inhibitor 1 prevents plasmin formation by inhibition of tissue plasminogen activator and urokinase [161].

7.4.7 Blood Clotting and Disease Progression

Several diseases scenarios such as impaired kidney function, diabetes, ischemic stroke, systemic inflammation, and sepsis are associated with an enhanced risk of thrombosis [162,163]. It has been assumed that abnormal fibrin structures delay fibrinolysis and promote the formation of circulating microclots [164]. Denser and more stable clots result from thinner fibrin fibers. These clots are characterized by diminished gel porosity and hindered diffusion of plasminogen and fibrinolytic agents [165–167].

Chemical modification is discussed as potential reason of the formation of dense fibrin clots. These alterations can result from the accompanying inflammatory process. However, glycation of fibrinogen does not contribute to formation of more thin fibers [167].

Oxidation of methionine residues of fibrinogen by hypochlorous acid forms thin fibrin fibers after activation by thrombin [164].

Blood of healthy individuals is free of any fibrin fibers. As evidenced by scanning electron microscopy, blood of patients of a wide range of diseases such as diabetes, Parkinson's disease, and rheumatoid arthritis contains deformed red blood cells embedded in fibrin fibers [168–171]. It is assumed that heme release from defective red blood cells and the appearance of unliganded iron contribute to pathological fibrin formation [32].

7.5 LINKS BETWEEN COMPLEMENT, COAGULATION, AND INFLAMMATION

7.5.1 The Kallikrein–Kinin Pathway: Activation by Coagulation Factor XIIa

On contact of blood with collagen and other negatively charged surfaces, the blood components coagulation factor XII, also known as Hageman factor, prekallikrein, and high molecular kininogen bind to these surfaces and interact with each other [138–140]. As a result, the cleavage of prekallikrein by factor XIIa leads to formation of kallikrein (Fig. 7.5). The latter protease activates additional XII molecules, cleaves kininogen under release of bradykinin, and converts plasminogen to plasmin. Bradykinin, the final product of the kallikrein–kinin pathway, is a potent proinflammatory agent [172]. It promotes

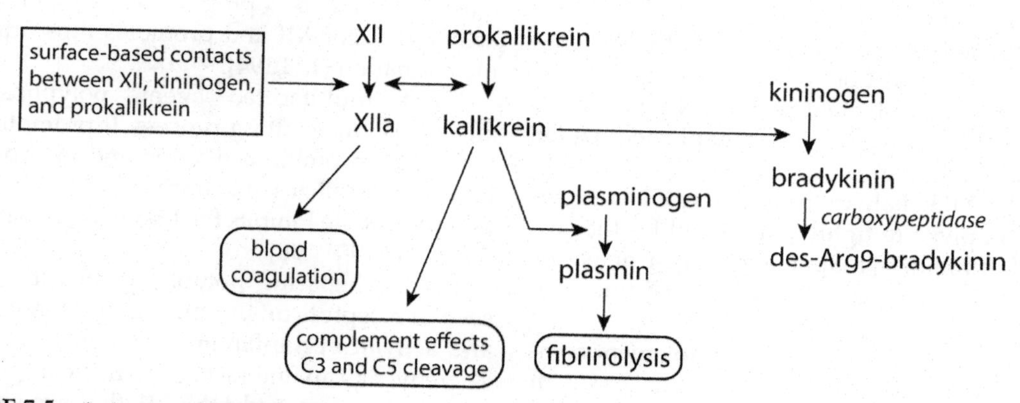

FIGURE 7.5 Activation of the contact system. This pathway is activated by assembling of the blood components factor XII, prekallikrein, and kininogen on collagen and other surfaces. Further explanations are given in the text.

vascular permeability and functions as vasodilator acting mainly via the constitutively expressed bradykinin receptor B_2 [173]. The nonapeptide bradykinin contributes also to pain and itching sensations [174–176]. The action of bradykinin is terminated by aminopeptidase P and angiotensin-converting enzyme transforming bradykinin into inactive metabolites.

Depletion of the C-terminal arginine in bradykinin by carboxypeptidase N results in formation of des-Arg9-bradykinin, a selective agonist to the bradykinin B_1 receptor [177]. The significance of this pathway rises when the angiotensin-converting enzyme is inhibited [178]. The bradykinin receptor B_1 is only expressed in inflamed and injured tissue [179,180]. This receptor plays a role in recruitment of neutrophils [181]. It is also involved in bradykinin effects during brain inflammations [182,183].

7.5.2 Contact System: Effects on the Complement System

The reverse interplay between factor XIIa and kallikrein affects also the complement system. Several factors of the contact-dependent coagulation pathway promote the cleavage of the complement factors C3 and C5 into the active forms C3a and C5a. The same holds for thrombin and plasmin [184]. Kallikrein itself cleaves C3 to yield both C3a and C3b [185].

Other interactions between both systems concern the activation of the protease C1s of the classical complement pathway by factor XII [186] and the cleavage of the complement factor B by kallikrein [187].

Factor XIIa, kallikrein, and the C1 complement complex are tightly controlled by the serpin C1 inhibitor [188]. Plasminogen activator inhibitor 1 and α_1-antitrypsin can additionally inhibit FXIIa.

Concerted activation of the complement components and bradykinin by the contact system can mediate severe inflammatory reactions and anaphylaxis.

7.5.3 Contact System: Activation by Mast Cell and Platelet Products

Heparin released from activated mast cells and basophils is able to interact with the coagulation factor XII and promotes, thus, the formation of bradykinin [189]. Mast cell heparin differs from the therapeutically used heparin by a higher negative charge density. This property is necessary for induction of bradykinin activation [189]. By the way, contamination of the anticoagulant drug heparin with oversulfated chondroitin sulfate was determined as the reason for lethal complications in some patients receiving heparin [190]. In factor XII activation, heparin exerts a dual function. Besides contact system activation, it binds antithrombin, which irreversibly inhibits factor XIIa. Activation of this factor by heparin follows a bell-shaped dose response [191].

Mast cell granules contain also the negatively charged chondroitin sulfate E, which is able to activate factor XIIa [192].

7.5.4 Manifold Actions of Polyphosphate

A similar activation of the contact system has been described for polyphosphate that is released from activated platelets [122] as well as from mast cells and basophils [193]. Polyphosphate activates factor XII and promotes thus bradykinin formation [123,194].

As a multifaceted agent, polyphosphate affects also the clotting process. It promotes factor X and thrombin activation and incorporates into and stabilizes the fibrin clot. In addition, polyphosphate inhibits the tissue factor pathway inhibitor [195–198].

Unlike heparin, polyphosphates binds to several receptors on the surface of inflamed cells and activates proinflammatory pathways [199]. Thereby, it can interact with nuclear proteins such as histones and HMGB1 that are secreted from activated immune cells. The presence of these nuclear components in plasma correlates with poor outcome in severe sepsis and cancer [200]. Interaction between polyphosphate and histone amplifies thrombin generation and proinflammatory signaling via RAGE receptors and purinergic P2Y1 receptors [201]. In endothelial cells, the key metabolic regulator mTOR is activated on polyphosphate-mediated signaling [202].

7.5.5 Some Examples of Inflammatory Diseases

The close association of contact, coagulation, and complement systems with platelet and mast cell degranulation can lead to anaphylactic reactions. Plasma levels for heparin, mast cell constituents, and bradykinin increase in severe anaphylaxis together with parameters for activation of the contact system [139,203]. Mast cell—derived heparin is known to initiate generation of bradykinin via contact activation of factor XII. Moreover, the severity of clinical symptoms during anaphylaxis correlates with heparin and bradykinin levels [203].

Hereditary angioedema is characterized by recurrent episodes of angioedema with manifestations in skin and mucosal surfaces. This disease is caused by reduced C1 inhibitor levels, genetic defects in C1 inhibitor, or hyperactive factor XII [140].

Activation of the kallikrein—kinin pathway via factor XII is also induced by hantaviruses [204] and herpes simplex virus 1 [205]. Different strains of hantaviruses can cause severe hemorrhagic fever with renal syndrome or hantavirus pulmonary syndrome. These syndromes are associated with systemic vascular leakage, hypotension, and shock [206]. Several new strategies have been proposed for therapeutic interventions against hantavirus-induced symptoms on basis of inhibitors of the contact system [204].

Gram-negative and Gram-positive bacteria as well as fungi bind also factor XII [207]. Although the activation of the contact system by microbes is part of immunological defense reactions, excessive bradykinin formation can cause edema and hypotension [139]. For example, enhanced bradykinin levels are found in septic patients infected with *Staphylococcus aureus* [208]. Contact system can also be activated by neutrophil-derived extracellular traps [209].

Factor H plays a central role in complement regulation by acceleration of convertase 3 decay in the alternative pathway of complement activation and as cofactor for factor I—mediated inactivation of C3b [210]. Mutations, polymorphisms, or deficiencies of factor H are associated with increased complement activation as evidenced in numerous disease states such as atypical hemolytic uremic syndrome, age-related macular degeneration, and membranoproliferative glomerulonephritis type II [211—213].

Some cancer cells and microbes developed immune evasion strategies by upregulating factor H in their microenvironment [210,214].

7.6 SUMMARY

During complex immune response, recruitment and activation of immune cells is additionally supported by acute-phase proteins, complement proteins, and by proteins of the coagulation and contact systems. These important mechanisms act supplementary to immune cells and help to find an adequate response against external and internal impacts.

Majority of proteins of these supplementary systems is synthesized in the liver. Adequate response of these systems depends on intact metabolic pathways in this organ. Acute-phase proteins assist in inflammatory immune mechanisms and help to protect tissues from immune cell-induced destructions. The complement system contributes to immune response by opsonization of microbial surfaces, antigen—antibody complexes, and defective targets labeled by the CRP as well as by formation of membrane attack complexes. Microbial, collagenous surfaces and negatively charged polymers activate the contact system with multiple effects on blood vessel permeability, inflammation, fibrinolysis, complement, and coagulation. The coagulation system itself is very crucial to prevent any loss of this fluid from leaky and injured blood vessel. This system can also be activated without induction of any inflammation.

Close interrelations exist between inflammation, acute-phase proteins, and activation of complement, coagulation, and contact systems. Dysregulations in these mechanisms affect immune response and favor the development of pathologies.

References

[1] I. Kushner, The phenomenon of the acute phase response, Ann. NY Acad. Sci. 389 (1982) 39–48.

[2] H. Baumann, J. Gauldie, The acute phase response, Immunol. Today 15 (1994) 74–80.

[3] H. Moshage, Cytokines and the hepatic acute phase response, J. Pathol. 181 (1997) 257–266.

[4] R.F. Ritchie, G.E. Palomaki, L.M. Neveux, O. Navolotskaia, T.B. Ledue, W.Y. Craig, Reference distributions for the negative acute-phase serum proteins, albumin, transferrin and transthyretin: a practical simple and clinically relevant approach in a large cohort, J. Clin. Lab. Anal. 13 (1999) 273–279.

[5] G. Tsirpanlis, P. Bagos, D. Ioannou, A. Bleta, I. Marinou, S. Chatzipanagiotou, C. Nicolaou, Serum albumin: a late-reacting negative acute-phase protein in clinically evident inflammation in dialysis patients, Nephrol. Dial. Transplant. 20 (2005) 658–659.

[6] Y. Ingenbleek, V. Young, Transthyretin (prealbumin) in health and disease: nutritional implications, Annu. Rev. Nutr. 14 (1994) 495–533.

[7] A. Koj, Biological functions of acute-phase proteins, in: A.H. Gordon, A. Koj (Eds.), The Acute Phase Response to Injury and Infection, Elsevier, Amsterdam, 1985, pp. 145–160.

[8] D.G. Hesse, K.J. Tracey, Y. Fong, K.R. Manogue, M.A. Palladino Jr., A. Cerami, G.T. Shires, S.F. Lowry, Cytokine appearance in human endotoxemia and primate bacteremia, Surg. Gynecol. Obstet. 166 (1988) 147–153.

[9] K.J. Tracey, Role of tumor necrosis factor in protein metabolism, in: N.C.R. Räihä (Ed.), Protein Metabolism During Infancy, Vevey/Raven Press, New York, 1994, pp. 229–242.

[10] C.A. Dinarello, Interleukin-1 and interleukin-1 antagonism, Blood 77 (1991) 1627–1652.

[11] F. Tacke, T. Luedde, C. Trautwein, Inflammatory pathways in liver homeostasis and liver injury, Clin. Rev. Allergy Immunol. 36 (2009) 4–12.

[12] M.W. Robinson, C. Harmon, C. O'Farrelly, Liver immunology and its role in inflammation and homeostasis, Cell. Mol. Immunol. 13 (2016) 267–276.

[13] M.B. Pepys, G.M. Hirschfeld, C-Reactive protein: a critical update, J. Clin. Investig. 111 (2003) 1805–1812.

[14] C. Garlanda, B. Bottazzi, A. Bastone, A. Mantovani, Pentraxins at the crossroad between innate immunity, inflammation, matrix deposition, and female fertility, Annu. Rev. Immunol. 23 (2005) 337–366.

[15] M.B. Pepys, M.I. Baltz, Acute phase proteins with special reference to C-reactive protein and related proteins (pentraxins) and serum amyloid A protein, Adv. Immunol. 34 (1983) 141–212.

[16] D. Real de Asúa, R. Costa, J.M. Galván, M.T. Filigheddu, D. Trujillo, J. Cadiñanos, Systemic AA amyloidosis: epidemiology, diagnosis, and management, Clin. Epidemiol. 6 (2014) 369–377.

[17] R.D. Ye, L. Sun, Emerging functions of serum amyloid A in inflammation, J. Leukoc. Biol. 98 (2015) 923–929.

[18] K.J. Frommherz, B. Faller, J.G. Bieth, Heparin strongly decreases the rate of inhibition of neutrophil elastase by α_1-proteinase inhibitor, J. Biol. Chem. 266 (1991) 15356–15362.

[19] J. Ermolieff, C. Boudier, A. Laine, B. Meyer, J.G. Bieth, Heparin protects cathepsin G against inhibition by protein proteinase inhibitors, J. Biol. Chem. 269 (1994) 29502–29508.

[20] C.C. Taggart, D. Cervantes-Laurean, G. Kim, N.G. McElvaney, N. Wehr, J. Moss, R.L. Levine, Oxidation of either methionine 351 or methionine 358 in alpha 1-antitrypsin causes loss of anti-neutrophil elastase activity, J. Biol. Chem. 275 (2000) 27258–27265.

[21] J. Ma, A. Yee, H.B. Brewer Jr., S. Das, H. Potter, Amyloid-associated proteins alpha 1-antichymotrypsin and apolipoprotein E promote assembly of Alzheimer beta-protein into filaments, Nature 372 (1994) 92–94.

[22] J. Duranton, C. Boudier, D. Belorgey, P. Mellet, J.G. Blieth, DNA strongly impairs the inhibition of cathepsin G by α_1-antichymotrypsin and α_1-proteinase inhibitor, J. Biol. Chem. 275 (2000) 3787–3792.

[23] D. Chiabrando, F. Vinchi, V. Fiorito, E. Tolosano, Haptoglobin and hemopexin in heme detoxification and iron recycling, in: F. Veas (Ed.), Acute Phase Proteins – Regulation and Functions of Acute Phase Proteins, Intech, Rijeka, Croatia, 2011, pp. 261–288.

[24] Z.L. Harris, A.P. Durley, T.K. Man, J.D. Gitlin, Targeted gene disruption reveals essential role of

ceruloplasmin in cellular iron efflux, Proc. Natl. Acad. Sci. U.S.A. 96 (1999) 10812–10817.

[25] A.V. Sokolov, K.V. Ageeva, M.O. Pulina, O.S. Cherkalina, V.R. Samygina, I.I. Vlasova, O.M. Panasenko, E.T. Zakharova, V.B. Vasilyev, Ceruloplasmin and myeloperoxidase in complex affect the enzymatic properties of each other, Free Radic. Res. 42 (2008) 989–998.

[26] K. Schmid, R.B. Nimberg, A. Kimura, J.P. Yamaguchi, J.P. Binette, The carbohydrate units of human plasma alpha-1-acid glycoprotein, Biochim. Biophys. Acta 492 (1977) 291–302.

[27] T. de Graaf, M. van der Stelt, M. Ambergen, W. van Dijk, Inflammation-induced expression of sialyl Lewis X-containing glycan structures on alpha 1-acid glycoprotein (orosomucoid) in human sera, J. Exp. Med. 177 (1993) 657–666.

[28] J. Epstein, Q. Eichbaum, S. Sheriff, R.A.B. Ezekowitz, The collectins in innate immunity, Curr. Opin. Immunol. 8 (1996) 29–35.

[29] C. Suankratay, C. Mold, Y. Zhang, L.A. Potempa, T.F. Lint, H. Gewurz, Complement regulation in innate immunity and the acute-phase response: inhibition of mannan-binding lectin-initiated complement cytolysis by C-reactive protein (CRP), Clin. Exp. Immunol. 113 (1998) 353–359.

[30] C. Gabay, I. Kushner, Acute-phase proteins and other systemic responses to inflammation, N. Engl. J. Med. 340 (1999) 448–454.

[31] J.T.B. Crowley, S. Zanardelli, C.K.N.K. Chion, D.A. Lane, The central role of thrombin in hemostasis, J. Thromb. Haemost. 5 (Suppl. 1) (2007) 95–101.

[32] E. Pretorius, N. Vermeulen, J. Bester, P. Lipinski, Novel use of scanning electron microscopy for detection of iron-induced morphological changes in human blood, Microsc. Res. Tech. 76 (2013) 268–271.

[33] A. Etzerodt, M. Kjolby, M.J. Nielsen, M. Maniecki, P. Svendsen, S.K. Moestrup, Plasma clearance of hemoglobin and haptoglobin in mice and effect of CD163 gene targeting disruption, Antioxidants Redox Signal. 18 (2013) 2254–2263.

[34] T. Lin, D. Maita, S.R. Thundivalappil, F.E. Riley, J. Hambsch, L.J. van Marter, H.A. Christou, L. Berra, S. Fagan, D.C. Christiani, H.S. Warren, Hemopexin in severe inflammation and infection: mouse models and human diseases, Crit. Care 19 (166) (2015) 1–8, https://doi.org/10.1186/s13054-015-0885-x.

[35] M.B. Pepys, M. Baltz, K. Gomer, A.J.S. Davies, M. Doenhoff, Serum amyloid P-component is an acute-phase reactant in the mouse, Nature 278 (1979) 259–261.

[36] B. Shine, F.C. de Beer, M.B. Pepys, Solid phase radioimmunoassays for C-reactive protein, Clin. Chim. Acta 117 (1981) 13–23.

[37] D. Velissaris, N. Pantzaris, I. Koniari, N. Koutsogiannis, V. Karamouzos, I. Kotroni, A. Skroumpelou, J. Ellul, C-Reactive protein and frailty in the elderly: a literature review, J. Clin. Med. Res. 9 (2017) 461–465.

[38] D.C.W. Lau, B. Dillon, H. Yan, P.E. Szmitko, S. Verma, Adipokines: molecular links between obesity and atherosclerosis, Am. J. Physiol. Heart Circ. Physiol. 288 (2005) H2031–H2041.

[39] T. Palosuo, T. Husman, J. Koistinen, K. Aho, C-Reactive protein in population samples, Acta Med. Scand. 220 (1986) 175–179.

[40] B. Young, M. Gleeson, A.W. Cripps, C-reactive protein: a critical review, Pathology 23 (1991) 118–124.

[41] H. Gewurz, C. Mold, J. Siegel, B. Fiedel, C-Reactive protein and the acute phase response, Adv. Intern. Med. 27 (1982) 345–372.

[42] M.-K. Chang, C.J. Binder, M. Torzewski, J.L. Witztum, C-Reactive protein binds to both oxidized LDL and apoptotic cells through recognition of a common ligand: phosphorylcholine of oxidized phospholipids, Proc. Nat. Acad. Sci. U.S.A. 99 (2002) 13043–13048.

[43] P.S. Hicks, L. Saunero-Nava, T.W. du Clos, C. Mold, Serum amyloid P component binds to histones and activates the classical complement pathway, J. Immunol. 149 (1992) 3689–3694.

[44] M.B. Pepys, S.E. Booth, G.A. Tennent, P.J.G. Butler, D.G. Williams, Binding of pentraxins to different nuclear structures: C-reactive protein binds to small nuclear ribonucleoprotein particles, serum amyloid P component binds to chromatin and nucleoli, Clin. Exp. Immunol. 97 (1994) 152–157.

[45] A. Mantovani, C. Garlanda, A. Doni, B. Bottazzi, Pentraxins in innate immunity: from C-reactive protein to the long pentraxin PTX3, J. Clin. Immunol. 28 (2008) 1–13.

[46] T.J. Holzer, K.M. Edwards, H. Gewurz, C. Mold, Binding of C-reactive protein to the pneumococcal capsule or cell wall results in differential localization of C3 and stimulation of phagocytosis, J. Immunol. 133 (1984) 1424–1430.

[47] J.M. Gould, J.N. Weiser, Expression of C-reactive protein in the human respiratory tract, Infect. Immun. 69 (2001) 1747–1754.

[48] F.J. Culley, R.A. Harris, P.M. Kaye, K.P. McAdam, J.G. Raynes, C-Reactive protein binds to a novel ligand on Leishmania donovani and increases uptake into human macrophages, J. Immunol. 156 (1996) 4691–4696.

[49] M.H. Kaplan, J.E. Volanakis, Interactions of C-reactive protein with the complement system. I. Consumption of human complement associated with the reaction of C-reactive protein with pneumococcal polysaccharide and with choline phosphatides, lecithin, and sphingomyelin, J. Immunol. 112 (1974) 2135–2147.
[50] J. Siegel, R. Rent, H. Gewurz, Interactions of C-reactive protein with the complement system. I. Protamine-induced consumption of complement in acute phase sera, J. Exp. Med. 140 (1974) 631–647.
[51] C.M. Uhlar, C.J. Burgess, P.M. Sharp, A.S. Whitehead, Evolution of the serum amyloid A (SAA) protein superfamily, Genomics 19 (1994) 228–235.
[52] C.L. Banka, T. Yuan, M.C. de Beer, M. Kindy, L.K. Curtiss, F.C. de Beer, Serum amyloid A (SAA): influence on HDL-mediated cellular cholesterol efflux, J. Lipid Res. 36 (1995) 1058–1065.
[53] D.R. van der Westhuyzen, F.C. de Beer, N.R. Webb, HDL cholesterol transport during inflammation, Curr. Opin. Lipidol. 18 (2007) 147–151.
[54] R. Kisilevsky, P.N. Manley, Acute-phase serum amyloid A: perspectives on its physiological and pathological roles, Amyloid 19 (2012) 5–14.
[55] I. Migeotte, D. Communi, M. Parmentier, Formyl peptide receptors: a promiscuous subfamily of G protein-coupled receptors controlling immune responses, Cytokine Growth Factor Rev. 17 (2006) 501–519.
[56] D. El-Kebir, L. József, T. Khreiss, W. Pan, N.A. Petasis, C.N. Serhan, J.G. Filep, Aspirin-triggered lipoxins override the apoptosis-delaying action of serum amyloid A in human neutrophils: a novel mechanism for resolution of inflammation, J. Immunol. 179 (2008) 616–622.
[57] L. Sun, H. Zhou, Z. Zhu, Q. Wang, Q. Liang, R.D. Ye, Ex vivo and in vitro effect of serum amyloid A in the induction of macrophage M2 markers and efferocytosis of apoptotic neutrophils, J. Immunol. 194 (2015) 4891–4900.
[58] N. Cheng, R. He, J. Tian, P.P. Ye, R.D. Ye, Cutting edge: TLR2 is a functional receptor for acute-phase serum amyloid A, J. Immunol. 181 (2008) 22–26.
[59] S. Sandri, D. Rodriguez, E. Gomes, H.P. Monteiro, M. Russo, A. Campa, Is serum amyloid A an endogenous TLR4 agonist? J. Leukoc. Biol. 83 (2008) 1174–1180.
[60] C. Shah, R. Hari-Dass, J.G. Raynes, Serum amyloid A is an innate immune opsonin for Gram-negative bacteria, Blood 108 (2006) 1751–1757.
[61] G.T. Westermark, M. Fändrich, P. Westermark, AA amyloidosis: pathogenesis and targeted therapy, Annu. Rev. Pathol. 10 (2015) 321–344.
[62] S. Jayaraman, D.L. Gantz, C. Haupt, O. Gursky, Serum amyloid A forms stable oligomers that disrupt vesicles at lysosomal pH and contribute to the pathogenesis of reactive amyloidosis, Proc. Natl. Acad. Sci. U.S.A. 114 (2017) E6507–E6515.
[63] E. Tyagi, T. Fiorelli, M. Norden, J. Padmanabhan, Alpha 1-antichymotrypsin, an inflammatory protein overexpressed in the brains of patients with Alzheimer's disease, induced tau hyperphosphorylation through c-Jun N-terminal kinase activation, Int. J. Alzheimer's Dis. (2013), ID 6060831, 1–11, https://doi.org/10.1155/2013/606083.
[64] J. Fleddermann, A. Pichert, J. Arnhold, Interaction of serine proteases from polymorphonuclear leucocytes with the cell surface and heparin, Inflammation 35 (2012) 81–88.
[65] C.T.N. Pham, Neutrophil serine proteases: specific regulators of inflammation, Nat. Rev. Immunol. 6 (2006) 541–550.
[66] C. Kantari, M. Pederzoli-Ribeil, O. Amir-Moazami, V. Gauson-Dorey, I. Cruz Mora, M.-C. Lecomte, M. Benhoume, V. Witko-Sarsat, Proteinase 3, the Wegener autoantigen, is externalized during neutrophil apoptosis: evidence for a functional association with phospholipid scramblase 1 and interference with macrophage phagocytosis, Blood 110 (2007) 4086–4095.
[67] S.R. Lammers, P.J. Kao, H.J. Qi, K. Hunter, C. Lamming, J. Albietz, S. Hofmeister, R. Mecham, K.R. Stenmark, R. Shandas, Changes in the structure-function relationship of elastin and its impact on the proximal pulmonary arterial mechanics of hypertensive calves, Am. J. Physiol. Heart Circ. Physiol. 295 (2008) H1451–H1459.
[68] C.C. Taggart, C.M. Greene, T.P. Carroll, S.J. O'Neill, N.G. McElvaney, Elastolytic proteases. Inflammation resolution and dysregulation in chronic infective lung disease, Am. J. Respir. Crit. Care Med. 171 (2005) 1070–1076.
[69] S.K. Brode, S.C. Ling, K.R. Chapman, Alpha-1 antitrypsin deficiency: a commonly overlooked cause of lung disease, Can. Med. Assoc. J. 184 (2012) 1365–1371.
[70] H. Bouriche, P. Salavei, J. Lessig, J. Arnhold, Differential effects of flavonols on inactivation of α_1-antitrypsin induced by hypohalous acids and the myeloperoxidase-hydrogen peroxide-halide system, Arch. Biochem. Biophys. 459 (2007) 137–142.
[71] J.H. Teckman, J.-K. An, K. Blomenkamp, B. Schmidt, D. Perlmutter, Mitochondrial autophagy and injury in the liver in α_1-antitrypsin deficiency, Am. J. Physiol. Gastrointest. Liver Physiol. 286 (2004) G851–G862.
[72] Y. Weinrauch, D. Drujan, S.D. Shapiro, J. Weiss, A. Zychlinsky, Neutrophil elastase targets virulence factors of enterobacteria, Nature 417 (2002) 91–94.

[73] A. Belaaouaj, Neutrophil elastase-mediated killing of bacteria: lessons from targeted mutagenesis, Microb. Infect. 4 (2002) 1259–1264.
[74] J. Duranton, C. Adam, J.G. Blieth, Kinetic mechanism of the inhibition of cathepsin G by α_1-antichymotrypsin and α_1-proteinase inhibitor, Biochemistry 37 (1997) 11239–11245.
[75] J. Travis, J. Bowen, R. Baugh, Human α_1-antichymotrypsin: interaction with chymotrypsin-like proteinases, Biochemistry 26 (1978) 5651–5656.
[76] N.A. Kalsheker, α_1-Antichymotrypsin, Int. J. Biochem. Cell Biol. 28 (1996) 961–964.
[77] C.R. Abraham, W.T. McGraw, F. Slot, R. Yamin, Alpha 1-antichymotrypsin inhibits A beta degradation in vitro and in vivo, Ann. NY Acad. Sci. 920 (2000) 245–248.
[78] J. Padmanabhan, M. Levy, D.W. Dickson, H. Potter, Alpha 1-antichymotrypsin, an inflammatory protein overexpressed in Alzheimer's disease brain, induces tau phosphorylation in neurons, Brain 129 (2006) 3020–3034.
[79] M.G. Dik, C. Jonker, C.E. Hack, J.H. Smit, H.C. Comijs, P. Eikelenboom, Serum inflammatory proteins and cognitive decline in elderly persons, Neurology 64 (2005) 1371–1377.
[80] N. Berkova, C. Gilbert, S. Goupil, J. Yuan, V. Korobko, P.H. Naccache, TNF-induced haptoglobin release from human neutrophils: pivotal role of the TNF p55 receptor, J. Immunol. 162 (1999) 6226–6232.
[81] G. Musci, Structure/function relationships in ceruloplasmin, Adv. Exp. Med. Biol. 448 (1999) 175–182.
[82] V.R. Samygina, A.V. Sokolov, G. Bourenkov, M.V. Petoukhov, M.O. Pulina, E.T. Zakharova, V.B. Vasilyev, H. Bartunik, D.I. Svergun, Ceruloplasmin: macromolecular assemblies with iron-containing acute phase proteins, PLoS One 8 (2013), 367145, 1–12, https://doi.org/10.1371/journal.pone.0067145.
[83] A.L. Chapman, T.J. Mocatta, S. Shiva, A. Seidel, B. Chen, I. Khalilova, M.E. Paumann-Page, G.N. Jameson, C.C. Winterbourn, A.J. Kettle, Ceruloplasmin is an endogenous inhibitor of myeloperoxidase, J. Biol. Chem. 288 (2013) 6465–6477.
[84] V.B. Vasilyev, A.M. Kachurin, N.V. Soronka, Dismutation of superoxide anion radicals by ceruloplasmin. Details of the mechanism, Biokhimija 53 (1988) 2051–2058.
[85] T. Fournier, N. Medjoubi-N, D. Porquet, Alpha-1-acid glycoprotein, Biochim. Biophys. Acta 1482 (2000) 157–171.
[86] J.P. Williams, M.R. Weiser, T.T. Pechet, L. Kobzik, F.D. Moore Jr., H.B. Hechtman, Alpha 1-acid glycoprotein reduces local and remote injuries after intestinal ischemia in the rat, Am. J. Physiol. 273 (1997) G1031–G1035.
[87] L. Rabehi, F. Ferriere, L. Saffar, L. Gattegno, Alpha 1-acid glycoprotein binds human immunodeficiency virus type 1 (HIV-1) envelope glycoprotein via N-linked glycans, Glycoconj. J. 12 (1995) 7–16.
[88] J.A. Schifferli, J.C. Ng, K. Peters, The role of complement and its receptor in the elimination of immune complexes, N. Engl. J. Med. 315 (1986) 488–495.
[89] B.P. Morgan, Regulation of the complement membrane attack pathway, Crit. Rev. Immunol. 19 (1999) 173–198.
[90] P.F. Zipfel, C. Skerka, Complement regulators and inhibitory proteins, Nat. Rev. Immunol. 9 (2009) 729–740.
[91] P.N. Nesargikar, B. Spiller, R. Chavez, The complement system: history, pathways, cascade and inhibitors, Eur. J. Microbiol. Immunol. 2 (2012) 103–111.
[92] J.V. Sarma, P.A. Ward, The complement system, Cell Tissue Res. 343 (2011) 227–235.
[93] M. Noris, G. Remuzzi, Overview of complement activation and regulation, Semin. Nephrol. 33 (2013) 479–492.
[94] T. Vorup-Jensen, S.V. Petersen, A.G. Hansen, K. Poulsen, W. Schwaeble, R.B. Sim, K.B.M. Reid, S.J. Davis, S. Thiel, J.C. Jensenius, Distinct pathways of mannan-binding lectin (MBL)- and C1-complex autoactivation revealed by reconstitution of MBL with recombinant MBL-associated serine protease-2, J. Immunol. 165 (2000) 2093–2100.
[95] M. Matsushita, Y. Endo, N. Hamasaki, T. Fujita, Activation of the lectin complement pathway by ficolins, Int. Immunopharmacol. 1 (2001) 359–363.
[96] M. Matsushita, Ficolins: complement-activating lectins involved in innate immunity, J. Innate Immunol. 2 (2010) 24–32.
[97] J.M. Thurman, V.M. Holers, The central role of the alternative complement pathway in human disease, J. Immunol. 176 (2006) 1305–1310.
[98] M.T. Ganter, K. Brohi, M.J. Cohen, Role of the alternative pathway in the early complement activation following major trauma, Shock 28 (2007) 29–34.
[99] F.C. Kimberley, B. Sivasankar, B.P. Morgan, Alternative roles of CD59, Mol. Immunol. 44 (2007) 73–81.
[100] M. Yamashina, E. Ueda, T. Kinoshita, T. Takami, A. Ojima, H. Ono, H. Tanaka, N. Kondo, T. Orii, N. Okada, H. Okada, K. Inoue, T. Kitani, Inherited complete deficiency of 20-kilodalton homologous restriction factor (CD95) as a cause of paroxysmal nocturnal hemoglobinuria, N. Engl. J. Med. 323 (1990) 1184–1189.

[101] N. Motoyama, N. Okada, M. Yamashina, H. Okada, Paroxysmal nocturnal hemoglobinuria due to hereditary nucleotide deletion in the HRF20 (CD59) gene, Eur. J. Immunol. 22 (1992) 2669–2673.

[102] A. Klos, A.J. Tenner, K.-O. Johswich, R.R. Ager, E.S. Reis, J. Köhl, The role of the anaphylatoxins in health and disease, Mol. Immunol. 46 (2009) 2753–2766.

[103] J.A. Ember, M.A. Jagels, T.E. Hugli, Characterization of complement anaphylatoxins and their biological responses, in: J.E. Volanakis, M.M. Frank (Eds.), The Human Complement System in Health and Disease, Dekker, New York, 1998, pp. 241–284.

[104] S.R. Barnum, C4a: an anaphylatoxin in name only, J. Innate Immun. 7 (2015) 333–339.

[105] J.A. Wingrove, R.G. DiScipio, Z. Chen, J. Potempa, J. Travis, T.E. Hugli, Activation of complement components C3 and C5 by a cysteine protease (gingipain-1) from *Porphyromonas (Bacteroides) gingivalis*, J. Biol. Chem. 267 (1992) 18902–18907.

[106] K. Maruo, T. Akaike, T. Ono, T. Okamoto, H. Maeda, Generation of anaphylatoxins through proteolytic processing of C3 and C5 by house dust mite protease, J. Allergy Clin. Immunol. 100 (1997) 253–260.

[107] H. Nitta, T. Iwamura, Y. Wada, A. Irie, H. Kobayashi, K. Okamoto, H. Baba, Production of C5a by ASP, a serine protease released from *Aeromonas sobria*, J. Immunol. 181 (2008) 3602–3608.

[108] M. Jusko, J. Potempa, A.Y. Karim, M. Ksiazek, K. Riesbeck, P. Garred, S. Eick, A.M. Blom, A metalloproteinase karilysin present in the majority of *Tannerella forsythia* isolates inhibits all pathways of the complement system, J. Immunol. 188 (2012) 2338–2349.

[109] S.M. Kanse, A. Gallenmueller, S. Zeerleder, F. Stephan, O. Rannou, S. Denk, M. Etscheid, G. Lochnit, M. Krueger, M. Huber-Lang, Factor VII-activating protease is activated in multiple trauma patients and generates anaphylatoxin C5a, J. Immunol. 188 (2012) 2858–2865.

[110] M. Huber-Lang, S. Denk, S. Fulda, E. Erler, M. Kalbitz, S. Weckbach, E.M. Schneider, M. Weiss, S.M. Kanse, M. Perl, Cathepsin D is released after severe tissue trauma in vivo and is capable of generating C5a in vitro, Mol. Immunol. 50 (2012) 60–65.

[111] Y. Fukuoka, H.Z. Xia, L.B. Sanchez-Muñoz, A.L. Dellinger, L. Escibano, L.B. Schwartz, Generation of anaphylatoxins by human beta-tryptase from C3, C4, and C5, J. Immunol. 180 (2008) 6307–6316.

[112] M. Perl, S. Denk, M. Kalbitz, M. Huber-Lang, Granzyme B: a new crossroad of complement and apoptosis, Adv. Exp. Med. Biol. 946 (2012) 135–146.

[113] K.W. Matthews, S.L. Mueller-Ortiz, R.A. Wetsel, Carboxypeptidase N: a pleiotropic regulator of inflammation, Mol. Immunol. 40 (2004) 785–793.

[114] K. Cianflone, Z. Xia, L.Y. Chen, Critical review of acylation-stimulating protein physiology in humans and rodents, Biochim. Biophys. Acta 1609 (2003) 127–143.

[115] A. Klos, E. Wende, K.J. Wareham, P.N. Monk, International union of basic and clinical Pharmacology. LXXXVII. Complement peptide C5a, C4a, and C3a receptors, Pharmacol. Rev. 65 (2013) 500–543.

[116] A. Bengtson, M. Heideman, Anaphylatoxin formation in sepsis, Arch. Surg. 123 (1998) 645–649.

[117] S. Palta, R. Saroa, A. Palta, Overview of the coagulation system, Ind. J. Anasth. 58 (2014) 515–523.

[118] G.R. Hetzel, C. Sucker, The heparins: all a nephrologist should know, Nephrol. Dial. Transplant. 20 (2005) 2036–2042.

[119] A. Finley, C. Greenberg, Review article: heparin sensitivity and resistance: management during cardiopulmonary bypass, Anesth. Analg. 116 (2013) 1210–1222.

[120] G.D. Boon, An overview of hemostasis, Toxicol. Pathol. 21 (1993) 170–179.

[121] K. Ghoshal, M. Bhattacharyya, Overview of platelet physiology: its hemostatic and nonhemostatic role in disease pathogenesis, Sci. World J. (2014), article 781857, 1–16, https://doi.org/10.1155/2014/781757.

[122] F.A. Ruiz, C.R. Lea, E. Oldfield, R. Docampo, Human platelet dense granules contain polyphosphate and are similar to acidocalcisomes of bacteria and unicellular eukaryotes, J. Biol. Chem. 279 (2004) 44250–44257.

[123] S.A. Smith, S.H. Choi, R. Davis-Harrison, J. Huyck, J. Boettcher, C.M. Rienstra, J.H. Morrissey, Polyphosphate exerts differential effects on blood clotting, depending on polymer size, Blood 116 (2010) 4353–4359.

[124] B. Dahlbeck, Coagulation and inflammation – close allies in health and disease, Semin. Immunopathol. 34 (2012) 1–3.

[125] R.J. Travers, S.A. Smith, J.H. Morrissey, Polyphosphate, platelets, and coagulation, Int. J. Lab. Hematol. 37 (2015) 31–35.

[126] J.E. Sadler, Biochemistry and genetics of von Willebrand factor, Annu. Rev. Biochem. 67 (1998) 395–424.

[127] Z. Li, K. Delaney, K.A. O'Brien, X. Du, Signaling during platelet adhesion and activation, Arterioscler. Thromb. Vasc. Biol. 30 (2010) 2341–2349.

[128] E.M. Golebiewska, A.W. Poole, Platelet secretion: from haemostasis to wound healing and beyond, Blood Rev. 29 (2015) 153–162.

[129] T. Bombeli, D.R. Spahn, Updates in perioperative coagulation: physiology and management of

thromboembolism and haemorrhage, Br. J. Anasth. 93 (2004) 275–287.
[130] T.S. Edgington, C.D. Dickinson, W. Ruf, The structural basis of function of the TF-VIIa complex in the cellular initiation of coagulation, Thromb. Haemost. 78 (1997) 401–405.
[131] N. Mackman, The role of tissue factor and factor VIIa in hemostasis, Anesth, Analgesia 108 (2009) 1447–1452.
[132] A.J. Chu, Tissue factor, blood coagulation, and beyond: an overview, Int. J. Inflamm. (2011), article 367284, 1–30, https://doi.org/10.4061/2011/367284.
[133] R.H. Thomas, Hypercoagulability syndromes, Arch. Intern. Med. 161 (2001) 2433–2439.
[134] A.J. Chu, Tissue factor mediates inflammation, Arch, Biochem. Biophys 440 (2005) 123–132.
[135] A.J. Chu, Tissue factor upregulation drives a thrombosis-inflammation circuit in relation to cardiovascular complications, Cell Biochem. Funct. 24 (2006) 173–192.
[136] Y. Hattori, M. Matsumura, K. Kasai, Vascular smooth muscle cell activation by C-reactive protein, Cardiovasc. Res. 58 (2003) 186–195.
[137] M. Levi, T. van der Poll, H. ten Cate, Tissue factor in infection and severe inflammation, Sem, Thromb. Hemost. 32 (2006) 33–39.
[138] B. Ghebrehiwet, A.P. Kaplan, K. Joseph, E.I. Peerschke, The complement and contact activation systems: partnership in pathogenesis beyond angioedema, Immunol. Rev. 274 (2016) 281–289.
[139] A.T. Long, E. Kenne, R. Jung, T.A. Fuchs, T. Renné, Contact system revisited: an interface between inflammation, coagulation, and innate immunity, J. Thromb. Haemost. 14 (2016) 427–437.
[140] L. Bender, H. Weidmann, S. Rose-John, T. Renné, A.T. Long, Factor XII-driven inflammatory reactions with implications for anaphylaxis, Front. Immunol. 8 (2017), article 1115, 1–11, https://doi.org/10.3389/fimmu.2017.01115.
[141] C. Maas, C. Oschatz, T. Renné, The plasma contact system 2.0, Semin. Thromb. Hemost. 37 (2011) 375–381.
[142] S. de Maat, C. Tersteeg, E. Herczenik, C. Maas, Tracking down contact activation – from coagulation in vitro to inflammation in vivo, Int. J. Lab. Hematol. 36 (2014) 374–381.
[143] M.W. Mosesson, Fibrinogen and fibrin structure and functions, J. Thromb. Haemost. 3 (2005) 1800–1814.
[144] M. Nesheim, Thrombin and fibrinolysis, Chest 124 (2003) 33S–39S.
[145] D.A. Meh, K.R. Siebenlist, M.W. Mosesson, Identification and characterization of the thrombin binding sites on fibrin, J. Biol. Chem. 271 (1996) 23121–23125.
[146] R. Asselta, S. Duga, M.L. Tenchini, The molecular basis of quantitative fibrinogen disorders, J. Thromb. Haemost. 4 (2006) 2115–2229.
[147] C. Fuss, J.C. Palmaz, E.A. Spargue, Fibrinogen: structure, function, and surface interaction, J. Vasc. Interv. Radiol. 12 (2001) 677–682.
[148] G.A. Tennant, S.O. Brennan, A.J. Stangou, J. O'Grady, P.N. Hawkins, M.B. Pepys, Human plasma fibrinogen is synthesized in the liver, Blood 109 (2007) 1971–1974.
[149] D.G. Chalupowicz, Z.A. Chowdhury, T.L. Bach, C. Barsigian, J. Martinez, Fibrin II induces endothelial cell capillary tube formation, J. Cell Biol. 130 (1995) 207–215.
[150] L.A. Sporn, L.A. Bunce, C.W. Francis, Cell proliferation on fibrin: modulation by fibrinopeptide cleavage, Blood 86 (1995) 1802–1810.
[151] T.M. Ordljin, J.R. Shainoff, S.O. Lawrence, P.J. Simpson-Haidaris, Thrombin cleavage enhances exposure of a heparin binding domain in the N-terminus of the fibrin beta chain, Blood 88 (1996) 2050–2061.
[152] T.L. Bach, C. Barsigian, D.G. Chalupowicz, D. Busler, C.H. Yaen, D.S. Grant, J. Martinez, VE-Cadherin mediates endothelial cell capillary tube formation in fibrin and collagen gels, Exp. Cell Res. 238 (1998) 324–334.
[153] D.J. Geer, S.T. Andreadis, A novel role of fibrin in epidermal healing: plasminogen-mediated migration and selective detachment of differentiated keratinocytes, J. Investig. Dermatol. 121 (2003) 1210–1216.
[154] S.J. Palumbo, K.E. Talamge, H. Liu, C.M. La Jeunesse, D.P. Witte, J.L. Degen, Plasminogen supports tumor growth through a fibrinogen-dependent mechanism linked to vascular patency, Blood 102 (2003) 2819–2827.
[155] R.I. Litvinov, D.A. Faizullin, Y.F. Zuev, J.W. Weisel, The α-helix to β-sheet transition in stretched and compressed hydrated fibrin clots, Biophys. J. 103 (2012) 1020–1027.
[156] M.A.A. Parry, X.C. Zhang, W. Bode, Molecular mechanisms of plasminogen activation: bacterial cofactors provide clues, Trends Biochem. Sci. 25 (2000) 53–59.
[157] G. Cesarman-Maus, K.A. Hajjar, Molecular mechanisms of fibrinolysis, Br. J. Haematol. 129 (2005) 307–321.
[158] J.P. Irigoyen, P. Muñoz-Cánoves, L. Montero, M. Koziczak, Y. Nagamine, The plasminogen activator system: biology and regulation, Cell. Mol. Life Sci. 56 (1999) 104–132.
[159] B.H. Shieh, J. Travis, The reactive site of human alpha$_2$-antiplasmin, J. Biol. Chem. 262 (1987) 6055–6059.

[160] S.J. Kolodziej, H.U. Klueppelberg, N. Nolasco, W. Ehses, D.K. Strickland, J.K. Stoops, Three dimensional structure of the human plasmin α_2-macroglobulin complex, J. Struct. Biol. 123 (1998) 124–133.

[161] M. Cesari, M. Pahor, R.A. Incalzi, Plasminogen activator inhibitor-1 (PAI-1): a key factor linking fibrinolysis and age-related subclinical and clinical conditions, Cardiovasc. Ther. 28 (2010) e72–e91.

[162] J.P. Colle, Z. Mishal, C. Lesty, M. Mirshahi, J. Peyne, A. Baumelou, A. Bensman, J. Soria, C. Soria, Abnormal fibrin clot architecture in nephrotic patients is related to hypofibrinolysis: influence of plasma biochemical modifications: a possible mechanism for the high thrombotic tendency? Thromb. Haemost. 82 (1999) 1482–1489.

[163] E. Pretorius, H. Steyn, M. Engelbrecht, A.C. Swanepoel, H.M. Oberholzer, Differences in fibrin fiber diameters in healthy individuals and thromboembolic ischemic stroke patients, Blood Coagul. Fibrinolysis 22 (2001) 696–700.

[164] K.M. Weigandt, N. White, D. Chung, E. Ellingson, Y. Wang, X. Fu, D.C. Pozzo, Fibrin clot structure and mechanics associated with specific oxidation of methionine residues in fibrinogen, Biophys. J. 103 (2012) 2399–2407.

[165] J.P. Collet, D. Park, C. Lesty, J. Soria, C. Soria, G. Motalescot, J.W. Weisel, Influence of fibrin network conformation and fibrin diameter on fibrinolysis speed: dynamic and structural approaches by confocal microscopy, Arterioscler. Thromb. Vasc. Biol. 20 (2000) 1354–1361.

[166] A. Undas, Fibrin clot properties and their modulation in thrombotic disorders, Thromb. Haemost. 112 (2014) 32–42.

[167] W. Li, J. Sigley, M. Pieters, C.C. Helms, C. Nagaswami, J.W. Weisel, M. Guthold, Fibrin fiber stiffness is strongly affected by fiber diameter, but not by fibrinogen glycation, Biophys. J. 110 (2016) 1400–1410.

[168] E. Pretorius, A.C. Swanepoel, A.V. Buys, N. Vermeulen, W. Duim, D.B. Kell, Eryptosis as a marker of Parkinson's disease, Aging 6 (2014) 788–819.

[169] E. Pretorius, J. Bester, N. Vermeulen, S. Alummoottil, P. Soma, A.V. Buys, D.B. Kell, Poorly controlled type 2 diabetes is accompanied by significant morphological and ultrastructural changes in both erythrocytes and in thrombin-generated fibrin: implications for diagnostics, Cardiovasc. Diabetol. 14 (2015), 30, 1–20, https://doi.org/10.1186/s12933-015-0192-5.

[170] O.O. Olumuyiwa-Akeredolu, P. Soma, A.V. Buys, L.K. Debusho, E. Pretorius, Characterizing pathology in erythrocytes using morphological and biophysical membrane properties: relation to impaired hemorheology and cardiovascular function in rheumatoid arthritis, Biochim. Biophys. Acta 1859 (2017) 2381–2391.

[171] D.B. Kell, E. Pretorius, No effects without causes: the iron dysregulation and dormant microbes hypothesis for chronic, inflammatory diseases, Biol. Rev. Camb. Philos. Soc. 93 (2018) 1518–1557.

[172] F. Muller, T. Renné, Novel roles for factor XII-driven plasma contact activation system, Curr. Opin. Hematol. 15 (2008) 516–521.

[173] K. Ishihara, M. Kamata, I. Hayashi, S. Yamashina, M. Majima, Roles of bradykinin in vascular permeability and angiogenesis in solid tumor, Int. Immunopharmacol. 2 (2002) 499–509.

[174] D.C. Manning, S.N. Raja, R.A. Meyer, J.N. Campbell, Pain and hyperanalgesia after intradermal injection of bradykinin in humans, Clin. Pharmacol. Ther. 50 (1991) 721–729.

[175] W. Koppert, P.W. Reeh, H.O. Handwerker, Conditioning of histamine by bradykinin alters responses of rat nociceptor and human itch sensation, Neurosci. Lett. 152 (1993) 117–120.

[176] K.J. Paterson, L. Zambreanu, D.L. Bennett, S.B. McMahon, Characterisation and mechanisms of bradykinin-evoked pain in man using iontophoresis, Pain 154 (2013) 782–792.

[177] F. Marceau, H.J. Hess, D.R. Rachvarov, The B1 receptors for kinins, Pharmacol. Rev. 50 (1998) 357–386.

[178] M. Cyr, Y. Lepage, C. Blais Jr., N. Gervais, M. Cugno, J.-L. Rouleau, A. Adam, Bradykinin and des-Arg^9-bradykinin metabolic pathways and kinetics of activation of human plasma, Am. J. Physiol. Heart Circ. Physiol. 281 (2001) H275–H283.

[179] J.B. Calixto, R. Medeiros, E.S. Fernandes, J. Ferreira, D.A. Cabrini, M.M. Campos, Kinin B1 receptors: key G-protein-coupled receptors and their role in inflammatory and painful processes, Br. J. Pharmacol. 143 (2004) 803–818.

[180] M. Hamza, X.M. Wang, A. Adam, J.S. Brahim, J.S. Rowan, G.N. Carmona, R.A. Dionne, Kinin B_1 receptors contributes to acute pain following minor surgery in humans, Mol. Pain 6 (12) (2010) 1–11, https://doi.org/10.1186/1744-8069-6-12.

[181] J. Duchene, F. Lecomte, S. Ahmed, C. Cayla, J. Pasquero, M. Bader, M. Peretti, A. Ahluwalia, A novel inflammatory pathway involved in leukocyte recruitment: role for the kinin B1 receptor and the chemokine CXCL5, J. Immunol. 179 (2007) 4849–4856.

[182] M. Austinat, S. Braeuninger, J.B. Pesquero, M. Brede, M. Bader, G. Stoll, T. Renné, C. Kleinschnitz, Blockade of bradykinin receptor B1 but not bradykinin receptor

B2 provides protection from cerebral infarction and brain edema, Stroke 40 (2009) 285–293.

[183] K. Asraf, N. Torika, A. Danon, S. Fleisher-Berkovich, Involvement of the bradykinin receptor B1 in microglial activation: in vitro and in vivo studies, Front. Endocrinol. 8 (2017), article 82, 1–9, https://doi.org/10.3389/fendo.2017.00082.

[184] U. Amara, M. Flierl, D. Rittirsch, A. Klos, H. Chen, B. Acker, U.B. Brückner, B. Nilsson, F. Gebhard, J.D. Lambris, M. Huber-Lang, Molecular intercommunication between the complement and coagulation systems, J. Immunol. 185 (2010) 5628–5636.

[185] S. Irmscher, N. Döring, L.D. Halder, E.A.H. Jo, I. Kopka, C. Dunker, I.D. Jacobsen, S. Luo, H. Slevogt, S. Lorkowski, N. Beyersdorf, P.F. Zipfel, C. Skerka, Kallikrein cleaves C3 and activates complement, J. Innate Immun. 10 (2018) 94–105.

[186] B. Ghebrehiwet, B.P. Randazzo, J.T. Dunn, M. Silverberg, A.P. Kaplan, Mechanism of activation of the classical pathway of complement by Hageman factor fragment, J. Clin. Investig. 71 (1983) 1450–1456.

[187] R.G. DiScipio, The activation of the alternative pathway C3 convertase by human plasma kallikrein, Immunology 45 (1982) 587–595.

[188] J. Bjorkvist, K.F. Nickel, E. Stavrou, T. Renné, In vivo activation and functions of the protease factor XII, Thromb. Haemast. 112 (2014) 868–875.

[189] C. Oschatz, C. Maas, B. Lecher, T. Jansen, J. Bjorkqvist, T. Tradler, R. Sedlmeier, P. Burfeind, S. Cichon, S. Hammerschmidt, W. Müller-Esterl, W.A. Wuillemin, G. Nilsson, T. Renné, Mast cells increase vascular permeability by heparin-initiated bradykinin formation in vivo, Immunity 34 (2011) 258–268.

[190] M. Guerrini, D. Beccati, Z. Shriver, A. Naggi, K. Viswanathan, A. Bisio, I. Capila, J.C. Lansing, S. Guglieri, B. Fraser, A. Al-Hakim, N.S. Gunay, Z. Zhang, L. Robinson, L. Buhse, M. Nasr, J. Woodcock, R. Langer, G. Venkataraman, R.J. Linhardt, B. Casu, G. Torri, R. Sasisekharan, Oversulfated chondroitin sulfate is a contaminant in heparin associated with adverse clinicals events, Nat. Biotechnol. 26 (2008) 669–675.

[191] T.K. Kishimoto, K. Viswanathan, T. Ganguly, S. Elankumaran, S. Smith, S. Pelzer, J.C. Lansing, N. Sriranganathan, G. Zhao, Z. Galcheva-Gargova, A. Al-Hakim, G.S. Bailey, B. Fraser, S. Roy, T. Rogers-Cotrone, L. Buhse, M. Whary, J. Fox, M. Nasr, G.J. Dal Pan, Z. Shriver, R.S. Langer, G. Ventakaraman, K.F. Austen, J. Woodcock, R. Sasisekharan, Contaminated heparin associated with adverse clinical events and activation of the contact system, N. Engl. J. Med. 358 (2008) 2457–2467.

[192] Y. Hojima, C.G. Cochrane, R.C. Wiggins, K.F. Austen, R.L. Stevens, In vitro activation of the contact (Hageman factor) system of plasma by heparin and condroitin sulfate E, Blood 63 (1984) 1453–1459.

[193] D. Moreno-Sanchez, L. Hernandez-Ruiz, F.A. Ruiz, R. Docampo, Polyphosphate is a novel proinflammatory regulator of mast cells and is located in acidocalcisomes, J. Biol. Chem. 287 (2012) 28435–28444.

[194] J.M. Gajsiewicz, S.A. Smith, J.H. Morrissey, Polyphosphate and RNA differentially modulate the contact pathway of blood clotting, J. Biol. Chem. 292 (2017) 1808–1814.

[195] S.A. Smith, N.J. Mutch, D. Baskar, P. Rohloff, R. Docampo, J.H. Morrissey, Polyphosphate modulates blood coagulation and fibrinolysis, Proc. Natl. Acad. Sci. U.S.A. 103 (2006) 903–908.

[196] S.A. Smith, J.H. Morrissey, Polyphosphate enhances fibrin clot structure, Blood 112 (2008) 2810–2816.

[197] N.J. Mutch, R. Engel, S. Uitte de Willige, H. Philippou, R.A. Ariens, Polyphosphate modifies the fibrin network and down-regulates fibrinolysis by attenuating binding of tPA and plasminogen to fibrin, Blood 115 (2010) 3980–3988.

[198] J.H. Morrissey, S.A. Smith, Polyphosphate as modulator of haemostasis, thrombosis and inflammation, J. Thromb. Haemost. 13 (2015) S92–S97.

[199] S.M. Hassanian, A. Avan, A. Ardeshirylajimi, Inorganic polyphosphate: a key modulator of inflammation, J. Thromb. Haemost. 15 (2016) 213–218.

[200] J. Xu, X. Zhang, R. Pelayo, M. Monestier, C.T. Ammollo, F. Semeraro, F.B. Taylor, N.L. Esmon, F. Lupu, C.T. Emson, Extracellular histones are major mediators of death in sepsis, Nat. Med. 15 (2009) 1318–1321.

[201] P. Dinarvand, S.M. Hassanian, S.H. Qureshi, C. Manithody, J.C. Eisenberg, L. Yang, A.R. Rezaie, Polyphosphate amplifies proinflammatory responses of nuclear proteins through interaction with receptor for advanced glycation end products and P2Y1 purinergic receptor, Blood 123 (2014) 935–945.

[202] S.M. Hassanian, P. Dinarvand, S.A. Smith, A.R. Rezaie, Inorganic polyphosphate elicits proinflammatory responses through activation of mTOR complexes 1 and 2 in vascular endothelial cells, J. Thromb. Haemost. 13 (2015) 860–871.

[203] A. Sala-Cunill, J. Bjorkqvist, R. Senter, M. Guilarte, V. Cardona, M. Labrador, K.F. Nickel, L. Butler, O. Luengo, P. Kumar, L. Labberton, A. Long, A. Di Gennaro, E. Kenne, A. Jämsä, T. Krieger, H. Schlüter, T. Fuchs, S. Flohr, U. Hassiepen, F. Cumin, K. McCrae, C. Maas, E. Stavrou, T. Renné, Plasma contact system activation drives anaphylaxis

in severe-mast cell-mediated allergic reactions, J. Allergy Clin. Immunol. 135 (2015) 1031–1043.

[204] S.L. Taylor, V. Wahl-Jensen, A.M. Copeland, P.B. Jahrling, C.S. Schmaljohn, Endothelial cell permeability during hantavirus infection involved factor XII-dependent increased activation of the kallikrein-kinin system, PLoS Pathog. 9 (2013), e1003470, 1–14, https://doi.org/10.1371/journal.ppat.1003470.

[205] E.S. Gershom, M.R. Sutherland, P. Lollar, E.L. Pryzdial, Involvement of the contact phase and intrinsic pathway in herpes simplex virus-initiated plasma coagulation, J. Thromb. Haemost. 8 (2010) 1037–1043.

[206] S.F. Khaiboullina, S.P. Morzunov, S.C. St Jeor, Hantaviruses: molecular biology, evolution and pathogenesis, Curr. Mol. Med. 5 (2005) 773–790.

[207] K.F. Nickel, T. Renné, Crosstalk of the plasma contact system with bacteria, Thromb. Res. 130 (Suppl. 1) (2012) S78–S83.

[208] E. Mattsson, H. Herwald, H. Cramer, K. Persson, U. Sjobring, L. Bjorck, *Staphylococcus aureus* induces release of bradykinin in human plasma, Infect. Immun. 69 (2001) 3877–3882.

[209] S. Oehmcke, M. Mörgelin, H. Herwald, Activation of the human contact system on neutrophil extracellular traps, J. Innate Immun. 1 (2009) 225–230.

[210] V.P. Ferreira, M.K. Pangburn, C. Cortés, Complement control protein factor H: the good, the bad, and the inadequate, Mol. Immunol. 47 (2010) 2187–2197.

[211] S.R. de Cordoba, E.G. de Jorge, Translational mini-review series on complement factor H: genetics and disease associations of human complement factor H, Clin. Exp. Immunol. 151 (2008) 1–13.

[212] V.M. Holers, The spectrum of complement alternative pathway-associated diseases, Immunol. Rev. 223 (2008) 300–316.

[213] M.C. Pickering, H.T. Cook, Translational mini-review series on complement factor H: renal diseases associated with complement factor H: novel insights from humans and animals, Clin. Exp. Immunol. 151 (2008) 210–230.

[214] J.D. Lambris, D. Ricklin, B.V. Geisbrecht, Complement evasion by human pathogens, Nat. Rev. Microbiol. 6 (2008) 132–142.

PART III

AGING PROCESSES AND DEVELOPMENT OF PATHOLOGICAL STATES

CHAPTER

8

Aging in Complex Multicellular Organisms

8.1 INTRODUCTION

Aging is an indispensable fundamental property of humans and animal species. It originates from destructive chemical and physical processes that are always present in living systems at all stages of the life span. Of course, processes of aging appear more evident in the second half of the lifetime, when the years of juvenile confidence are unrecoverable over. There is a gradual, and sometimes more abrupt, loss of different physiological functions with increasing age. Many diseases appear predominantly in elderly persons. Finally, each individual will die. However, it is unpredictable when a given person dies and under what circumstances this death takes place.

Miracles of aging have been fascinating scientists since ancient times. About 300 different hypothesis and theories have been proposed to explain the enigma aging [1]. However, none of them gives an all-encompassing statement generally accepted by the scientific community, what is aging. Sometimes, these efforts are driven by the misbelief to find any wondrous recipe against aging. The everlasting "fountain of youth" is a recurring motive in the commercially driven area of antiaging products.

Similar to life, aging processes are very multifaceted in appearance. Existing theories of aging can be roughly divided into genetic, neuroendocrine, and damage-associated theories despite considerable overlap between them. In this chapter, advantages and disadvantages of the main theories will be shortly outlined. However, it is not the objective of this book to provide a comprehensive overview of all aspects of aging. Finally, the link between aging and thermodynamics will be deepened to underscore how processes of cell and tissue damage and accumulation of damaged material promote irreversible alterations in the aging organism.

8.2 GENES AND AGING

8.2.1 Premature Aging Due to Gene Defects

The primary sequence of proteins is encoded in the genes. This information provides the basis for the enormous complexity of biological structures. The knowledge about mutations of selected single genes, which can shorten the life span of individuals and induce their premature aging, underlies the significance of genes in the aging process.

Cell and Tissue Destruction
https://doi.org/10.1016/B978-0-12-816388-7.00008-5

TABLE 8.1 Genetically Based Human Diseases Associated With Premature Aging.

Disease	Defective Gene	Defective Protein	References
Hutchinson–Gilford syndrome (progeria)	LMNA gene	Lamin A, lamin C	[2,3]
Restrictive dermopathy	LMNA gene, ZMPSTE24 (prelamin-processing enzyme)	Lamin A	[4]
Werner syndrome	WRN gene	DNA helicase	[5]
Trichothiodystrophy	ERCC2, ERCC3, GTF2H5	Reduced DNA repair, increased photosensitivity	[6,7]
Cockayne syndrome	Cross-complementing gene 8 and 6	CSA protein, CSB protein, defects in transcription-coupled nucleotide excision repair	[8]
Bloom syndrome	BLM gene	Helicase, suppression of inappropriate homologous recombination	[9]
Xeroderma pigmentosum	Several genes such as DDB2, ERCC2, ERCC3, ERCC4, ERCC5, XPA, XPC	Defects in nucleotide excision repair, no exposure to ultraviolet light	[10,11]
Wiedemann–Rautenstrauch syndrome	Unknown	Apparently defects in DNA repair	[12]

Some rare human diseases exist, in which mutation of a single gene causes premature aging, sometimes already during childhood or before birth. These diseases are summarized as progeroid syndromes. Selected examples of these diseases are given in Table 8.1.

In the Hutchinson–Gilford syndrome, also known as progeria, the defective nuclear protein lamin A impairs the repair of DNA damage [2,3]. Intact lamins maintain the level of proteins playing a key role in the repair of DNA double-strand breaks. These scaffold proteins determine the shape and integrity of the nucleus acting at the inner nuclear membrane [13]. Further, laminopathies are restrictive dermopathy, muscular dystrophy, neuropathy, and cardiomyopathy [4,14–16].

Other diseases of this group are characterized by defects in proteins involved in DNA repair such as mutations in genes encoding helicase [5,9] or avoiding the nucleotide excision repair [8,11]. Of course, numerous other examples are known, in which malfunctions, mutations, or knockout of certain genes affect the aging process.

Taken together, genetic disorders associated with premature aging of individuals are mainly caused by insufficient repair of damaged DNA. These diseases originate either from defects in some genes encoding proteins involved in DNA repair or proteins responsible for the stability of the nucleus.

8.2.2 Life Span Prolongation: Effects of Genes

Otherwise, some genes are known to be responsible for longevity. These data are mainly based on numerous experimental investigations on small organisms and animals, in particular on yeast, nematode (*Caenorhabditis elegans*), drosophila, and mice. In these species, a life span–modulating effect was associated with some daf genes or their mammalian orthologs, the forkhead box O (FOXO) gene [17–21]. These genes are important for regulation of nutrient uptake to cells and growth processes. The corresponding growth hormone/insulin-like growth factor 1 (IGF-1) signaling pathway also plays an important role in human longevity [22,23]. This pathway is described in more detail in Section 8.3.2.

Other studies imply a role of sirtuins in aging delay [24–26]. These proteins are activated by fasting and function as energy sensors during metabolism. Sirtuins are involved in nucleotide excision repair and other DNA repair processes. Their overexpression in transgenic mice extends the life span [27].

Poly(ADP-ribose) polymerase 1 (PARP-1), which catalyzes the synthesis of poly(ADP-ribose), also plays a role in DNA repair processes and contributes to genomic maintenance at low DNA damage. Otherwise, this nuclear enzyme is involved in apoptosis and necrosis induction under inflammatory conditions. Effects of PARP-1 on the aging process are under discussion [28,29].

8.2.3 Life Span Prolongation by Caloric Restriction

The hypothesis has been established that limitation of food intake extends the life span and delays the onset of age-related diseases [30]. On different species, a relationship between caloric restriction and prolongation of lifetime could be demonstrated [31–34].

A lower level of 8-hydroxy-2′-deoxyguanosine, an indicator of DNA damage, was observed in several rodents on caloric restriction [35–37]. It is assumed that this restriction reduces the generation of reactive species [38,39]. The mechanism of how caloric restriction acts and prolongs the life span of animals remains unsolved. In some mutants of the nematode *C. elegans*, it has been shown that caloric restriction and prolongation of life span are associated with elevated autophagy as well as smaller cell and body size [40,41].

Although the effect of caloric restriction is clearly shown in animal models, it remains puzzling whether caloric restriction prolongs the life span in humans.

Critical evaluation of experimental data about caloric restriction revealed two main problems [42]. First, the effect of lifetime extension by caloric restriction is not universal over different species and is even not observed within different strains of the same species. Second, in most cases, control animals were usually fed at libitum, a condition leading to overweight and more disease-susceptible animals.

8.2.4 Human Longevity

It is obvious that human life expectancy has considerably been prolonged for the last 150 years in modern countries. Responsible factors are continuous amelioration of living conditions, delivery of clean and safe nutrient resources, reduction of infectious diseases, general improvement of health care, and health-conscious lifestyle. Thus, nongenetic factors mainly contribute to the observed increase in human longevity in contrast to genetic factors that comprise about 25% of the variation of human life span as shown by family studies [43,44].

Other longevity determining factors are lifestyle-based and linked to the upregulation of some genes promoting survival. For example, the regular uptake of the green tea polyphenol epigallocatechin gallate by rats increased their lifetime and favored the expression of sirtuin1, superoxide dismutase, glutathione peroxidase, and most of all FoxO3 [45]. Beneficial effects of polyphenols associated with an increase of FoxO3 gene expression are also demonstrated on humans and on human cells [46–48].

8.2.5 Role of Telomeres

Repetitive nucleotide sequences, the telomeres, are found at each end of chromosomes in eukaryotes. They protect the end of chromosomes from neighboring chromosomes and from deterioration. These sequences are formed by telomerase [49]. Telomeres play an important role in the action of DNA polymerases as the binding place for the primer.

Immortal cells, which express telomerase, are characterized by stable telomeres. Germ cells, stem cells, and T lymphocytes are immortal as they maintain telomere length. In the overwhelming amount of cancer cells, telomerase is upregulated [50,51]. All other cell types do not express telomerase. In these cells, telomeres are shortened with each cell division leading to a finite number of 50–60 replication cycles [52]. As a result, terminally arrested cells are formed with altered physiology [53].

In addition to telomerases, other yet unknown factors determine whether a cell line becomes immortal or not [54].

Telomere shortening is discussed as a factor contributing to age-related diseases and increasing risk of cancer [55,56]. However, further studies are necessary to apply determination of telomere lengths as a suitable biomarker of aging. Although many different factors are known affecting the telomere length, cumulative effects of these parameters over the lifetime remain unknown. Moreover, there is no consensus about validation criteria for a biomarker of aging [57].

8.2.6 Genetic Theories of Aging

Data about the role of genes clearly show a close link between shortened or prolonged life span and enhancement or delay of cell and tissue damage. On this basis, numerous genetically associated theories of aging have been formulated.

According to the DNA damage theory, the aging process results from accumulation of unrepaired DNA. This theory is supported by known mutations impairing DNA repair and causing thus a premature aging. Otherwise, the overexpression of genes or higher activity of some proteins involved in DNA repair can prolong the life span.

Mitochondrial DNA, which is not protected like nuclear DNA by histones, has 10- to 20-fold higher level of oxidized bases than that in nuclei [58,59]. As mitochondria are highly important for ATP synthesis in many cells and mitochondrial proteins are encoded by mitochondrial DNA, the mitochondrial theory of aging was proposed [60,61]. This theory suggests that mutations in mitochondrial DNA are responsible for aging.

The somatic mutation theory of aging focusses on the accumulation of genetic mutations in somatic cells [62]. The increasing inability of stem cells to replenish tissues with novel differentiated cells is the main message of the stem cell theory of aging [63,64]. The effect of telomere shortening is the central point in the telomere theory of aging [52].

Despite the huge impact of these theories, they describe mainly characteristic features of the aging process. However, answers on the questions why these changes occur and which mechanisms are behind these alterations remain open. Theories favoring genetic reasons of aging are not isolated from the other existing theories. A closer look shows that they are also interrelated to neuroendocrine and most of all to damage-associated theories.

8.3 NEUROENDOCRINE THEORIES OF AGING

8.3.1 Regulation of Growth

In humans, general growth processes are under control of the hypothalamus–pituitary–somatic axis. The pituitary gland releases the growth hormone (also known as somatotropin) that acts stimulating on growth, cell reproduction, and cell regeneration. Growth hormone stimulates either directly cell division of chondrocytes in cartilage or indirectly replication processes in a wide variety of tissues via the formation of the IGF-1 [65].

All multicellular species originate from a single cell, the fertilized egg cell. At the beginning, cell divisions dominate and the novel organism grows notably. After reaching an optimum size for a given species, general growth processes are diminished or completely stopped. Blood levels of the growth hormone and the downstream hormone, IGF-1, decline with increasing age [66–68].

In blood, IGF-1 forms predominantly a complex with IGF-binding proteins (IGFBP), whereby most IGF-1 forms a ternary complex with IGFBP-3 and the acid-labile subunit [69]. This complex formation protects IGF-1 from premature degradation.

Application of growth hormone as an antiaging therapy is questionable and critically reviewed [70]. Maybe, the administration of low levels of growth hormone and IGF-1 can improve protein synthesis and some age-related functions. However, these low doses can also cause increase of insulin resistance, risk of metabolic syndrome, and risk of certain cancers. Other adverse effects are edema formation, arthralgias, and carpal tunnel syndrome [70].

8.3.2 The Insulin-Like Growth Factor 1 Signaling Pathway

IGF-1 is a high-affine agonist to the transmembrane IGF-1 receptor [69]. Signaling by IGF-1 initiates the phosphoinositide 3-kinase/Akt/protein kinase B (PKB) and the mitogen-activated protein kinase pathways [71]. Other factors enhancing these growth- and proliferation-driven pathways are insulin, growth factors, sonic hedgehog protein, and calmodulin [72–75]. In regulation of protein synthesis, important downstream targets of Akt/PKB are factors controlling the activity of mammalian target of rapamycin (mTOR), glycogen synthase kinase 3β, and the forkhead box O (FoxO) transcription factors [71].

In cells, the catalytic subunit mTOR, a serine/threonine protein kinase, forms together with subcellular components two structurally distinct complexes, mTOR complex 1 (mTORC1, also known as mTOR (raptor)) and mTOR complex 2 (mTORC2, other name mTOR (rictor)). The complex mTORC1, which is inhibited by rapamycin, functions as a sensor for the nutrient, energy, and redox state and controls, thus, protein synthesis [76]. This complex is activated by Akt/PKB by inhibition of repressive factors [71]. Active mTORC1, which is localized on lysosomal surfaces, inhibits the eukaryotic initiation factor 4E-binding proteins 1 and 2, repressors of mRNA translation, and phosphorylates the ribosomal S6 kinases 1 and 2, which stimulates protein synthesis through activation of ribosomes and promotes ribosome biogenesis [76–79].

The second complex, mTORC2, which is mainly insensitive to rapamycin and inhibited by this agent only under certain conditions and prolonged exposure, is an important regulator of the cytoskeleton [80]. It contributes also to full activation of the receptors for IGF-1 and insulin [81] and full activation of Akt/PKB [82] by phosphorylation of critical residues.

Activated Akt/PKB promotes protein synthesis as well by inhibition of glycogen synthase kinase 3β, which inhibits β-catenin, a cofactor for protein synthesis [83]. This protein kinase acts inhibitory on FoxO genes, which are involved in protein breakdown processes [84].

Main players in promotion of cellular protein synthesis during IGF-1 and insulin signaling are presented in Fig. 8.1.

Activation of mTORC1 is further supported by energy metabolites such as high glycogen yield and high ratio between ATP/

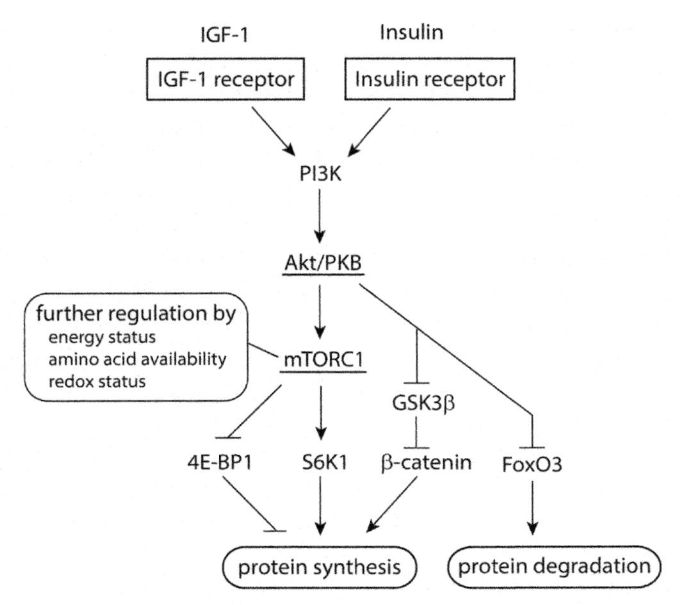

FIGURE 8.1 Simplified scheme of the IGF-1/insulin signaling pathway. Activation of the IGF-1 receptor or the insulin receptor by their respective ligands can induce processes of protein synthesis. Important regulatory proteins of this cascade are the kinases Akt/PKB and mTORC1. Abbreviations: *Akt/PKB*, protein kinase B; *FoxO*, forkhead box O; *GSK3β*, glycogen synthase kinase 3β; *IGF-1*, insulin-like growth factor 1; *mTORC1*, mammalian target of rapamycin complex 1; *PI3K*, phosphoinositide 3-kinase; *S6K*, ribosomal S6 kinase; *4E-BP*, eukaryotic initiation factor 4E-binding protein. Further explanations are given in the text.

(ADP + AMP) and by high amino acid concentrations, which favor assembling of mTORC1 to lysosomal membranes [76,85]. Accumulation of cysteine oxidation products promotes mTORC1 activation as well [86]. Otherwise, the inactivation of mTORC1 on lysosomal damage favors autophagy [87].

8.3.3 Growth Factor Signaling and Longevity

Individuals with an attenuated growth hormone/IGF-1 signaling pathway live longer and have a reduced risk for developing diabetes, cancer, or stroke [22,88–90]. A low level of circulating IGF-1 is also beneficial to human aging and confers protection against age-related pathologies [91,92].

Of the four human FoxO genes, FoxO3 is the major gene associated with longevity [48]. Selected single-nucleotide polymorphisms in or near intron 2 of this gene contribute to longevity [43,93,94].

These facts and data from laboratory animals underscore the significant role of the growth hormone/IGF-1 axis in aging process and development of age-dependent diseases. However, the present knowledge about modulating factors is limited.

8.4 DAMAGE ACCUMULATION THEORIES

8.4.1 Free Radical Theory of Aging

A central role in damage-associated theories of aging plays the free radical theory. The intracellular accumulation of any dioxygen-derived free radicals is thought to be responsible for any harm to cells and tissues. This theory was first formulated in 1957 by D. Harman [95] and revisited several times [60,96]. A critical survey of this theory is given in the excellent book of Barry Halliwell and John M. C. Gutteridge *"Free Radicals in Biology and Medicine"* [97].

Originally, putative cell and tissue damage was associated with overproduction of free radicals [95]. This was extended to reactive species in general according to the increasing knowledge of reactive species. Indeed, numerous studies demonstrate enhanced production of reactive species in senescent animals and humans [98–103].

Unequivocally, the damage of biological material by reactive species including free radicals is important during aging. However, the question raises are reactive species really the reason for aging processes or, what is more likely, is their enhanced production the consequence of underlying physiological alterations such as the presence of chronic inflammatory processes, deficits in energy supply, or limited capacity of protecting systems.

8.4.2 Mitochondria and Aging

Dysfunctional mitochondria are an important source for reactive species. This relationship provides the basis for the formulation of the mitochondrial free radical theory of aging [61]. It is well known that accumulation of reactive species disturbs mitochondrial integrity and function (see Sections 4.5.3–4.5.6). Damage of mitochondrial DNA, which is less well protected than nuclear DNA, by reactive species has been proposed as well as the existence of a vicious cycle of a putative positive feedback mechanism between damaging of mitochondrial DNA and generation of reactive species [60,104]. Otherwise, point DNA mutations do not affect largely to life span in mice [105,106]. A detailed analysis revealed little evidence for the existence of the aforementioned positive feedback mechanism [107,108].

According to the epigenetic oxidative shift theory of aging, physical inactivity of seniors should be associated with a drastic increase of mitochondrial $O_2^{\bullet -}$ production due to an impaired function of mitochondria [109].

Mitochondria are able to adapt to enhanced production of reactive species caused by physical exercise or caloric restriction. This adaptive response, known as mitochondrial hormesis or mitohormesis, is the basis of the hormesis theory of aging [110]. This theory was primarily developed on the basis of observations on yeast, fruit flies, nematodes, and mice. In humans, short-term exercise-mediated increase of reactive species induces enhanced expression of several enzymes involved in insulin sensitivity and antioxidative defense [111]. Contrary to the free radical theory of aging, mitochondrial-induced reactive species are assumed to delay the aging process as demonstrated by abrogation of these reactive species by antioxidants, which impair the health-promoting and lifetime-extending effects [110].

8.4.3 Inflammation Theories of Aging

Enhanced formation of reactive species is also observed during extensive immune reactions and inflammatory processes. Many developing pathologies are accompanied by a chronic inflammatory state. This inflammation is also regarded as a driving force for cell and tissue damage, whereby not only reactive species but also different proteolytic activities may contribute to destruction of biological material. In the inflammation theory of aging, these underlying processes are assumed to be responsible for age-dependent worsening of structural and functional parameters. In this regard, the term "inflammageing" was created [112].

The significance of inflammatory processes in the pathogenesis of age-related diseases is supported by enhanced levels of proinflammatory markers in circulating blood of older individuals [113–116]. Moreover, numerous genetic variants of inflammatory mediators and microRNAs can contribute to age-related inflammatory disease processes [117].

Several origins of inflammageing are discussed. Among them are visceral obesity [118,119], alteration in gut microbiota and age-related changes in gut permeability [120], accumulation of senescent cells in tissues [121,122], increased formation of damage-associated molecular patterns [123], and infections [117,124].

On a similar basis, the immune theory of aging suggests that the molecular and cellular defense mechanisms become limited with increasing age [112,125]. This would explain that the incidence of tumors and infections with pathogens increases in elder persons.

8.5 AGING AS A CONSEQUENCE OF GROWTH LIMITATION

8.5.1 Main Consequences from Aging Theories

Contemplating a closer look on aging, it becomes evident that processes of aging are tightly associated with mechanisms of cell and tissue damage. In more detail, aging processes are characterized by the increasing inability of the organism to reverse and to repair spontaneous deteriorations. It is not useful to search for a single source such as a certain gene or a given damage pattern as the reason explaining all facets of aging process. Aging is a very complex and multifactorial process affecting different levels of functional organization in multicellular organisms [126–128].

On cellular level, the accumulation of deteriorated molecules disturbs most of all the barrier function of membranes, leads to enhanced consumption of ATP and ATP depletion [129,130], and finally leads to cell death. Under ATP depletion, necrotic cell death dominates over apoptosis [131,132]. Increased intracellular Ca^{2+} levels and long-lasting overexcitation of cells are further consequences of disturbed metabolic functions [133–135]. For example, this is evident for muscle damage by strong overexertion [136] or neuronal loss by glutamate-induced overexcitation [137–139]. Another cellular deterioration concerns the increasing inability to repair some DNA damages [140].

On systemic level, consequences of the enhanced accumulation of unwanted defects and destructions are more complex. First of all, serious damage of biological material is mainly linked to the increasing inability of protecting systems to compensate any potentially toxic components. These systems have a limited capacity to counterregulate unfavorable alterations. With increasing age, the capacity of protecting systems is often restricted due to energy deficiency. Protective mechanisms and repair reactions are energy-consuming processes. Thus, some defects may accumulate with time contributing thus to long-lasting inflammations and pathologies.

A second problem is the inadequate supply with nutrients and dioxygen at local areas. This deficiency also favors cell and tissue damage. In addition, there is an increasing lack of energy substrates to compensate these destructive processes. Because of decreased blood flow, some local areas become ischemic, and the risk of cardiovascular complication increases [141,142]. Although certain regulatory mechanisms exist to counterregulate a mild ischemia, a strong and long-lasting ischemic state is associated with serious damage and increasing cell loss. A dangerous situation is also a transient ischemic condition followed by reperfusion [141,143]. Some metabolites can accumulate during the ischemic phase and contribute then to damage after sufficient dioxygen supply at the beginning of reperfusion phase.

Third, diminished local energy supply and blood circulation can induce altered values for homeostatic parameters in tissues, organs, and the whole organism [144,145]. Any deviation of the desired value of a homeostatic parameter induces an energy-consuming process of counterregulation. It is expected that with increasing age, deviations from desired values occur more frequent and heavier.

To sum up, all these age-dependent alterations on cellular and systemic level require more and more energy with increasing age for compensation of deviations of desired homeostatic parameters. This energy portion increases on the cost of energy that will be used for specific processes in cells, tissues, organs, and on the levels of the whole organism. Of course, it is possible to slowdown aging processes by avoiding conditions enhancing damage formation, by permanent moderate training of the protecting systems most of all of the immune system. Numerous suggestions are given in scientific and popular literature for healthy lifestyle and optimal nutrition.

8.5.2 Thermodynamics and Aging

When we look on processes of the development of living species from a thermodynamic point of view, it appears that cells have to permanently grow and divide, if they will successfully overcome the accumulation of unwanted waste products (see Chapter 1). Only in this way, the high order and low entropy of biological material will be maintained. This is the main consequence of the coexistence of genetically encoded, enzymatically driven reactions favoring the formation of high-ordered living structures on the one hand and unwanted chemical side reactions and physical processes disturbing in any way the integrity of biological material on the other hand.

Aging processes are unknown so far for bacteria and other unicellular species. The lifetime of an individual bacterium is limited. It ends either by cell division or by cell damage. In these unicellular species, any low-level destructive events will be efficiently compensated by augmentation of cell mass. It remains unclear whether during the process of cell division the deteriorated molecules predominate in one of the novel cells or they are equally distributed to both newly formed cells.

If we take a cell culture or a bioreactor, we can produce a large amount of biomass under certain optimum conditions in a given time. Nonoptimum circumstances will slowdown the growth in a cell culture. Some bacteria and other species can survive unfavorable conditions by forming endospores, microbial cysts, and related structures, in which the metabolic processes are slowed down or stopped [146,147]. In cell cultures, the application of very low temperatures stops metabolic processes and enables the use of these cells at later times.

Although for unicellular species the growth condition is always fulfilled under the assumption of sufficient supply with nutrients and energy substrates, in a growing multicellular organism, there is a continuous decrease of the general growth rate with time. Thus, a sufficient high growth rate is only given during embryonic phase, after birth as well as during childhood and juvenile age. In adult organisms, the general growth is very low or negligible.

In humans and higher animals, processes of cell and tissue damage can affect all stages of the development of this multicellular organism. These processes become more evident in adults after the organism reached its optimum size. In this period, the accumulation of any deteriorated molecules can be compensated only by the preexisting or inducible protective repair and replacement systems but not by growth processes. Thus, processes of aging and senescence result from insufficient compensation of cell and tissue damage in nongrowing elderly organisms.

8.5.3 Energy Distribution in Elderly Organisms

As a result of the stepwise worsening of cellular and physiological parameters with increasing age, more and more equivalents of energy in form of ATP are required to maintain basic homeostatic parameters in cells, tissues, and the whole organism. We can roughly divide the produced energy into a first part necessary to ensure physiological and metabolic processes, t. e. to maintain characteristic structures and homeostatic parameters of a living system, and into a second part that is used for specific purposes of the organism such as active movement, sensing the environment, intake of food, providing any physical or mental work, and so on. In the elderly organism, there is a permanent shift between these two portions of produced energy to the disadvantage of the second part. Hence, the physical and mental power declines with increasing age. In addition to this age-related shift in energy distribution, there is also general deficit of produced energy in aging individuals. The changed energy balance disturbs the ability of protecting systems to compensate any perturbations, leads to alterations in homeostatic parameters, and provides the basis for the development of numerous health complications and diseases typical of advanced age.

This thermodynamic view explains well the enormous variety of aging phenomena and is also consistent with many of the aforementioned theories of aging. These theories direct the focus on certain special aspects playing a role in the whole aging process. In truth, these aspects represent characteristic features of aging but not the reason. Aging originates from the limited ability of nongrowing complex organisms to compensate always existing processes of cell and tissue damage.

How can an elderly individual adapt to and manage with this stepwise increasing energy deficit and worsening of physiological functions? First, typical risk factors favoring aging and disease progression should be avoided such as inactivity, smoking, inadequate alcohol consumption, and unhealthy diet. A healthy lifestyle, well-balanced diet, and physical and mental activity considerably improve the physiological status even in older persons and shift the inevitable decline in physiological parameters to higher ages.

Second, the organism can gradually adapt some homeostatic values to altered conditions to reduce the energy demand for regulatory processes [144,145]. For instance, this is seen in the increase of blood pressure in some patients. Another example is the transient increase of body temperature during some severe infections. To the same category of adaptations belongs the decrease of general metabolic rate in some disease states.

8.5.4 Blood Pressure and Aging

One of the important physiological parameters that is under strict control in humans is the blood pressure. This value depends on arterial stiffness, sensitivity of baroreceptors, responsiveness of the sympathetic nervous system, electrolyte metabolism, and the renin–aldosterone system. Several studies state an increase of the systolic and diastolic arterial pressure in adults with increasing age [148–152]. Most of all, an enhanced arterial stiffness is responsible for this increase [150,153]. Arteriosclerotic alterations and calcification lead to a higher arterial stiffness. As a result, pulse waves are accelerated and earlier reflected from the arterioles toward the heart [154]. Thus, the systolic blood pressure and pulse pressure rise. Molecular details of these alterations are under discussion. Among main risk factors for hypertension are the modern lifestyle in developed countries, diabetes, smoking, and missing mobility. Arterial stiffness and hypertension provide the basis for stroke, myocardial infarction, kidney disease, loss of mental power, and other disturbances [155–157]. The prevalence of these complications rises with increasing age of individuals.

With increasing age, the variability of blood pressure increases too [158]. Some individuals are prone to orthostatic and postprandial hypotension with the consequences of syncope and falls [159].

Several factors determine the intensity of blood flow through tissues. The vasodilatory agent NO widens the blood vessels and diminishes the blood pressure, whereby the action of NO depends on its bioavailability. In the blood vessel wall, NO is produced by the endothelial nitric oxide synthase (eNOS) [160]. The physiological target of NO is the guanylate cyclase in adjacent vascular smooth muscles cells that produce the muscle relaxant cyclic guanosine monophosphate. A diminished activity of NO can be caused by enhanced hemolysis rate and/or rhabdomyolysis [161,162]. Both released hemoglobin and myoglobin are targets for NO.

8.6 CONCLUDING REMARK

Degradation, removal, and replacement of dysfunctional biological material, cell constituents, cells, and tissues occur regularly in living organisms. The complex action of protective systems, waste-elimination processes, and permanent de novo synthesis of proteins and other biomolecules in combination with cell divisions and growth processes ensures the long-term survival of biological structures. In aging multicellular organisms, there is only a limited contribution of growth processes to maintain this tissue homeostasis. Hence, cellular wastes and injured material accumulate step by step in close association with a gradual worsening of physiological parameters.

The percentage of the total amount of produced energy to maintain major physiological values on a more or less constant level rises with increasing age. In return, decreasing amount of energy is available for organism-specific purposes in advanced age. Of course, it is unpredictable which functional constraints develop first on aging. Damage patterns and physiological worsening are very multifaceted. From this point of view, the aforementioned theories of aging mirror the dazzling array of mechanisms occurring in elderly persons. Each of these theories and hypotheses focusses on a certain phenomenon, which can actually emerge in some individuals, arguing that this aspect represents the real reason of aging. This generalization fails due to a one-sided perspective.

Aging is an indispensable property of each complex multicellular organism. As outlined in Section 1.4.2, there are three main strategies to protect organisms against destructions of biological material: (1) by inhibition and repair mechanisms, (2) by removal and replacement mechanisms, and (3) by growth processes in combination with cell divisions. In full-grown organisms, the growth rate is very low or negligible. Consequently, growth processes do not contribute to maintain the high order and low entropy in adult individuals. Protective repair and replacement mechanisms can slow down the aging process. However, they cannot prevent aging. The same is valid for all aspects concerning a healthy lifestyle.

Taken together, aging as a fundamental property of multicellular organisms originates from the increasing inability of organisms to manage with processes of cell and tissue destruction.

References

[1] Z.A. Medvedev, An attempt at a rational classification of theories of aging, Biol. Rev. 65 (1990) 375–398.

[2] A. De Sandre-Giovannoli, R. Bernard, P. Cau, C. Navarro, J. Amiel, I. Boccaccio, S. Lyonnet, C.L. Stewart, A. Munnich, M. Le Merrer, N. Lévy, Lamin A truncation in Hutchinson-Gilford progeria, Science 300 (2003) 2055.

[3] M. Eriksson, W.T. Brown, L.B. Gordon, M.W. Glynn, J. Singer, L. Scott, M.R. Erdos, C.M. Robbins, T.Y. Moses, P. Berglund, A. Dutra, E. Pak, S. Durkin, A.B. Csoka, M. Boehnke, T.W. Glover, F.S. Collins, Recurrent de novo point mutations in lamin A cause Hutchinson-Gilford progeria syndrome, Nature 423 (2003) 293–298.

[4] C.L. Navarro, A. De Sandre-Giovannoli, R. Bernard, I. Boccaccio, A. Boyer, D. Genevieve, S. Hadj-Rabia, C. Gaudy-Marqueste, H.S. Smitt, P. Vabres, L. Faivre, A. Verloes, T. van Essen, E. Flori, R. Hennekam, F.A. Beemer, N. Laurent, M. Le Merrer, P. Cau, N. Levy, Lamin A and ZMPSTE24 (FACE-1) defects cause nuclear disorganization and identity restrictive dermopathy as a lethal neonatal laminopathy, Hum. Mol. Genet. 13 (2004) 2493–2503.

[5] S. Huang, L. Lee, N.B. Hanson, C. Lenaerts, H. Hoehn, M. Poot, C.D. Rubin, D.F. Chen, C.C. Yang, H. Juch, T. Dorn, R. Spiegel, E.A. Oral, M. Abid, C. Battisti, E. Lucci-Cordisco, G. Neri, E.H. Steed, A. Kidd, W. Isley, D. Showalter, J.L. Vittone, A. Konstantinow, J. Ring, P. Meyer, S.L. Wenger, A. von Herbay, U. Wollina, M. Schuelke, C.R. Huizenga, D.F. Leistritz, G.M. Martin, I.S. Mian, J. Oshima, The spectrum of WRN mutations in Werner syndrome patients, Hum. Mutat. 27 (2006) 558–567.

[6] P.H. Itin, A. Sarasin, M.R. Pittelkow, Trichothiodystrophy: update on the sulfur-deficient brittle hair syndromes, J. Am. Acad. Dermatol. 44 (2001) 891–920.

[7] S. Faghri, D. Tamura, K.H. Kraemer, J.J. Digiovanna, Trichothiodystrophy: a systematic review of 112 published cases characterizes a wide spectrum of clinical manifestations, J. Med. Genet. 45 (2008) 609–621.

[8] A. Komatsu, S. Suzuki, T. Inagaki, K. Yamashita, K. Hashizume, A kindred with Cockayne syndrome caused by multiple splicing variants of the CSA gene, Am. J. Med. Genet. 128A (2004) 67–71.

[9] C.F. Cheok, C.Z. Bachrati, K.L. Chan, C. Ralf, L. Wu, I.D. Hickson, Roles of the Bloom's syndrome helicase in the maintenance of genome stability, Biochem. Soc. Trans. 33 (2005) 1456–1459.

[10] K.H. Kraemer, M.M. Lee, J. Scotto, Xeroderma pigmentosum. Cutaneous, ocular, and neurological abnormalities in 830 published cases, Arch. Dermatol. 123 (1987) 241–250.

[11] U.R. Hengge, S. Emmert, Clinical features of xeroderma pigmentosum, Adv. Exp. Med. Biol. 637 (2008) 10–18.

[12] H.V. Toriello, Wiedemann-Rautenstrauch syndrome, J. Med. Genet. 27 (1990) 256–257.

[13] T.A. Dittmer, T. Misteli, The lamin protein family, Genome Biol. 12 (2011), 222, 1–14, https://doi.org/10.1186/gb-m2011-12-5-222.

[14] H.J. Woman, G. Bonne, "Laminopathies": a wide spectrum of human diseases, Exp. Cell Res. 313 (2007) 2121–2133.

[15] G. Captur, E. Arbustini, G. Bonne, P. Syrris, K. Mills, K. Wahbi, S.S. Mohiddin, W.J. McKenna, S. Petit, Lamin and the heart, Heart 104 (2017) 468–479.

[16] A. Barateau, N. Vadrot, P. Vicart, A. Ferreiro, M. Mayer, D. Héron, C. Vigouroux, B. Buendia, A novel lamin A mutant responsible for congenital muscular dystrophy causes distinct abnormalities of the cell nucleus, PLoS One 12 (2017), e0169189, 1–18, https://doi.org/10.1371/journal.pone.0169189.

[17] C. Kenyon, J. Chang, E. Gensch, A. Rudner, R. Tabtiang, A *C. elegans* mutant that lives twice as long as wild type, Nature 366 (1994) 461–464.

[18] S. Gottlieb, G. Ruvkun, Daf-2, daf-6, daf-16 and daf-23: genetically interacting genes controlling Dauer formation in *Caenorhabditis elegans*, Genetics 137 (1994) 107–120.

[19] H.M. Brown-Borg, K.E. Borg, C.J. Meliska, A. Bartke, Dwarf mice and the ageing process, Nature 384 (1996) 33.

[20] Y. Ikeno, R.T. Bronson, G.B. Hubbard, S. Lee, A. Bartke, Delayed occurrence of fatal neoplastic diseases in ames dwarf mice: correlation to extended longevity, J. Gerontol. A Biol. Sci. Med. 58 (2003) 291–296. Brown-Borg HM 1.

[21] C. Kenyon, The plasticity of ageing, Nature 464 (2010) 504–512.

[22] Y. Suh, G. Atzmon, M.O. Cho, D. Hwang, B. Liu, D.J. Leahy, N. Barzilai, P. Cohen, Functionally significant insulin-like growth factor 1 receptor mutations in centenarians, Proc. Natl. Acad. Sci. U.S.A. 105 (2008) 3438–3442.

[23] S. Milman, G. Atzmon, D.M. Huffman, J. Wan, J.P. Crandall, P. Cohen, N. Brazilai, Low insulin-like growth factor-1 level predicts survival in humans with exceptional longevity, Aging Cell 13 (2014) 769–771.

[24] L. Guarante, Sirtuins in aging and disease, Cold Spring Harbor Symp. Quant. Biol. 72 (2007) 483–488.

[25] M. Watroba, D. Szukiewicz, The role of sirtuins in aging and age-related diseases, Adv. Med. Sci. 61 (2016) 52–62.

[26] W. Grabowska, E. Sikora, A. Bielak-Zmijewska, Sirtuins, a promising target in slowing down the ageing process, Biogerontology 18 (2017) 447–476.

[27] S.J. Mitchell, A. Martin-Montalvo, E.M. Mercken, H.H. Palacios, T.M. Ward, G. Abulwerdi, R.K. Minor, G.P. Vlasuk, J.L. Ellis, D.A. Sinclair, J. Dawson, D.B. Allison, Y. Zhang, K.G. Becker, M. Bernier, R. de Cabo, The SIRT1 activator SRT1720 extends lifespan and improves health of mice fed a standard diet, Cell Rep. 6 (2014) 836–843.

[28] T.S. Piskunova, M.N. Yurova, A.I. Osvyannikov, A.V. Semenchenko, M.A. Zabezhinski, I.G. Popovich, Z.-Q. Wang, V.N. Anisimov, Deficiency in poly(ADP-ribose) polymerase-1 (PARP-1) accelerates aging and spontaneous carcinogenesis in mice, Curr. Gerontol. Geriatr. Res. (2008), article 754190, 1–11, https://doi.org/10.1155/2008/754190.

[29] A. Mangerich, A. Bürkle, Pleiotropic cellular functions of PARP1 in longevity and aging: genome maintenance meets inflammation, Oxid. Med. Cell. Longev. (2012), article 321653, 1–19, https://doi.org/10.1155/2012/321653.
[30] T.B. Osborne, L.B. Mendel, E.L. Ferry, The effect of retardation of growth upon the breeding period and duration of life of rats, Science 45 (1917) 294–295.
[31] A.M. Holehan, B.J. Merry, The experimental manipulation of aging by diet, Biol. Rev. 61 (1986) 329–368.
[32] B.P. Yu, Aging and oxidative stress: modulation by dietary restriction, Free Radic. Biol. Med. 21 (1996) 321–326.
[33] E.J. Masoro, Overview of caloric restriction and ageing, Mech. Ageing Dev. 126 (2005) 913–922.
[34] L. Partridge, M.D. Piper, W. Mair, Dietary restriction in Drosophila, Mech. Ageing Dev. 126 (2005) 938–950.
[35] G.E. Holmes, C. Bernstein, H. Bernstein, Oxidative and other DNA damages as the basis of aging: a review, Mutat. Res. 275 (1992) 305–315.
[36] R.S. Sohal, S. Agarwal, M. Candas, M.J. Forster, H. Lal, Effect of age and caloric restriction on DNA oxidative damage in different tissues of C57BL/6 mice, Mech. Ageing Dev. 76 (1994) 215–224.
[37] T. Kaneko, S. Tahara, M. Matsuo, Retarding effect of dietary restriction on the accumulation of 8-hydroxy-2′-deoxyguanosine in organs of Fischer 344 rats during aging, Free Radic. Biol. Med. 23 (1997) 76–81.
[38] R.S. Sohal, A. Dubey, Mitochondrial oxidative damage, hydrogen peroxide release and aging, Free Radic. Biol. Med. 16 (1994) 621–626.
[39] R. Gredilla, A. Sanz, M. Lopez-Torres, G. Barja, Caloric restriction decreases mitochondrial free radical generation at complex I and lowers oxidative damage to mitochondrial DNA in the rat heart, FASEB J. 15 (2001) 1589–1591.
[40] C. Mörck, M. Pilon, *C. elegans* feeding defective mutants have shorter body lengths and increased autophagy, BioMed Cent. Dev. Biol. 6 (2006), 39, 1–11, https://doi.org/10.1186/1471-213X-6-39.
[41] C. Mörck, M. Pilon, Addenda, Caloric restriction and autophagy in *Caenorhabditis elegans*, Autophagy 3 (2007) 51–53.
[42] R.S. Sohal, M.J. Forster, Caloric restriction and the aging process: a critique, Free Radic. Biol. Med. 73 (2014) 366–382.
[43] A.R. Brooks-Wilson, Genetics of healthy aging and longevity, Hum. Genet. 132 (2013) 1323–1338.
[44] G. Passarino, F. De Rango, A. Montesanto, Human longevity: genetics or lifestyle? It takes two to tango, Immun. Ageing 13 (2016), 12, 1–6, https://doi.org/10.1186/s12979-016-0066-z.
[45] Y. Niu, L. Na, R. Feng, L. Gong, Y. Zhao, Q. Li, Y. Li, C. Sun, The phytochemical, EGCG, extends lifespan by reducing liver and kidney damage and improving age-associated inflammation and oxidative stress in healthy rats, Aging Cell 12 (2013) 1041–1049.
[46] M. Stefani, M.A. Markus, R.C. Lin, M. Pinese, I.W. Dawes, B.J. Morris, The effect of resveratrol on a cell model of human aging, Ann. NY Acad. Sci. 1114 (2007) 407–418.
[47] Y. Zeng, H. Chen, T. Ni, R. Ruan, L. Feng, C. Nie, L. Cheng, Y. Li, W. Tao, J. Gu, K.C. Land, A. Yashin, Q. Tan, Z. Yang, L. Bolund, H. Yang, E. Hauser, D.C. Willcox, B.J. Willcox, X.L. Tian, J.W. Vaupel, GxE interactions between FOXO genotypes and tea drinking are significantly associated with cognitive disability at advanced ages in China, J. Gerontol. A Biol. Sci. Med. Sci. 70 (2015) 426–433.
[48] B.J. Morris, D.C. Willcox, T.A. Donlon, B.J. Willcox, FOXO3: a major gene for human longevity – a mini review, Gerontology 61 (2015) 515–525.
[49] J. Lingner, T.R. Hughes, A. Shevchenko, M. Mann, V. Lundblad, T.R. Cech, Reverse transcriptase motifs in the catalytic subunit of telomerase, Science 276 (1997) 561–567.
[50] J.W. Shay, W.E. Wright, Telomerase activity in human cancer, Curr. Opin. Oncol. 8 (1996) 66–71.
[51] S.E. Artandi, Telomeres, telomerase, and human disease, N. Engl. J. Med. 355 (2006) 1195–1197.
[52] L. Hayflick, The limited in vitro lifetime of human diploid cell strains, Exp. Cell Res. 37 (1965) 614–636.
[53] J. Campisi, Cellular senescence and cell death, in: P.S. Timiras (Ed.), Physiological Basis of Aging and Geriatrics, third ed., CRC, Boca Raton, 2003, pp. 47–59.
[54] T.M. Bryan, A. Englezou, J. Gupta, S. Bacchetti, R.R. Reddel, Telomere elongation on immortal human cells without detectable telomerase activity, EMBO J. 14 (1995) 4240–4248.
[55] H. Jiang, Z. Ju, K.L. Rudolph, Telomere shortening and ageing, Z. Gerontol. Geriatr. 40 (2007) 314–324.
[56] M.A. Shammas, Telomeres, lifestyle, cancer and aging, Curr. Opin. Clin. Nutr. Metab. Care 14 (2011) 28–34.
[57] K.A. Mather, A.F. Jorm, R.A. Parslow, H. Christensen, Is telomere length a biomarker of aging? A review, J. Gerontol. A Biol. Sci. Med. Sci. 66A (2011) 202–213.
[58] C. Richter, J.W. Park, B.N. Ames, Normal oxidative damage to mitochondrial and nuclear DNA is extensive, Proc. Natl. Acad. Sci. U.S.A. 85 (1988) 6465–6467.
[59] B.N. Ames, Endogenous oxidative DNA damage, aging, and cancer, Free Radic. Res. Commun. 7 (1989) 121–128.

[60] D. Harman, The biological clock: the mitochondria? J. Am. Geriatr. Soc. 20 (1972) 145–147.
[61] J. Miquel, A.C. Economos, J. Fleming, J.E. Johnson Jr., Mitochondrial role in cell aging, Exp. Gerontol. 15 (1980) 575–591.
[62] K.B. Beckman, B.N. Ames, The free radical theory of aging matures, Physiol. Res. 78 (1998) 547–581.
[63] J.A. Smith, R. Daniel, Stem cells and aging: a chicken-or-egg issue? Aging Dis. 3 (2012) 260–268.
[64] A. Behrens, J.M. van Deursen, K.L. Rudolph, B. Schumacher, Impact of genomic damage and ageing on stem cell function, Nat. Cell Biol. 16 (2014) 201–207.
[65] A. Vijayakumar, S. Yakar, D. LeRoith, The intricate role of growth hormone in metabolism, Front. Endocrinol. 2 (2011), article 32, 1–11, https://doi.org/10.3389/fendo.2011.00032.
[66] D. Rudman, M.H. Kutner, C.M. Rogers, M.F. Lubin, G.A. Fleming, R.P. Bain, Impaired growth hormone secretion in the adult population: relation to age and adiposity, J. Clin. Investig. 67 (1981) 1361–1369.
[67] A.A. Abbasi, P.J. Drinka, D.E. Mattson, D. Rudman, Low circulating levels of insulin-like growth factors and testosterone in chronically institutionalized elderly men, J. Am. Geriatr. Soc. 41 (1993) 975–982.
[68] A. Giustina, J.D. Veldhuis, Pathophysiology of the neuroregulation of growth hormone secretion in experimental animals and the human, Endocr. Rev. 19 (1998) 717–797.
[69] J.M. Holly, C.M. Perks, Insulin-like growth factor physiology: what we have learned from human studies, Metab. Clin. North Am. 41 (2012) 249–263.
[70] F.R. Sattler, Growth hormone in the aging male, Best Pract. Res. Clin. Endocrinol. Metabol. 27 (2013) 541–555.
[71] C.M. Taniguchi, B. Emanuelli, C.R. Kahn, Critical nodes in signaling pathways: insight into insulin action, Nat. Rev. Mol. Cell Biol. 7 (2006) 85–96.
[72] H.Y. Man, Q. Wang, W.Y. Lu, W. Ju, G. Ahmadian, L. Liu, S. d'Souza, T.P. Wong, C. Taghibiglou, J. Lu, L.E. Becker, L. Pei, M.P. Wymann, J.F. MacDonald, Y.T. Wang, Activation of PI3-kinase is required for AMPA receptor insertion during LTP of mEPSCs in cultural hippocampal neurons, Neuron 38 (2003) 611–624.
[73] J. Peltier, A. O'Neill, D.V. Schaffer, PI3K/Akt and CREB regulate adult neural hippocampal progenitor proliferation and differentiation, Dev. Neurobiol. 67 (2007) 1348–1361.
[74] V.A. Rafalski, A. Brunet, Energy metabolism in adult neural stem cell fate, Prog. Neurobiol. 93 (2011) 182–203.
[75] L. Ojeda, J. Gao, K.G. Hooten, E. Wang, J.R. Thonhoff, T.J. Dunn, T. Gao, P. Wu, Critical role of PI3K/Akt/GSK3β in motoneuron specification from human neural stem cells in response to FGF2 and EGF, PLoS One 6 (2011), e23414, 1–14, https://doi.org/10.1371/journal.pone.0023414.
[76] N. Hay, N. Sonnenberg, Upstream and downstream of mTOR, Genes Dev. 18 (2004) 1926–1945.
[77] X.M. Ma, J. Blenis, Molecular mechanisms of mTOR-mediated translational control, Nat. Rev. Mol. Cell Biol. 10 (2009) 307–318.
[78] N. Sonnenberg, A.G. Hinnebusch, Regulation of translation initiation in eukaryotes: mechanisms and biological targets, Cell 136 (2009) 731–745.
[79] M. Brian, E. Bilgen, C.F. Diana, Regulation and function of ribosomal protein s6 kinase (s6k) within motor signaling networks, Biochem. J. 441 (2012) 1–21.
[80] D.D. Sarbassov, S.M. Ali, D.H. Kim, D.A. Guertin, R.R. Latek, H. Erdjument-Bromage, P. Tempst, D.M. Sabatini, Rictor, a novel partner of mTOR, defines a rapamycin-insensitive and raptor-independent pathway that regulates the cytoskeleton, Curr. Biol. 14 (2004) 1296–1302.
[81] Y. Yin, H. Hua, M. Li, S. Liu, Q. Kong, T. Shao, J. Wang, Y. Luo, Q. Wang, T. Luo, Y. Jiang, mTORC2 promotes type I insulin-like growth factor receptor and insulin receptor activation through the tyrosine kinase activity of mTOR, Cell Res. 26 (2016) 46–65.
[82] D.D. Sarbassov, D.A. Guertin, S.M. Ali, D.M. Sabatini, Phosphorylation and regulation of Akt/PKB by the rictor-mTOR complex, Science 307 (2005) 1098–1101.
[83] D.A. Cross, D.R. Alessi, P. Cohen, M. Andjelkovich, B.A. Hemmings, Inhibition of glycogen synthase kinase-3 by insulin mediated by protein kinase B, Nature 378 (1995) 785–789.
[84] H. Tran, A. Brunet, E.C. Griffith, M.E. Greenberg, The many forks in foxo's road, Sci. STKE 2003 (172) (2003) 1–5, https://doi.org/10.1126/stke.2003.172.re5.
[85] C. Tokunaga, K. Yoshino, K. Yonezawa, mTOR integrates amino acid- and energy-sensing pathways, Biochem. Biophys. Res. Commun. 313 (2004) 443–446.
[86] S. Yoshida, S. Hong, T. Suzuki, S. Nada, A.M. Mannan, J. Wang, M. Okada, K.-L. Guan, K. Inoki, Redox regulates mammalian target of rapamycin complex 1 (mTORC1) activity by modulating the TSC1/TSC2-Rheb GTPase pathway, J. Biol. Chem. 286 (2011) 32651–32660.
[87] S. Chauhan, S. Kumar, A. Jain, M. Ponpuak, M.H. Mudd, T. Kimura, S.W. Choi, R. Peters, M. Mandell, J.A. Bruun, T. Johansen, V. Deretic, TRIMs and galectins globally cooperate and TRIM16 and galectin-3 co-direct autophagy in endomembrane damage homeostasis, Dev. Cell 39 (2016) 13–27.

[88] J. Guevara-Aguirre, P. Balasubramanian, M. Guevara-Aguirre, M. Wei, F. Madia, C.W. Cheng, D. Hwang, A. Martin-Montalvo, J. Saavedra, S. Ingles, R. de Cabo, P. Cohen, V.D. Longo, Growth hormone receptor deficiency is associated with a major reduction in pro-aging signaling, cancer, and diabetes in humans, Sci. Transl. Med. 3 (2011), 70ra13, 1–16, https://doi.org/10.1126/scitranslmed.3001845.

[89] R. Steuerman, O. Shevah, Z. Laron, Congenital IGF1 deficiency tends to confer protection against post-natal development of malignancies, Eur. J. Endocrinol. 164 (2011) 485–489.

[90] C. Tazearslan, J. Hunag, N. Barzilai, Y. Suh, Impaired IGF1R signaling in cells expressing longevity-associated human IGF1R alleles, Aging Cell 10 (2011) 551–554.

[91] E. van der Spoel, M.P. Rozing, J.J. Houwing-Duistermaat, P.E. Slagboom, M. Beekman, A.J. de Craen, R.G. Westendorp, D. van Heernst, Association analysis of insulin-like growth factor-1 axis parameters with survival and functional status in nonagenarians of the Leiden longevity study, Aging 7 (2015) 956–963.

[92] S. Milman, D.M. Huffman, N. Barzilai, The somatotropic axis in human aging: framework for the current state of knowledge and future research, Cell Metabol. 23 (2016) 980–989.

[93] J.M. Murabito, R. Yuan, K.L. Lunetta, The search for longevity and healthy aging genes: insights from epidemiological studies and samples of long-lived individuals, J. Gerontol. A Biol. Sci. Med. Sci. 67 (2012) 470–479.

[94] T.A. Donlon, J.D. Curb, Q. He, J.S. Grove, K.H. Masaki, B. Rodriguez, A. Elliott, D.C. Willcox, B.J. Willcox, FOXO3 gene variants and human aging: coding variants may not be key players, J. Gerontol. A Biol. Sci. Med. Sci. 67 (2012) 1132–1139.

[95] D. Harman, Aging: a theory based on free radical and radiation chemistry, J. Gerontol. 2 (1957) 298–300.

[96] T. Finkel, N.J. Holbrook, Oxidants, oxidative stress and the biology of ageing, Nature 408 (2000) 239–247.

[97] B. Halliwell, J.M.C. Gutteridge, Free Radicals in Biology and Medicine, fifth ed., Oxford Press, 2015.

[98] M. Sawada, J.C. Carlson, Changes in superoxide radical and lipid peroxide formation in the brain, heart and liver during the lifetime of the rat, Mech. Ageing Dev. 41 (1987) 125–137.

[99] A.J. Donato, I. Eskurza, A.E. Silver, A.S. Levy, G.L. Pierce, P.E. Gates, D.R. Seals, Direct evidence of endothelial oxidative stress with aging in humans: relation to impaired endothelium-dependent dilation and upregulation of nuclear factor-kappa B, Circ. Res. 100 (2007) 1659–1666.

[100] A. Jacobson, C. Yan, Q. Gao, T. Rincon-Skinner, A. Rivera, J. Edwards, A. Huang, G. Kaley, D. Sun, Aging enhances pressure-induced arterial superoxide formation, Am. J. Physiol. Heart Circ. Physiol. 293 (2007) H1344–H1350.

[101] B. Lener, R. Kozeil, H. Pircher, E. Hutter, R. Greussing, D. Herndler-Brandstetter, M. Hermann, H. Unterluggauer, P. Jansen-Durr, The NADPH oxidase Nox4 restricts the replicative lifespan of human endothelial cells, Biochem. J. 423 (2009) 353–374.

[102] L. Rodriguez-Manas, M. El-Assar, S. Vallejo, P. Lopez-Dorgia, J. Solis, R. Petidier, M. Montes, J. Nevado, M. Castro, C. Gomez-Guerrero, C. Peiro, C.F. Sanchez-Ferrer, Endothelial dysfunction in aged humans is related with oxidative stress and vascular inflammation, Aging Cell 8 (2009) 226–238.

[103] I. Afanas'ev, Signaling and damaging functions of free radicals in aging – free radical theory, hormesis, and TOR, Aging Dis. 1 (2010) 75–88.

[104] M.L. Hamilton, H. van Remmen, J.A. Drake, H. Yang, Z.M. Guo, K. Kewitt, C.A. Walter, A. Richardson, Does oxidative damage to DNA increase with age? Proc. Natl. Acad. Sci. U.S.A. 98 (2001) 10469–10474.

[105] M. Vermulst, J.H. Bielas, G.C. Kujoth, W.C. Ladiges, P.S. Rabinovich, T.A. Prolla, L.A. Loeb, Mitochondrial point mutations do not limit the natural lifespan in mice, Nat. Genet. 39 (2007) 540–543.

[106] A. Hiona, C. Leeuwenburgh, The role of mitochondrial DNA mutations in aging and sarcopenia: implications for the mitochondria vicious cycle theory of aging, Exp. Gerontol. 43 (2008) 24–33.

[107] J. Gruber, S. Schaffer, B. Halliwell, The mitochondrial free radical theory of ageing – where do we stand? Front. Biosci. 13 (2008) 6554–6579.

[108] S.K. Poovathingal, J. Gruber, B. Halliwell, R. Gunawan, Stochastic shift in mitochondrial DNA point mutations: a novel perspective ex silico, PLoS Comput. Biol. 5 (2009), e1000572, 1–12, https://doi.org/10.1371/journal.pcbi.1000572.

[109] G.J. Brewer, Epigenetic oxidative redox shift (EORS) theory of aging unifies the free radical and insulin signaling theories, Exp. Gerontol. 45 (2010) 173–179.

[110] M. Ristow, K. Zarse, How increased oxidative stress promotes longevity and metabolic health: the concept of mitochondrial hormesis (mitohormesis), Exp. Gerontol. 45 (2010) 410–418.

[111] M. Ristow, K. Zarse, A. Oberbach, N. Klöting, M. Birringer, M. Kiehntopf, M. Stumvoll, C.R. Kahn, M. Blüher, Antioxidants prevent health-promoting effects of physical exercise in humans, Proc. Natl. Acad. Sci. U.S.A. 106 (2009) 8665–8670.

[112] C. Franceschi, M. Bonafè, S. Valensin, F. Olivieri, M. De Luca, E. Ottaviani, G. De Benedictis, Inflamm-aging – an evolutionary perspective on immunosenescence, Ann. N. Y. Acad. Sci. 908 (2000) 244–254.
[113] H.J. Cohen, C.F. Pieper, T. Harris, K.M. Rao, M.S. Currie, The association of plasma IL-6 levels with functional disability in community-dwelling elderly, J. Gerontol. A Biol. Sci. Med. Sci. 52 (1997) M201–M208.
[114] R. Gerli, D. Monti, O. Bistoni, A.M. Mazzone, G. Peri, A. Cossarizza, M. Di Gioacchino, M.E. Cesarotti, A. Doni, A. Mantovani, C. Franceschi, R. Paganelli, Chemokines, sTNF-Rs and sCD30 serum levels in healthy aged people and centenarians, Mech. Ageing Dev. 121 (2000) 37–46.
[115] L. Ferrucci, R.D. Semba, J.M. Guralnik, W.B. Ershler, S. Bandinelli, K.V. Patel, K. Sun, R.C. Woodman, N.C. Andrews, R.J. Cotter, T. Ganz, E. Nemeth, D.L. Longo, Proinflammatory state, hepcidin, and anemia in older persons, Blood 115 (2010) 3810–3816.
[116] A.B. Newman, J.L. Sanders, J.R. Kizer, R.M. Boudreau, M.C. Odden, A. Zeki Al Hazzouri, A.M. Arnold, Trajectories of function and biomarkers with age: the CHS all stars study, Int. J. Epidemiol. 45 (2016) 1135–1145.
[117] L. Ferrucci, E. Fabbri, Inflammageing: chronic inflammation in ageing, cardiovascular disease, and frailty, Nat. Rev. Cardiol. 15 (2018) 505–522.
[118] B. Vandanmagsar, Y.H. Youm, A. Ravussin, J.E. Galgani, K. Stadler, R.L. Mynatt, E. Ravussin, J.M. Stephens, V.D. Dixit, The NLPR3 inflammasome instigates obesity-induced inflammation and insulin resistance, Nat. Med. 17 (2011) 179–189.
[119] D. Frasca, B.B. Blomberg, R.T. Paganelli, Aging, obesity, and inflammatory age-related diseases, Front. Immunol. 8 (2017), article 1745, 1–10, https://doi.org/10.3389/fimmu.2017.01745.
[120] H. Shapiro, C.A. Thaiss, M. Levy, E. Elinav, The cross talk between microbiota and the immune system: metabolites take center stage, Curr. Opin. Immunol. 30 (2014) 54–62.
[121] Y. Liu, H.K. Sanoff, H. Cho, C.E. Burd, C. Torrice, J.G. Ibrahim, N.E. Thomas, N.E. Sharpless, Expression of p161NK4a in peripheral blood T-cells is a biomarker of human aging, Aging Cell 8 (2009) 439–448.
[122] M.E. Waaijer, W.E. Parish, B.H. Strongitham, D. van Heemst, P.E. Slagboom, A.J. de Craen, J.M. Sedivy, R.G. Westendorp, D.A. Gunn, A.B. Maier, The number of p161NK4a positive cells in human skin reflects biological age, Aging Cell 11 (2012) 722–725.
[123] R. Medzhitov, C.A. Janeway Jr., Decoding the patterns of self and nonself by the innate immune system, Science 296 (2002) 298–300.
[124] M.J. Cannon, D.S. Schmid, T.B. Hyde, Review of cytomegalovirus seroprevalence and demographic characteristics associated with infection, Rev. Med. Virol. 20 (2010) 202–213.
[125] C. Franceschi, Cell proliferation and cell death in the aging process, Aging Clin. Exp. Res. 1 (1989) 3–13.
[126] A. Kowald, T.B. Kirkwood, A network theory of ageing: the interactions of defective mitochondria, aberrant proteins, free radicals and scavengers in the ageing process, Mutat. Res. 316 (1996) 209–236.
[127] C. Franseschi, S. Valentin, M. Bonafè, G. Paolisso, A.I. Yashin, D. Monti, G. De Benedictis, The network and remodeling theories of aging: historical background and new perspectives, Exp. Gerontol. 35 (2000) 879–896.
[128] B.T. Weinert, P.S. Timiras, Theories of aging, J. Appl. Physiol. 95 (2003) 1706–1716.
[129] N. Miyoshi, H. Oubrahim, P.B. Chock, E.R. Stadtman, Age-dependent cell death and the role of ATP in hydrogen peroxide-induced apoptosis and and necrosis, Proc. Natl. Acad. Sci. U.S.A. 103 (2006) 1727–1731.
[130] C.D. Wiley, J. Campisi, From ancient pathways to aging cells – connecting metabolism and cellular senescence, Cell Metabol. 23 (2016) 1013–1021.
[131] Y. Eguchi, S. Shimizu, Y. Tsujimoto, Intracellular ATP levels determine cell death fate by apoptosis or necrosis, Cancer Res. 57 (1997) 1835–1840.
[132] A. Wochna, E. Niemczyk, C. Kurano, M. Masaoka, A. Majczak, J. Kedzior, E. Slominska, M. Lipinski, T. Wakabayashi, Role of mitochondria in the switch mechanisms of the cell death mode from apoptosis to necrosis – studies on rho0 cells, J. Electron. Microsc. 54 (2005) 127–138.
[133] G.E. Gibson, C. Peterson, Calcium and the aging nervous system, Neurobiol. Aging 8 (1987) 329–343.
[134] C. Chinopoulos, V. Adam-Vizi, Calcium, mitochondria, and oxidative stress in neuronal pathology, FEBS J. 273 (2006) 433–450.
[135] M.P. Mattson, Calcium and neurodegeneration, Aging Cell 6 (2007) 337–350.
[136] G.L. Close, A. Kayani, A. Vasilaki, A. McArdle, Skeletal muscle damage with exercise and aging, Sports Med. 35 (2005) 413–427.
[137] E. Salinska, W. Danysz, J.W. Lazarewicz, The role of excitotoxicity in neurodegeneration, Folia Neuropathol. 43 (2005) 322–329.
[138] M.M. Fan, L.A. Raymond, N-Methyl-D-aspartate (NMDA) receptor function and excitotoxicity in Huntington's disease, Prog. Neurobiol. 81 (2007) 272–293.

[139] U. Wojda, E. Salinka, J. Kuznicki, Calcium ions in neuronal degeneration, IUBMB Life 60 (2008) 575–590.

[140] S. Maynard, E. Fei Fang, M. Scheibye-Knudsen, D.L. Croteau, V.A. Bohr, DNA damage, DNA repair, aging, and neurodegeneration, Cold Spring Harb. Perspect. Med. 5 (2015), a025130, 1–18, https://doi.org/10.1101/cshperspect.a025130.

[141] M. Liu, P. Zhan, M. Chen, W. Zhang, L. Yu, X.-C. Yang, Q. Fan, Aging might increase myocardial ischemia/reperfusion-induced apoptosis in humans and rats, Age 34 (2012) 621–632.

[142] J.B. Strait, E.G. Lakatta, Aging-associated cardiovascular changes and their relationship to heart failure, Heart Fail. Clin. 8 (2012) 143–164.

[143] E.J. Lesnefsky, S. Moghaddas, B. Tandler, J. Kerner, C.L. Hoppel, Mitochondrial dysfunction in cardiac disease: ischemia-reperfusion, aging, and heart failure, J. Mol. Cell. Cardiol. 33 (2001) 1065–1089.

[144] K.J.A. Davies, Adaptive homeostasis, Mol. Aspect. Med. 49 (2016) 1–7.

[145] L.C.D. Pomatto, K.J.A. Davies, The role of declining adaptive homeostasis in ageing, J. Physiol. 24 (2017) 7275–7309.

[146] J. Errington, Regulation of endospore formation in Bacillus subtilis, Nat. Rev. Microbiol. 1 (2003) 117–126.

[147] S. van Vliet, Bacterial dormancy: how to decide when to wake up, Curr. Biol. 25 (2015) R753–R755.

[148] J.M. Kotchen, H.E. McKean, T.A. Kotchen, Blood pressure trends with aging, Hypertension 4 (Suppl. III) (1982) 128–134.

[149] I. Ferreira, R.J. van de Laar, M.H. Prins, J.W. Twisk, C.D. Stehouwer, Carotid stiffness in young adults: a life-course analysis of its early determinants: the Amsterdam growth and health longitudinal study, Hypertension 59 (2012) 54–61.

[150] M. AlGhatrif, J.B. Strait, C.H. Morrell, M. Canepa, J. Wright, P. Elango, A. Scuteri, S.S. Najjar, L. Ferrucci, E.G. Lakatta, Longitudinal trajectories of arterial stiffness and the role in blood pressure: the Baltimore longitudinal study of aging, Hypertension 62 (2013) 934–941.

[151] C. Rosano, N. Watson, Y. Chang, A.B. Newman, H.J. Aizenstein, Y. Du, V. Venkatraman, T.B. Harris, Aortic pulse wave velocity predicts focal white matter hyperintensities in a biracial cohort of older adults, Hypertension 61 (2013) 160–165.

[152] T.W. Buford, Hypertension and aging, Ageing Res. Rev. 26 (2016) 96–111.

[153] Z. Sun, Aging, arterial stiffness, and hypertension, Hypertension 65 (2015) 252–256.

[154] H.Y. Lee, B.H. Oh, Aging and arterial stiffness, Circ. J. 74 (2010) 2257–2262.

[155] A. Karras, J.P. Haymann, E. Bozec, M. Metzger, C. Jacquot, G. Maruani, P. Houillier, M. Froissart, B. Stengel, P. Guardiola, S. Laurent, P. Boutouyrie, M. Briet, Nephro test study group, Large artery stiffening and remodeling are independently associated with all-cause mortality and cardiovascular events in chronic kidney disease, Hypertension 60 (2012) 1451–1457.

[156] J. Hashimoto, S. Ito, Aortic stiffness determines diastolic blood flow reversal in the descending thoracic aorta: potential implication for retrograde embolic stroke in hypertension, Hypertension 62 (2013) 542–549.

[157] D.W. Kitzman, D.M. Herrington, P.H. Brubaker, J.B. Moore, J. Eggebeen, M.J. Haykowsky, Carotid arterial stiffness and its relationship to exercise intolerance in older patients with heart failure and preserved ejection fraction, Hypertension 61 (2013) 112–119.

[158] E. Pringle, C. Philips, L. Thijs, C. Davidson, J.A. Staessen, P.W. de Leeuw, M. Jaaskivi, C. Nachev, G. Parati, E.T. O'Brien, J. Tuomilehto, J. Webster, C.J. Bulpitt, R.H. Fagard, Syst-EU investigators, Systolic blood pressure variability as a risk factor for stroke and cardiovascular mortality in the elderly hypertensive population, J. Hypertens. 21 (2003) 2251–2257.

[159] E. Pinto, Blood pressure and ageing, Postgrad. Med. J. 83 (2007) 109–114.

[160] I. Fleming, R. Busse, Molecular mechanisms involved in the regulation of the endothelial nitric oxide synthase, Am. J. Physiol. Regul. Integr. Comp. Physiol. 284 (2003) R1–R12.

[161] C.D. Reiter, X. Wang, J.E. Tanus-Santos, N. Hogg, R.O. Cannon III, A.N. Schechter, M.T. Gladwin, Cell-free hemoglobin limits nitric oxide bioavailability in sickle-cell disease, Nat. Med. 8 (2002) 1383–1389.

[162] R.P. Rother, L. Bell, P. Hillman, M.T. Gladwin, The clinical sequelae of intravascular hemolysis and extracellular plasma hemoglobin, J. Am. Med. Assoc. 293 (2005) 1653–1662.

CHAPTER

9

Cell and Tissue Destruction in Selected Disorders

9.1 INTRODUCTION

Many disease scenarios are accompanied by a more or less massive impairment of cell and tissue functions. In this chapter, the association between damage of biological structures and disease pathogenesis will be highlighted for selected disorders. It is not the attention to give a complete overview of all possible diseases. Numerous external and internal impacts can affect physiological parameters and lead to inflammatory and adaptive responses. Long-term chronic diseases are generally favored by impaired immune functions, immunosuppressive states, limitation and exhaustion of natural protective systems, and disturbances in key physiological parameters.

Normally in description of disease pathogenesis, typical features of a disorder are in the main focus and on how to treat this disease. Here, the main attention is directed on the role of cell and tissue destruction in the development and maintenance of chronic disease states. In this chapter, selected disorders are depicted, which can affect all stages of lifetime but dominate clearly in elderly persons. Chapter 10 is devoted to diseases leading to impairment of the kidney, liver, and spleen functions and finally to septic conditions and multiple organ failure.

9.2 DISTURBANCES OF THE CARDIOVASCULAR SYSTEM

9.2.1 Risk Factors for CardioVascular Diseases

The fluid tissue/blood delivers dioxygen and nutrients to all parts of our body. At the same time, carbon dioxide and metabolic and waste products are transported by blood to the lungs, liver, and kidneys, where they are released from the organisms. Furthermore, immune cells are supplied by circulating blood to inflammatory sites, and endocrine hormones are forwarded to target cells and tissues. Hence, an intact cardiovascular system is of utmost importance for normal functioning of all living functions in humans and animals.

Disturbances in blood vessels and heart function are widespread reasons for health problems, in particular in elder persons. Among common risk factors for cardiovascular diseases are physical inactivity, obesity, unbalanced diet, smoking, excessive alcohol consumption, air pollution, enhanced blood pressure, hyperglycemia, hyperlipidemia, and increased age. Here, the main focus is directed on those physiological factors, which provide the basis for disturbances in the cardiovascular systems. These factors are

Cell and Tissue Destruction
https://doi.org/10.1016/B978-0-12-816388-7.00009-7

increase in blood pressure, loss of blood vessel elasticity, enhanced intravascular hemolysis, internal hemorrhage, and inflammatory processes affecting blood vessels.

As a result, numerous disturbances arise in physiological cardiovascular functions with changed homeostatic parameters and manifold impacts on the metabolism of affected tissues. These impacts concern the development of mild, severe, or strong ischemic conditions in tissues, diminished blood flow through atherosclerotic vessels, and disturbances in wound healing.

9.2.2 Hypertension

Long-term hypertension is a very common risk factor for cardiovascular complications and diseases. Among them are coronary artery disease, stroke, heart failure, peripheral vascular disease, disturbance of heart rhythm, and functional impacts on other organs such as the brain, kidney, and eye [1–3].

Under normal conditions, blood pressure is well regulated via arterial and low-pressure baroreceptors acting as sensors and transmitting signals to the nervous and endocrine system [4]. The renin–angiotensin–aldosterone system is important for the long-term adjustment of arterial pressure and normal functioning of the kidneys [5]. The steroid hormone aldosterone contributes to maintenance of osmotic pressure in blood by regulating sodium and potassium concentrations ions via reabsorption or excretion of ions and water in the kidney [6].

Functioning blood pressure regulation is crucial for a laminar flow of blood through large blood vessels. Laminar blood flow safes energies and avoids disturbances of endothelial layers by turbulences and sheer forces. Atherosclerotic plaques are initially formed at locations, where turbulences in blood flow and an oscillatory sheer stress may arise [7,8].

In most cases of patients, primary or essential hypertension results from lifestyle factors (e.g., inactivity, smoking, excess weight, unbalanced diet) or genetic reasons. The so-called secondary hypertension is mainly a consequence of the kidney or endocrine disorders [9].

There are several reasons of disturbances in blood flow causing an increased blood pressure. Among them are the changes in the blood vessel wall and the reduced bioavailability of nitric monoxide. These alterations can lead to hemorrhage formation, intravascular hemolysis, local damages, and inflammations in the endothelium. Of course, fine mechanisms of these deteriorations are under discussion, and it is impossible to depict a precise mechanism for each form of damage.

9.2.3 Intravascular Hemolysis and Rhabdomyolysis

Numerous disease scenarios are associated with an enhanced intravascular hemolysis [10–13]. Common causes for increased intravascular hemolysis are genetic hemoglobinopathies [14–16], malaria [17,18], diabetes mellitus [19], and severe traumata [20,21].

Different physical and chemical factors such as osmotic stress, sheer stress, the presence of lytic poisons, some secreted microbial components from Gram-positive bacteria, autoantibodies, elevated oxidative processes in membranes of red blood cells, burn-associated necrosis, and hemorrhagic conditions promote release of hemoglobin from red blood cells [11–13]. Blood transfusion with stored red blood cells [19,22] and therapeutic procedures requiring extracorporeal assistance such as cardiopulmonary bypass and mechanical heart valve–induced anemia [12] are further reasons for increased intravascular hemolysis.

Damage of skeletal muscle cells is associated with release of myoglobin and other constituents of muscle cells into blood. Rhabdomyolysis can be caused by intense exercise of muscles, alcohol and illicit drug abuses, improper use of certain medications, infections, electrical injury, heat stroke, and prolonged immobilization [23,24].

On release, hemoglobin and myoglobin are rapidly oxidized by nitrogen monoxide. This diminishes the bioavailability of nitrogen monoxide with adverse effects on blood pressure and blood circulation [10,11,22]. In blood, both heme proteins are bound by haptoglobin and removed from circulation [13,25]. In case of haptoglobin diminution or exhaustion, unbound hemoglobin and myoglobin are subjected to glomerular filtration and reabsorption and contribute to kidney damage [13,26–28].

Oxidized free hemoglobin and myoglobin are the source for free heme (ferric protoporphyrin IX) [29,30], a highly cytotoxic molecule (see Section 4.4.2) that forms a tight complex with hemopexin, and is removed as heme–hemopexin complex from circulation [31].

Excessive hemolytic and rhabdomyolytic events are very problematic as the protecting haptoglobin and hemopexin are rapidly exhausted and cytotoxic free heme accumulates. Free heme is involved in the pathogenesis of atherosclerosis and oxidative damage of lipoproteins [32,33], kidney, and liver toxicity [22,34,35] and proinflammatory activation of endothelial cells via toll-like receptor 4 (TLR4) signaling [36,37]. However, free heme is a weak initiator of inflammation via TLR4 in contrast to LPS [38]. These authors negate an inflammation-promoting role of free heme under in vivo conditions. A very dangerous property of free heme is the induction of novel hemolytic events from unperturbed red blood cells [39,40].

Free heme is able to activate the alternative complement pathway during the acute phase of malaria and other hemolytic disorders [41–43]. On endothelial cells, this interaction induces the release of the prothrombotic von Willebrand factor [43].

9.2.4 Internal Hemorrhage

Rupture of small blood vessels can lead to internal hemorrhage at various locations in our body. The loss of large amount of blood is a serious medical emergency and can cause hemorrhagic shock, cardiac arrest, and death [44]. Here, the focus will be directed on potential consequences of small, localized internal bleedings, which remain often unrecognized. In particular, those bleedings will be considered where the escaped blood remains in circumjacent tissue such as intracranial hemorrhage [45]. A special form of hemorrhage is the intratumor hemorrhage (Section 9.8.4).

Among reasons for internal bleeding are high blood pressure, deficient blood clotting, traumata, aneurysms, surgery, alcohol, and drug abuse. The most common risk factor of intracerebral hemorrhage is the uncontrolled hypertension [46,47]. With increasing age, a progressive loss of elasticity of blood vessels and reduced arterial compliance are reported [48]. Moreover, deficient lysine and proline hydroxylation in collagen weakens collagen linkage to vessel walls and contributes to diminished resistance of vasculature as seen in ascorbate deficiency [49–52].

In newly formed internal hemorrhage areas, the blood is not flowing and loses dioxygen and glucose with increasing time. Subsequently, naturally protecting systems are exhausted. As a result, intravascular hemolysis of red blood cells is observed. Finally, the cytotoxic free heme will be formed after exhaustion of haptoglobin and hemopexin. This molecule can affect neighbored cells, leading to increased oxidative processes and finally to necrotic cell death in near surroundings. Small hemorrhagic infiltrations remain often undetected. The great danger of these infiltrations is the induction of numerous unrecognized destructions in adjacent tissues.

Nonspecific bleeding can also be associated with vitamin C deficiency in patients with plasma levels of ascorbic acid lower than $0.6\ \text{mg mL}^{-1}$ [53]. The risk of low vitamin C levels increased with advanced age, prolonged illness, heavy smoking and ethanol drinking, poor diet, psychiatric disturbances, in dialysis patients, or after surgery [53–55]. Vitamin C deficiency impairs wound healing as well [56]. As cofactor of lysyl and prolyl hydroxylases, ascorbic acid contributes to stability of collagen and adjacent blood vessels [51,57].

9.2.5 Atherosclerosis

In atherosclerotic arteries and arterioles, blood flow is progressively disturbed due to plaque formation on the vessel wall and diminished lumen. In plaque formation, the involvement of lipoproteins, endothelial cells, immune cells, and others has been described in numerous publications [58–62]. Without any doubt, processes of lipid peroxidation in lipoproteins and plaque cells play a role in the development of plaque deposits [63]. Nevertheless, it remains unknown how atherosclerosis is initiated.

Among the hypothesis about the molecular reason of atherosclerosis are disturbances in the lipoprotein metabolism. Excess of cholesterol, high level of low-density lipoprotein, and diminished level of lipid antioxidants contribute to atherosclerosis [64]. This hypothesis is supported by the fact that in rare case of hereditary alterations in cholesterol metabolism these patients are prone to develop atherosclerosis [65].

Another hypothesis pays attention on enhanced systemic blood pressure and the resulting turbulences in some regions of aorta and arteries. These oscillating turbulences mediate a shear stress and can damage the endothelial layer of blood vessels and attract, thus, immune cells. This hypothesis is favored by the fact that atherosclerotic lesions are predominantly located in regions where turbulences are more likely [7,8].

As immune cells are progressively attracted to endothelial lesions, most of all neutrophils as cells of the first line of defense, products of these cells are assumed to be involved in plaque formation. Special attention has been focused on myeloperoxidase, which is released from these cells [66]. Myeloperoxidase easily attaches to negatively charged surface epitopes. It produces hypochlorous acid and promotes numerous substrate oxidations. Moreover, active myeloperoxidase was detected in atherosclerotic plaques [67,68].

Free heme is another likely candidate to be involved in initiation of atherosclerotic plaques. This hydrophobic agent activates endothelial cells via TLR4 signaling [37]. Otherwise, this molecule is easily inserted into hydrophobic areas of lipoproteins, membranes, and proteins, where it can catalyze oxidative reactions [32,33]. This hypothesis about a central role of free heme in the pathogenesis of atherosclerosis is well linked to the other aforementioned reasons. The formation of free heme is closely interconnected to the diminished bioavailability of NO and the enhancement of blood pressure [11]. Moreover, free heme contributes to oxidative processes, endothelial cell damage, and immune cell activation (Section 4.4.2).

Another aspect merits attention. Wild-type mice do not develop atherosclerosis. To use murine models of atherosclerosis, mice strains with special genetic defects or transgenic animals have been developed such as apolipoprotein E-deficient mice or ApoE*3-Leiden transgenic mice [69,70]. The reasons for failure of wild-type mice to develop atherosclerosis remains unknown at present. There are fundamental differences in regulation of blood flow between man and mouse [71]. Under stress situations, mice can upregulate the NO production by inducible NO synthase by three to four orders of magnitude, t.e. up to 10,000-fold [72,73]. In humans, there is only a small upregulation of NO production. Further, in mice, both the heme-protecting proteins haptoglobin and hemopexin are acute-phase proteins. In humans, only the haptoglobin level is two- to threefold enhanced under stress. Human hemopexin is not an acute-phase protein [74]. Thus, mice have a much greater potential to resist against consequences of intravascular hemolysis. A further aspect is also highly important. In humans, the TLR4 receptor on endothelial and other cells responds very sensitive against its ligand lipopolysaccharide (LPS). By this mechanism, a low burden of LPS released from Gram-negative bacteria can be recognized. Mice cells are much more insensitive against LPS by about one order of magnitude. It is unknown if this different responsiveness of human and murine cells against LPS is also valid for other TLR4 ligands such a free heme.

9.2.6 Ischemia, Thrombosis, and Total Blockade of the Blood Flow

Diminished supply of dioxygen and nutrients leads to ischemic states in tissues. Numerous adaptations are known to mild stages of hypoxia (see Section 4.6.3). Clinical problems may arise from reperfusion after ischemia, leading to cell destruction due to enhanced formation of reactive species (Section 4.6.5). Complications result also from severe and prolonged hypoxic states that are accompanied by necrotic cell death.

A critical situation can occur in patients prone to thrombophilia. In combination with disturbed blood flow and endothelial cell injury, the danger of development of thrombosis increases [75]. Activation of the coagulation system, platelet adherence, erosion or rupture of an atherosclerotic plaque, major surgery, increasing age, and prolonged traveling have an enhanced risk of thrombosis in addition to the common risk factors for cardiovascular diseases [76,77].

Thrombosis may concern both veins and arteries. Characteristic features of arterial and venous thrombosis are given in Table 9.1. While arterial thrombosis damages less-supplied tissues by ischemia and necrosis, venous thrombosis leads mainly to congestion [78,79].

Full blockade of blood vessels is a serious clinical incidence with major implications to patients. A blockade of blood vessels can principally occur at all places in the organisms. Preferred appearances are cardiac infarction and stroke in brain vessels for arterial thrombosis and pulmonary embolism and deep venous thrombosis for blockade of venous vessels.

Arterial thrombi are derived from ruptured atherosclerotic plaques. They are rich in platelets [80]. In therapy of acute thrombotic events, the main target of administered drugs is platelets. In patients prone to cardiovascular disease, these medications are also used prophylactically [81]. Cyclooxygenase inhibitors, ADP receptor antagonists, inhibitors of protease-activated receptor-1, integrin inhibitors, and activators of fibrinolysis are applied as antiplatelet drugs [82]. Insertion of drug-eluting stents in the occluded coronary artery is also used in patients with coronary artery disease [83]. A further

TABLE 9.1 Characterization of Arterial and Venous Thrombosis.

Feature	Arterial Thrombosis	Venous Thrombosis
Origin of thrombus	Rupture of atherosclerotic plaque	Changes in blood composition, diminished blood flow, vessel wall alterations
Consequences of thrombus formation	Ischemia, necrosis	Congestion, in severe cases, ischemia and necrosis
Thrombus constituents	Rich in platelets	Rich in fibrin and red blood cells
Clinical manifestation	Myocardial infarct, stroke	Deep vein thrombosis, pulmonary embolism
Drug therapy	Antiplatelet drugs, anticoagulants	Anticoagulants

therapeutic option is the use of anticoagulants such as vitamin K antagonists such as warfarin and heparins [84].

Deep vein thrombosis occurs predominantly in the large veins of the legs. If a venous thrombus travels to the lungs, it plugs a pulmonary artery and causes pulmonary embolism. Anticoagulants are applied to protect and treat deep vein thrombosis and pulmonary embolism [85].

9.2.7 Disturbances of Wound Healing

Wound healing is a complex dynamic process to restore skin and adjacent tissue after injury. This process can be subdivided into the four overlapping phases: hemostasis, inflammation, proliferation, and tissue remodeling [86,87]. All events of each phase should occur well-regulated and within a specific time frame.

Any disturbances in these processes lead to a delayed wound healing or nonhealing chronic wounds such as ulcers. Reasons of impaired healing of wounds can be manifold. Often the inflammatory stage lasts too long and is insufficiently resolved [88]. Among factors affecting wound healing are disturbed blood circulation, neuropathy, immunosuppression, high age, physical inactivity, smoking, alcohol consumption, infections, obesity, diabetes, and others [89]. With increasing age, all phases of wound healing can be disturbed leading to an increased risk of inflammatory reaction and bacterial colonization [90,91].

Persistence of these wounds is favored by prolonged hypoxia, insufficient perfusion, inadequate angiogenesis, hyperglycemia, neuropathy, and disbalance of reactive species and matrix metalloproteases [92–95]. Majority of chronic wounds can be divided into venous ulcers, pressure ulcers, and diabetic ulcers, and a small group of wounds resulting from arterial ischemia [96,97]. Foot ulcers of diabetic patients are a typical example of nonhealing chronic wounds (see Section 9.3.7). Disturbances in blood flow, ischemia, and hypoxia in adjacent tissues are common to all kinds of chronic wounds [96,98]. From this background, it is not surprising that chronic wounds are associated with a prolonged inflammatory state [99], dysregulated functions of neutrophils [100], and bacterial colonization [101]. These factors further delay wound healing. Predisposition to scarring and neoplastic progression is enhanced in retarded healing of ulcers [99,102].

In chronic wounds, the tight balance between production and degradation of molecules is shifted toward degradation processes. Chronic wounds are rich in elastase from invading neutrophils [103,104]. Several wound dressing materials were developed to inhibit elastase and matrix metalloproteases in chronic wounds [105–107]. In chronic ulcera, migration of neutrophils is disturbed. They are scattered around the wound [100,108].

To sum up, in chronic ulcers, the inflammatory state is insufficiently resolved. As outlined in Section 6.3.7, an inflammatory response is associated with release and generation of cell and tissue-damaging agents by activated immune cells. Their impact on destruction increases with a diminished efficiency of protective mechanisms. Thus, persistence of chronic wounds is mainly favored by already exhausted or low-capacity protecting systems of the host.

One serious problem is the accumulation of noncirculating blood in the wound region. Similar to hemorrhages, this blood undergoes alterations in its physicochemical properties with the most serious consequences of release of a sufficient percentage of hemoglobin from red blood cells. Subsequently, the protecting haptoglobin and hemopexin are exhausted at these loci, and free heme accumulates. Of course, these processes occur more intensively in patients, which are already prone to intravascular hemolysis like in patients having severe diabetes mellitus. Free heme is a potent inducer of novel hemolytic events in yet unperturbed red blood cells [39,40] and contributes to oxidative damage in lipid matrices. This mechanism can also play a role in disturbed wound healing.

9.2.8 Vasculitis

Destruction of blood vessels by inflammatory agents takes place in vasculitis. This group of disorder can affect arteries and veins. Mainly neutrophils, but also leukocytes (eosinophils, macrophages, lymphocytes) contribute to vessel damage.

In relation of the affected vessels, these disorders are classified into large vessel vasculitis, medium vessel vasculitis, and small vessel vasculitis [109]. Further categories of this consensus classification are variable vessel vasculitis, single-organ vasculitis, and vasculitis associated with systemic disease or probable etiology [110].

Here, I will stress only the antineutrophil cytoplasmic antibody (ANCA)–associated vasculitis, which represents a small vessel vasculitis. In ANCA-associated vasculitis, autoantibodies against the neutrophil enzymes myeloperoxidase (MPO-ANCA) and proteinase 3 (PR3-ANCA) are formed [111]. As both proteins become expressed on the surface of activated neutrophils, these antibodies are directed against neutrophils attached to the vessel epithelium.

As a result of these attacks, necrotizing vasculitis develops. Prominent forms of ANCA-associated vasculitis are glomerulonephritis [112] and vasculitis of the upper and lower respiratory tract [113].

Conventional therapy of ANCA-associated vasculitis with cyclophosphamide and long-term treatment with glucocorticoids have numerous side effects [114]. Moreover, relapses of the disease are common. Depletion of antibody-producing B cells by rituximab is a further strategy in therapy of this disease [114,115].

9.3 DIABETES MELLITUS AND COMPLICATIONS

9.3.1 Disturbed Glucose Metabolism

Potential reasons for the development of diabetes mellitus are addressed in Sections 4.7.5–4.7.9. The same holds for the development of insulin resistance. Here, the main focus is directed on complications of diabetes mellitus and disease scenarios that result from disturbed metabolism in diabetic patients. The main reason for the development of these pathologies is the enhanced glucose level that disturbs the cell metabolism in noninsulin-dependent cells. Among these cells are red blood cells, endothelial cells, and neurons. These cells are equipped with insulin-independent transporters for glucose uptake [116–118].

Several dietary factors promote insulin resistance and favor development of diabetes mellitus such as elevated blood and tissue levels of free fatty acids and triglycerides, excess body weight, high dietary levels of fructose or sucrose, and long-lasting physical inactivity [119,120].

9.3.2 Hyperglycemia and Red Blood Cells

The accumulation of glucose in red blood cells under hyperglycemic conditions favors not only the formation of glycated products such as hemoglobin A1c (HbA1c) and others but also the formation of sorbitol by aldose reductase [121,122]. In contrast to glucose, sorbitol cannot leave the cell. Increased sorbitol level disturbs the cellular osmolarity and enhances the stiffness of red blood cells [123]. These cells have a reduced ability to resist to mechanical deformations, a property that is highly dispensable for passage of red blood cells through narrow capillaries [124–126]. Moreover, in diabetic microangiopathy, a thickening of basal membrane and hence reduction of capillary diameter occur

[127]. This further disturbs the blood flow and contributes to occlusion and ischemia.

Osmotically stressed red blood cells are prone to intravascular hemolysis [125]. There is a clear relationship between hemolysis degree and fasting plasma glucose concentrations as well as the HbA1c value in patients suffering from diabetes mellitus type 2 [128].

Hyperglycemic red blood cells are prone to intravascular hemolysis. In consequence, the bioavailability of nitrogen monoxide is disturbed, the blood pressure increases, and tissues are less supplied with dioxygen and nutrients. In severe cases, cytotoxic free heme is formed. A dreadful activity of free heme is the induction of novel hemolytic events [39,40]. This amplifies substantially any toxic effects of this agent. For example, free hemoglobin is known to contribute to kidney insufficiency (see Section 10.1.4).

Although serum levels of HbA1c is a widely used indicator for the mean glucose level during the last 3 months, this value highly depends on the lifetime of red blood cells being on average 120 d. Diabetes patients with accompanying disease scenarios with a shortened lifetime of these cells show often low HbA1c levels. This underestimation of HbA1c values is found in patients with cirrhotic liver [129,130], hemoglobinopathies [131–133], severe hemolysis and hemolytic anemia [134–136], splenomegaly [137], chronic renal failure [138,139], and after acute blood loss [140]. Otherwise, falsely enhanced HbA1c values may occur under conditions that prolong the lifetime of red blood cells or decrease the turnover of these cells [137]. Prolonged iron deficiency anemia results, for example, in enhanced HbA1c values [141].

9.3.3 Hyperglycemia and Endothelial Cells

On hyperglycemia, vascular endothelial cells change their properties to an inflammatory phenotype [142,143], resulting in accumulation of different advanced glycation end products and formation of reactive species [144,145]. Moreover, free heme promotes signaling over TLR4 receptors in these cells [37]. These alterations are accompanied by reduced bioavailability of nitric monoxide, decreased vasorelaxation, impaired fibrinolytic activities, and reorganization of cellular cytoskeleton [143]. A thickening of the capillary endothelial cells' basal lamina is observed in diabetes patients [146–148]. Taken together, hyperglycemia promotes a worsening of endothelial and vascular functions.

Endothelial cells exposed to high levels of glucose enhance the production and release of the von Willebrand factor, which is able to initiate thrombotic events [149]. In patients with diabetes mellitus, enhanced blood levels of von Willebrand factor are highly predictive of adverse cardiovascular events [150,151].

Disturbance of endothelial functions by high glucose levels is associated with enhanced thrombotic events such as acute coronary syndrome, stroke, and peripheral vascular disease, which often lead to a lethal outcome among diabetic patients [152].

9.3.4 Hyperglycemia and Neurons

Pronounced hyperglycemia in association with critical illness leads to activation of microglia, followed by impaired astrocyte functions, and neuronal damage in frontal cortex and hippocampus. These data were obtained from analysis of nonsurviving critically ill patients and animal experiments [153].

In patients with elevated glucose levels, the risk raises for development of Alzheimer's disease and dementia [154–156]. Hyperglycemia favors an increase of monomeric amyloid β (Aβ) in the hippocampal interstitial fluid and enhanced lactate levels indicating an increase of neuronal activity. These alterations were mediated by closure of ATP-sensitive potassium channels [157,158].

9.3.5 Diabetes Mellitus and Its Complications

Many disease complications are reported for diabetic patients. Typical disorders are retinopathy, nephropathy, neuropathy, and diabetic foot syndrome. Late consequences of these diseases are fatal to patients. Kidney dysfunction can result in the need of dialysis.

Vision can be disturbed with retinal detachment due to microvascular alteration and retinal ischemia. Apparently, altered rheological properties of both leukocytes and red blood cells contribute to disturbance of retinal blood flow [126,142,159,160]. In strong cases of retinopathy, blindness results [127,161].

In diabetic patients, the pain sensing can be disturbed especially in those regions that are far from the brain [162,163]. Most of all feet are subjected by neuropathy. On this basis, some small injuries on feet are badly recognized, and the development of chronic wounds is favored in diabetic foot syndrome. The combination of neuropathy, chronic foot ulcers, and disturbed blood circulation is often the reason for below-knee amputations (Section 9.3.7).

Although the molecular reasons for these complications are under discussion, it is strikingly that in all diabetes-associated pathologies, disturbed functions of red blood, endothelial, and other cells are involved, which accumulate glucose in an uncontrolled fashion.

9.3.6 Therapeutic Options

Because of the enormous distribution of diabetes, numerous therapeutic options exist. Mild forms of diabetes mellitus type 2 are treated with different antidiabetic drugs, while insulin administration is the therapy of choice for diabetes mellitus type 1 and severe forms of type 2 diabetes. These therapies are directed to reduce and maintain blood glucose and HbA1c values on a low, safety level. In addition, consequent diet, regular physical activity, reduction of body weight, and a general healthy lifestyle are highly recommended to avoid and delay the onset of comorbidities.

It is not the objective of this book to describe the existing therapeutic options in the treatment of patients with diabetes.

9.3.7 Foot Ulcers in Diabetic Patients

Diabetic foot syndrome is a common comorbidity of diabetes mellitus characterized by persistent wounds, further complications such as gangrened toes and osteomyelitis, as well as a high risk for below-knee amputation [164]. In close association with a neuropathy, angiopathy, hypoxia, and disturbed immune functions, small wounds or pressure points on feet remain often unrecognized and lead to development of ulcera [165,166]. These wounds are resistant against conventional therapies.

In therapy of nonhealing foot ulcera of diabetic patients, it has been demonstrated that the intravenous administration of the chlorite-based isotonic drug solution WF10 significantly improves the wound healing in affected patients [167,168]. In addition, enhanced glucose and HbA1c levels are significantly reduced to nonpathological levels under adjuvant application of WF10 [169]. As free heme is inactivated by chlorite and loses its ability to initiate novel hemolytic events in unperturbed red blood cells [40], this drug prevents efficiently cytotoxic activities of free heme. A replacement of senescent, dysfunctional red blood cells, which are enriched with glucose, by newly produced red blood cells is also discussed as a reason for improvement of blood parameters in WF10-treated patients [169]. These data indicate that the administration of WF10 diminishes cytotoxic effects resulting from hemolysis together with improvement of microcirculation and glucose utilization in affected tissues. This approach should also be useful to other comorbidities of diabetes mellitus.

9.4 NEURODEGENERATIVE DISORDERS

9.4.1 Overexcitation of Cells

The progressive loss of vital brain cells observed in several pathologies can result from overexcitation of these cells. Several molecular scenarios are known supporting this hypothesis.

Glutamate, a major excitatory neurotransmitter, is a known excitotoxin. Increased extracellular glutamate concentrations result from insufficient removal of glutamate from synaptic cleft and after brain or spinal cord injury [170,171]. Ischemia is another condition, which promotes the release of glutamate and aspartate from cells.

Increased extracellular levels of glutamate can excessively stimulate glutamate receptors. These receptors are spread around glutaminergic synapses on the presynaptic and postsynaptic terminals and adjacent astrocytes [172,173]. The latter cells are known to release glutamate as well [174]. Furthermore, oligodendrocytes and oligodendroglia also express glutamate receptors [175].

Activation of neuronal receptors by excessive glutamate raises the Ca^{2+} influx above critical values [176–178]. Increased intracellular Ca^{2+} level activates mitochondria, enhances the generation of reactive species, depletes ATP, promotes stress to endoplasmic reticulum, upregulates neuronal nitric monoxide synthase, favors the release of lysosomal enzymes, and drives cells into apoptosis [179,180]. In addition, excess glutamate inhibits the cystine uptake by cells on reversal of the cystine–glutamate antiporter [181].

Control over extracellular level of glutamate and other excitatory amino acids is achieved by special amino acid transporters that clear these molecules from the extracellular space [182]. Transporters of the excitatory amino acid transporter family are localized in the plasma membrane of neurons and glial cells. These antiporter systems are driven by an electrochemical gradient of sodium ions [183,184]. The cystine–glutamate antiporter is another type of plasma membrane glutamate transporters [185]. Vesicular glutamate transporters move glutamate from cell cytoplasm into synaptic vesicles. These transporters depend on a proton gradient [182,186]. The concerted activity of plasma membrane and intracellular glutamate transporters allows an efficient recycling of this neurotransmitter for repeated release from synapses [187].

Glutamate-induced excitotoxicity plays a role in Alzheimer's disease, Parkinson's disease, Huntington's disease, amyotrophic lateral sclerosis, and multiple sclerosis [188,189].

9.4.2 Macroautophagy and Fibrillation in Neurodegenerative Diseases

Accumulation of fibrillated proteins is a hallmark of neuronal tissue in Alzheimer's and other neurodegenerative diseases (see Section 5.6.6). Molecular mechanisms leading to fibrillation are largely unknown. It is also unclear if the formation of amyloid fibrils is the reason or consequence of neurodegenerative disorders.

In Alzheimer's disease, Aβ is the main component of large extracellular deposits in the brain [190,191]. Aβ is derived from processing of the amyloid precursor protein (APP), an integral protein found in plasma and endosomal membranes. Other amyloid fibrils such as mutant huntingtin, α-synuclein, or phosphorylated τ protein are derived from cytosolic proteins. In neurodegenerative pathologies, all neurotoxic peptide fragments result from processing via the endosomal–lysosomal pathway or the autophagic pathway of removal of dysfunctional cell organelles. Both pathways converge at the final lysosomal stage [192].

The macroautophagic pathway is highly important for removal of dysfunctional proteins and damaged organelles in hepatocytes, neurons, and myocytes [193–196]. It is the only known mechanism for degradation of large aggregates and organelles in eukaryotic cells. In neurodegenerative pathologies, the autophagic pathway does not function properly [197–199].

Impaired lysosomal clearance contributes to the accumulation of Aβ and thus to the pathogenesis of Alzheimer's disease [200,201]. On mouse models of Alzheimer's disease, defective lysosomal clearance was evaluated by selective inhibition of lysosomal enzymes [202,203]. Disturbances in autophagy also play a role in Parkinson's disease [204–206].

Intriguingly, activation of macroautophagy in cultured cells by the disaccharide trehalose diminishes the formation of cytosolic aggregates derived from huntingtin [207,208], α-synuclein [207,208], and τ-protein [208–210]. Similarly, trehalose reduces the secretion of Aβ and decreases the degradation of APP [211].

In experimental models of Parkinson's disease, further agents have been applied to stimulate autophagy such as metformin, rapamycin, lithium, valproate, carbamazepine, and dietary polyphenols [212]. The most serious problem in translating these findings into clinical practice is the limited selectivity of these agents [212]. In addition, macroautophagy is also involved in tumor pathogenesis by tumor-suppressing or tumor-promoting activities [213,214]. Thus, stimulation of autophagy can induce detrimental effects. A concerted balance between protective and detrimental effects is therefore of utmost importance for potential therapeutic applications of autophagy-stimulating substances [215].

9.4.3 Alzheimer's Disease: State of Art

Alzheimer's disease is characterized by a progressive loss of neurons and synapses in the cerebral cortex and other brain regions. This widespread disease is associated with gradually worsening of cognitive functions and development of dementia. Despite intense research, no consensus exists about molecular reasons. In addition, at present, it is impossible to stop or reverse this disease by drugs. Many proposed medications against Alzheimer's disease failed.

Known risk factors for Alzheimer's disease are genetic predisposition, advanced age, head injuries, depression, hypertension, diabetes, smoking, hypercholesterolaemia, physical and mental inactivity, unsocial lifestyle, and unbalanced diet [216–218].

Numerous hypotheses have been formulated in the past to explain typical features of this disease. According to the cholinergic hypothesis, decline in the neurotransmitter acetylcholine provides the basis of this disease. However, any drugs to treat this deficiency have no or only small effects [219,220]. Currently, the acetylcholinesterase inhibitors tacrine, rivastigmine, galantamine, and donepezil are applied to improve cognitive problems, however, with only marginal benefit [221]. The drug memantine, a noncompetitive *N*-methyl-D-aspartate (NMDA) receptor antagonist, inhibits the overstimulation by glutamate and acts as noncompetitive antagonist at nicotinic acetylcholine and serotonin (5-hydroxytryptamine)-3 receptors. Its benefit is small in the therapy of Alzheimer's disease [222,223].

Accumulation of Aβ and hyperphosphorylated τ protein in affected brain regions are central items in the amyloid hypothesis and tau hypothesis [224,225]. Despite the proposed role of oligomeric Aβ or Aβ plaques in the pathogenesis of Alzheimer's disease, no clinical improvement is found by drugs that inhibit β- or γ-secretases, the responsible enzymes for Aβ formation.

Further hypothesis about Alzheimer's disease concern the poor functioning of the blood–brain barrier [226], disturbed homeostasis of metal ions [227], and dysfunctions of oligodendrocytes [228,229].

9.4.4 Alzheimer's Disease: Accumulation of Amyloid β

In nervous system, APP is crucial for synapse development, maintenance at the neuromuscular junction, and calcium homeostasis for synaptic transmission [230,231]. This protein is also involved in iron and copper regulation [232,233]. An increased expression of APP under hypoxic and ischemic conditions is thought to act neuroprotective [234,235]. Otherwise, this upregulation can result in enhanced Aβ formation by concomitant activation of β-secretase or reduced levels of neprilysin and endothelin-converting enzyme-1, both of which degrade amyloid [235,236].

Intracellular degradation of APP starts either with the action of α-secretase or β-secretase followed by γ-secretase. All secretases are localized at different loci within cells [237]. The dominating neuroprotective α-secretase pathway yields nonamyloidogenic fragments. α-Secretase acts predominantly on APP in the plasma membrane, whereas β-secretase cleaves this protein in early endosomes [237]. Huntingtin-associated protein 1 interacts with APP and favors traffic of this protein toward plasma membrane [238].

Aβ, which consists mainly of 40 or 42 amino acid residues, results from sequential cleavage of β- and γ-secretase on the precursor protein [239]. The Aβ42 form is more fibrillogenic and thus related to Alzheimer's disease [240]. Human and mouse Aβ differ in three amino acid positions making the human variant more prone to formation of fibril aggregates. In human Aβ, Arg (instead of Gly in the mouse peptide), Tyr (for Phe), and His (for Arg) are found at positions 5, 10, and 13, respectively [241]. His13 can bind zinc, which is required for initiation of fibrillogenesis [242]. Moreover, mouse APP is better protected by its sequence against β-secretase activity [243].

Soluble Aβ, mainly as Aβ40, is found in plasma, cerebrospinal fluid, and brain interstitial fluid [244]. Soluble Aβ oligomers can interact with membranes and form apparently ion channels in membranes allowing an unregulated Ca^{2+} influx into neurons [245,246].

9.4.5 Alzheimer's Disease: Oxidative Stress and Amyloid β-Heme Complexes

Several lines of evidence show that affected regions in brain of Alzheimer's patients are subjected to an enhanced oxidative stress. In Alzheimer's disease, there is an increased expression of heme oxygenase 1 especially in hippocampal and cortical neurons and astrocytes [247,248]. Hence bilirubin and iron, products of heme oxygenase 1, are also enhanced in patient's material [249,250]. This indicates an enhanced heme release and degradation. Although thought to be an antioxidant protein, heme oxygenase exhibits also prooxidant activities by release of catalytically active iron when internal iron stores are overloaded [251,252]. Heme oxygenase 1 promotes macroautophagy, a bulk degradative pathway for demolition of damaged organelles in lysosomes [253]. In Alzheimer patients, the mitochondrial complex IV declines [254,255]. Some proteins of this complex contain the unique heme form heme-α, which is rate-limiting to the formation of complex IV [256]. Free heme is apparently involved in inhibition of muscarinic acetylcholine receptors in Alzheimer's disease [257,258].

Human Aβ and free heme tightly bind to each other [259–261]. Heme binding is more intense with human Aβ than with the rodent form [261]. Inside cells this complex formation is assumed to deplete exchangeable heme (also known as regulatory heme), a process that induces heme synthesis and iron uptake [260,262]. The aggregation of Aβ is accelerated in vitro by copper, zinc, and iron ions [263]. Copper and zinc ions are found in Aβ amyloid deposits taken from postmortem brain tissue [264,265]. The metal ion chelator clioquinol, which can permeate through the blood–brain barrier, inhibits both Aβ aggregation and plaque formation in a transgenic human APP mouse model of Alzheimer's disease [266,267].

Heme–Aβ interaction inhibits the in vitro aggregation of Aβ40 and Aβ42 [268]. Otherwise, the human Aβ–heme complex exerts a weak peroxidase-like activity [260,269,270]. In the presence of H_2O_2, this complex oxidizes serotonin and 3,4-dihydroxyphenylalanine [260]. Oxidation of Tyr-10 in human Aβ–heme complexes or by metal ions contributes to formation of stable dityrosine linkages between adjacent Aβ molecules. This oxidation is favored by copper ions and acidic pH values [270–274]. Dityrosine formation promotes aggregation of Aβ via stabilization of peptide dimers [274,275].

In summary, formation of human Aβ aggregates in Alzheimer's disease is most likely favored by oxidative processes, the presence of catalytically active metal ions, and disturbances in the removal of defective cell organelles by autophagy.

9.4.6 Parkinson's Disease

Death of neurons in substantia nigra affects the motor system and dopamine secretion in patients with Parkinson's disease [276]. In addition, astrocytes die also in substantia nigra, whereas the number of microglia cells increases [277]. Dopamine depletion is associated with hypokinesia [278].

In diseased neurons, so-called Lewy bodies accumulate. They are composed of insoluble fibrillar α-synuclein bound to ubiquitin [276,279]. Again, it is unknown whether toxicity originates from aggregates or from oligomers of α-synuclein.

α-Synuclein is predominantly expressed in substantia nigra and some other regions. It is found in the presynaptic termini, where it is essential to normal functioning of neurotransmission [206]. Autophagy and the ubiquitin proteasomal system are involved in the clearance of α-synuclein. Defects in these mechanisms have been associated with disease progression [204–206]. Moreover, overexpression of wild-type α-synuclein inhibits autophagy in mammalian cells and transgenic mice [280].

9.4.7 Huntington's Disease

In Huntington's disease, aggregated huntingtin protein causes death of brain cells that leads to lack of coordination, unstable gait, decline of mental abilities, and development of dementia [281]. Exact functions of huntingtin are unclear for brain cells. Apparently, it plays a functional role in cytoskeletal assembling, vesicle trafficking, and endocytosis [282–284].

Huntington's disease is associated with mutants of the gene, which codes for the protein huntingtin. In this gene, the DNA base sequence cytosine-adenine-guanine is repeated multiple times. This base sequence encodes the amino acid glutamine. If the number of these repeats is too high in the gene, an unstable protein is produced that promotes mitochondrial dysfunction and oxidative stress [285–287]. Huntingtin proteins with elongated polyglutamine tract are susceptible to aggregation [288]. Mutant huntingtin aggregates increase in cells when autophagy is impaired [289].

9.4.8 Amyotrophic Lateral Sclerosis

The death of neurons controlling voluntary muscle activity occurs in patients with ALS. Stiff muscles, muscle twitching, weakness, difficulty in speaking, swallowing, and breathing are symptoms and complications of this progressive paralytic disease [290].

In cytoplasm of motoneurons of patients with ALS, inclusion bodies are found composed of aggregated misfolded proteins. Mutations in distinct genes are responsible for formation of these deposits. These genes can be clustered in three categories, protein homeostasis, RNA homeostasis, and cytoskeletal dynamics [290,291]. The most common mutated gene is *C9orf72* that encodes a protein, which is involved in nuclear and endosomal membrane trafficking and autophagy. In this mutated gene, a noncoding hexanucleotide repeat is largely expanded [292].

Despite extensive research, only the drugs riluzole and edaravone slightly improve the survival in ALS patients [293,294].

9.4.9 Multiple Sclerosis

In multiple sclerosis, the insulating myelin cover of nerve cells is progressively damaged in the brain and spinal cord. This disturbs the transition of action potentials along axons and leads to numerous neurological symptoms. Failures in immune response and defects in myelin-producing cells are thought to be the underlying mechanisms of this disease [295,296]. Genetic and environmental factors and viral infections are discussed as potential reasons of multiple sclerosis [297,298].

Demyelination and inflammation are typical features of multiple sclerosis [299]. $CD8^+$ T cells contribute to these lesions after entering the defective blood–brain barrier [300]. They recognize myelin as foreign substance. The ongoing inflammation disturbs the myelin sheath and irreversibly damages axons [301].

Several oral drugs and monoclonal antibodies are used as medication in multiple sclerosis [298].

9.5 AUTOIMMUNE DISORDERS

9.5.1 Formation and Cytotoxic Action of Autoantibodies

The formation of autoantibodies against host's own proteins and other components is serious problem in autoimmune diseases (Sections 6.7.1 and 6.7.2). About 100 different disease scenarios are known, in which autoantibodies are involved [302]. Common autoimmune diseases are diabetes mellitus type 1, multiple sclerosis, rheumatoid arthritis, systemic lupus erythematosus, celiac disease, inflammatory bowel disease, psoriasis, and celiac disease.

During their maturation, self-reactive T cells and B cells are eliminated by central tolerance, peripheral tolerance, and the action of T regulatory cell–derived cytokines (Section 6.6.7). Some self-reactive B and T cells, which can escape these tolerance mechanisms and being normally unable to induce autoimmune processes, can be triggered by genetic predisposition or environmental factors to be involved in autoimmunity reactions [303].

Once formed, autoantibodies can be involved in tissue damage by antibody-dependent cell-mediated cytotoxicity, complement activation, formation of immune complexes, activation or blocking cell surface receptors, and binding to extracellular molecules [304–306]. In addition, autoreactive T cells contribute also to tissue damage [307].

9.5.2 Rheumatoid Arthritis

In patients with rheumatoid arthritis, joints of hands and other parts of the body are typically swollen, warm, and painful. Sometime, other organs are affected [308].

Several autoantibodies are found in patients with rheumatoid arthritis. Rheumatoid factor is an autoantibody that is directed against the Fc region of human immunoglobulin G [309]. These antibodies are also found in other autoimmune diseases such as systemic lupus erythematosus or Sjögren syndrome [310].

Other antibodies are directed against citrullinated proteins. Citrulline represents a nonstandard amino acid. It is generated from arginine by peptidylarginine deiminase most likely in apoptotic neutrophils [311]. Insufficient clearance of apoptotic neutrophils in inflamed synovium can promote the formation of citrullinated targets [312].

Anticarbamylated protein antibodies are directed against proteins containing homocitrulline. This posttranslational modification is favored by the reaction between cyanide with lysine [313]. Among factors increasing the cyanide level are smoking, kidney disease, and inflammation [314]. Antiacetylated protein antibodies are a further group of antibodies found in about 40% of patients with rheumatoid arthritis [315].

In serum, autoantibodies are already detected several years before the onset of rheumatoid arthritis [316,317]. Several autoantibody-mediated factors are discussed playing a role in joint destruction. Immunocomplexes with autoantibodies especially those with anticitrullinated autoantibodies can boost the inflammation by stimulating innate immune cells [318]. In inflamed synovia, these autoantibodies contribute to complement activation [319]. In rheumatoid arthritis patients, the formation of extracellular traps from neutrophils is enhanced both in circulating blood and in synovial fluid. These traps contain citrullinated residues. A vicious cycle between a nonresolving inflammation is proposed as anticitrullinated antibodies mediate trap formation from neutrophils (NETosis) and the traps exhibit several proinflammatory activities [320].

9.5.3 Systemic Lupus Erythematosus

A typical autoimmune disease is lupus erythematosus, which affects many parts of the body. Common symptoms are swollen and painful joints, red rash on the skin, chest pain, mouth ulcers, mild fever, feeling of fatigue, swollen lymph nodes, and hair loss. However, symptoms can vary widely and impede the right diagnosis [321].

A delayed phagocytosis of apoptotic cell material with resulting secondary necrosis is discussed as reason of this disease. Otherwise, apoptotic bodies can be captured by antigen-presenting cells [321,322].

Different antinuclear antibodies are found in patients with systemic lupus erythematosus [323,324]. Components of nucleus acids are major autoantigens in this disease. They contribute via toll-like receptor 7 and 9 signaling to plasma cell differentiation and increase thus circulating autoreactive B cells [324].

9.6 DISEASES OF THE RESPIRATORY TRACT

9.6.1 Chronic Obstructive Pulmonary Disease

Typical symptoms of chronic obstructive pulmonary disease (COPD) are shortness of breath, productive cough, and sputum production [325]. In this disease, poor airflow and the inability to breathe out fully contribute to altered lung functions. The affected lungs contain damaged bronchioles and alveoli. Walls of alveoli can be destroyed with the appearance of fewer larger alveoli. The number of terminal bronchioles is drastically reduced. Bronchioles can be filled with mucus. Thus, emphysema and obstructive bronchiolitis are characteristic to COPD.

Persistent chronic inflammations of lung tissue are associated with tissue destruction and fibrosis [326,327]. Destruction starts most likely on terminal bronchioles [328]. An insufficient inactivation of neutrophil elastase by α_1-antitrypsin is also discussed as a reason of tissue damage in COPD [329,330].

Inhaled irritants (e.g., tobacco smoke, pollution, toxic airborne particles) can induce an inflammatory response under involvement of neutrophils, macrophages, and sometimes eosinophils and T cells. Neutrophil-driven inflammation is a hallmark of acute exacerbations of COPD [331–333]. Persistent and acute bacterial colonization is a further factor in the pathogenesis of COPD [334].

9.6.2 Asthma

Recurrent episodes of wheezing, shortness of breath, chest tightness, and coughing are typical to asthma. Both genetic and environmental factors have been associated with the development of asthma [335,336].

Asthma is characterized by a chronic inflammation in the region of bronchi and bronchioles. As a consequence, smooth muscles surrounding these airways show an increased contractibility, and narrowing of the airways results. Asthma combines different pathogenic mechanisms, immunological characteristics, and responsiveness to allergens. Besides eosinophilic asthma, there are noneosinophilic asthma forms. In addition, inhaled glucocorticoids can differentially affect inflammatory granulocytes. This complicates the assessment of primary pathology [337].

Asthma and COPD are different obstructive inflammatory diseases, which can prevail together in some patients.

9.7 DISEASES OF THE DIGESTIVE SYSTEM

9.7.1 Inflammatory Bowel Disease

In small intestine and colon, long-lasting inflammatory conditions are known as inflammatory bowel disease. This group of diseases can further be specified as Crohn's disease and ulcerative colitis. Crohn's disease affects mainly the whole digestive tract from the mouth to anus with preference to the small intestine and colon, while ulcerative colitis dominates in the colon and rectum. Common properties of both diseases are abdominal pain, diarrhea, rectal bleeding, spasms, and weight loss [338,339]. Different comorbidities can result from bowel disease such as arthritis, deep vein thrombosis, pyoderma gangrenosum, primary sclerosing cholangitis, and nonthyroidal illness syndrome [338].

It is challenging to separate clearly Crohn's disease and ulcerative colitis by clinical signs and symptoms from each other. In ulcerative colitis, feces are often mucus-like and mixed with blood, and tenesmus is more common. Porridge-like defecation is found in Crohn's disease. It contains sometimes excess of fat.

9.7.2 Peculiarities of Immune Defense in the Gut

The intestinal epithelial layer separates the sterile host tissues from the gut microbiota, an intense populated microbial habitat [340]. In intact intestinal epithelium, both immunotolerance and immunostimulatory mechanisms contribute in a well-balanced regulation to maintain the integrity of this barrier. Immune tolerance comprises not only components of the gut microbiota but also orally administrated food antigens. Whereas tolerance to gut bacteria is restricted to the colon without attenuation of systemic responses, tolerance to food antigens comprises both local and systemic unresponsiveness [341]. This systemic tolerance to food components is referred to as oral tolerance.

Composition of the intestinal microbiota is crucial for maintenance of gut homeostasis. Despite large variety of microorganisms, *Streptococcus* and *Lactobacillus* species dominate in the microbiota of the small intestine. *Bacteroides, Clostridium, Fusobacterium,* and *Bifidobacterium* are the prevailing species in the colon and rectum [342]. In inflammatory bowel disease, the diversity of microbiota decreases and the number of Gram-negative bacteria increases [343].

Expression of toll-like receptors is downregulated at the apical side on enterocytes of the epithelium, but markedly enhanced at the basolateral side [344]. Although permanently exposed to PAMPs from commensal bacteria, no activation of immune functions results. Pattern recognition receptors on intestinal epithelial cells contribute to immune tolerance to beneficial microbes, but their signaling is activated in the presence of pathogenic microorganisms [345,346]. The fine-tuning of this discrimination remains unknown.

Further, intestinal antigens are transferred via microfold cells (M cells) of the epithelium to nearby residing dendritic cells [347]. The liver is apparently a further tolerogenic site for gut antigen [348]. These mechanisms contribute to downregulation and anergy induction in antigen-specific $CD4^+$ and $CD8^+$ T cells [349]. Adaptations of T regulatory cells to gut microbiota are also important for immune tolerance mechanisms [350].

9.7.3 Specific Functions of the Intestinal Epithelium

The intestinal epithelium consists of specialized differentiated cell types (Table 9.2), which fulfill specific functions in absorption of nutrients and protection against microbe invasion and immune tolerance [351]. Enterocytes take up nutrients, water, ions, vitamins, and others from the gut lumen [352].

Goblet cells secrete glycosylated mucins that form a mucus layer on the epithelium as a natural barrier preventing microbial invasion [353]. In the small intestine, this barrier is a thin mucus layer to allow the penetration of food substances [357]. In contrast, a double mucus layer covers the colon epithelium with an inner denser layer [358]. In the mucus layer, the C-type lectin RegIIIγ promotes spatial segregation of bacteria from the intestinal epithelial surface [359]. Another immunological important function of goblet cells concerns the secretion of special molecules to initiate a T_h2 response on nematode infection [360].

Paneth cells are involved in protection and functional modulation of epithelial stem cells [354]. These cells are characterized by large secretory granules, which are rich in basic peptides [361]. On stimulation by acetylcholine receptor agonists and toll-like receptor agonists, they release antimicrobial agents such as defensins, lysozyme, TNF-α, and phospholipase A_2. This arsenal of secretory molecules is directed against a broad spectrum of pathogens such as Gram-positive and Gram-negative bacteria, fungi, and some viruses [354]. The most abundant human α-defensin HD5 is also involved in regulation of the composition of colonizing microbiota [362]. Stress situations can induce autophagy in Paneth cells [363].

M cells take up antigens by selective endocytosis from the gut lumen and deliver them to intraepithelial antigen-presenting cells. The later cells migrate to lymph nodes to initiate an immune response [355].

Different neuroendocrine cells are spread as single cells throughout the intestinal tract. They

TABLE 9.2 Cell Types of the Intestinal Epithelium and Their Functions.

Cell Type	Main Functions	Peculiarities	References
Enterocytes	Uptake of foodstuffs, ions, and water from the gut lumen		[352]
Goblet cells	Generation and release of glycosylated mucins	Mucus formation as an important regulative against microbe invasion	[353]
Paneth cells	Protection and regulation of intestinal stem cells, secretion of antimicrobial agents	Lifetime up to 23 d	[354]
Microfold cells	Antigen uptake from the gut lumen	Short or no microvilli on their surface, thin mucus layer	[355]
Neuroendocrine cells	Secretion of gastrointestinal hormones		[356]

secrete about 30 gastrointestinal hormones and regulatory peptides that are essential for regulation of gut functions and transmission of systemic effects [356].

9.7.4 Inflammatory Response in Inflammatory Bowel Disease

Several alterations in the integrity of the epithelial layer and immunity contribute to the pathogenesis of inflammatory bowel disease. Increased epithelial permeability can be mediated by genetic defects, T cell–induced disruption of tight junction protein, or enteric neuron dysfunction [364–367]. Antigen recognition and processing is disturbed in patients with inflammatory bowel disease [368]. Dendritic cells incorrectly recognize commensal bacteria and induce thus proinflammatory T cell responses. In addition, patient's epithelial cells might be also function as antigen-presenting cells [369]. Moreover, activated T cells resist apoptosis induction in patients with Crohn's disease [370]. The balance between T helper cells and T regulatory cells is shifted toward the first group [371]. Overstimulation of the sympathetic nerve acts proinflammatory, promotes mast cell degranulation, and alters expression of tight junction proteins [372,373].

There are contradictory data about the dominating T effector cells in the pathogenesis of inflammatory bowel disease. The focus is mainly directed either on T_h17 cells to play a central role [374,375] or on T_h1 cells [376,377]. Otherwise, higher levels of T_h2 cytokines were reported in patients with ulcerative colitis [378].

9.8 CANCER

9.8.1 Energy Metabolism and Hypoxia in Cancer Tissue

Many types of tumors are characterized by lower partial pressure of dioxygen (pO_2) in comparison with the surrounding unperturbed tissue. In human tumors, median values of pO_2 were determined to be in the range from 0.3% to 4.2%, whereas pO_2 values were 3.4%–6.8% in the healthy counterparts [379].

Two main reasons are discussed for the deficient supply of O_2 in tumors. Tumor vasculature is often of very poor quality that disturbs the adequate supply with O_2. A lymphatic drainage is insufficient. Moreover, tumor vessels are characterized by fluctuations in the interstitial pressure, intermittent vascular collapse, and, thus, phases of transient acute hypoxia [380,381].

The second reason concerns the development of a tolerance of tumor cells against hypoxia. Tumor cells are able to switch their metabolism to dominance of glycolysis in energy production. This allows tumors to better survive hypoxic conditions. This tolerance includes changes in the expression of hypoxia-inducible factor 1 toward a constitutive expression pattern and changes in genes controlling HIF expression [382,383]. The enhanced demand of glucose is used for imaging of cancerous tissue by positron emission tomography after accumulation of ^{18}F-labeled glucose in tumors [384].

Of course, there are great variations in the O_2 supply between individual tumors [385]. A profound low oxygenation is found in many prostate and pancreatic tumors [386–388].

There are several lines of evidence that tumor malignancy increases with higher degree of hypoxia [389–391]. For example, in cervical carcinoma, the survival rate decreases with the impact of hypoxia independent of the applied therapy [392].

9.8.2 Lactate Production and Acidosis by Tumors

Both normoxic and hypoxic cancer cells produce large amounts of lactate due to the dominance of glycolysis. Lactate produced by cancer cells contributes to immune escape of T cells. As glycolysis is the dominating pathway for production of energy equivalents in activated T cells too, lactate released by cancer cells hinders immune cells to release their lactate. In consequence, cytotoxic T cells are dampened in their activity [393]. Lactate concentration in tumor tissues correlates with incidence of metastases and reduces the overall survival of cancer patients [394,395].

Tumor cell–derived lactate affects also the metabolism of macrophages. A subset of macrophages, the tumor-associated macrophages, resides in cancer tissue such as in breast cancer. Lactate drives these macrophages into an antiinflammatory M2 subtype [396]. Accumulation of macrophages in tumors is associated with poor clinical outcome [397]. In and around tumors, M2 macrophages promote immunosuppression and proangiogenic activity [398].

Tumor cells produce up to 40 times more lactate than other cells [399]. Macrophages residing in the tumor microenvironment contribute also to lactate production [400]. Lactate suppresses antitumor immunity, increases angiogenesis, and supports tumor progression [396,401]. It favors the release of hyaluronan by adjacent fibroblasts. A hyaluronan cover around carcinomas promotes cell growth, invasion, and metastasis [402]. Further, lactate enhances the resistance of tumors to radiation and chemotherapy. It acts as an antioxidant and is assumed to scavenge reactive species generated during these therapies [403].

Shift to glycolysis in tumor cells is also associated with a high acid load in the tumor microenvironment. This extracellular acidosis correlates with the lower dioxygen level [404]. It facilitates tumor invasion, promotes multidrug resistance, and inhibits the activity of natural killer cells and cytotoxic T cells [393,405–407].

9.8.3 Immunosuppression by Tumors

In addition to the aforementioned mechanisms to manipulate host's immune system, tumor cells contribute also by secretion of active component to tumor surveillance, progression, and metastasis formation.

Many types of tumors express high levels of transforming growth factor β (TGF-β) [408,409]. This pleiotropic cytokine is able to inhibit functional response of natural killer cells and cytotoxic T cells, those cell types that directly attack and kill tumor cells, as well as dampen the activation of many other immune cells such as neutrophils, mast cell, and dendritic cells [410,411]. TGF-β is also known to promote tumor angiogenesis [412,413]. High levels of TGF-β in tumors correlate with a poor clinical outcome [414–416]. The development of inhibitors of TGF-β is a strategy to dampen immunosuppressive activities of TGF-β [410].

Interleukin-10 (IL-10) is known to promote antiinflammatory activity and immune suppression (Section 6.3.5). However, there are conflicting data about its effects in cancer patients. Both tumor-promoting and tumor-inhibiting effects are reported for IL-10 [417,418].

Indoleamine 2,3-dioxygenase 1 is upregulated in many tumors and utilized to promote tumor spread and survival [419,420]. This heme enzyme catalyzes the conversion of tryptophan into *N*-formylkynurenine. Products of the tryptophan metabolism give rise to immunosuppression by inducing differentiation of T regulatory cells and apoptosis of T effector cells [421–423]. These products dampen also natural killer cell functions [424,425].

9.8.4 Intratumoral Hemorrhages

Rupture of blood vessels, which occurs mostly because of tumor invasion, but also as a result of surgery, biopsies, radiotherapy, and chemotherapy, can lead to formation of hemorrhages. This is a common clinical manifestation in patients with the lung, bladder, gastric, or colorectal cancers. Intratumor hemorrhage has been associated with a poor clinical outcome for affected patients [426]. The presence of intratumoral hemorrhage in glioblastoma multiforme is associated with upregulation of growth factors promoting angiogenesis and tumor growth [427].

In cancer mouse models, tumor growth and proliferation are promoted by inoculation of cancer cells with red blood cells, red blood cells lysates, and hemoglobin. In addition, heme oxygenase 1, the NFκB pathway, and proangiogenic factors are upregulated. These effects are not found at low doses of red blood cells and hemoglobin [426]. Enhanced NFκB coincides with increased chemoresistance [426,428]. Intratumor lysed blood cells are an important factor for recruitment of macrophages. Intratumor and peritumor macrophages are polarized into the M2 phenotype [426].

It has also been assumed that the hemoglobin level in tumors can serve as a prognostic parameter before, during, and after radiation therapy [429,430]. In cancer patients, anemia correlates with a poor clinical result [431].

Besides rupture of red blood cells, platelets are also related to tumor pathogenesis. Platelets prevent cytotoxicity of killer cells, promote angiogenesis, and induce metastasis via epithelial–mesenchymal transition [432–434].

9.8.5 Therapeutic Approaches

Surgery, chemotherapy, and radiotherapy are often utilized as therapeutic approaches against cancer. A major challenge for drug-based therapies is the great heterogeneity of cancer. Moreover, tumors adapt in different ways to immune pressure through reducing antigens, expressing of immunoinhibitory molecules, and establishing an immunosuppressive microenvironment [435]. A personalized immune-based therapy is required for cancer patients.

A number of different inhibitors and agents for immunotherapy and combinative applications are currently under investigation as potential anticancer therapies [435–437].

Several approaches have been introduced to increase tumor oxygenation. Limitations of these methods result mainly from adaptations of tumor cells to altered conditions [379]. The application of hypoxia-activated prodrugs is a novel approach to target tumor cells. These drugs are only metabolized in hypoxic cells [438,439]. A combined therapy consisting of hypoxia-activated prodrugs, radiotherapy, and standard cytotoxic chemotherapy is recommended [379].

9.9 CONCLUDING REMARKS

Any deviations from homeostatic parameters and disturbance in the integrity of tissues as well as the inability of the organism to counterregulate them in an appropriate time are associated with development of disease states. There is a large variety of diseases, which affect different organs and sometimes the whole organism and highly disturb the quality of life. In the frame of this book, it is quite impossible to consider all potential disease scenarios. Therefore, the focus has been directed on most common widespread diseases that can principally affect all phases of life, but dominate in elderly individuals.

Despite great variations in their characteristics, there are some common features in mechanisms, appearances, and properties of these disorders. All diseases are accompanied by a more or less intense cell and tissue damage. In chronic diseases, this damage can result from the initiating impact such as injury or microbial activity, but in most cases damage results from activated immune cells. An imbalance between immune cell damage and host's protective mechanisms favors these deteriorations. The resulting disease scenarios are driven by repeated cycles of immune cell activation. In other words, a long-persisting chronic inflammation is the outcome. In strong cases, this inflammation can lead to severe damage with pronounced necrotic areas.

The question arises under which conditions a chronic inflammation develops. An adequate answer is hard to give. Hypoxia, energy, and nutrient deficits impede normal physiological functions in cells and tissues and aggravate the functioning of protective systems. Thus, an additional damaging impact will more seriously disturb the integrity in tissues, which are already weakened by existing deficits.

A long-lasting chronic inflammation is associated with a suppressed immune response. An immunocompromised state favors markedly the development of chronic diseases. As immunosuppression increases in elderly people, it is not surprising that cardiovascular disturbances, diabetes mellitus, neurodegenerative disorders, cancer, and other pathologies dominate in persons with advanced age. As a result of an immunodeficiency, infections with opportunistic microbes and the activation of silent viruses can occur. In addition, different comorbidities can develop on the basis of already existing disorders. In seriously ill patients, often several disease scenarios are observed at the same time.

Addressing risk factors for individual diseases, there are also some of them very widespread to all pathologies. Prevalent risk factors are smoking, alcohol and drug abuse, obesity, inactivity, improper diet, and environmental factors. Of course, many diseases are also favored by genetic predisposition. However, here, it is necessary to differentiate which genes are involved in a certain clinical picture.

And two further aspects are very common in the description of the aforementioned diseases. Often it is written that the cause of the disease is unknown. And secondly, no medication exists that can cure the disease. Despite immense research work of physicians and other scientists, these two statements manifest a dilemma in science. Main reasons of chronic diseases are not the genes or metabolic pathways or specific cellular functions. These factors can enhance or dampen the pathological process, but not induce. The main reason of chronic diseases is the physical inability of complex multicellular organisms (such as the human organism) to hold the order of their structural elements on a high level over the time despite the presence of efficient repair and replacement mechanisms.

Numerous beneficial recommendations around a healthy lifestyle are given to prevent the development of life-threatening diseases. Anyway, aging and the appearance of diseases in advanced age are indispensable. These processes are an essential part of life, and it is impossible to escape them.

References

[1] D.T. Lackland, M.A. Weber, Global burden of cardiovascular disease and stroke: hypertension at the core, Can. J. Cardiol. 31 (2015) 569–571.

[2] I. Hermanodena, E. Duron, J.S. Vidal, O. Hanon, Treatment options and considerations for hypertensive patients to prevent dementia, Expert Opin. Pharmacother. 18 (2017) 989–1000.

[3] D.H. Lau, S. Nattel, J.M. Kalman, P. Sanders, Modifiable risk factors and atrial fibrillation, Circulation 136 (2017) 583–596.

[4] P.B. Raven, M.W. Chapleau, Blood pressure regulation XI: overview and future research directions, Eur. J. Appl. Physiol. 114 (2014) 579–586.

[5] M.A. Sparks, S.D. Crowley, S.B. Gurley, M. Mirotsou, T.M. Coffman, Classical renin-angiotensin system in kidney physiology, Comp. Physiol. 4 (2014) 1201–1228.

[6] A.D. Struthers, T.M. MacDonald, Review of aldosterone- and angiotensin II-induced target organ damage and prevention, Cardiovasc. Res. 61 (2004) 663–670.

[7] C.M. Prado, S.G. Ramos, J. Elias Jr., M.A. Rossi, Turbulent blood flow plays an essential localizing role in the development of atherosclerotic lesions in experimentally induced hypercholesterolaemia in rats, Int. J. Exp. Med. 89 (2008) 72–80.

[8] C.M. Warboys, N. Amini, A. de Luca, P.C. Evans, The role of blood flow in determining the sites of atherosclerotic plaques, F1000 Med. Rep. 3 (2001), 5, 1–8, https://doi.org/10.3410/M3-5.

[9] N.R. Poulter, D. Prabhakaran, M. Caulfield, Hypertension, Lancet 386 (2015) 801–812.

[10] C.D. Reiter, X. Wang, J.E. Tanus-Santos, N. Hogg, R.O. Cannon III, A.N. Schechter, M.T. Gladwin, Cell-free hemoglobin limits nitric oxide bioavailability in sickle-cell disease, Nat. Med. 8 (2002) 1383–1389.

[11] R.P. Rother, L. Bell, P. Hillman, M.T. Gladwin, The clinical sequelae of intravascular hemolysis and extracellular plasma hemoglobin, J. Am. Med. Assoc. 293 (2005) 1653–1662.

[12] R. Gozzelino, V. Jeney, M.P. Soares, Mechanisms of cell protection by heme oxygenase-1, Annu. Rev. Pharmacol. Toxicol. 50 (2010) 323–354.

[13] D. Chiabrando, F. Vinchi, V. Fiorito, E. Tolosano, Haptoglobin and hemopexin in heme detoxification and iron recycling, in: F. Veas (Ed.), Acute Phase Proteins – Regulation and Functions of Acute Phase Proteins, Intech, Rijeka, Croatia, 2011, pp. 261–288.

[14] S. Sarkar, D. Prakash, R.K. Marwaha, G. Garewal, L. Kumar, S. Singhi, B.N. Walia, Acute intravascular haemolysis in glucose-6-phosphate dehydrogenase deficiency, Ann. Trop. Paediatr. 13 (1993) 391–394.

[15] C. Parker, Paroxysmal nocturnal hemoglobinuria, Curr. Opin. Hematol. 19 (2012) 141–148.

[16] G.J. Kato, M.H. Steinberg, M.T. Gladwin, Intravascular hemolysis and the pathophysiology of sickle cell disease, J. Clin. Investig. 127 (2017) 750–760.

[17] A. Ferreira, J. Balla, V. Jeney, G. Balla, M.P. Soares, A central role for free heme in the pathogenesis of experimental cerebral malaria: the missing link? Nat. Med. 13 (2007) 703–710.

[18] R. Fendel, C. Brandts, A. Rudat, A. Kreidenweiss, C. Steur, I. Appelmann, B. Ruehe, P. Schröder, W.E. Berdel, P.G. Kremsner, B. Mordmüller, Hemolysis is associated with low reticulocyte production index and predicts blood transfusion in severe malarial anemia, PLoS One 5 (2010), e10038, 1–8, https://doi.org/10.1371/journal.pone.0010038.

[19] W.G. Land, Transfusion-related acute lung injury: the work of DAMPs, Transfus. Med. Hemother. 40 (2013) 3–13.

[20] D.J. Loegering, Hemolysis following thermal injury and depression of reticuloendothelial system phagocytic function, J. Trauma 21 (1981) 130–134.

[21] Y. Hua, R.F. Keep, J.T. Hoff, G. Xi, Brain injury after intracerebral hemorrhage: the role of thrombin and iron, Stroke 38 (2007) 759–762.

[22] D.J. Schaer, P.W. Buehler, A.I. Alayash, J.D. Belcher, G.M. Vercelotti, Hemolysis and free hemoglobin revisited: exploring hemoglobin and hemin scavengers as a novel class of therapeutic proteins, Blood 121 (2013) 1276–1284.

[23] J.M. Sauret, G. Marinides, G.K. Wang, Rhabdomyolysis, Am. Fam. Phys. 65 (2002) 907–912.

[24] J.D. Hunter, K. Gregg, Z. Damani, Rhabdomyolysis, Cont. Educ. Anaesth. Crit. Care Pain 6 (2006) 141–143.

[25] M. Kristiansen, J.H. Graversen, C. Jacobsen, O. Sonne, H.J. Hoffmann, S.K. Law, S.K. Moestrup, Identification of the haemoglobin scavenger receptor, Nature 409 (2001) 198–201.

[26] N. Petejova, A. Martinek, Acute renal failure due to rhabdomyolysis and renal replacement therapy: a critical review, Crit. Care 18 (2014), 224, 1–8, https://doi.org/10.1188/cc13897.

[27] I.C. Vermeulen Windsant, N.C.J. de Wit, J.T.C. Sertorio, A.A. van Bijnen, Y.M. Ganushchak, J.H. Heijmans, J.E. Tanus-Santos, M.J. Jacobs, J.G. Maessen, W.A. Buurman, Hemolysis during cardiac surgery is associated with increased intravascular nitric oxide consumption and perioperative kidney and intestinal tissue damage, Front. Physiol. 5 (2014), 340, 1–9, https://doi.org/10.3389/fphys.2014.00340.eCollection 2014.

[28] N. Tabibzadeh, C. Estournet, S. Placier, J. Perez, H. Bilbault, A. Girshovich, A. Vandermeersch, C. Jouanneau, E. Letavenier, N. Hammoudi, F. Lionnet, J.P. Haymann, Plasma heme-induced renal toxicity is related to capillary rarefaction, Sci. Rep. 7 (2017), 40156, 1–8, https://doi.org/10.1038/srep40156.

[29] H.F. Bunn, J.H. Jandl, Exchange of heme among hemoglobins and between hemoglobin and albumin, J. Biol. Chem. 243 (1968) 465–475.

[30] J. Balla, G.M. Vercellotti, V. Jeney, A. Yachie, Z. Varga, J.W. Eaton, G. Balla, Heme, heme oxygenase and ferritin in vascular endothelial cell injury, Mol. Nutr. Food Res. 49 (2005) 1030–1043.

[31] V. Hvidberg, M.B. Maniecki, C. Jacobson, P. Hojrup, H.J. Moller, S.K. Moestrup, Identification of the receptor scavenging hemopexin-heme complexes, Blood 106 (1999) 2572–2579.

[32] G. Balla, H.S. Jacob, J.W. Eaton, J.D. Belcher, G.M. Vercelotti, Hemin: a possible physiological mediator of low density lipoprotein oxidation and endothelial injury, Arterioscler. Thromb. 11 (1991) 1700–1711.

[33] V. Jeney, J. Balla, A. Yachie, Z. Varga, G.M. Vercellotti, J.W. Eaton, G. Balla, Pro-oxidant and cytotoxic effects of circulating heme, Blood 100 (2002) 879–887.

[34] T. Lin, F. Sammy, H. Yang, S. Thundivalappil, J. Hellman, K.C. Tracey, H.S. Warren, Identification of hemopexin as an anti-inflammatory factor that inhibits synergy of hemoglobin with HMGB1 in sterile and infectious inflammation, J. Immunol. 189 (2012) 2017–2022.

[35] J.W. Deuel, C.A. Schaer, F.S. Boretti, L. Opitz, I. Garcia-Rubio, J.H. Baek, D.R. Spahn, P.W. Buehler, D.J. Schaer, Hemoglobinuria-related acute kidney injury is driven by intrarenal oxidative reactions triggering a heme toxicity response, Cell Death Dis. 7 (2016), e2064, 1–12, https://doi.org/10.1038/cddis.2015.392.

[36] R.T. Figueiredo, P.L. Fernandez, D.S. Mourao-Sa, B.N. Porto, F.F. Dutra, L.S. Alves, M.F. Oliviera, P.L. Oliviera, A.V. Graca-Souza, M.T. Bozza, Characterization of heme as activator of toll-like receptor 4, J. Biol. Chem. 282 (2007) 20221–20229.
[37] J.D. Belcher, C. Chen, J. Nguyen, L. Milbauer, F. Abdulla, A.I. Alayash, A. Smith, K.A. Nath, R.P. Hebbel, G.M. Vercelotti, Heme triggers TLR4 signaling leading to endothelial cell activation and vaso-occlusion in murine sickle cell disease, Blood 123 (2014) 377–390.
[38] F. Vallelian, C.A. Schaer, J.W. Deutel, G. Ingoglia, R. Humar, P.W. Buehler, D.J. Schaer, Revisiting the putative role of heme as a trigger of inflammation, Pharmacol. Res. Perspect. 2018 (2018), e00392, 1–15, https://doi.org/10.1002/prp2.392.
[39] S. Kumar, U. Bandyopadhyay, Free heme toxicity and its detoxification systems in human, Toxicol. Lett. 157 (2005) 175–188.
[40] J. Flemmig, D. Schlorke, F.-W. Kühne, J. Arnhold, Inhibition of the heme-induced hemolysis of red blood cells by the chlorite-based drug WF10, Free Radic. Res. 50 (2016) 1386–1395.
[41] C. Mold, J.D. Tamerius, G. Philips Jr., Complement activation during painful crisis in sickle cell anemia, Clin. Immunol. Immunopathol. 76 (1995) 314–320.
[42] A.W. Pawluczkowycz, M.A. Lindorfer, J.N. Waitumbi, R.P. Taylor, Hematin promotes complement alternative pathway-mediated deposition of C3 activation fragments on human erythrocytes: potential implications for the pathogenesis of anemia in malaria, J. Immunol. 179 (2007) 5543–5552.
[43] M. Frimat, F. Tabarin, J.D. Dimitrov, C. Poitou, L. Halbwachs-Mecarelli, V. Fremeaux-Bacchi, L.T. Roumenia, Complement activation by heme as a secondary hit for atypical hemolytic uremic syndrome, Blood 122 (2013) 282–292.
[44] G. Gutierrez, H.D. Reines, M.E. Wulf-Gutierrez, Clinical review: hemorrhagic shock, Crit. Care 8 (2004) 373–381.
[45] J.A. Caceres, J.N. Goldstein, Intracranial hemorrhage, Emerg. Med. Clin. N. Am. 30 (2012) 771–794.
[46] M.J. Areisen, S.P. Claus, G.J. Rinkel, A. Algra, Risk factors for intracerebral hemorrhage in the general population: a systemic view, Stroke 34 (2003) 2060–2065.
[47] C.K. Dastur, W. Yu, Current management of spontaneous intracerebral haemorrhage, Stroke Vasc. Neurol. (2017), e000047, 1–9, https://doi.org/10.1136/svn-2016-000047.
[48] B. Jani, C. Rajkumar, Ageing and vascular ageing, Postgrad. Med. J. 82 (2006) 357–362.
[49] M.J. Barnes, B.J. Constable, E. Kodicek, Studies in vivo on the biosynthesis of collagen and elastin in ascorbic acid-deficient Guinea pigs, Biochem. J. 113 (1969) 387–397.
[50] S. Murad, D. Grove, K.A. Lindberg, G. Reynolds, A. Sivarajah, S.R. Pinnell, Regulation of collagen synthesis by ascorbic acid, Proc. Natl. Acad. Sci. U.S.A. 78 (1981) 2879–2882.
[51] L. de Jong, S.P. Albrecht, A. Kemp, Prolyl 4-hydroxylase activity in relation to the oxidation state of enzyme-bound iron. The role of ascorbate in peptidyl proline hydroxylation, Biochim. Biophys. Acta 704 (1982) 326–332.
[52] J.A. May, F.E. Harrison, Role of vitamin C in the function of the vascular endothelium, Antioxid. Redox Signal 19 (2013) 2068–2083.
[53] R.A. Jacob, in: M.E. Shils, J.A. Olson, M. Shike (Eds.), Modern Nutrition in Health and Disease, eighth ed., Lea & Febinger, Philadelphia, 1994, pp. 432–448.
[54] J.B. Reuler, V.C. Broudy, T.G. Cooney, Adult scurvy, J. Am. Med. Assoc. 253 (1985) 805–807.
[55] T.H. Blee, T.H. Cogbill, P.J. Lambert, Hemorrhage associated with vitamin C deficiency in surgical patients, Surgery 131 (2002) 408–412.
[56] A. Bikker, J. Wielders, R. van Loo, M. Loubert, Ascorbic acid deficiency impairs wound healing in surgical patients. Four case reports, Int. J. Surg. Open 2 (2016) 15–18, https://doi.org/10.1016/j.ijso.2016.02.009.
[57] N. Boyera, I. Galey, B.A. Bernard, Effect of vitamin C and its derivatives on collagen synthesis and cross-linking by normal human fibroblasts, Int. J. Cosmet. Sci. 20 (1998) 151–158.
[58] A.J. Lusis, Atherosclerosis, Nature 407 (2000) 233–241.
[59] G. Wick, M. Knoflach, Q. Xu, Autoimmune and inflammatory mechanisms in atherosclerosis, Annu. Rev. Atheroscl. 22 (2004) 361–403.
[60] I. Tabas, G. Garcia-Cardeña, G.K. Owens, Recent insights into the cellular biology of atherosclerosis, J. Cell Biol. 209 (2015) 13–22.
[61] S.C. Bergheanu, M.C. Bodde, J.W. Jukema, Pathophysiology and treatment of atherosclerosis, Current view and future perspective on lipoprotein modification treatment, Neth. Heart J. 25 (2017) 231–242.
[62] A. Gisterå, G. Hansson, The immunology of atherosclerosis, Nat. Rev. Nephrol. 13 (2017) 368–380.
[63] D. Steinberg, S. Parthasarathy, T.E. Carew, J.C. Khoo, J.L. Witztum, Beyond cholesterol: modification of low-density lipoprotein that increases its atherogenicity, N. Engl. J. Med. 320 (1989) 915–924.

[64] B.A. Ference, H.N. Ginsberg, I. Graham, K.K. Ray, C.J. Packard, E. Bruckert, R.A. Hegele, R.M. Krauss, F.J. Raal, H. Schunkert, G.F. Watts, J. Borén, S. Fazio, J.D. Horton, L. Masana, S.J. Nicholls, B.G. Nordestgaard, B. van de Sluis, M.-R. Taskinen, L. Tokgözoğlu, U. Landmesser, U. Laufs, O. Wiklund, J.K. Stock, M.J. Chapman, A.L. Catapano, Low-density lipoproteins cause atherosclerotic cardiovascular disease. 1. Evidence from genetic, epidemiologic, and clinical studies. A consensus statement from the European atherosclerosis society consensus panel, Eur. Heart J. 38 (2017) 2459–2472.

[65] A.C. Goldberg, P.N. Hopkins, P.P. Toth, C.M. Ballantyne, D.J. Rader, D.J. Robinson, S.R. Daniels, S.S. Gidding, S.D. de Feranti, M.K. Ito, M.P. McGowan, P.M. Moriarty, W.C. Cromwell, J.L. Ross, P.E. Ziajka, National lipid association expert panel on familial hypercholesterolemia, Familial hypercholesterolemia: screening, diagnosis and management of pediatric and adult patients: clinical guidance from the national lipid association expert panel on familial hypercholesterolemia, J. Clin. Lipidol. 5 (Suppl.) (2011) S1–S8.

[66] E. Malle, G. Marsche, J. Arnhold, M.J. Davies, Modification of low-density lipoprotein by myeloperoxidase-derived oxidants and reagent hypochlorous acid, Biochim. Biophys. Acta 1761 (2006) 392–415.

[67] A. Daugherty, J.L. Dunn, D.L. Rateri, J.W. Heinecke, Myeloperoxidase, a catalyst for lipoprotein oxidation, is expressed in human atherosclerotic lesions, J. Clin. Investig. 94 (1994) 437–444.

[68] S.L. Hazen, J.W. Heinecke, 3-Chlorotyrosine, a specific marker of myeloperoxidase-catalyzed oxidation, is markedly elevated in low density lipoprotein isolated from human atherosclerotic intima, J. Clin. Investig. 99 (1997) 2075–2081.

[69] J. Jawien, P. Nastalek, R. Korbut, Mouse models of experimental atherosclerosis, J. Physiol. Pharmacol. 55 (2004) 503–517.

[70] S. Zadellar, R. Kleemann, L. Verschuren, J. de Vries-van der Weij, J. van der Hoorn, H.M. Princen, T. Kooistra, Mouse models for atherosclerosis and pharmaceutical modifiers, Arterioscler. Thromb. Vasc. Biol. 27 (2007) 1706–1721.

[71] J. Zschaler, D. Schlorke, J. Arnhold, Differences in innate immune response between man and mouse, Crit. Rev. Immunol. 34 (2014) 433–454.

[72] G.D. Barish, M. Downes, W.A. Alaynick, R.T. Yu, C.B. Ocampo, A.L. Bookout, D.J. Mangelsdorf, R.M. Evans, A nuclear receptor atlas: macrophage activation, Mol. Endocrinol. 19 (2005) 2466–2477.

[73] M. Schneemann, G. Schoeden, Macrophage biology and immunology: man is not a mouse, J. Leukoc. Biol. 81 (2007) 579.

[74] T. Lin, D. Maita, S.R. Thundivalappil, F.E. Riley, J. Hambsch, L.J. van Marter, H.A. Christou, L. Berra, S. Fagan, D.C. Christiani, H.S. Warren, Hemopexin in severe inflammation and infection: mouse models and human diseases, Crit. Care 19 (2015), 166, 1–8, https://doi.org/10.1186/s13054-015-0885-x.

[75] G.D. Lowe, Virchow's triad revisited: abnormal flow, Pathophysiol. Haemostasis Thrombosis 33 (2003) 455–457.

[76] M. Cushman, Epidemiology and risk factors for venous thrombosis, Semin. Hematol. 44 (2007) 62–69.

[77] E. Previtali, P. Bucciarelli, S.M. Passamonti, I. Martinelli, Risk factors for venous and arterial thrombosis, Blood Transfus. 9 (2011) 120–138.

[78] G. Lippi, M. Franchini, G. Targher, Arterial thrombus formation in cardiovascular disease, Nat. Rev. Cardiol. 8 (2011) 502–512.

[79] A.S. Wolberg, F.R. Rosendaal, J.I. Weitz, I.H. Jaffer, G. Agnelli, T. Baglin, N. Mackman, Venous thrombosis, Nat. Rev. Dis. Prim. 1 (2015), article 15006, 1–17, https://doi.org/10.1038/nrdp.2015.6.

[80] B. Savage, F. Almus-Jacobs, Z.M. Ruggeri, Specific synergy of multiple substrate-receptor interactions in platelet thrombus formation under flow, Cell 94 (1998) 657–666.

[81] T.A. Meadows, D.L. Bhatt, Clinical aspects of platelets inhibitors and thrombus formation, Circ. Res. 100 (2007) 1261–1275.

[82] N. Mackman, Triggers, targets and treatment for thrombosis, Nature 451 (2008) 914–918.

[83] M.N. Babapulle, L. Joseph, P. Belisle, J.M. Brophy, A.M. Eisenberg, A hierarchical Bayesian meta-analysis of randomized clinical trials of drug-eluting stents, Lancet 364 (2004) 583–591.

[84] J.I. Weitz, L.A. Linkins, Beyond heparin and warfarin: the new generation of anticoagulants, Expert Opin. Investig. Drugs 16 (2007) 271–282.

[85] M.B. Streiff, G. Agnelli, J.M. Connors, M. Crowther, S. Eichinger, R. Lopes, R.D. McBane, S. Moll, J. Ansell, Guidance for the treatment of deep vein thrombosis and pulmonary treatment, J. Thromb. Thrombolysis 41 (2016) 32–67.

[86] A. Gosain, L.A. DiPietro, Aging and wound healing, World J. Surg. 28 (2004) 321–326.

[87] G.C. Gartner, S. Werner, Y. Barrandon, M.T. Longaker, Wound repair and regeneration, Nature 453 (2008) 314–321.

[88] R. Snyder, Treatment of nonhealing ulcers with allografts, Clin. Dermatol. 23 (2005) 388–395.

[89] S. Guo, L.A. DiPietro, Factors affecting wound healing, J. Dent. Res. 89 (2010) 219–229.
[90] M. Cáceres, A. Oyarzún, P.C. Smith, Defective wound-healing in aging gingival tissue, J. Dent. Res. 93 (2014) 691–697.
[91] P.C. Smith, M. Cáceres, C. Martinez, A. Oyarzún, J. Martinez, Gingival wound healing: an essential response disturbed by aging? J. Dent. Res. 94 (2015) 395–402.
[92] A.A. Tandara, T.A. Mustoe, Oxygen in wound healing – more than a nutrient, World J. Surg. 28 (2004) 294–300.
[93] A.M. Vincent, J.W. Russell, P. Low, E.L. Feldman, Oxidative stress in the pathogenesis of diabetic neuropathy, Endocr. Rev. 25 (2004) 612–628.
[94] K. Woo, E.A. Ayello, R.G. Sibbald, The edge effect: current therapeutic options to advance the wound edge, Adv. Skin Wound Care 20 (2007) 99–117.
[95] R.G. Sibbald, K.Y. Woo, The biology of chronic foot ulcers in persons with diabetes, Diabetes Metab. Res. Rev. 24 (2008) 25–30.
[96] T. Mustoe, Understanding chronic wounds: a unifying hypothesis on their pathogenesis and implications for therapy, Am. J. Surg. 187 (2004) S65–S70.
[97] R. Zhao, H. Liang, E. Clarke, C. Jackson, M. Xue, Inflammation in chronic wounds, Int. J. Mol. Sci. 17 (2016), 2085, 1–14, https://doi.org/10.3390/ijms17122085.
[98] S. Schreml, R.M. Szeimies, L. Prantl, S. Karrer, M. Landthaler, P. Babilas, Oxygen in acute and chronic wound healing, Br. J. Dermatol. 163 (2010) 257–268.
[99] S.A. Emring, T. Krieg, J.M. Davidson, Inflammation in wound repair: molecular and cellular mechanisms, J. Investig. Dermatol. 127 (2007) 514–525.
[100] G.S. Ashcroft, S.J. Mills, J.J. Ashworth, Ageing and wound healing, Biogerontology 3 (2002) 337–345.
[101] E.J. Mudge, Recent accomplishments in wound healing, Int. Wound J. 12 (2005) 4–9.
[102] W.M. van der Veer, M.C. Bloemen, M.M. Ulrich, G. Molema, P.P. van Zuijlen, E. Middelkoop, F.B. Niessen, Potential cellular and molecular causes of hypertrophic scar formation, Burns 35 (2009) 15–29.
[103] F. Grinell, M. Zhu, Identification of neutrophil elastase as the proteinase in burn wound fluid responsible for degradation of fibronectin, J. Investig. Dermatol. 103 (1994) 155–161.
[104] D.R. Yager, S.M. Chen, S.I. Ward, O.O. Olutoye, R.F. Diegelmann, I.K. Cohen, Ability of chronic wound fluids to degrade peptide growth factors is associated with increased levels of elastase activity and diminishes levels of proteinase inhibitors, Wound Repair Regen. 5 (1997) 23–32.
[105] J.V. Edwards, P. Howley, I.K. Cohen, In vitro inhibition of human neutrophil elastase by oleic acid albumin formulations from derivatized cotton wound dressings, Int. J. Pharm. 284 (2004) 1–12.
[106] U. Schönfelder, M. Abel, C. Wiegand, D. Klemm, P. Elsner, U.C. Hipler, Influence of selected dressings on PMN elastase in chronic wound fluid and their antioxidative potential in vivo, Biomaterials 26 (2005) 6664–6673.
[107] J.V. Edwards, P.S. Howley, Human neutrophil elastase and collagenase sequestration with phosphorylated cotton wound dressings, J. Biomed. Mater. Res. 83A (2007) 446–454.
[108] C. Qing, The molecular biology of wound healing & non-healing wound, Chin. J. Traumatol. 20 (2017) 189–193.
[109] J.C. Jennette, R.J. Falk, P.A. Bacon, N. Basu, M.C. Cid, F. Ferrario, L.F. Flores-Suarez, W.L. Gross, L. Guillevin, E.C. Hagen, G.S. Hoffman, D.R. Jayne, C.G.M. Kallenberg, P. Lamprecht, C.A. Langford, R.A. Luqmani, A.D. Mahr, E.L. Matteson, P.A. Merkel, S. Ozen, C.D. Pusey, N. Rasmussen, A.J. Rees, D.G.I. Scott, U. Specks, J.H. Stone, K. Takahashi, R.A. Watts, 2012 Revised international Chapel Hill consensus conference nomenclature of vasculitides, Arthritis Rheum. 65 (2013) 1–11.
[110] J.C. Jennette, Overview of the 2012 revised international Chapel Hill consensus conference nomenclature of vasculitides, Clin. Exp. Nephrol. 17 (2013) 603–606.
[111] J.C. Jennette, R.J. Falk, P. Hu, H. Xiao, Pathogenesis of anti-neutrophil cytoplasmic autoantibody associated small vessel vasculitis, Annu. Rev. Pathol. Mech. Dis. 8 (2013) 139–160.
[112] J.C. Jennette, P.H. Nachman, ANCA glomerulonephritis and vasculitis, Clin. J. Am. Soc. Nephrol. 12 (2017) 1680–1691.
[113] K. Greenan, D. Vassallo, R. Chinnadurai, J. Ritchie, K. Shepard, D. Green, A. Ponnusamy, S. Sinha, Respiratory manifestations of ANCA-associated vasculitis, Clin. Respir. J. 12 (2018) 57–61.
[114] M. Yates, R. Watts, ANCA-associated vasculitis, Clin. Med. 17 (2017) 60–64.
[115] M. McClure, S. Gopaluni, D. Jayne, B cell therapy in ANCA-associated vasculitis: current and emerging treatment options, Nat. Rev. Rheumatol. 14 (2018) 580–591.
[116] M. Mueckler, C. Caruso, S.A. Baldwin, M. Panico, I. Blench, H.R. Morris, W.J. Allard, G.E. Lienhard, H.F. Lodish, Sequence and structure of human glucose transporter, Science 229 (1985) 941–945.
[117] F. Maher, S.J. Vannucci, I.A. Simpson, Glucose transport proteins in brain, FASEB J. 8 (1994) 1003–1011.

[118] S.J. Vannucci, F. Maher, I.A. Simpson, Glucose transporter proteins in brain: delivery of glucose to neurons and glia, Glia 21 (1997) 2–21.
[119] D.M. Muoio, C.B. Newgard, Molecular and metabolic mechanisms of insulin resistance and β-cell failure in type 2 diabetes, Nat. Rev. Mol. Cell Biol. 9 (2008) 193–205.
[120] V.T. Samuel, G.I. Shulman, Mechanisms for insulin resistance: common threads and missing links, Cell 148 (2012) 852–871.
[121] A. Gugliucci, Glycation as the glucose link to diabetic complications, J. Am. Osteopath. Assoc. 100 (2000) 621–634.
[122] M. Brownlee, The pathophysiology of diabetic complications. A unifying mechanism, Diabetes 54 (2005) 1615–1625.
[123] S. Shin, Y.-H. Ku, J.-X. Ho, Y.-K. Kim, J.-S. Suh, M. Singh, Progressive impairment of erythrocyte deformability as indicator of microangiopathy in type 2 diabetes mellitus, Clin. Hemorheol. Microcirc. 36 (2007) 253–261.
[124] N. Babu, M. Singh, Influence of hyperglycemia on aggregation, deformability and shape parameters of erythrocytes, Clin. Hemorheol. Microcirc. 31 (2004) 273–280.
[125] C.-M. Kung, Z.-I. Tseng, H.-L. Wang, Erythrocyte fragility with level of glycosylated hemoglobin in type 2 diabetic patients, Clin. Hemorheol. Microcirc. 43 (2009) 345–351.
[126] R. Agarwal, T. Smart, J. Nobre-Cardosa, C. Richards, R. Bhatnagar, A. Tufail, D. Shima, P.H. Jones, C. Pavesio, Sci. Rep. 6 (2016), 15873, 1–12, https://doi.org/10.1038/srep15873.
[127] N. Cheung, P. Mitchell, T.Y. Wong, Diabetic retinopathy, Lancet 376 (2010) 124–136.
[128] G. Lippi, M. Mercadanti, R. Aloe, G. Targher, Erythrocyte mechanical fragility is increased in patients with type 2 diabetes, Eur. J. Intern. Med. 23 (2012) 150–153.
[129] T. Lahousen, K. Hegenbarth, R. Ille, R.W. Lipp, R. Krause, R.R. Little, W.J. Schnedl, Determination of glycated hemoglobin in patients with advanced liver disease, World J. Gastroenterol. 10 (2004) 2284–2286.
[130] J. Nadelson, S.K. Satapathy, S. Nair, Glycated hemoglobin levels in patients with decompensated cirrhosis, Int. J. Endocrinol. 2016 (2016), article 8390210, 1–8, https://doi.org/10.1155/2016/8390210.
[131] W.J. Schnedl, A. Liebminger, R.E. Roller, R.W. Lipp, G.J. Krejs, Hemoglobin variants and determinations of glycated hemoglobin (HbA1c), Diabetes Metab. Res. Rev. 17 (2001) 94–98.
[132] R.R. Little, W.L. Roberts, A review of variant hemoglobins interfering with hemoglobin A1c measurements, J. Diabetes Sci. Technol. 3 (2009) 446–451.
[133] R.F. Arakaki, B. Changcharoen, Glycemic assessment in a patient with Hb Leiden and type 2 diabetes, AACE Clin. Case Rep. 2 (2016) e307–e310.
[134] D.B. Sacks, D.E. Bruns, D.E. Goldstein, N.K. Maclaren, J.M. McDonald, M. Parrott, Guidelines and recommendations for laboratory analysis in the diagnosis and management of diabetes mellitus, Clin. Chem. 48 (2002) 436–472.
[135] G. Lum, Artefactually low hemoglobin A_{1c} in a patient with hemolytic anemia, Lab. Med. 41 (2010) 267–270.
[136] N. Aggarwal, A.K. Rai, Y. Kupfer, S. Tessler, Immeasurable glycosylated haemoglobin: a marker for severe haemolysis, Br. Med. J. Case Rep. 2013 (2013) 1–2, https://doi.org/10.1136/bcr-2013-200307.
[137] M.S. Radin, Pitfalls in hemoglobin A1c measurement: when results may be misleading, J. Gen. Intern. Med. 29 (2013) 388–394.
[138] J. Ly, R. Marticorena, S. Donnelly, Red blood cell survival in chronic renal failure, Am. J. Kidney Dis. 44 (2004) 715–719.
[139] J.N. Brown, D.W. Kemp, K.R. Brice, Class effect of erythropoietin therapy on hemoglobin A(1c) in a patient with diabetes mellitus and chronic kidney disease not undergoing hemodialysis, Pharmacotherapy 29 (2009) 468–472.
[140] T.M. Reynolds, W.S.A. Smellie, P. Twomey, Glycated haemoglobin (HbA_{1c}) monitoring, Br. Med. J. 333 (2006) 586–588.
[141] K. Bhardwaj, S.K. Sharma, N. Rajpal, A. Sachdev, Effect of iron deficiency anemia on hemoglobin A1c levels, Ann. Clin. Lab. Res. 4 (2016), 123, 1–7, https://doi.org/10.21767/2386-5180.1000123.
[142] E. Ernst, A. Matrai, Altered red and white blood cell rheology in type II diabetes, Diabetes 35 (1986) 1412–1415.
[143] D. Popov, Endothelial cell dysfunction in hyperglycemia: phenotypic change, intracellular signaling modification, ultrastructural alteration, and potential clinical outcomes, Int. J. Diabetes Mellitus 2 (2010) 189–195.
[144] M.A. Potenza, S. Gagliardi, C. Nacci, M.R. Carratu, M. Montagnani, Endothelial dysfunction in diabetes: from mechanisms to therapeutic targets, Curr. Med. Chem. 16 (2009) 94–112.
[145] A. Hirose, T. Tanikawa, H. Mori, Y. Okada, Y. Tanaka, Advanced glycation end products increase endothelial permeability through RAGE/Rho signaling pathway, FEBS Lett. 584 (2010) 61–66.

[146] D. Popov, Towards understanding the mechanisms of impeded vascular function in the diabetic kidney, in: M. Simionescu, A. Sima, D. Popov (Eds.), Cellular Dysfunction in Atherosclerosis and Diabetes – Reports from Bench to Bedside, Publishing House Rom. Acad., Bucharest, 2004, pp. 244–259.

[147] J.S. Estrella, R.N. Nelson, B.K. Sturges, K.M. Vernau, D.C. Williams, R.A. LeCouteur, G.D. Shelton, A.P. Mizisin, Endoneurial microvascular pathology in feline diabetic neuropathy, Microvasc. Res. 75 (2008) 403–410.

[148] B.B. Dokken, The pathophysiology of cardiovascular disease and diabetes: beyond blood pressure and lipids, Diabetes Spectr. 21 (2008) 160–165.

[149] Y. Xiang, J. Cheng, D. Wang, X. Hu, Y. Xie, J. Stitham, G. Atteya, J. Du, W.H. Tang, S.H. Lee, K. Leslie, G. Spollett, Z. Liu, E. Herzog, R.I. Herzog, J. Lu, K.A. Martin, J. Hwa, Hyperglycemia repression of miR-24 coordinately upregulates endothelial cell expression and secretion of von Willebrand factor, Blood 125 (2015) 3377–3387.

[150] P.H. Whincup, J. Danesh, M. Walker, L. Lennon, A. Thomson, P. Appleby, A. Rumley, G.D. Lowe, Von Willebrand factor and coronary heart disease: prospective study and meta-analysis, Eur. Heart J. 23 (2002) 1764–1770.

[151] D.S. Frankel, J.B. Meigs, J.M. Massaro, P.W. Wilson, C.J. O'Donnell, R.B. D'Agostino, G.H. Tofler, Von Willebrand factor, type 2 diabetes mellitus, and risk of cardiovascular disease: the Framingham offspring study, Circulation 118 (2008) 2533–2539.

[152] M. Laakso, J. Kuusisto, Insulin resistance and hyperglycemia in cardiovascular disease development, Nat. Rev. Endocrinol. 10 (2014) 293–302.

[153] R. Sonneville, H.M. den Hertog, F. Güiza, J. Gunst, I. Derese, P.J. Wouters, J.-P. Brouland, A. Polito, F. Gray, F. Chrétien, P. Charlier, D. Annane, T. Sharshar, G. Van den Berghe, I. Vanhorebeek, Impact of hyperglycemia on neuropathological alterations during critical illness, J. Clin. Endocrinol. Metab. 97 (2012) 2113–2123.

[154] A. Ott, R.P. Stolk, F. van Harskamp, H.A. Pols, A. Hofman, M.M. Breteler, Diabetes mellitus and the risk of dementia: the Rotterdam study, Neurology 53 (1999) 1937–1942.

[155] P.K. Crane, R. Walker, R.A. Hubbard, G. Li, D.M. Nathan, H. Zheng, S. Haneuse, S. Craft, T.J. Montine, S.E. Kahn, W. McCormick, S.M. McCurry, J.D. Bowen, E.B. Larsen, Glucose levels and risk of dementia, N. Engl. J. Med. 369 (2013) 540–548.

[156] C.-C. Huang, C.-M. Chung, H.-B. Leu, L.-Y. Lin, C.C. Chiu, C.-Y. Chiang, P.-H. Huang, T.-J. Chen, S.-J. Lin, J.-W. Chen, W.-L. Chan, Diabetes mellitus and the risk of Alzheimer's disease: a nationwide population-based study, PLoS One 9 (2014), e87095, 1–7, https://doi.org/10.1371/journal.pone.0087095.

[157] C.G. Nichols, K-ATP channels as molecular sensors of cellular metabolism, Nature 440 (2006) 470–476.

[158] S.L. Macauley, M. Stanley, E.E. Caesar, S.A. Yamada, M.E. Raichle, R. Perez, T.E. Mahan, C.L. Sutphen, D.M. Holtzman, Hyperglycemia modulates extracellular amyloid-β concentrations and neuronal activity in vivo, J. Clin. Investig. 125 (2015) 2463–2467.

[159] I. Vermes, E.T. Steinmetz, L.J. Zeyen, E.A. van der Veen, Rheological properties of white blood cells are changed in diabetic patients with microvascular complications, Diabetologia 30 (1987) 434–436.

[160] Z. Pecsvarady, T.C. Fisher, C.H. Darwin, A. Fadók, T.S. Maqueda, M.F. Saad, H.J. Meiselman, Decreased polymorphonuclear leukocyte deformability in NIDDM, Diabetes Care 17 (1994) 57–63.

[161] R.L. Engerman, Pathogenesis of diabetic retinopathy, Diabetes 38 (1989) 1203–1206.

[162] V. Bansal, J. Kalita, U.K. Misra, Diabetic neuropathy, Postgrad. Med. J. 86 (2006) 95–100.

[163] L.M. Román-Pintos, G. Villegas-Rivera, A.D. Rodriguez-Carrizalez, A.G. Miranda-Diaz, E.G. Cordona-Muñoz, Diabetic polyneuropathy in type 2 diabetes mellitus: inflammation, oxidative stress, and mitochondrial function, J. Diabetes Res. 2016 (2016), article 3425617, 1–16, https://doi.org/10.1155/2016/3425617.

[164] L.R. Braun, W.A. Fisk, H. Lev-Tov, R.S. Kirsner, R.R. Isseroff, Diabetic foot ulcer: an evidence-based treatment update, Am. J. Clin. Dermatol. 15 (2014) 267–281.

[165] R.E. Pecoraro, J.H. Ahroni, E.J. Boyko, V.L. Stensel, Chronology and determinants of tissue repair in diabetic lower-extremity ulcers, Diabetes Care 40 (1991) 1305–1313.

[166] K. Jonsson, J.A. Jensen, W.H. Goodson III, H. Scheunenstuhl, J. West, H.W. Hopf, T.K. Hunt, Tissue oxygenation, anemia, and perfusion in relation to wound healing in surgical patients, Ann. Surg. 214 (1991) 605–613.

[167] N. Yingsakmongkol, P. Maraprygsavan, P. Sukosit, Effect of WF10 (immunokine) on diabetic foot ulcer therapy: a double-blind, randomized, placebo-controlled trail, J. Foot Ankle Surg. 50 (2011) 635–640.

[168] N. Yingsakmongkol, Clinical outcome of WF10 adjunct to standard treatment of diabetic foot ulcers, J. Wound Care 22 (2013) 1–6.

[169] P. Maraprygsavan, J. Mongkulsuk, J. Arnhold, F.-W. Kuehne, The chlorite-based drug WF10 constantly reduces hemoglobin A1c values and improves glucose control in diabetes patients with severe foot syndrome, J. Clin. Transl. Endocrinol. 4 (2016) 53–58.

[170] M. Ankakrona, J.M. Dypbukt, E. Bonfoco, B. Zhivotovsky, S. Orrenius, S.A. Lipton, P. Nicotera, Glutamate-induced neuronal death: a succession of necrosis or apoptosis depending on mitochondrial function, Neuron 15 (1995) 961–973.

[171] C.E. Hulsebosch, B.C. Hains, E.D. Crown, S.M. Carlton, Mechanisms of chronic central neuropathic pain after spinal cord injury, Brain Res. Rev. 60 (2009) 202–213.

[172] D.D. Schoepp, P.J. Conn, Metabotropic glutamate receptors in brain function and pathology, Trends Pharmacol. Sci. 14 (1993) 13–20.

[173] H.S. Engelmann, A.B. MacDermott, Presynaptic ionotropic receptors and control of transmitter release, Nat. Rev. Neurosci. 5 (2004) 135–145.

[174] P.G. Haydon, G. Carmignoto, Astrocyte control of synaptic transmission and neurovascular coupling, Physiol. Rev. 86 (2006) 1009–1031.

[175] A.M. Butt, Neurotransmitter-mediated calcium signaling in oligodendrocyte physiology and pathology, Glia 54 (2006) 666–675.

[176] D.W. Choi, Calcium-mediated neurotoxicity: relationship to specific channel types and role in ischemic damage, Trends Neurosci. 11 (1988) 465–469.

[177] R. Bullock, A. Zauner, J.J. Woodward, J. Myseros, S.C. Choi, J.D. Ward, A. Marmarou, H.F. Young, Factors affecting amino acid release following severe human head injury, J. Neurosurg. 89 (1998) 507–518.

[178] K. Gupta, G.E. Hardingham, S. Chandran, NMDA receptor-dependent glutamate excitotoxicity in human embryonic stem cell-derived neurons, Neurosci. Lett. 543 (2013) 95–100.

[179] Y. Wang, Z.H. Qin, Molecular and cellular mechanisms of excitotoxic neuronal death, Apoptosis 15 (2010) 1382–1402.

[180] A.A. Kritis, E.G. Stamoula, K.A. Paniskaki, T.D. Vavilis, Researching glutamate-induced cytotoxicity in different cell lines: a comparative/collective analysis/study, Front. Cell. Neurosci. 9 (2015), article 91, 1–18, https://doi.org/10.3389/fncell.2015.00091.

[181] F.N. Soria, A. Pérez-Samatin, A. Martin, K.B. Gona, J. Llop, B. Szczupak, J.C. Chara, C. Matute, M. Domercq, Extrasynaptic glutamate release through cystine/glutamate antiporter contributes to ischemic damage, J. Clin. Investig. 124 (2014) 3645–3655.

[182] Y. Shigeri, R.P. Seal, K. Shimamoto, Molecular pharmacology of glutamate transporters, EAATs and VGLUTs, Brain Res. Rev. 45 (2004) 250–265.

[183] S. Holmseth, H.A. Scott, K. Real, K.P. Lehre, T.B. Leergaard, J.G. Bjaalie, N.C. Danbolt, The concentrations and distributions of three C-terminal variants of the GLT1 (EAAT2; slc1a2) glutamate transporter protein in rat brain tissue suggest differential regulation, Neuroscience 162 (2009) 1055–1071.

[184] P. Rao, M.M. Yallapu, Y. Sari, P.B. Fisher, S. Kumar, Designing novel nanoformulations targeting glutamate transporter excitatory amino acid transporter 2: implications in treating drug addiction, J. Pers. Nanomed. 1 (2015) 3–9.

[185] J. Lewerenz, S.J. Hewett, Y. Huang, M. Lambros, P.W. Gout, P.W. Kalivas, A. Massie, I. Smolders, A. Methner, M. Pergande, S.B. Smith, V. Ganapathy, P. Maher, The cysteine/glutamate antiporter system x_c^- in health and disease: from molecular mechanisms to novel therapeutic opportunities, Antioxidants Redox Signal. 18 (2013) 522–555.

[186] S. Naito, T. Ueda, Adenosine triphosphate-dependent uptake of glutamate into protein I-associated synaptic vesicles, J. Biol. Chem. 258 (1983) 696–699.

[187] J.Y. Zou, F.T. Crews, TNFα potentiates glutamate neurotoxicity by inhibiting glutamate uptake in organotypic brain slice cultures: neuroprotection by NFκB inhibition, Brain Res. 1034 (2005) 11–24.

[188] S.A. Lipton, Paradigm shift in the neuroprotection by NMDA receptor blockade: memantine and beyond, Nat. Rev. Drug Discov. 5 (2006) 160–170.

[189] X.-x. Dong, Y. Wang, Z.-h. Qin, Molecular mechanisms of excitotoxicity and their relevance to pathogenesis of neurodegenerative diseases, Acta Pharmacol. Sin. 30 (2009) 379–389.

[190] R. Riek, D.S. Eisenberg, The activities of amyloids from a structural perspective, Nature 539 (2016) 227–235.

[191] M.A. Walti, F. Ravotti, H. Arai, C.G. Glabe, J.S. Wall, A. Bockmann, P. Guntert, B.H. Meier, R. Riek, Atomic-resolution structure of a disease-relevant Abeta(1-42) amyloid fibril, Proc. Natl. Acad. Sci. U.S.A. 113 (2016) E4976–E4984.

[192] B. Boland, D.A. Smith, D. Mooney, S.S. Jung, D.M. Walsh, F.M. Platt, Macroautophagy is not directly involved in the metabolism of amyloid precursor protein, J. Biol. Chem. 285 (2010) 37415–37426.

[193] T. Hara, K. Nakamura, M. Matsui, A. Yamamoto, Y. Nakahara, R. Suzuki-Migishima, M. Yokoyama, K. Mishima, I. Saito, H. Okano, N. Mizushima, Suppression of basal autophagy in neural cells causes neurodegenerative disease in mice, Nature 441 (2006) 885–889.

[194] M. Komatsu, Q.J. Wang, G.R. Holstein, V.L. Friedrich Jr., J. Iwata, E. Kominami, B.T. Chait, K. Tanaka, Z. Yue, Essential role for autophagy protein Atg7 in the maintenance of axonal homeostasis and the prevention of axonal degeneration, Proc. Natl. Acad. Sci. U.S.A. 104 (2007) 14489–14494.

[195] A. Nakai, O. Yamaguchi, T. Takeda, Y. Higuchi, S. Nikoso, M. Taniike, S. Omiya, I. Mizote, Y. Matsumura, M. Asahi, K. Nishida, M. Hori, N. Mizushima, K. Otsu, The role of autophagy in cardiomyocytes in the basal state and in response to hemodynamic stress, Nat. Med. 13 (2007) 619–624.

[196] Y. Feng, D. He, Z. Yao, D.J. Klionsky, The machinery of macroautophagy, Cell Res. 24 (2014) 24–41.

[197] M. Kundu, C.B. Thompson, Autophagy: basic principles and relevance to disease, Annu. Rev. Pathol. 3 (2008) 427–455.

[198] P.A. Jaeger, T. Wyss-Corey, All-you-ca-eat: autophagy in neurodegeneration and neuroprotection, Mol. Neurodegener. 4 (2009), 16, 1–22, https://doi.org/10.1186/1750-1326-4-16.

[199] R. Banerjee, M.F. Beal, B. Thomas, Autophagy in neurodegenerative disorders: pathogenetic roles and therapeutic implications, Trends Neurosci. 33 (2010) 541–549.

[200] J.W. Steele, E. Fan, Y. Kelahmetoglu, Y. Tian, V. Bustos, Modulation of autophagy as a therapeutic target for Alzheimer's disease, Postdoc J. 1 (2013) 21–34.

[201] A. Zare-shahabadi, E. Masliah, G.,V.,W. Johnson, N. Rezael, Autophagy in Alzheimer's disease, Rev. Neurosci. 26 (2015) 385–395.

[202] D.S. Yang, P. Stavrides, P.S. Mohan, S. Kaushik, A. Kumar, M. Ohno, S.D. Schmidt, D.W. Wesson, U. Bandyopadhyay, Y. Jiang, M. Pawlik, C.M. Peterhoff, A.J. Yang, D.A. Wilson, P.S. George-Hyslop, D. Westawey, P.M. Mathews, E. Levy, A.M. Cuervo, R.A. Nixon, Reversal of autophagy dysfunction in the TgCRND8 mouse model of Alzheimer's disease ameliorates amyloid pathologies and memory deficits, Brain 134 (2011) 258–277.

[203] D.S. Yang, P. Stavrides, P.S. Mohan, S. Kaushik, A. Kumar, M. Ohno, S.D. Schmidt, D.W. Wesson, U. Bandyopadhyay, Y. Jiang, M. Pawlik, C.M. Peterhoff, A.J. Yang, D.A. Wilson, P.S. George-Hyslop, D. Westawey, P.M. Mathews, E. Levy, A.M. Cuervo, R.A. Nixon, Therapeutic effects of remediating autophagy failure in a mouse model of Alzheimer disease by enhancing lysosomal proteolysis, Autophagy 7 (2011) 788–789.

[204] M.A. Lynch-Day, K. Mao, K. Wang, M. Zhao, D.J. Klionsky, The role of autophagy in Parkinson's disease, Cold Spring Harbor Perspect. Med. 2 (2012), a009357, 1–13, https://doi.org/10.1101/cshperspect.a009357.

[205] L. Zhang, Y. Dong, X. Xu, Z. Xu, The role of autophagy in Parkinson's disease, Neural Regen. Res. 7 (2012) 141–145.

[206] L. Stefanis, α-Synuclein in Parkinson's disease, Cold Spring Harb. Perspect. Med. 4 (2012), a009399, 1–23, https://doi.org/10.1101/cshperspect.a009399.

[207] S. Sarkar, J.E. Davies, Z. Huang, A. Tunnacliffe, D.C. Rubinsztein, Trehalose, a novel mTOR-independent autophagy enhancer, accelerates the clearance of mutant huntingtin and α-synuclein, J. Biol. Chem. 282 (2007) 5641–5652.

[208] M.J. Casarejos, R.M. Solano, A. Gómez, J. Perucho, J.G. de Yébenes, M.A. Mena, The accumulation of neurotoxic proteins, induced by proteasome inhibition, is reverted by trehalose, an enhancer of autophagy, in human neuroblastoma cells, Neurochem. Int. 58 (2011) 512–520.

[209] U. Krüger, Y. Wang, S. Kumar, E.M. Mandelkow, Autophagic degradation of tau in primary neurons and its enhancement by trehalose, Neurobiol, Aging 33 (2012) 2291–2305.

[210] V. Schaeffer, L. Lavenir, S. Ozcelik, M. Tolnay, D.T. Winkler, M. Goedert, Stimulation of autophagy reduces neurodegeneration in a mouse model of human tauopathy, Brain 135 (2012) 2169–2177.

[211] N.T. Tien, I. Karaca, I.Y. Tamboli, J. Walter, Trehalose alters subcellular trafficking and the metabolism of the Alzheimer-associated amyloid precursor protein, J. Biol. Chem. 291 (2016) 10528–10540.

[212] T.E. Moors, J.J.M. Hoozemans, A. Ingrassia, T. Beccari, L. Parnetti, M.-C. Chartier-Harlin, W.D.J. van de Berg, Therapeutic potential of autophagy-enhancing agents in Parkinson's disease, Mol. Neurodegener. 12 (2017), 11, 1–18, https://doi.org/10.1186/s13024-017-0154-3.

[213] E. White, Deconvoluting the context-dependent role for autophagy in cancer, Nat. Rev. Canc. 12 (2012) 401–410.

[214] R. Amaravadi, A.C. Kimmelmann, E. White, Recent insights into function of autophagy in cancer, Genes Dev. 30 (2016) 1913–1930.

[215] R.W. Button, S. Luo, D.C. Rubinsztein, Autophagic activity in neuronal cell death, Neurosci. Bull. 31 (2015) 382–394.

[216] C. Rosendorff, M.S. Beeri, J.M. Silverman, Cardiovascular risk factors for Alzheimer's disease, Am. J. Geriatr. Cardiol. 16 (2007) 143–149.

[217] C. Patterson, J.W. Feightner, A. Garcia, G.Y. Hsiung, C. MacKnight, A.D. Sadovnick, Diagnosis and treatment of dementia: 1. Risk assessment and primary prevention of Alzheimer disease, Can. Med. Assoc. J. 178 (2008) 548–556.

[218] C. Ballard, S. Gauthier, A. Corbett, C. Brayne, D. Aarsland, E. Jones, Alzheimer's disease, Lancet 377 (2011) 1019–1031.

[219] P.T. Francis, A.M. Palmer, M. Snape, G.K. Wilcock, The cholinergic hypothesis of Alzheimer's disease: a review of progress, J. Neurol. Neurosurg. Psychiatry 66 (1999) 137–147.

[220] A. Martorana, Z. Esposito, G. Koch, Beyond the cholinergic hypothesis: do current drugs work in Alzheimer's disease, CNS Neurosci. Ther. 16 (2010) 235–245.

[221] M. Mehta, A. Adem, M. Sabbagh, New acetylcholinesterase inhibitors for Alzheimer's disease, Int. J. Alzheimer's Dis. 2012 (2012), article 728983, 1–8, https://doi.org/10.1155/2012/728983.

[222] M.A. Rogawski, G.L. Wenk, The neuropharmacological basis for the use of memantine in the treatment of Alzheimer's disease, CNS Drug Rev. 9 (2003) 275–308.

[223] D.M. Robinson, G.M. Keating, Memantine: a review of its use in Alzheimer's disease, Drugs 66 (2006) 1515–1534.

[224] J. Hardy, D. Allsop, Amyloid deposition as a central event in the aetiology of Alzheimer's disease, Trends Pharmacol. Sci. 12 (1991) 383–388.

[225] A. Mudher, S. Lovestone, Alzheimer's disease – do tauists and Baptists finally shake hands? Trends Neurosci. 25 (2002) 22–26.

[226] R. Deane, B.V. Zlokovic, Role of the blood-brain barrier in the pathogenesis of Alzheimer's disease, Curr. Alzheimer Res. 4 (2007) 191–197.

[227] H. Xu, D.I. Finkelstein, P.A. Adlard, Interactions of metals and apolipoprotein E in Alzheimer's disease, Front. Aging Neurosci. 6 (2014), article 121, 1–7, https://doi.org/10.3389/fnagi.2014.00121.

[228] G. Bartzokis, Alzheimer's disease as homeostatic responses to age-related myelin breakdown, Neurobiol. Aging 32 (2011) 1341–1371.

[229] Z. Cai, M. Xiao, Oligodendrocytes and Alzheimer's disease, Int. J. Neurosci. 126 (2016) 97–104.

[230] J.H. Caldwell, M. Klevanski, M. Saar, U.C. Müller, Roles of the amyloid precursor protein family in the peripheral nervous system, Mech. Dev. 130 (2013) 433–446.

[231] J.-N. Octave, N. Pierrot, S. Ferao Santos, N.N. Nalivaeva, A.J. Turner, From synaptic spines to nuclear signaling: nuclear and synaptic actions of the amyloid precursor protein, J. Neurochem. 126 (2013) 183–190.

[232] S.A. Bellingham, D.K. Lahiri, B. Maloney, S. La Fontaine, G. Multhaup, J. Camakaris, Copper depletion down-regulates expression of the Alzheimer's disease amyloid-beta precursor protein gene, J. Biol. Chem. 279 (2004) 20378–20386.

[233] J.A. Duce, A. Tsatsanis, M.A. Cater, S.A. James, E. Robb, K. Wikhe, S.L. Leong, K. Perez, T. Johansen, M.A. Greenough, H.H. Cho, D. Galatis, R.D. Moir, C.L. Masters, C. McLean, R.E. Tanzi, R. Cappai, K.J. Barnham, G.D. Ciccotosto, J.T. Rogers, A.I. Bush, Iron-export ferrioxidase activity of β-amyloid precursor protein is inhibited by zinc in Alzheimer's disease, Cell 142 (2010) 857–867.

[234] K. Baiden-Amissah, U. Joashi, R. Blumberg, H. Mehmet, A.D. Edwards, P.M. Cox, Expression of amyloid precursor protein (β-APP) in the neonatal brain following hypoxic ischemic injury, Neuropathol. Appl. Neurobiol. 24 (1998) 346–352.

[235] N.N. Nalivaeva, L. Fisk, E.G. Kochkina, S.A. Plesneva, I.A. Zhuravin, E. Babusikova, D. Dobrota, A.J. Turner, Effect of hypoxia/ischemia and hypoxic preconditioning/reperfusion on expression of some amyloid-degrading enzymes, Ann. N. Y. Acad. Sci. 1035 (2004) 21–33.

[236] N.N. Nalivaeva, N.D. Belyaev, D.I. Lewis, A.R. Pickles, N.Z. Makova, D.I. Bagrova, N.M. Dubrovskaya, S.A. Plesneva, I.A. Zhuravin, A.J. Turner, Effect of sodium valproate administration on brain neprilysin expression and memory in rats, J. Mol. Neurosci. 46 (2012) 569–577.

[237] R. van der Kant, L.S.B. Goldstein, Cellular function of the amyloid precursor protein from development to dementia, Dev. Cell 32 (2015) 502–515.

[238] G.-Z. Yang, M. Yang, Y. Lim, J.-J. Lu, T.-H. Wang, J.-G. Qi, J.-H. Zhong, X.-F. Zhou, Huntingtin associated protein 1 regulates trafficking of the amyloid precursor protein and modulates beta levels in neurons, J. Neurochem. 122 (2012) 1010–1022.

[239] T. Hartmann, S.C. Bieger, B. Brühl, P.J. Tienari, N. Ida, D. Allsop, G.W. Roberts, C.L. Masters, C.G. Dotti, K. Unsicker, K. Bayreuther, Distinct sites of intracellular production for Alzheimer's disease A beta 40/42 amyloid peptides, Nat. Med. 3 (1997) 1016–1020.

[240] Y.I. Yin, B. Bassit, L. Zhu, X. Yang, C. Wang, Y.M. Li, γ-Secretase substrate concentration modulates the Aβ42/Aβ40 ratio. Implications for Alzheimer disease, J. Biol. Chem. 282 (2007) 23639–23644.

[241] P.E. Fraser, J.T. Nguyen, H. Inouye, W.K. Surewicz, D.J. Selkoe, M.B. Podlisny, D.A. Kirschner, Fibril formation by primate, rodents, and Dutch-hemorrhagic analogues of Alzheimer amyloid β-protein, Biochemistry 31 (1992) 10716–10723.

[242] L. Hong, T.M. Carducci, W.D. Bush, C.G. Dudzik, G.L. Millhauser, J.D. Simon, Quantification of the binding properties of Cu^{2+} to the amyloid beta peptide: coordination spheres for human and rat peptides and implication on Cu^{2+}-induced aggregation, J. Phys. Chem. B 114 (2010) 11261–11271.

[243] B. De Strooper, M. Simons, G. Multhaup, F. van Leuven, K. Beyreuther, C.G. Dotti, Production of

intracellular amyloid-containing fragments in hippocampal neurons expressing human amyloid precursor protein and protection against amyloidogenesis by subtle amino acid substitutions in the rodent sequence, EMBO J. 14 (1995) 4932–4938.

[244] J. Ghiso, B. Frangione, Amyloidosis and Alzheimer's disease, Adv. Drug Deliv. Rev. 54 (2002) 1539–1551.

[245] N. Arispe, E. Rojas, H.B. Pollard, Alzheimer disease amyloid beta protein forms calcium channels in bilayer membranes: blockade by tromethamine and aluminum, Proc. Natl. Acad. Sci. U.S.A. 90 (1993) 567–571.

[246] G.M. Shankar, S. Li, T.H. Mehta, A. Garcia-Munoz, N.E. Shepardson, I. Smitt, F.M. Brett, M.A. Farrell, M.J. Rowan, C.A. Lemere, C.M. Regan, D.M. Walsh, B.L. Sabatini, D.J. Selkoe, Amyloid-beta protein dimers isolated directly from Alzheimer's brain impair synaptic plasticity and memory, Nat. Med. 14 (2008) 837–842.

[247] M.A. Smith, R.K. Kutty, P.L. Richey, S.-D. Yan, D. Stern, G.J. Chader, B. Wiggert, R.B. Petersen, G. Perry, Heme oxygenase-1 is associated with the neurofibrillary pathology of Alzheimer's disease, Am. J. Pathol. 145 (1994) 42–47.

[248] H. Schipper, S. Cisse, E. Stopa, Expression of heme oxygenase-1 in the senescent and Alzheimer-diseased brain, Ann. Neurol. 37 (1995) 758–768.

[249] J.R. Connor, B.S. Snyder, J.L. Beard, R.E. Fine, E.J. Mufson, Regional distribution of iron and iron-regulatory proteins in the brain in aging and Alzheimer's disease, J. Neurosci. Res. 31 (1992) 327–335.

[250] T. Kimpara, A. Takeda, T. Yamaguchi, H. Arai, N. Okita, S. Takase, H. Sasaki, Y. Itoyama, Increased bilirubins and their derivatives in cerebrospinal fluid in Alzheimer's disease, Neurobiol. Aging 21 (2000) 551–554.

[251] S.W. Ryter, R.M. Tyrrell, The heme synthesis and degradation pathways: role in oxidant sensitivity: heme oxygenase has both pro-and antioxidant properties, Free Radic. Biol. Med. 28 (2000) 289–309.

[252] W. Song, H. Su, S. Song, H.K. Paudel, H.M. Schipper, Over-expression of heme oxygenase-1 promotes oxidative mitochondrial damage in rat astroglia, J. Cell. Physiol. 185 (2000) 80–86.

[253] H. Zukor, W. Song, A. Lieberman, J. Mui, H. Vali, C. Fillebeen, K. Pantopoulos, T.D. Wu, J.L. Guerquin-Kern, H.M. Schipper, HO-1-mediated macroautophagy: a mechanism for unregulated iron deposition in aging and degenerating neural tissues, J. Neurochem. 109 (2009) 776–791.

[254] I. Maurer, S. Zierz, H.J. Möller, A selective defect of cytochrome c oxidase is present in brain of Alzheimer disease patients, Neurobiol. Aging 21 (2000) 455–462.

[255] D.A. Cottrell, E.L. Blakely, M.A. Johnson, P.G. Inge, D.M. Turnbull, Mitochondrial enzyme-deficient hippocampal neurons and choroidal cell in AD, Neurology 57 (2001) 260–264.

[256] G.C. Steffens, R. Biewald, G. Buse, Cytochrome c oxidase is a three-copper, two-heme-A-protein, Eur. J. Biochem. 164 (1987) 295–300.

[257] H.D. Venters Jr., L.E. Bonilla, T. Jensen, H.P. Garner, E.Z. Bordayo, M.M. Najarian, T.A. Ala, R.P. Mason, W.H. Frey 2nd, Heme from Alzheimer's brain inhibits muscarinic receptor binding via thiyl radical generation, Brain Res. 764 (1997) 93–100.

[258] J.R. Fawcett, E.Z. Bordayo, K. Jackson, H. Liu, J. Peterson, A. Svitak, W.H. Frey 2nd, Inactivation of the human brain muscarinic acetylcholine receptor by oxidative damage catalyzed by low molecular weight endogenous inhibitor from Alzheimer's brain is prevented by pyrophosphate analogs, bioflavonoids and other antioxidants, Brain Res. 950 (2002) 10–20.

[259] H. Atamna, W.H. Frey II, A role for heme in Alzheimer's disease: heme binds amyloid β and has altered metabolism, Proc. Natl. Acad. Sci. U.S.A. 101 (2004) 11153–11158.

[260] H. Atamna, K. Boyle, Amyloid-β peptide binds with heme to form a peroxidase: relationship to the cytopathologies of Alzheimer's disease, Proc. Natl. Acad. Sci. U.S.A. 103 (2006) 3381–3386.

[261] H. Atamna, W.H. Frey II, N. Ko, Human and rodent amyloid-β peptides differentially bind heme: relevance to the human susceptibility to Alzheimer's disease, Arch. Biochem. Biophys. 487 (2009) 59–65.

[262] H. Atamna, Heme binding to amyloid-beta peptide: mechanistic role in Alzheimer's disease, J. Alzheimer's Dis. 10 (2006) 255–266.

[263] A.I. Bush, Copper, zinc, and the metallobiology of Alzheimer disease, Alzheimers Dis. Assoc. Disord. 17 (2003) 147–150.

[264] R.A. Cherny, J.T. Legg, C.A. McLean, D.P. Fairlie, X. Huang, C.S. Atwood, K. Beyrether, R.E. Tanzi, C.L. Masters, A.I. Bush, Aqueous dissolution of Alzheimer's disease Aβ amyloid deposits by biometal depletion, J. Biol. Chem. 274 (1999) 23223–23228.

[265] J. Dong, C.S. Atwood, V.E. Anderson, S.L. Siedlak, M.A. Smith, G. Perry, P.R. Carey, Biochemistry 42 (2003) 2768–2773.

[266] R.A. Cherny, C.S. Atwood, M.E. Xilinas, D.N. Gray, W.D. Jones, C.A. McLean, K.J. Barnham, I. Volitakis, F.W. Fraser, Y. Kim, X. Huang, L.E. Goldstein, R.D. Moir, J.T. Lim, K. Beyreuther, H. Zheng, R.E. Tanzi, C.L. Masters, A.I. Bush, Treatment with a copper-zinc chelator markedly and rapidly inhibits beta-amyloid accumulation in Alzheimer's disease transgenic mice, Neuron 30 (2001) 665–676.

[267] A.I. Bush, C.L. Masters, R.E. Tanzi, Copper, β-amyloid, and Alzheimer's disease: tapping a sensitive connection, Proc. Natl. Acad. Sci. U.S.A. 100 (2003) 11193–11194.

[268] D. Howlett, P. Cutler, S. Heales, P. Camilleri, Hemin and related porphyrins inhibit β-amyloid aggregation, FEBS Lett. 417 (1997) 249–251.

[269] N. Lu, J. Li, R. Tian, Y.-Y. Peng, Key roles of Arg^5, Tyr^{10} and His residues in Aβ-heme peroxidase: relevance to Alzheimer's disease, Biochem. Biophys. Res. Commun. 452 (2014) 676–681.

[270] E. Chiziane, H. Telemann, M. Krueger, J. Adler, J. Arnhold, A. Alia, J. Flemmig, Free heme and amyloid-β: a fatal liaison in Alzheimer's disease, J. Alzheimer's Dis. 61 (2018) 963–984.

[271] C.S. Atwood, R.D. Moir, X. Huang, R.C. Scarpa, N.M.E. Bacarra, D.M. Romano, M.A. Hartshorn, R.E. Tanzi, A.I. Bush, Dramatic aggregation of Alzheimer Aβ by Cu(II) is induced by conditions representing physiological acidosis, J. Biol. Chem. 273 (1998) 12817–12826.

[272] J.C. Yoburn, W. Tian, J.O. Brower, J.S. Novick, C.G. Glabe, D.L. Van Vranken, Dityrosine cross-linked Aβ peptides: fibrillary β-structure in Aβ(1-40) is conductive to formation of dityrosine cross-links but a dityrosine cross-link in Aβ(8-14) does not induce β-structure, Chem. Res. Toxicol. 16 (2003) 531–535.

[273] C.S. Atwood, G. Perry, H. Zeng, Y. Kato, W.D. Jones, K.-Q. Ling, X. Huang, R.D. Moir, D. Wang, L.M. Sayre, M.A. Smith, S.G. Chen, A.I. Bush, Copper mediates dityrosine cross-linking of Alzheimer's amyloid-β, Biochemistry 43 (2003) 560–568.

[274] Y.K. Al-Hilaly, T.L. Williams, M. Stewart-Parker, L. Ford, E. Skaria, M. Cole, W.G. Bucher, K.L. Morris, A.A. Sada, J.R. Thorpe, L.C. Serpell, A central role for dityrosine crosslinking of amyloid-β in Alzheimer's disease, Acta Neuropathol. Commun. 1 (2013), 83, 1–17, https://doi.org/10.1186/2051-5960-1-83.

[275] N. Lu, J. Li, R. Tian, Y.-Y. Peng, Key roles for tyrosine 10 in Aβ-heme complexes and its relevance to oxidative stress, Chem. Res. Toxicol. 28 (2015) 365–372.

[276] C.A. Davie, A review of Parkinson's disease, Br. Med. Bull. 86 (2008) 109–127.

[277] D.V. Dickson, Neuropathology of movement disorders, in: E. Tolosa, J.J. Jankovic (Eds.), Parkinson's Disease and Movement Disorders, Lippincott Williams & Wilkins, 2007, pp. 271–283.

[278] M.E. Domellöf, E. Elgh, L. Forsgren, The relation between cognition and motor dysfunction in drug-naïve newly diagnosed patients with Parkinson's disease, Mov. Disord. 26 (2011) 2183–2189.

[279] W.J. Schulz-Schaeffer, The synaptic pathology of alpha-synuclein aggregation in dementia with Lewy bodies, Parkinson's disease and Parkinson's disease dementia, Acta Neuropathol. 120 (2010) 131–143.

[280] A.R. Winslow, C.-W. Chen, S. Corrochano, A. Acevedo-Arozena, D.E. Gordon, A.A. Peden, M. Lichtenberg, F.M. Menzies, B. Ravikumar, S. Imarisio, S. Brown, C.J. O'Kane, D.C. Rubinsztein, α-Synuclein impairs macroautophagy: implications for Parkinson's disease, J. Cell Biol. 190 (2010) 1023–1037.

[281] P. Dayalu, R.L. Albin, Huntington's disease: pathogenesis and treatment, Neurol. Clin. 33 (2015) 101–114.

[282] M. DiFiglia, E. Sapp, K. Chase, C. Schwarz, A. Meloni, C. Young, E. Martin, J.P. Vonsattel, R. Carraway, S.A. Reeves, Huntingtin is a cytoplasmic protein associated with vesicles in human and rat brain neurons, Neuron 14 (1995) 1075–1081.

[283] J. Velier, M. Kim, C. Schwarz, T.W. Kim, E. Sapp, K. Chase, N. Aronin, M. DiFiglia, Wild-type and mutant huntingtins function in vesicle trafficking in the secretory and endocytic pathways, Exp. Neurol. 152 (1998) 34–40.

[284] G. Hoffner, P. Kahlem, P. Dijan, Perinuclear localization of huntingtin as a consequence of its binding to microtubules through the interaction with beta-tubulin: relevance to Huntington's disease, J. Cell Sci. 115 (2002) 941–948.

[285] F.O. Walker, Huntington's disease, Lancet 369 (2007) 218–228.

[286] S. Ayala-Peña, Role of oxidative DNA damage in mitochondrial dysfunction and Huntington's disease pathogenesis, Free Radic. Biol. Med. 62 (2013) 102–110.

[287] Z. Liu, T. Zhou, A.C. Ziegler, P. Dimitrion, L. Zuo, Oxidative stress in neurodegenerative diseases: from molecular mechanisms to clinical applications, Oxid. Med. Cell. Longev. 2017 (2017), article 2525967, 1–11, https://doi.org/10.1155/2017/2525967.

[288] J. Labbadia, R.I. Morimoto, Huntington's disease: underlying molecular mechanisms and emerging concepts, Trends Biochem. Sci. 38 (2013) 378–385.

[289] B. Ravikumar, R. Duden, D.C. Rubinsztein, Aggregate-prone proteins with polyglutamine and polyalanine expansions are degraded by autophagy, Hum. Mol. Genet. 11 (2002) 1107–1117.

[290] R.H. Brown, A. Al-Chalabi, Amyotrophic lateral sclerosis, N. Engl. J. Med. 377 (2017) 162–172.

[291] O.M. Peters, M. Ghasemi, R.H. Brown Jr., Emerging mechanisms of molecular pathology in ALS, J. Clin. Investig. 125 (2015) 1767–1779.

[292] M. DeJesus-Hernandez, I.R. Mackenzie, B.F. Boeve, A.L. Boxer, M. Baker, N.J. Rutherford, A.M. Nicholson, N.A. Finch, H. Flynn, J. Adamson, N. Kouri, A. Wojtas, P. Sengdy, G.Y. Hsiung, A. Kanydas, W.W. Seeley, K.A. Josephs, G. Coppola, D.H. Geschwind, Z.K. Wszolek, H. Feldman, D.S. Knopman, R.C. Petersen, B.L. Miller, D.W. Dickson, K.B. Boylan, N.R. Graff-Radford, R. Rademakers, Expanded GGGGCC hexanucleotide repeat in noncoding region of C9ORF72 causes chromosome 9p-linked FTD and ALS, Neuron 72 (2011) 245–256.

[293] G. Bensimon, L. Lacomblez, V. Meininger, ALS/riluzole study group, A controlled trial of riluzole in amyotrophic lateral sclerosis, N. Engl. J. Med. 330 (1994) 585–591.

[294] M. Paspe Cruz, Edaravone (Radicava). A novel neuroprotective agent for the treatment of amyotrophic lateral sclerosis, Pharmacol. Therapeut. 43 (2018) 25–28.

[295] J. Nakahara, M. Maeda, S. Aiso, N. Suzuki, Current concepts in multiple sclerosis: autoimmunity versus oligodendrogliopathy, Clin. Rev. Allergy Immunol. 42 (2012) 26–34.

[296] R. Dobson, G. Giovannoni, Multiple sclerosis – a review, Eur. J. Neurol. 26 (2019) 27–40.

[297] A. Compston, A. Coles, Multiple sclerosis, Lancet 359 (2002) 1221–1231.

[298] A. Ascherio, K.L. Munger, Environmental risk factors for multiple sclerosis. Part I: the role of infections, Ann. Neurol. 61 (2007) 288–299.

[299] D. Karussis, The diagnosis of multiple sclerosis and the various related demyelinating syndromes: a critical review, J. Autoimmun. 48–49 (2014) 134–142.

[300] H. Lassmann, Pathology and disease mechanisms in different stages of multiple sclerosis, J. Neurol. Sci. 333 (2013) 1–4.

[301] B.D. Trapp, J. Peterson, R.M. Ransohoff, R. Rudick, S. Mörk, L. Bö, Axonal transection in the lesions of multiple sclerosis, N. Engl. J. Med. 338 (1998) 278–285.

[302] L. Wang, F.-S. Wang, M.E. Gershwin, Human autoimmune diseases: a comprehensive update, J. Intern. Med. 278 (2015) 369–395.

[303] S. Panda, J.L. Ding, Natural antibodies bridge innate and adaptive immunity, J. Immunol. 194 (2015) 13–20.

[304] K. Ohishi, M. Kanoh, H. Shinomiya, Y. Hitsumoto, S. Utsumi, Complement activation by cross-linked B cell-membrane IgM, J. Immunol. 154 (1995) 3173–3179.

[305] P. Rodien, A.M. Madec, J. Ruf, F. Rajas, H. Bornet, P. Carayon, J. Orgiazzi, Antibody-dependent cell-mediated cytotoxicity in autoimmune thyroid disease: relationship to antithyroperoxidase antibodies, J. Clin. Endocrinol. Metab. 81 (1996) 2595–2600.

[306] A. Lleo, P. Invernizzi, B. Gao, M. Podda, M.E. Gershwin, Definition of human autoimmunity-autoantibodies versus autoimmune disease, Autoimmun. Rev. 9 (2010) A259–A266.

[307] A. Davidson, B. Diamond, Autoimmune diseases, N. Engl. J. Med. 345 (2001) 340–350.

[308] C. Turesson, W.M. O'Fallon, C.S. Crowson, S.E. Gabriel, E.L. Matteson, Extra-articular disease manifestations in rheumatoid arthritis: incidence, trends and risk factors over 46 years, Ann. Rheum. Dis. 62 (2003) 722–727.

[309] V. Jones, P.C. Taylor, R.K. Jacoby, T.B. Wallington, Synovial synthesis of rheumatoid factors and immune complex constituents in early arthritis, Ann. Rheum. Dis. 43 (1984) 235–239.

[310] V.P. Nell, K.P. Machold, T.A. Stamm, G. Eberl, H. Heinzl, M. Uffmann, J.S. Smolen, G. Steiner, Autoantibody profiling as early diagnostic and prognostic tool for rheumatoid arthritis, Ann. Rheum. Dis. 64 (2005) 1731–1736.

[311] E.R. Vossenaar, A.J. Zendman, W.J. van Venrooij, G.J. Pruijn, PAD, a growing family of citrullinated enzymes: genes, features and involvement in disease, Bioassay 25 (2003) 1106–1118.

[312] C.C. Reparon-Schuijt, W.J. van Esch, C. van Kooten, G.A. Schellekens, B.A. de Jong, W.J. van Venrooij, F.C. Breedveld, C.L. Verweij, Secretion of anti-citrulline-containing peptide antibody by B lymphocyte in rheumatoid arthritis, Arthritis Rheum. 44 (2001) 41–47.

[313] Z. Wang, S.J. Nicholls, E.R. Rodriguez, O. Kommu, S. Hörkkö, J. Barnard, W.F. Reynolds, E.J. Topol, J.A. DiDonato, S.L. Hazen, Protein carbamylation links inflammation, smoking, uremia and atherogenesis, Nat. Med. 13 (2007) 1176–1184.

[314] V.F.A.M. Derksen, T.W.J. Huizinga, D. van der Woude, The role of autoantibodies in the pathophysiology of rheumatoid arthritis, Semin. Immunopathol. 39 (2017) 437–446.

[315] M. Juarez, H. Bang, F. Hammar, U. Reimer, B. Dyke, I. Sahbudin, C.D. Buckley, B. Fisher, A. Filer, K. Raza, Identification of novel vimentin antibodies in patients with early inflammatory arthritis, Ann. Rheum. Dis. 75 (2016) 1099–1107.

[316] M.M. Nielen, D. van Schaardenburg, H.W. Reesink, R.J. van de Stadt, I.E. van der Horst-Bruinsma, T. de Gast, M.R. Habibuw, J.P. Vandenbroucke, B.A. Dijkmans, Specific autoantibodies precede thy symptoms of rheumatoid arthritis: a study of serial measurements in blood donors, Arthritis Rheum. 50 (2004) 380–386.

[317] J. Shi, L.A. van de Stadt, E.W. Lewarht, T.W. Huizinga, D. Hamann, D. van Schaardenburg, R.E. Toes, L.A. Trouw, Anti-carbamylated protein (anti-CarP) antibodies precede the onset of rheumatoid arthritis, Ann. Rheum. Dis. 73 (2014) 780–783.

[318] C. Clavel, L. Nogueira, L. Laurent, C. Iobagui, C. Vincent, M. Sebbag, G. Serre, Induction of macrophage secretion to tumor necrosis factor α through Fcγ receptor IIa engagement by rheumatoid arthritis-specific autoantibodies to citrullinated proteins complexed with fibrinogen, Arthritis Rheum. 58 (2008) 678–688.

[319] L.A. Trouw, E.M. Haisma, E.W. Lewarht, D. van der Woude, A. Ioan-Fascinay, M.R. Daha, T.W. Huizinga, R.E. Toes, Anti-cyclic citrullinated peptide antibodies from rheumatoid arthritis patients activate complement via both the classical and alternative pathways, Arthritis Rheum. 60 (2009) 1923–1931.

[320] R. Khandpur, C. Carmona-Rivera, A. Vivekanandan-Giri, A. Gizinski, S. Yalavarthi, J.S. Knight, S. Friday, S. Li, R.M. Patel, V. Subramanian, P. Thompson, P. Chen, D.A. Fox, S. Pennathur, M.J. Kaplan, NETs are a source of citrullinated autoantigens and stimulate inflammatory responses in rheumatoid arthritis, Sci. Transl. Med. 5 (2013), 178ra40, 1–23, https://doi.org/10.1126/scitranslmed.3005580.

[321] L. Lisnevskaia, G. Murphy, D. Isenburg, Systemic lupus erythematosus, Lancet 384 (2014) 1878–1888.

[322] U.S. Gaipl, L.E. Munoz, G. Grossmayer, K. Lauber, S. Franz, K. Sarter, R.E. Voll, T. Winkler, A. Kuhn, J. Kalden, P. Kern, M. Herrmann, Clearance deficiency and systemic lupus erythematosus (SLE), J. Autoimmun. 28 (2007) 114–121.

[323] I. Fatal, N. Shental, D. Mevorach, J.-M. Anaya, A. Livneh, P. Langevitz, G. Zandman-Goddard, R. Pauzner, M. Lerner, M. Blank, M.-E. Hincapie, U. Gafter, Y. Naparstek, Y. Shoenfeld, E. Domany, I.R. Cohen, An antibody profile of systemic lupus erythematosus detected by antigen microarray, Immunology 130 (2010) 337–343.

[324] S. Han, H. Zhuang, S. Shumyak, L. Yang, W.H. Reeves, Mechanisms of autoantibody production in systemic lupus erythematosus, Front. Immunol. 6 (2015), article 228, 1–7, https://doi.org/10.3389/fimmu.2015.00228.

[325] M. Decramer, W. Janssens, M. Miravitlles, Chronic obstructive pulmonary disease, Lancet 379 (2012) 1341–1351.

[326] T.A. Wynn, T.R. Ramalingam, Mechanisms of fibrosis: therapeutic translation for fibrotic disease, Nat. Med. 18 (2012) 1028–1040.

[327] J.C. Hogg, A brief review of chronic obstructive pulmonary disease, Can. Respir. J. 19 (2012) 381–384.

[328] J.E. McDonough, R. Yuan, M. Suzuki, N. Seyednejad, W.M. Elliott, P.G. Sanchez, A.C. Wright, W.B. Gefter, L. Litzky, H.O. Coxson, P.D. Paré, D.D. Sin, R.A. Pierce, J.C. Woods, A.M. McWilliams, J.R. Mayo, S.C. Lam, J.D. Cooper, J.C. Hogg, Small-airway obstruction and emphysema in chronic obstructive pulmonary disease, N. Engl. J. Med. 365 (2011) 1567–1575.

[329] C.C. Taggart, C.M. Greene, T.P. Carroll, S.J. O'Neill, N.G. McElvaney, Elastolytic proteases. Inflammation resolution and dysregulation in chronic infective lung disease, Am. J. Respir. Crit. Care Med. 171 (2005) 1070–1076.

[330] S.K. Brode, S.C. Ling, K.R. Chapman, Alpha-1 antitrypsin deficiency: a commonly overlooked cause of lung disease, Can. Med. Assoc. J. 184 (2012) 1365–1371.

[331] S. Gompertz, C. O'Brien, D.L. Bayley, S.L. Hill, R.A. Stockley, Changes in bronchial inflammation during acute exacerbations of chronic bronchitis, Eur. Respir. J. 17 (2001) 1112–1119.

[332] G. Paone, V. Conti, A. Vestry, A. Leone, G. Puglisi, F. Benassi, G. Brunetti, G. Schmid, I. Cammarella, C. Terzano, Analysis of sputum markers in the evaluation of lung inflammation and functional impairment in symptomatic smokers and COPD patients, Dis. Markers 31 (2011) 91–100.

[333] K. Hoenderdos, A. Condliffe, The neutrophil in chronic obstructive pulmonary disease, too little, too late of too much, too soon? Am. J. Respir. Cell Mol. Biol. 48 (2013) 531–539.

[334] H. Desai, K. Eschberger, C. Wrona, L. Grove, A. Agrawal, B. Grant, J. Yin, G.I. Parameswaran, T. Murphy, S. Sethi, Bacterial colonization increases daily symptoms in patients with chronic obstructive pulmonary disease, Ann. Am. Thorac. Soc. 11 (2014) 303–309.

[335] F.D. Martinez, Genes, environments, development and asthma: a reappraisal, Eur. Respir. J. 29 (2007) 179–184.

[336] R.L. Miller, S.M. Ho, Environmental epigenetics and asthma: current concepts and call for studies, Am. J. Respir. Crit. Care Med. 177 (2008) 567–573.

[337] L. Borish, The immunology of asthma: asthma phenotypes and their implications for personalized treatment, Ann. Allergy Asthma Immunol. 117 (2016) 108–114.

[338] D.C. Baumgart, W.J. Sandborn, Crohn's disease, Lancet 380 (2012) 1590–1605.

[339] R. Ungaro, S. Mehandru, P.B. Allen, L. Peyrin-Biroulet, J.F. Colombel, Ulcerative colitis, Lancet 389 (2017) 1756–1770.
[340] The human microbiome project consortium, Structure, function and diversity of the healthy human microbiome, Nature 486 (2012) 207–214.
[341] O. Pabst, A.M. Mowat, Oral tolerance to food protein, Mucosal Immunol. 5 (2012) 232–239.
[342] P.B. Eckburg, E.M. Bik, C.N. Bernstein, E. Purdon, L. Dethlefsen, M. Sargent, S.R. Gill, K.E. Nelson, D.A. Relman, Diversity of the human intestinal microbial flora, Science 308 (2005) 1635–1638.
[343] J. Qin, R. Li, J. Raes, M. Aramugam, K.S. Burgdorf, C. Manichanh, T. Nielsen, N. Pons, F. Leverenz, T. Yamada, D.R. Mende, J. Li, J. Xu, S. Li, J. Cao, B. Wang, H. Liang, H. Zheng, Y. Xie, J. Tap, P. Lepage, M. Bertalan, J.M. Batto, T. Hansen, D. Le Paslier, A. Linneberg, H.B. Nielsen, E. Pelletier, P. Renault, T. Sicheritz-Ponten, K. Turner, H. Zhu, C. Yu, S. Li, M. Jian, Y. Zhou, Y. Li, X. Zhang, S. Li, N. Qin, H. Yang, J. Wang, S. Brunak, J. Doré, F. Guamer, K. Kristiansen, O. Pedersen, J. Parkhill, J. Weissenbach, MetaHIT Consortium, P. Bork, S.D. Ehrlich, J. Wang, A human gut microbial gene catalogue established by metagenomic sequencing, Nature 464 (2010) 59–65.
[344] M.T. Abreu, Toll-like receptor signalling in the intestinal epithelium: how bacterial recognition shapes intestinal function, Nat. Rev. Immunol. 10 (2010) 131–144.
[345] J.L. Round, S.K. Mazmanian, The gut microbiota shapes intestinal immune responses during health and disease, Nat. Rev. Immunol. 9 (2009) 313–323.
[346] H. Chu, S.K. Mazmanian, Innate immune recognition of the microbiota promote host-microbial symbiosis, Nat. Immunol. 14 (2013) 668–675.
[347] F.G. Chirdo, O.R. Millington, H. Beacock-Sharp, A.M. Mowat, Immunomodulatory dendritic cells in intestinal lamina propria, Eur. J. Immunol. 35 (2005) 1831–1840.
[348] D.A. Chistiakov, Y.V. Bobryshev, E. Kozarov, I.A. Sobenin, A.N. Orekhov, Intestinal mucosal tolerance and impact of gut microbiota to mucosal tolerance, Front. Microbiol. 5 (2015), article 781, 1–9, https://doi.org/10.3389/fmicb.2014.00781.
[349] B. Dubois, G. Joubert, M. Gomez de Agüero, M. Guanovic, A. Goubier, D. Kaiserlian, Sequential role of plasmacytoid dendritic cells and regulatory T cells in oral tolerance, Gastroenterology 137 (2009) 1019–1028.
[350] O.J. Harrison, F.M. Powrie, Regulatory T cells and immune tolerance in the intestine, Cold Spring Harbor Perspect. Biol 5 (2013), a018341, 1–17, https://doi.org/10.1101/cshperspect.a018341.
[351] L.G. van der Flier, H. Clevers, Stem cells, self-renewal, and differentiation in the intestinal epithelium, Annu. Rev. Physiol. 71 (2009) 241–260.
[352] N. Miron, V. Cristea, Enterocytes: active cells in tolerance to food and microbial antigens in the gut, Clin. Exp. Immunol. 167 (2011) 405–412.
[353] M.E.V. Johansson, G.C. Hansson, Mucus and the coblet cell, Dig. Dis. 31 (2013) 305–309.
[354] H.C. Clevers, C.L. Bevins, Paneth cells: maestros of the small intestinal crypts, Annu. Rev. Physiol. 75 (2013) 289–311.
[355] H. Miller, J. Zhang, R. KuoLee, G.B. Patel, W. Chen, Intestinal M cells: the fallible sentinels? World J. Gastroenterol. 13 (2007) 1477–1486.
[356] C. Stemini, L. Anselmi, E. Rozengurt, Enteroendocrine cells: a site of taste in gastrointestinal chemosensing, Curr. Opin. Endocrinol. Diabetes Obes. 15 (2008) 73–78.
[357] M.E.V. Johansson, D. Ambort, T. Pelaseyed, A. Schütte, J.K. Gustafsson, A. Ermund, D.B. Subramani, J.M. Holmén-Larsson, K.A. Thomsson, J.H. Bergström, S. van der Post, A.M. Rodriguez-Piñeiro, H. Sjövall, M. Backström, G.C. Hansson, Composition and functional role of the mucus layers in the intestine, Cell. Mol. Life Sci. 68 (2011) 3635–3641.
[358] M.E. V Johansson, J.M. Holmén-Larsson, G.C. Hansson, The two mucus layers of colon are organized by the MUC2 mucin, whereas the outer layer is a legislator of host-microbial interactions, Proc. Natl. Acad. Sci. U.S.A. 108 (Suppl. 1) (2011) 4659–4665.
[359] S. Vaishnava, M. Yamamoto, K.M. Severson, K.A. Ruhn, X. Yu, O. Koren, R. Ley, E.K. Wakeland, L.V. Hooper, The antibacterial lectin RegIIIγ promotes the spatial segregation of microbiota and host in the intestine, Science 334 (2011) 255–258.
[360] D.R. Herbert, J.Q. Yang, S.P. Hogan, K. Groschwitz, M. Khodoun, A. Munitzu, T. Orekov, C. Perkins, Q. Wang, F. Brombacher, J.F. Urban Jr., M.E. Rothenberg, F.D. Finkelman, Intestinal epithelial cell secretion of RELM-β protects against gastrointestinal worm infection, J. Exp. Med. 206 (2009) 2947–2957.
[361] E.M. Porter, C.L. Bevins, D. Ghosh, T. Ganz, The multifaceted Paneth cell, Cell. Mol. Life Sci. 59 (2002) 156–170.
[362] N.H. Salzman, K. Hung, D. Haribhai, H. Chu, J. Karlsson-Sjöberg, E. Amir, P. Teggatz, M. Barman, M. Hayward, D. Eastwood, M. Stoel, Y. Zhou, E. Sodergren, G.M. Weinstock, C.L. Bevins, C.B. Williams, N.A. Bos, Enteric defensins are essential regulators of intestinal microbial ecology, Nat. Immunol. 11 (2010) 76–83.

[363] T.S. Stappenbeck, J.D. Rioux, A. Mizoguchi, T. Saitoh, A. Huett, A. Darfeuille-Michaud, T. Wileman, N. Mizushima, S. Carding, S. Akira, M. Parkes, R.J. Xavier, Crohn disease: a current perspective on genetics, autophagy, and immunity, Autophagy 7 (2011) 355–374.

[364] J.D. Soderholm, G. Olaison, K.H. Peterson, L.E. Franzén, T. Lindmark, M. Wirén, C. Tagesson, R. Sjödahl, Augmented increase in tight junction permeability by luminal stimuli in the non-inflamed ileum of Crohn's disease, Gut 50 (2002) 307–313.

[365] Y. Sun, B.M. Fihn, H. Sjovall, M. Jodal, Enteric neurons modulate the colonic permeability response to luminal bile acids in rat colon in vivo, Gut 53 (2004) 362–367.

[366] F. Heller, P. Florian, C. Bojarski, J. Richter, M. Christ, B. Hillenbrand, J. Mankertz, A.H. Gitter, N. Bürgel, M. Fromm, M. Zeitz, I. Fuss, W. Strober, J.D. Schulzke, Interleukin-13 is the key effector Th2 cytokine in ulcerative colitis that affects epithelial tight junctions, apoptosis, and cell restitution, Gastroenterology 129 (2005) 550–564.

[367] S. Buhner, C. Buning, J. Genschel, K.I. Kling, D. Herrmann, A. Dignass, I. Kuechler, S. Krueger, H.H. Schmidt, H. Lochs, Genetic basis for increased intestinal permeability in families with Crohn's disease: role of CARD15 3020insC mutation? Gut 55 (2006) 342–347.

[368] D.C. Baumgart, S.R. Carding, Inflammatory bowel disease: cause and immunobiology, Lancet 369 (2007) 1627–1640.

[369] A. Nakazawa, I. Dotan, J. Brimnes, M. Allez, L. Shao, F. Tsushima, M. Azuma, L. Mayer, The expression and function of costimulatory molecules B7H and B7-H1 on colonic epithelial cells, Gastroenterology 126 (2004) 1347–1357.

[370] K. Ina, J. Itoh, K. Fukushima, K. Kusugami, T. Yamauchi, K. Kyokane, A. Imada, D.G. Binion, A. Musso, G.A. West, G.M. Dobrea, T.S. McCormick, E.G. Lapetina, A.D. Levine, C.A. Ottaway, C. Fiocchi, Resistance of Crohn's disease T cells to multiple apoptosis signals is associated with Bcl-2/Bax mucosal imbalance, J. Immunol. 163 (1999) 1081–1090.

[371] B. Martin, A. Banz, B. Bienvenu, C. Cordier, N. Dautigny, C. Bécourt, B. Lucas, Suppression of $CD4^+$ T lymphocyte effector functions by $CD4^+CD25^+$ cells in vivo, J. Immunol. 172 (2004) 3391–3398.

[372] R. Furlan, S. Ardizzone, L. Palazzolo, A. Rimoldi, F. Perego, F. Barbic, M. Bevilacqua, L. Vago, G. Bianchi Porro, A. Malliani, Sympathetic overactivity in active ulcerative colitis, Am. J. Physiol. Regul. Integr. Comp. Physiol. 290 (2006) R224–R232.

[373] B. Bonaz, Is-there a place for vagus nerve stimulation in inflammatory bowel disease? Bioelectr. Med. 4 (2018) 4, 1–9, https://doi.org/10.1186/s42234-018-0004-9.

[374] M. Leppkes, C. Becker, I.I. Ivanov, S. Hirth, S. Wirtz, C. Neufert, S. Pouly, A.J. Murphy, D.M. Valenzuela, G.D. Yancopoulos, B. Becher, D.R. Littman, M.F. Neurath, RORγ-expressing Th17 cells induce murine chronic intestinal inflammation via redundant effects of IL-17A and IL17-F, Gastroenterology 136 (2009) 257–267.

[375] S.-Y. Lee, S.H. Lee, E.-J. Yang, E.-K. Kim, J.-K. Kim, D.-Y. Shin, M.-L. Cho, Metformin ameliorates inflammatory bowel disease by suppression of the STAT3 signaling pathway and regulation of the between Th17/Treg balance, PLoS One 10 (2015), e0135858, 1–12, https://doi.org/10.1371/journal.pone.0315858.

[376] G. Montelone, F. Trapasso, T. Parrello, L. Baincone, A. Stella, R. Iuliano, F. Luzza, A. Fusco, F. Pallone, Bioactive IL-18 expression is up-regulated in Crohn's disease, J. Immunol. 163 (1999) 143–147.

[377] D.K. Podolsky, Inflammatory bowel disease, N. Engl. J. Med. 347 (2002) 417–429.

[378] A. Di Sabatino, P. Biancheri, L. Rovedatti, T.T. Macdonald, C.R. Corazza, New pathogenic paradigms in inflammatory bowel disease, Inflamm. Bowel Dis. 18 (2012) 368–371.

[379] S.R. McKeown, Defined normoxia, physoxia and hypoxia in tumours – implications for treatment response, Br. J. Radiol. 87 (2014), 20130676, 1–12, https://doi.org/10.1259/brj.20130676.

[380] M.A. Konerding, E. Fait, A. Gaumann, 3D microvascular architecture of pre-cancerous lesions and invasive carcinomas of the colon, Br. J. Canc. 84 (2001) 1354–1362.

[381] D.W. Siemann, The unique characteristics of tumor vasculature and preclinical evidence for its selective disruption by tumor-vascular disrupting agents, Cancer Treat Rev. 37 (2011) 63–74.

[382] G.L. Semenza, Hypoxia-inducible factor 1 and cancer pathogenesis, IUBMB Life 60 (2008) 591–597.

[383] J. Casavaugh, K.M. Lounsbury, Hypoxia-mediated biological control, J. Cell. Biochem. 112 (2011) 735–744.

[384] G.J. Kelloff, J.M. Hoffman, B. Johnson, H.I. Scher, B.A. Siegel, E.Y. Cheng, B.D. Cheson, J. O'Shaughnessy, K.Z. Guyton, D.A. Mankoff, L. Shankar, S.M. Larson, C.C. Sigman, R.L. Schilsky, D.C. Sullivan, Progress and promise of FDG-PET imaging for cancer patient management and oncologic drug development, Clin. Cancer Res. 11 (2005) 2785–2808.

[385] P. Vaupel, M. Höckel, A. Mayer, Detection and characterization of tumor hypoxia using pO_2 histography, Antioxidants Redox Signal. 9 (2007) 1221–1235.

[386] A.C. Koong, V.K. Mehta, Q.T. Le, G.A. Fisher, D.J. Terris, J.M. Brown, A.J. Bastidas, M. Vierra, Pancreatic tumors show high level of hypoxia, Int. J. Radiat. Oncol. Biol. Phys. 48 (2000) 919–922.

[387] B. Movsas, J.D. Chapman, A.L. Hanlon, E.M. Horwitz, W.H. Pinover, R.E. Greenberg, C. Stobbe, G.E. Hanks, Hypoxia in human prostate carcinoma: an Eppendorf pO2 study, Am. J. Clin. Oncol. 24 (2001) 458–461.

[388] C. Parker, M. Milosevic, A. Toi, J. Sweet, T. Panzarella, R. Bristow, C. Catton, P. Catton, J. Crook, M. Gospodarowicz, M. McLean, P. Warde, R.P. Hill, Polarographic electrode study of tumor oxygenation in clinically localized prostate cancer, Int. J. Radiat. Oncol. Biol. Phys. 58 (2004) 750–757.

[389] C.N. Coleman, J.B. Mitchell, K. Camphausen, Tumor hypoxia: chicken, egg, or a piece of the farm? J. Clin. Oncol. 20 (2002) 610–615.

[390] P. Okunieff, I. Ding, P. Vaupel, M. Höckel, Evidence for and against hypoxia as the primary cause of tumor aggressiveness, Adv. Exp. Med. Biol. 510 (2003) 69–75.

[391] J.L. Spivak, The anemia of cancer: death by a thousand cuts, Nat. Rev. Canc. 5 (2005) 543–555.

[392] M. Höckel, K. Schlenger, B. Aral, M. Mitze, U. Schaffer, P. Vaupel, Association between tumor hypoxia and malignant progression in advanced cancer of the uterine cervix, Cancer Res. 56 (1996) 4509–4515.

[393] K. Fischer, P. Hoffmann, S. Voelkl, N. Meidenbauer, J. Ammer, M. Edinger, E. Gottfired, S. Schwarz, G. Rothe, S. Hoves, K. Renner, B. Timischl, A. Mackensen, L. Kunz-Schughart, R. Andreesen, S.W. Krause, M. Kreutz, Inhibitory effect of tumor cell-derived lactic acid on human T cells, Blood 109 (2007) 3812–3819.

[394] S. Walenta, M. Wetterling, M. Lehrke, G. Schwickert, K. Sundfør, E.K. Rofstad, W. Mueller-Klieser, High lactate levels predict likelihood of metastases, tumor recurrence, and restricted patient survival in human cervical cancers, Cancer Res. 60 (2000) 916–921.

[395] S. Walenta, T. Schroeder, W. Mueller-Klieser, Lactate in solid malignant tumors: potential basis of a metabolic classification in clinical oncology, Curr. Med. Chem. 11 (2004) 2195–2204.

[396] X. Mu, W. Shi, Y. Xu, C. Xu, T. Zhao, B. Geng, J. Yang, J. Pan, S. Hu, C. Zhang, J. Zhang, C. Wang, J. Shen, Y. Che, Z. Liu, Y. Lv, H. Wen, Q. You, Tumor-derived lactate induces M2 macrophage polarization via the activation of the ERK/STAT3 signaling pathway in breast cancer, Cell Cycle 17 (2018) 428–438.

[397] J. Choi, J. Gyamfi, H. Jang, J.S. Koo, The role of tumor-associated macrophage in breast cancer biology, Histol. Histopathol. 33 (2018) 133–145.

[398] A.A. Mantovani, A. Sica, S. Sozzani, P. Allavena, A. Vecchi, M. Locati, The chemokine system in diverse forms of macrophage activation and polarization, Trends Immunol. 25 (2004) 677–686.

[399] E. Holm, E. Hagmuller, U. Staedt, G. Schlickeiser, H.J. Günther, H. Leweling, M. Tokus, H.B. Kollmar, Substrate balances across colonic carcinomas in humans, Cancer Res. 55 (1995) 1373–1378.

[400] S. Romero-Garcia, M.M. Moreno-Altamirano, H. Prado-Garcia, F.J. Sánchez-Garcia, Lactate contribution to the tumor microenvironment: mechanisms, effects on immune cells and therapeutic relevance, Front. Immunol. 7 (2016), article 52, 1–11, https://doi.org/10.3389/fimmu.2016.00052.

[401] F. Végran, R. Boidot, C. Michiels, P. Sonveaux, O. Feron, Lactate influx through the endothelial cell monocarboxylate transporter MCT1 supports an NFκB/IL-8 pathway that drives tumor angiogenesis, Cancer Res. 71 (2011) 2550–2560.

[402] T. Chamnee, P. Ontong, N. Itano, Hyaluronan: a modulator of the tumor microenvironment, Cancer Lett. 375 (2016) 20–30.

[403] U.G.A. Sattler, W. Mueller-Klieser, The anti-oxidant capacity of tumor glycolysis, Int. J. Radiat. Biol. 85 (2009) 963–971.

[404] J. Chiche, C. Brahimi-Horn, J. Pouysségur, Tumor hypoxia induces a metabolic shift causing acidosis: a common feature in cancer, J. Cell Mol. Med. 14 (2010) 771–794.

[405] R. Martinez-Zaguilan, E.A. Seftor, R.E. Seftor, Y.W. Chu, R.J. Gillies, M.J. Hendrix, Acidic pH enhances the invasive behavior of human melanoma cells, Clin. Exp. Metastasis 14 (1996) 176–186.

[406] A. Lardner, The effect of extracellular pH on immune function, J. Leukoc. Biol. 69 (2001) 522–530.

[407] C. Sauvant, M. Nowak, C. Wirth, B. Schneider, A. Riemann, M. Gekle, O. Thews, Acidosis induces multi-drug resistance in rat prostate cancer cells (AT1) in vitro and in vivo by increasing the activity of the p-glycoprotein via activation of p38, Int. J. Cancer 123 (2008) 2532–2542.

[408] B.A. Teicher, Malignant cells, directors of the malignant process: role of transforming growth factor-β, Cancer Metastasis Rev. 20 (2001) 133–143.

[409] M. Dong, G.C. Blobe, Role of transforming growth factor-β in hematologic malignancies, Blood 107 (2006) 4589–4596.

[410] S.H. Wrzesinski, Y.Y. Wan, R.A. Flavell, Transforming growth factor-β and the immune response: implications for anticancer therapy, Clin. Cancer Res. 13 (2007) 5262–5270.

[411] B. Bierie, H.L. Moses, Transforming growth factor beta (TGF-β) and inflammation in cancer, Cytokine Growth Factor Rev. 21 (2010) 49–59.

[412] G.C. Blobe, W.P. Schiemann, H.F. Lodish, Role of transforming growth factor β in human disease, N. Engl. J. Med. 342 (2000) 1350–1358.

[413] A. Galliher, W. Schiemann, β_3 Integrin and Src facilitate transforming growth factor-β mediated induction of epithelial-mesenchymal transition in mammary epithelial cells, Breast Cancer Res. 8 (2006), R42, 1–16, https://doi.org/10.1186/bcr1524.

[414] S. Gorsch, V. Memoli, T. Stukel, L.I. Gold, B.A. Arrick, Immunohistochemical staining for transforming growth factor β1 associates with disease progression in human breast cancer, Cancer Res. 52 (1992) 6949–6952.

[415] E. Friedman, L.I. Gold, D. Klimstra, Z.S. Zeng, S. Winawer, A. Cohen, High levels of transforming growth factor β1 correlate with disease progression in human colon cancer, Cancer Epidemiol. Biomark. Prev. 4 (1995) 549–554.

[416] H. Saito, S. Tsujitani, S. Oka, A. Kondo, M. Ikeguchi, M. Maeta, N. Kaibara, An elevated serum level of transforming growth factor-β1 (TGF-β1) significantly correlates with lymph node metastasis and poor prognosis in patients with gastric carcinoma, Anticancer Res. 20 (2000) 4489–4493.

[417] S. Mocellin, F.M. Marincola, H.A. Young, Interleukin-10 and the immune response against cancer: a counterpoint, J. Leukoc. Biol. 78 (2005) 1043–1051.

[418] K.L. Dennis, N.R. Blatner, F. Gounair, K. Khazaie, Current status of IL-10 and regulatory T-cells in cancer, Curr. Opin. Oncol. 25 (2013) 637–645.

[419] S. Löb, A. Königsrainer, D. Zieker, B.L.D.M. Brücher, H.G. Rammensee, G. Opelz, P. Terness, Ido1 and Ido2 are expressed in human tumors: levo- but not dextro-1-methyl tryptophan inhibits tryptophan metabolism, Cancer Immunol. Immunother. 58 (2009) 153–157.

[420] J. Godin-Ethier, L.-A. Hanafi, C.A. Piccirillo, R. Lapointe, Indoleamine 2,3-dioxygenase expression in human cancer: clinical and immunologic perspectives, Clin. Cancer Res. 17 (2011) 6985–6991.

[421] A.L. Mellor, D.B. Keskin, T. Johnson, P. Chandler, D.H. Munn, Cells expressing indoleamine 2,3-dioxygenase inhibit T cell responses, J. Immunol. 168 (2002) 3771–3776.

[422] D.H. Munn, A.L. Mellor, Indoleamine 2,3 dioxygenase and metabolic control of immune responses, Trends Immunol. 34 (2013) 137–143.

[423] N. van Baren, B.J. Van den Eynde, Tryptophan-degrading enzymes in tumoral immune resistance, Front. Immunol. 6 (2015), 34, 1–9, https://doi.org/10.3389/fimmu.2015.00034.

[424] G. Pietra, M. Vitale, L. Moretta, M.C. Mingari, How melanoma cells inactivate NK cells, OncoImmunology 1 (2012) 974–975.

[425] D. Wang, Y. Saga, H. Mizukami, N. Sato, H. Nonaka, H. Fujiwara, Y. Takai, S. Machida, O. Takikawa, K. Ozawa, M. Suzuki, Indoleamine-2,3-dioxygenase, an immunosuppressive enzyme that inhibits natural killer cell function, as a useful target for ovarian cancer therapy, Int. J. Oncol. 40 (2012) 929–934.

[426] T. Yin, S. He, X. Liu, W. Jiang, T. Ye, Z. Lin, Y. Sang, C. Su, Y. Wan, G. Shen, X. Ma, M. Yu, F. Guo, Y. Liu, L. Li, Q. Hu, Y. Wang, Y. Wei, Extravascular red blood cells and hemoglobin promote tumor growth and therapeutic resistance as endogenous danger signals, J. Immunol. 194 (2015) 429–437.

[427] B. Kaya, O. Çiçek, F. Erdi, S. Findik, Y. Karatas, H. Esen, F. Keskin, E. Kalkan, Intratumoral hemorrhage-related differences in the expression of vascular endothelial growth factor, basic fibroblast growth factor and thioredoxin reductase 1 in human glioblastoma, Mol. Clin. Oncol. 5 (2006) 343–346.

[428] B.B. Aggarwal, Nuclear factor-κB, Cancer Cell 6 (2004) 203–208.

[429] R. Tarnawski, K. Skladowski, B. Maciejewski, Prognostic value of hemoglobin concentration in radiotherapy for cancer of supraglottic larynx, Int. J. Radiat. Oncol. Biol. Phys. 38 (1997) 1007–1011.

[430] M. Grogan, G.M. Thomas, I. Melamed, F.L. Wong, R.G. Pearcey, P.K. Joseph, L. Portelance, J. Crook, K.D. Jones, The importance of hemoglobin levels during radiotherapy for carcinoma of the cervix, Cancer 86 (1999) 1528–1536.

[431] S.J.-P. Van Belle, What is the value of hemoglobin as a prognostic and predictive factor in cancer? EJC Suppl. 2 (2004) 11–19.

[432] J. Kisucka, C.E. Butterfield, D.G. Duda, S.C. Eichenberger, S. Saffaripour, J. Ware, Z.M. Ruggeri, R.K. Jain, J. Folkman, D.D. Wagner, Platelets and platelet adhesion support angiogenesis while preventing excessive hemorrhage, Proc. Natl. Acad. Sci. U.S.A. 103 (2006) 855–860.

[433] M. Labelle, S. Begum, R.O. Hynes, Direct signalling between platelets and cancer cells induces an epithelial-mesenchymal-like transition and promotes metastasis, Cancer Cell 20 (2011) 576–590.

[434] T. Placke, M. Örgel, M. Schaller, G. Jung, H.G. Rammensee, H.G. Kopp, H.R. Saith, Platelet-derived MHC class I confers a pseudonormal phenotype to cancer cells that subverts the antitumor reactivity of natural killer immune cells, Cancer Res. 72 (2012) 440–448.
[435] G.L. Beatty, W.L. Gladney, Immune escape mechanisms as a guide for cancer immunotherapy, Clin. Cancer Res. 21 (2015) 687–692.
[436] P. Gotwals, S. Cameron, D. Cipolletta, V. Cremasco, A. Crystal, B. Hewes, B. Mueller, S. Quarantino, C. Sabatos-Peyton, L. Petruzzelli, J.A. Engelmann, G. Dranoff, Prospects for combining targeted and conventional cancer therapy with immunotherapy, Nat. Rev. Canc. 17 (2017) 286–301.
[437] L. Hornyák, N. Dobos, G. Koncz, Z. Karányi, D. Páll, Z. Szabo, G. Halmos, L. Székvölyi, The role of indoleamine-2,3-dioxygenase in cancer development, diagnostics, and therapy, Front. Immunol. 9 (2018), 151, 1–8, https://doi.org/10.3389/fimmu.2018.00151.
[438] P. Michieli, Hypoxia, angiogenesis and cancer therapy. To breathe or not to breathe? Cell Cycle 8 (2009) 3291–3296.
[439] W.R. Wilson, M.P. Hay, Targeting hypoxia in cancer therapy, Nat. Rev. Canc. 11 (2011) 393–410.

CHAPTER

10

Organ Damage and Failure

10.1 ELIMINATION OF WASTES AND XENOBIOTICS FROM THE ORGANISM

In tissues and organs, metabolic processes generate numerous end products and wastes that are released into blood and should be detoxified and eliminated from the organism. These important functions are mainly performed by the kidney, liver, and spleen. Of course, lungs and skin participate also in elimination of some metabolic products.

Otherwise, different drugs or more general xenobiotics can be taken up via the intestine, skin, or lungs and are transported by the blood. Their metabolism is usually divided into three main phases, modification, conjugation, and further modification and excretion [1,2]. The major organ for these biotransformations is the liver. However, some reactions can also take place in other organs and tissues. Elimination of transformed xenobiotics occurs mainly via the kidney. Thus, in addition to their main functions, intact liver and kidneys are very crucial for detoxification and excretion of metabolic products, drugs, and others.

In this chapter, functional disturbances of the kidney, liver, and spleen are highlighted. Failure of kidney and liver has dramatical consequences for the functioning of the organism. Development of sepsis and multiple organ dysfunction syndrome are further topics of this chapter.

10.2 KIDNEY DYSFUNCTIONS

10.2.1 Homeostatic Conditions in Intact Kidney

Kidneys are composed of multiple independently working units, the nephrons, which permanently filter the blood to eliminate waste products and to concentrate them in the urine. In the human organism, around two million nephrons are generally contained in both kidneys [3]. On average, 1 L of urine is produced per day by an adult person.

In a nephron, blood is first filtered flowing through a branched capillary network in the glomerulus. Each capillary is surrounded by three distinctive layers forming a sieve that filters blood components [4]. The resulting ultrafiltrate flows through the whole length of the nephron, where many filtered substances are reabsorbed, and others secreted from adjacent blood capillaries. In other words, complex exchange processes occur between the filtrate and blood. Finally, filtrate from nephrons passes through collecting tubules to form urine under reabsorption of water. Details about renal physiology are given in physiological textbooks.

To fulfill these clearance functions, a constant flow rate of blood through glomeruli should be ensured. Regulation of renal blood flow is of utmost importance to maintain a more or less constant glomerular filtration rate (GFR). In

Cell and Tissue Destruction
https://doi.org/10.1016/B978-0-12-816388-7.00010-3

healthy humans, renal blood flow is constant at a systolic blood pressure in the range from 70 to 130 mmHg [5]. Two regulatory mechanisms provide this homeostasis. Elevated blood pressures are counterregulated by contraction of afferent arterioles, while reduced pressure values induce a NaCl-based tubuloglomerular feedback mechanism that dilates the afferent arteriole [5].

With these functions, kidneys participate in control of blood osmolarity, acid–base balances, maintenance of electrolyte concentration, and removal of toxins. Another, nonnephron-based kidney function is the secretion of calcitriol, erythropoietin, and renin.

10.2.2 Estimated Glomerular Filtration Rate

Several parameters are used to describe physiological kidney functions. An often applied parameter is the estimated GFR. It is defined as the fluid volume filtered from glomerular capillaries into the Bowman's capsule related to a time unit. In humans, this value cannot be measured directly. It is calculated from clearance measurements or serum levels of defined markers subjected to glomerular filtration [6].

Clearance of exogenous substances such as inulin, iohexol, ^{51}Cr-EDTA, ^{99}Tc-diethylenetriaminepentaacetic acid, and ^{125}I-labeled iothalamate generates reliable GFR values [7]. However, these methods are improper for routine use. Instead serum creatinine is widely applied as an indirect crude marker to assess GFR despite the presence of numerous factors affecting this value [8,9]. Other parameters used for determination of GFR are blood urea nitrogen and cystatin C [10,11].

As creatinine clearance depends on muscle mass, age, sex, and other parameters, the Cockcroft–Gault formula is often used for calculation of GFR to consider some of these factors [12].

Diminished value of GFR may result from a reduced number of nephrons. Otherwise, the filtration rate of a single nephron can be increased by an enhanced glomerular capillary pressure or by glomerular hypertrophy, compensating, thus, the loss of nephrons [13].

10.2.3 Chronic Kidney Disease

A gradual loss of kidney functions appears over a period of many months and years and is without symptoms at the beginning. The onset of kidney disturbances is often not recognized as unperturbed parts of kidney can compensate deteriorated ones. Unspecific symptoms of chronic kidney disease are lethargy, itch, or loss of appetite. Risk factors for the chronic kidney disease are hyperglycemia (diabetes mellitus), hypertension, dyslipidemia, smoking, glomerulonephritis, and polycystic kidney disease.

On the basis of the estimated GFR and markers of kidney damage, kidney function is divided into five stages [14]. Normal kidney function corresponds to stage G1 and is characterized by a GFR value higher than 90 mL min^{-1} 1.73 m^{-2} and the absence of proteinuria. In higher stages, there is a stepwise worsening of kidney parameters.

According to international guidelines, chronic kidney disease is defined as a decreased kidney function with GFR less than 60 mL min^{-1} 1.73 m^{-2} or the presence of one or more markers of kidney damage, or both criteria, of at least 3 months [14,15]. The following markers of kidney damage are listed: albuminuria, i.e., an albumin to creatinine ratio higher than 30 mg g^{-1}, urinary sediment abnormality, electrolyte or other abnormality due to tubular disorder, abnormalities on histology, structural abnormalities detected by imaging, and history of a kidney transplantation [15].

Kidney failure (stage G5), also known as end-stage kidney disease, is defined by a GFR value lower than 15 mL min^{-1} 1.73 m^{-2} [15]. In this stage, kidneys are unable to function. Both acute causes such as low blood pressure, blockade of the urinary tract, certain drugs, and excessive damage of red blood and muscle cells as well as chronic reasons such as hyperglycemia, hypertension, and some kidney diseases can result in kidney failure.

In biopsy samples, chronic kidney disease appears as glomerular sclerosis, tubular atrophy, and interstitial fibrosis. The final stage of damage is renal fibrosis. Because of impaired kidney function, uremic retention solutes may accumulate in the body and exert adverse effects. Functional systemic consequences of chronic kidney disease are inflammation, immune dysfunction, anemia, platelet dysfunction, bleeding, disruption of gut functions, altered drug metabolism, and further progression of chronic kidney disease [16,17]. Cardiovascular disease, cancer, and mineral bone disease are promoted by kidney insufficiency [18–21].

10.2.4 Acute Kidney Injury by Hemoglobin and Myoglobin

Kidney can clear from circulation of large quantities of free hemoglobin and myoglobin. This uptake can result in acute kidney injury. Examples are massive release of hemoglobin from infected red blood cells during malaria [22] or after major aortic or cardiac surgery [23,24]. Crushing injury to skeletal muscle is associated with muscle swelling and/or neurological disturbances in the affected regions. Systemic manifestations after crush injury are known as crush syndrome. Thereby, systemic effects are related to traumatic rhabdomyolysis, i.e., to release of myoglobin, creatinine, creatine kinase, phosphate, potassium, and others from injured skeletal muscle [25,26].

Excessive intravascular hemolysis and rhabdomyolysis lead to exhaustion of protecting haptoglobin and hemopexin and generate free hemoglobin, free myoglobin, and the highly cytotoxic ferric protoporphyrin IX (free heme) (see Sections 4.4.1 and 4.4.2). Free hemoglobin and myoglobin are filtered through the glomerular barrier in the kidney and reabsorbed by proximal tubular cells, where they contribute to kidney damage [24,27–31].

In tubular cells, low amounts of free hemoglobin and free myoglobin can be detoxified by intracellular heme oxygenase. However, in the presence of higher amounts of these heme agents, heme oxygenase exerts prooxidative activities and contributes to cell damage [30].

Several mechanisms contribute to hemoglobin-induced damage of tubular cells. It releases the cytotoxic free heme. The latter agent and other porphyrins are known to block proteasome activity by binding to high-affinity binding sites within the 26S catalytic unit of the proteasome [32,33]. In turn, proteasome inhibition results in accumulation of misfolded proteins and triggers the unfolded protein response [34]. Heme exposure to murine embryonic fibroblasts and human proximal tubule cells as well as transfusion of guinea pigs with aged blood clearly revealed enhanced oxidative stress and unfolded protein response in cells [30,33]. Within exposed cells, oxidative stress is derived from redox reactions of free iron ions generated by heme oxygenase reaction and/or from oxidative reactions in lipid matrices catalyzed by inserted free heme. This stress results in accumulation of lipid peroxidation and cross-linking products and depletion of glutathione and ATP and promotes ferroptosis [30,35].

10.3 LIVER DAMAGE

10.3.1 Modular Organization of Liver

Similar to the kidney, the liver is also composed of single modules. Thus, the disposal of some modules can be better compensated by the other ones. As in kidney injury, the deficit of some hepatic units remains often unrecognized. Microscopic cross section shows a hexagonal arrangement of basic hepatic units. In each unit, incoming blood flows from peripheral portal vein branches and branches of the hepatic artery through capillaries surrounded by hepatocytes to the central vein of this units. Bile duct and lymphatic vessels and vagus nerve branches are further constituents of these units.

Liver produces numerous enzymes necessary for digestion and further conversion of absorbed foodstuffs. Other proteins produced by the liver are involved in systemic immune functions and becomes activated when blood escapes from circulation. Detoxification of metabolic end products and xenobiotics are further tasks of liver cells. With these roles, liver fulfill systemic regulatory functions of utmost importance.

10.3.2 Nonalcoholic Fatty Liver Disease

In nonalcoholic fatty liver disease, liver accumulates an increasing amount of lipids. Risk factors for nonalcoholic hepatic steatosis are diabetes, obesity, and advanced age. Increased intake of dietary fat is a main reason of this disease. Increased lipolysis from the fat cells, insulin resistance, and hepatic de novo lipogenesis contribute also to lipid accumulation [36–38]. Several metabolic dysfunctions such as insulin resistance, dyslipidemia, and cardiovascular disturbances are related to fatty liver disease [39,40].

The balance between lipid acquisition and lipid disposal is disturbed in hepatic steatosis. However, molecular details contributing to this mismatch are not fully elucidated. Several mechanisms concerning uptake of circulating lipids, de novo lipogenesis, fatty acid oxidation, and export of lipids via very low-density lipoproteins can be involved in altered hepatic lipid metabolism [41].

Nonalcoholic fatty liver disease encompasses nonalcoholic steatohepatitis as well, which includes inflammation, recruitment of immune cells, and damage of hepatocytes [42]. This inflammation contributes to mitochondrial dysfunction including decrease of glutathione, diminished expression of glutathione peroxidase and superoxide dismutase, increased formation of reactive species, increased lipid oxidation, and damage of mitochondrial DNA [43–45]. Peroxisomes are also suggested as a source of reactive species during fatty acid oxidation in nonalcoholic steatohepatitis [46].

10.3.3 Other Kinds of Liver Disease

More than 100 different kinds of liver disease are known. In addition to the nonalcoholic fatty liver disease, the most common liver diseases are hepatitis, alcoholic fatty liver disease, liver cancer, primary sclerosing cholangitis, Budd–Chiari syndrome, and some hereditary diseases such as Wilson's disease, α_1-antitrypsin deficiency, glycogen storage disease type II, and Gilbert's syndrome.

Hepatitis may be caused by heavy alcohol consumption, autoimmune processes, certain drugs, and toxins, but in most cases, virus infections are the reason for liver inflammation [47–49].

In the liver, alcohol is mainly metabolized via the consecutive action of alcohol dehydrogenase and acetaldehyde dehydrogenase. In both enzymatic steps, NAD is reduced to $NADH^+$. An increased $NADH^+$/NAD ratio seriously affects the metabolism of carbohydrates and lipids and leads to development of liver steatosis [50,51].

10.3.4 Liver Cirrhosis as an Ultimate Stage of Liver Diseases

Liver cirrhosis is a final stage of chronic liver disease, where the liver is progressively damaged by replacement of normal liver tissue by scar tissue due to fibrosis. This disease starts usually without any symptoms. At more advanced stages of liver damage, patients become tired, weak, itchy, and have swollen legs, yellow skin, and fluid accumulation in the abdomen. Complications are abdominal infections, hepatic encephalopathy, bleeding from gastroesophageal varices, hepatorenal syndrome, liver cancer, and unconsciousness [52]. Main causes of liver cirrhosis are improper lifestyle, alcohol abuse, metabolic factors, and virus infections.

Cirrhosis is subdivided in several stages. In the absence of any complications (stages 1a, 1b, and 2), a compensated cirrhosis is existent. Higher stages (stages 3–5) are associated with complications and collectively named decompensated cirrhosis [52,53]. The transition from the compensated to the decompensated phase is mainly driven by portal hypertension and the presence of clinical complications [54].

Reduction of portal hypertension by pharmacological and other therapies diminishes the risk of further disease progression [55–57].

The cause of fibrosis during liver cirrhosis remains unclear. Normal parenchyma is step by step substituted by connective tissue [58]. The growth of hepatocytes is dysregulated and results in formation of regenerative nodules, dysplastic nodules, and hepatocellular carcinoma [59]. Apparently, hepatic stellate cells play a crucial role in fibrogenesis as these liver-specific pericytes undergo a transformation into myofibroblasts on activation by profibrogenic cytokines such as transforming growth factor β1 (TGFβ1) and growth factors [58,60,61]. Activated stellate cells migrate into necrotic areas, where they produce a variety of extracellular matrix molecules [62]. The involvement of portal fibroblasts and hepatocytes in liver fibrosis is also discussed [58,59,63,64].

During development of cirrhosis, the liver loses its elasticity and becomes stiffer. Increase in liver stiffness correlates with histologically assessed fibrosis stage independent of the underlying liver disease [65]. At early stages of cirrhosis a reduction of stiffness is possible, when the promoting cause is eliminated [66–68]. Liver stiffness coincides also with increased sinusoidal pressure in noncirrhotic livers [66]. According to the sinusoidal pressure hypothesis of cirrhosis development [69], an increase in sinusoidal pressure is regarded as the primary cause for matrix deposition. In the healthy liver, there is a low pressure difference of about 3 mmHg between the incoming portal vein and the outflowing caval vein [70]. This pressure difference can increase as a result of inflammatory processes, osmotic swelling, or outflow barriers and enhances the liver stiffness [71–73]. The sinusoidal pressure hypothesis explains also the irreversibility of fibrosis in advanced cirrhosis. The increase of sinusoidal pressure above a certain limit is accompanied by an increasing supply of blood to the liver via the hepatic artery. This arterialization of blood flow [74,75] exposes the liver to enhanced mechanical stress and further promotes liver damage and fibrosis [69].

10.3.5 Acute Liver Failure

While chronic liver failure is caused by developing cirrhosis, acute liver failure can develop in a patient without previous liver disease.

Several serious consequences arise from the inability of the liver to perform its usual synthetic and metabolic functions. Some toxic products accumulate such as ammonia. Acute liver failure starts with jaundice, general weakness, and can eventually lead to mental problems ranging from mild confusion to coma. Hepatic encephalopathy results from ammonia, which permeates through the blood–brain barrier and is metabolized by astrocytes. This induces enhanced formation of glutamine, osmotic effects, and brain edema [76,77]. Other liver waste products such as mercaptans, short-chain fatty acids, and phenol contribute to hepatic encephalopathy and disturbed neurotransmission too [78].

Disturbance in synthesis of coagulation factors and inhibitors of coagulation and fibrinolysis favors thrombocytopenia and thus the risk of bleeding [79]. Other deleterious consequences of liver failure are kidney failure, sepsis, electrolyte disturbances, hypotension, tissue hypoxia, lactate acidosis, and lung injury [79–81].

Acute liver toxicity results mainly from overdoses of some medications such as paracetamol, excessive alcohol consumption, viral hepatitis, and mushroom toxins.

10.4 SPLEEN DAMAGE

10.4.1 Physiological Functions of Spleen

The spleen acts as a blood filter and removes permanently senescent red blood cells from circulation. A reserve of blood is stored in the spleen and used to compensate blood loss in case of hemorrhagic shock. In the spleen, the globin part of from captured hemoglobin is degraded to amino acids, whereas heme is metabolized to bilirubin, which is finally removed by the liver, and heme iron is recycled [82].

Besides its role in hematopoiesis, the spleen contains numerous lymphocytes and myeloid cells, which are involved in antibody synthesis and removal of antibody-coated bacteria and antibody-coated blood cells. The spleen represents the largest lymphoid organ in the body. Resection of the spleen increases the predisposition to certain infections [83–85]. In asplenic patients, sepsis can occur with any bacteria, virus, fungi, or protozoan; however, infections dominate with encapsulated bacteria such as *Streptococcus pneumoniae*, *Haemophilus influenza*, and *Neisseria meningitidis* [86,87]. Other immune functions of spleen concern the synthesis of immunoglobulin M and opsonins such as tuftsin and properdin [88].

The spleen is also a large reservoir for monocytes. These cells are distributed from spleen to inflammatory sites, where they maturate into macrophages and dendritic cells and fulfill specific functions [89].

10.4.2 Spleen and Liver Cirrhosis

The spleen is connected with the liver by the portal vein. During progression of liver cirrhosis, spleen-derived immune cells and cytokine may further promote the inflammatory process in the injured liver. Otherwise, portal hypertension can cause a congestion in the portal system and subsequently gives rise to splenomegaly and hypersplenism [90,91]. In the enlarged spleen, the mTOR signaling pathway is activated [92].

A role of splenic macrophage-derived TGFβ1 has been proposed in hepatic fibrogenesis [93,94]. Other cross talks between the spleen and liver concern effects on the hepatic immune microenvironment [95].

10.5 SEPSIS

10.5.1 Systemic Inflammatory Response Syndrome versus Sepsis

Systemic inflammatory response syndrome (SIRS) is a generalized inflammatory reaction of the whole organism without identification of an initiating source. A noninfectious systemic inflammatory reaction can be caused by burns, severe injuries, severe organ damage and inflammations, anaphylactic reactions, intense loss of blood, and ischemic conditions after stroke or cardiac infarction. Often a certain pathogen can be detected in blood of patients with a generalized inflammatory syndrome. Then this infectious SIRS corresponds to sepsis. Noninfectious SIRS and sepsis are very dangerous clinical conditions that can lead to a progressive multiple organ failure and lethal outcome.

Originally, SIRS and sepsis concepts were closely linked to each other [96]. The 1991 consensus conference differentiates SIRS according the degree of organ damage and failure into SIRS, sepsis, severe sepsis, and septic shock. This sepsis concept was mainly based on the assumption of an infection-mediated inflammation [96].

A permanent refinement of the SIRS/sepsis concept took place over the last decades [97,98]. By now different pro- and antiinflammatory responses were identified in sepsis as well as the involvement of nonimmunological pathways [99–101].

At present sepsis is defined as life-threatening organ dysfunction caused by a dysregulated host response to infection [102]. This definition does not include the former SIRS criteria and inflammatory variables due to the inadequate specificity and sensitivity of these criteria. In other words, the focus is now directed on organ dysfunction caused in response to an infection [103]. Septic shock is defined as a subset of sepsis, in which underlying circulatory and cellular/metabolic abnormalities are profound enough to substantially increase mortality [102]. Sepsis is diagnosed using a sequential organ failure assessment scoring system [102]. This helps to predict the clinical outcomes of critically ill patients [104].

The existence of an aberrant or dysregulated systemic host response and the presence of organ dysfunction are key points that differentiate sepsis from infection.

Sepsis occurs most frequently in immunocompromised persons, such as injured, diseased, and elderly people or newborns [105–107]. The use of immunosuppressive agents increases also this risk of sepsis.

10.5.2 Pathogens in Sepsis

Sepsis may originate from an acute infection by exogenous highly virulent microbes. In most cases, however, sepsis-mediating microorganisms are members of the healthy human microbiome settling on skin and mucous surfaces. Their activation is consistent with the impaired host immunity. Among these commensal and mutualistic microbes or fungi are *Staphylococcus aureus*, *S. pneumoniae*, *Escherichia coli*, *Pseudomonas aeruginosa*, *Klebsiella pneumoniae*, *Acinetobacter* spp., *Enterococcus* spp., *Stenotrophomonas* spp., *Candida albicans*, and other *Candida* spp. [108–111].

Similarly, in immunocompromised patients with sepsis, some latent viruses can be reactivated. This reactivation concerns cytomegalovirus, herpes simplex virus, and some other viruses [112–114].

Several hypotheses try to explain the development of sepsis [115,116]. The healthy immune system avoids efficiently any invasion of microbes from skin and mucous surfaces. In injured and immunocompromised patients, limitation of this protection allows distribution of microbes by circulating blood and invasion into formally sterile tissues. According to the classical hypothesis, sepsis results as an accidental sequela of diminished host defense. The microbiome mutiny hypothesis is based on the assumption of a phenotype switch of involved microorganisms from a low to a high virulence phenotype as well as the existence of a microbial sensing mechanism to assess the health status of the host [115].

10.5.3 Experimental Models of Sepsis

To deepen the knowledge of molecular mechanisms of sepsis, several experimental models have been established. This allows the investigation of basic mechanisms in more detail, and the testing of different drug interventions on these models. The disadvantage of these models concerns the fact that they represent only a special aspect of the large variability of sepsis events. Moreover, previously healthy animals are often used in these models. In humans, sepsis develops predominantly in already ill immunocompromised individuals [105–107].

To mimic infectious state, lipopolysaccharides or a variety of different bacteria are injected intraperitoneally or intravenously into laboratory animals [117,118]. These models with a bolus injection of pathogens or endotoxins badly reflect the steady populating with microbes in septic persons [119]. Moreover, unlike humans, mice tolerate much better toll-like receptor agonists [120–122].

Inclusion of bacteria into fibrin clots and implantation of these clots into peritoneal cavity of mice provides a better control about bacterial quantities in the body [123].

The cecal ligation and puncture model of sepsis is widely applied [124,125]. This model expresses better characteristic features of human sepsis such as pro- and antiinflammatory phases, contribution of ischemic tissue to immune dysfunction, and detection of DAMPs. This model allows the testing of antibiotics and other drugs. Another approach for producing abdominal sepsis is the Colon Ascendens Stent Peritonitis [126].

In septic mice, contribution of neutrophils was analyzed in the pathogenesis of multiple organ failure caused by cecal ligation and puncture [127]. Depletion of neutrophils before or at the onset of sepsis enhanced substantially bacteremia and did not affect the level of kidney and liver dysfunction. In contrast, neutrophil depletion 12 h after cecal ligation and puncture resulted in decreased bacterial load, reduced liver and kidney damage, diminished levels of proinflammatory and antiinflammatory cytokines, and improved survival. At the onset of sepsis, neutrophils are crucial to regulate the level of pathogens. At later phases, these cells promote harmful effects and contribute to cell and tissue damage [127].

10.5.4 Immunosuppression in Sepsis

The time course of sepsis can be roughly divided into an initial hyperinflammatory phase followed by a second hypoinflammatory, immunosuppressive phase [128]. This corresponds to the timescale of sepsis seen in experimental models on animals.

In acute sepsis forms, the first phase can be characterized by a massive cytokine storm associated with shock, high fever, and multiple organ failure [129]. Determining factors of this cytokine-driven hyperinflammation are pathogen virulence, bacterial load, age, genetic factors, and most of all host's comorbidities. A previously healthy person is more prone to this form of sepsis than an immunocompromised individual with manifold existing chronic inflammatory processes.

Hence, a serious problem in human sepsis pathology is that the first proinflammatory phase is often weakly expressed, covered by antiinflammatory and immunosuppressive mechanisms, and therefore not recognized. This concerns mostly elderly people, who suffer from different chronic comorbidities and whose immune system is in a compromised state [130]. Then sepsis appears mainly as an immunosuppressive disorder.

There are some similarities in immunosuppressive mechanisms in sepsis and cancer such as increasing secretion or production of interleukin-10 (IL-10), T regulatory cells, myeloid suppressor cells, and programmed cell death protein 1 (PD-1) and its ligand PD-L1 [131–133]. Binding of PD-L1 to PD-1 induces gene expression of IL-10 in monocytes and impairs thus $CD4^+$ T cell activation [134]. The crucial role of immunosuppression in sepsis is evidenced by a substantial loss of immune cells, most notably $CD4^+$ and $CD8^+$ T cells, B cells, and dendritic cells as shown by postmortem analysis of patients [135–137]. Otherwise, the percentage of T regulatory cells increases in septic patients [138]. The dominance of opportunistic infections is a further indicator for immunodeficiency during sepsis.

In activated immune cells, aerobic glycolysis dominates for generation of energy equivalents (Section 6.2.6). Lactate as a product of glycolysis is related to severity and outcome of sepsis [139,140]. The present sepsis guideline (sepsis-3) recommends that persistence of serum lactate concentration higher than 2 mmol L^{-1} should be included as new criterion defining septic shock [102]. High serum lactate concentrations are predictor of mortality, while reduced lactate levels indicate an improved clinical outcome [141–144]. Lactate contributes to immunodeficiency in sepsis by driving macrophages into the suppressive M2 phenotype [145–147], inhibition of dendritic cell maturation [148], and affecting T cell functions [149]. Again, lactate contributes to immunosuppression in cancer as well (see section 9.8.2).

10.5.5 Neutrophil Functions During Sepsis

Assessment of neutrophil functions from sepsis patients is challenging. There are also numerous conflicting results in literature. In most cases, neutrophil characterization concerns the time after the onset of sepsis. Only few data about neutrophils are related to late phases of sepsis and septic shock, when immunosuppression is the hallmark of sepsis.

Neutrophils from septic patients and laboratory animals show a delayed apoptosis [150]. Oxidative burst and phagocytosis are increased by these neutrophils. However, the chemotactic movement of these cells is strongly inhibited [151]. Thus, they can contribute to tissue damage distant from infection sites [152,153].

There are an increasing number of immature neutrophils in septic shock patients. These immature cells exhibit a reduced chemotactic movement [151,154], reduced oxidative burst, show a decreased intracellular expression of lactoferrin and myeloperoxidase, and decreased expression of CD10 and CD16. Increased percentage of immature neutrophils is associated with higher risk of death after septic shock [151].

10.5.6 Tissue Damage and Sepsis

The strong relationship between sepsis and immunodeficiency affects also the general health status of patients. These patients are often in a hypoxic state and suffer from different metabolic and energy deficits. Hence, it is not surprising that not only immune functions but also other protective mechanisms are on the limit. This promotes a more or less uncontrolled increase in processes of cell and tissue damage. The spectrum of tissue-damaging mechanisms can widely vary in dependence on the state of individual protective mechanism [155,156]. Thus, it is not surprising that no consensus exists, which mechanisms contribute mainly to damage.

Invading microorganisms are involved in damaging reaction. Other key players in sepsis-induced cell and tissue damage are immune cells, in particular neutrophils, which can accumulate in large number in inflamed and infected tissues. Proteolytic enzymes and reactive species of these cells promote destructive reactions. Antigens and damage-associated molecular patterns (DAMPs) released from necrotic tissues can further attract and activate immune cells. Intravascular hemolysis and rhabdomyolysis are sources for free hemoglobin, free myoglobin, and ferric protoporphyrin IX (free heme). These agents are involved in damaging and oxidative reactions, too. A dysregulation in the sequestration of metal ions can also occur. Dysfunctional mitochondria, necrotic cells, defective heme, and other proteins can function as source for catalytically active metal ions. All these agents may contribute to cell and tissue damage in sepsis.

A potentiation of some damaging agents and reactions occurs in sepsis. This is caused by a considerable decline or exhaustion of the corresponding protective mechanisms that favor the uncontrolled rise of active counterparts. In sepsis, the uncertainty of destructive mechanisms results from the fact that it is very challenging to predict, which of the protective systems is exhausting first. This highly depends on the local situation in affected organs, on the energy and immune status of patients, on the individual state of protective systems, on the existing comorbidities, and on genetic predisposition. Postulating that most damage in septic tissues and organs is induced by a defined agent, whose antagonizing opponent exhausts first, this would provide a rationale to enhance therapeutically protection against this agent. To implement this therapeutic approach, a personalized analysis of the status of protective mechanisms is mandatory.

10.5.7 Therapeutic Options

From the background of dominating immunodeficiency in many patients with sepsis, it is not surprising that numerous trials of diverse anticytokine and antiinflammatory drugs were inefficient and reduced sometimes survival rates [157,158]. Moreover, an increased risk for nosocomial infections and late-stage mortality remains for those patients who survived the hyperinflammatory phase [159,160].

Considering strong immunosuppression in most patients with sepsis, a personalized treatment with immunostimulatory agents is highly recommended on the basis of laboratory and clinical data. Suitable immunostimulatory drugs are recombinant proteins on the basis of interferon γ, GM-CSF, or G-CSF [128]. Other potential therapeutic immunostimulatory approaches are the use of interleukin-7 [161], interleukin-15 [162], the application of anti-PD-L1 and anti-PD-1 antibodies [133], and the use of anticytotoxic T-lymphocyte antigen-4 antibodies [163].

In addition, several approaches of intensive care medicine are applied to critically ill patients.

10.6 DEVASTATING CONSEQUENCES OF SEPSIS

10.6.1 Septic Shock

Septic shock is associated with a dangerous decline in blood pressure, impaired tissue perfusion, and abnormal cell metabolism. Infections can cause septic shock also without a previous sepsis [157]. Further complications of septic shock are multiple organ dysfunction syndrome and death.

A mean arterial pressure below 60–65 mmHg is associated with a poor outcome [164]. Many vital organs such as the brain, kidney, heart, liver, and gut have a lower pressure threshold in the range between 65–80 mmHg for autoregulation of perfusion [165,166]. Patients with arterial hypertension may have higher autoregulatory thresholds [167].

Several factors contribute to the development of sepsis and septic shock. Inflammatory response to pathogens is associated with endothelial dysfunction and damage of blood vessels [168]. Neutrophil products and neutrophil extracellular traps are involved in this damage. In turn, the coagulation, complement, and contact systems are activated and red blood cells become hemolytic. The resulting hypercoagulable state can by accompanied by exhaustion of some components of the clotting cascade [169]. Then a transition occurs from coagulopathy to disseminated intravascular coagulation. These mechanisms contribute to vasodilation as seen in septic shock [168,170].

10.6.2 Multiple Organ Dysfunction Syndrome

Multiple organ dysfunction syndrome is regarded as the final stage of severe sepsis. In critically ill patients, usually two or more organs are affected. Homeostasis cannot be maintained without intervention.

Different alterations of functional activities contribute to development of multiple organ dysfunction syndrome. These pathogenic mechanisms include immune dysfunction, endothelial activation, microcirculatory dysfunction, mitochondrial damage, increase of cell death, hemolysis, hypoperfusion, cellular hypermetabolism thrombosis, hypoxia, and uncontrolled distribution of pathogens [171]. Altogether, the underlying mechanisms are rather complex and can widely differ from patient to patient.

The gut is assumed to play a critical role in sepsis development [172]. In critically ill patients, some markers are found in patient's blood for enterocyte damage [173]. Damage of gut epithelium can favor translocation of intestinal bacteria into the organism [174]. Another hypothesis supports the circulation of proinflammatory factors from the gut via the mesenteric lymphatic system [175].

Acute kidney injury is a common complication in sepsis. Kidney dysfunction is characterized by low or no urine output, disturbance in electrolyte compositions, or volume overload. It is induced by several ongoing mechanisms such as endothelial damage, inflammation, coagulation-induced disturbed microcirculation, and induced responses to injury. These mechanisms culminate in disturbed renal microcirculation [176,177].

Dysfunctional liver shows disruption of protein synthesis, diminished release of clotting factors, and impaired bilirubin metabolism. Again, multifactorial events contribute to liver failure. Besides infection, liver dysfunction can be caused by an uncontrolled inflammatory response, disturbed microcirculation, and therapeutic side effects [178,179].

In the lungs, severe septic can be associated with an acute respiratory distress syndrome. Lung damage results either directly from pneumonia and gastric aspiration or indirectly from other infections or traumata. Defective endothelial cells of pulmonary capillaries and disturbances in the alveolar epithelial layer contribute also to respiratory distress syndrome [180].

In sepsis-induced cardiac dysfunction, overstimulation can occur by sympathetic nerve by endogenously enhanced and exogenously administrated catecholamines. A reduced ejection fraction of septic heart is followed by systolic and diastolic dysfunction. Proinflammatory cytokines, mitochondrial dysfunction, enhanced generation of reactive species, dysregulation of β-adrenergic signaling, and cardiomyocyte apoptosis contribute to heart failure due to sepsis [181–183].

Excessive microglial activation, impaired cerebral perfusion, dysfunction of the blood–brain barrier, and altered neurotransmission are key features of brain dysfunction in sepsis. Both inflammatory and noninflammatory processes contribute to these functional alterations. As in other sepsis-induced organ dysfunctions, pathophysiology is rather complex [184].

10.7 CONCLUDING REMARKS

The same conclusions that have been drawn about the pathogenesis of chronic inflammatory diseases in the previous chapter (Section 9.9) are also valid for the regarded kidney, liver, and spleen disorders. This imbalance between cell and tissue damage and host's protective mechanism is further shifted toward the destruction side in sepsis and complications of sepsis. In these life-threatening clinical events, exhaustion of protective mechanisms seriously worsens the general condition of affected patients.

References

[1] E. Croom, Chapter three: metabolism of xenobiotics of human environments, Prog. Mol. Biol. Transl. Sci. 112 (2012) 31–88.

[2] D.K. Patel, D.J. Sen, Xenobiotics: an essential precursor for living systems, Am. J. Adv. Drug Deliv. 1 (2013) 262–270.

[3] V.G. Puelles, W.E. Hoy, M.D. Hughson, B. Diouf, R.N. Douglas-Denton, J.F. Bertram, Glomerular number and size variability and risk for kidney disease, Curr. Opin. Nephrol. Hypertens. 20 (2011) 7–15.

[4] R.P. Scott, S.E. Quaggin, The cell biology of renal filtration, J. Cell Biol. 209 (2015) 199–210.

[5] M. Burke, M.R. Pabbidi, J. Farley, R.J. Roman, Molecular mechanisms of renal blood flow autoregulation, Curr. Vasc. Pharmacol. 12 (2014) 845–858.

[6] A.S. Levey, L.A. Inker, J. Coresh, GFR estimation: from physiology to public health, Am. J. Kidney Dis. 63 (2014) 820–834.

[7] I. Soveri, U.B. Berg, J. Björk, C.G. Elinder, A. Grubb, I. Mejare, G. Sterner, S.E. Bäck, SBU GFR review group, measuring GFR: a systematic review, Am. J. Kidney Dis. 64 (2014) 411–424.
[8] V. Rigalleau, C. Lasseur, C. Perlemoine, N. Barthe, C. Raffatin, C. Liu, P. Chauveau, L. Bailet-Blanco, M.C. Beauvieux, C. Combe, H. Gin, Estimation of glomerular filtration rate in diabetic subjects, Diabetes Care 28 (2005) 838–843.
[9] C. White, A. Akbari, N. Hossain, L. Dinh, G. Filler, N. Lapage, G.A. Knoll, Estimation of glomerular filtration rate in kidney transplantation: a comparison between serum creatinine and cystatin C based methods, J. Am. Soc. Nephrol. 16 (2005) 3736–3770.
[10] J.L. Lyman, Blood urea nitrogen and creatinine, Emerg. Med. Clin. N. Am. 4 (1986) 223–233.
[11] L.A. Stevens, C.H. Schmid, T. Greene, L. Li, G.J. Beck, M.M. Joffe, M. Froissart, J.W. Kusek, Y.L. Zhang, J. Coresh, A.S. Levey, Factors other than glomerular filtration rate affect serum cystatin C levels, Kidney Int. 84 (2009) 164–173.
[12] D.W. Cockroft, M.H. Gault, Prediction of creatinine clearance from serum creatinine, Nephron 16 (1976) 31–41.
[13] A.S. Levey, L.A. Inker, Assessment of glomerular filtration rate in health and disease: a state of the art review, Clin. Pharmacol. Ther. 102 (2017) 405–419.
[14] Kidney Disease: Improving Global Outcomes (KDIGO) CKD World Group, KDIGO 2012. Clinical practice guideline for the evaluation and management of chronic kidney disease, Chapter 1: definition and classification of CKD, Kidney Int. Suppl. 3 (2013) 19–62.
[15] A.C. Webster, E.V. Nagler, R.L. Morton, P. Masson, Chronic kidney disease, Lancet 389 (2017) 1238–1252.
[16] R. Vanholder, U. Baurmeister, P. Brunet, G. Cohen, G. Glorieux, J. Jankowski, A bench to bedside view of uremic toxins, J. Am. Soc. Nephrol. 19 (2008) 863–870.
[17] H.J. Anders, K. Andersen, B. Stecher, The intestinal microbiota, a leaky gut, and abnormal immunity in kidney disease, Kidney Int. 83 (2013) 1010–1016.
[18] V. Perkovic, C. Verdon, T. Ninomiya, F. Barzi, A. Cass, A. Patel, M. Jardine, M. Gallagher, F. Turnball, J. Chalmers, J. Craig, R. Huxley, The relationship between proteinuria and coronary risk: a systematic review and meta-analysis, PLoS Med. 5 (2008), e207, 1–10, https://doi.org/10.1371/journal.pmed.0050207.
[19] G. Wong, A. Hayen, J.R. Chapman, A.C. Webster, J.J. Wang, P. Mitchell, J.C. Craig, Association of CKD and cancer risk in older people, J. Am. Soc. Nephrol. 20 (2009) 1341–1350.
[20] S. Iff, J.C. Craig, R. Turner, J.R. Chapman, J.J. Wang, P. Mitchell, G. Wong, Reduced estimated GFR and cancer mortality, Am. J. Kidney Dis. 63 (2014) 23–30.
[21] Y.C. Hou, C.L. Lu, K.C. Lu, Mineral bone disorders in chronic kidney disease, Nephrology 23 (2018) 88–94.
[22] M. Tombe, Images in clinical medicine. Hemoglobinuria with malaria, N. Engl. J. Med. 358 (2008) 1837.
[23] I.C. Vermeulen Windsant, M.G. Snoeijs, S.J. Hanssen, S. Altintas, J.H. Heijmans, T.A. Koeppel, G.W. Schurink, W.A. Buurman, M.J. Jacobs, Hemolysis is associated with acute kidney injury during major aortic surgery, Kidney Int. 77 (2010) 913–920.
[24] I.C. Vermeulen Windsant, N.C.J. de Wit, J.T.C. Sertorio, A.A. van Bijnen, Y.M. Ganushchak, J.H. Heijmans, J.E. Tanus-Santos, M.J. Jacobs, J.G. Maessen, W.A. Buurman, Hemolysis during cardiac surgery is associated with increased intravascular nitric oxide consumption and perioperative kidney and intestinal tissue damage, Front. Physiol. 5 (2014), 340, 1–9, https://doi.org/10.3389/fphys.2014.00340.eCollection 2014.
[25] R. Vanholder, M.S. Sever, E. Erek, N. Lameire, Acute renal failure related to the crush syndrome: towards an era of seismo-nephrology? Nephrol. Dial. Transplant. 15 (2000) 1517–1521.
[26] A. Genthon, S.R. Wilcox, Crush syndrome: a case report and review of the literature, J. Emerg. Med. 46 (2014) 313–319.
[27] M.J. Tracz, J. Alam, K.A. Nath, Physiology and pathophysiology of heme: implications for kidney disease, J. Am. Soc. Nephrol. 18 (2007) 414–420.
[28] D. Chiabrando, F. Vinchi, V. Fiorito, E. Tolosano, Haptoglobin and hemopexin in heme detoxification and iron recycling, in: F. Veas (Ed.), Acute Phase Proteins – Regulation and Functions of Acute Phase Proteins, Intech, Rijeka, Croatia, 2011, pp. 261–288.
[29] N. Petejova, A. Martinek, Acute renal failure due to rhabdomyolysis and renal replacement therapy: a critical review, Crit. Care 18 (2014), 224, 1–8, https://doi.org/10.1188/cc13897.
[30] J.W. Deuel, C.A. Schaer, F.S. Boretti, L. Opitz, I. Garcia-Rubio, J.H. Baek, D.R. Spahn, P.W. Buehler, D.J. Schaer, Hemoglobinuria-related acute kidney injury is driven by intrarenal oxidative reactions triggering a heme toxicity response, Cell Death Dis. 7 (2016), e2064, 1–12, https://doi.org/10.1038/cddis.2015.392.
[31] N. Tabibzadeh, C. Estournet, S. Placier, J. Perez, H. Bilbault, A. Girshovich, A. Vandermeersch, C. Jouanneau, E. Letavenier, N. Hammoudi, F. Lionnet, J.P. Haymann, Plasma heme-induced renal toxicity is related to capillary rarefaction, Sci. Rep. 7 (2017), 40156, 1–8, https://doi.org/10.1038/srep40156.

[32] A.M. Santoro, M.C. Lo Giudice, A. D'Urso, R. Lauceri, R. Purrello, D. Milardi, Cationic porphyrins are reversible proteasome inhibitors, J. Am. Chem. Soc. 134 (2012) 10451–10457.
[33] F. Vallelian, J.W. Deuel, L. Opitz, C.A. Schaer, M. Puglia, M. Lönn, W. Engelsberger, S. Schauer, E. Kamaukhova, D.R. Spahn, R. Stocker, P.W. Buehler, D.J. Schaer, Proteasome inhibition and oxidative reactions disrupt cellular homeostasis during heme stress, Cell Death Differ. 22 (2015) 597–611.
[34] M. Schröder, R.J. Kaufman, The mammalian unfolded protein response, Annu. Rev. Biochem. 74 (2005) 730–789.
[35] A. Linkermann, R. Skouta, N. Himmerkus, S.R. Mulay, C. Dewitz, F. De Zen, A. Prokai, G. Zuchtriegel, F. Krombach, P.S. Welz, R. Weinlich, T. Vanden Berghe, P. Vandenabeele, M. Pasparakis, M. Bleich, J.M. Weinberg, C.A. Reichel, J.H. Bräsen, U. Kunzendorf, H.J. Anders, B.R. Stockwell, D.R. Green, S.D. Krautwald, Synchronized renal tubular cell death involved ferroptosis, Proc. Natl. Acad. Sci. U.S.A. 111 (2014) 16836–16841.
[36] K.L. Donnelly, C.I. Smith, S.J. Schwarzenberg, J. Jessurun, M.D. Boldt, E.J. Parks, Sources of fatty acids stored in liver and secreted via lipoproteins in patients with nonalcoholic fatty liver disease, J. Clin. Investig. 115 (2005) 1343–1351.
[37] C. Postic, J. Girard, Contribution of de novo fatty acid synthesis to hepatic steatosis and insulin resistances: lessons from genetically engineered mice, J. Clin. Investig. 118 (2008) 829–838.
[38] E. Fabbrini, B.S. Mohammed, F. Magkos, K.M. Korenblat, B.W. Patterson, S. Klein, Alterations in adipose tissue and hepatic lipid kinetics in obese men and women with nonalcoholic fatty liver disease, Gastroenterology 134 (2008) 424–431.
[39] E. Fabbrini, F. Magkos, Hepatic steatosis as a marker of metabolic dysfunction, Nutrients 7 (2015) 4995–5019.
[40] D.H. Ipsen, P. Tveden-Nyborg, J. Lykkesfeldt, Dyslipidemia: obese or not obese – that is not the question, Curr. Obes. Rep. 5 (2016) 405–412.
[41] D.H. Ipsen, J. Lykkesfeldt, P. Tveden-Nyborg, Molecular mechanisms of hepatic lipid accumulation in non-alcoholic fatty liver disease, Cell. Mol. Life Sci. 75 (2018) 3313–3327.
[42] R. Loomba, N. Chalasani, The hierarchical model of NAFLD: prognostic significance of histologic features in NASH, Gastroenterology 149 (2015) 278–281.
[43] A.J. Sanyal, C. Campbell-Sargent, F. Mirshahi, W.B. Rizzo, M.J. Contos, R.K. Sterling, V.A. Luketic, M.L. Shiffman, J.N. Clore, Nonalcoholic steatohepatitis: association of insulin resistance and mitochondrial abnormalities, Gastroenterology 120 (2001) 1183–1192.
[44] S. Dasarathy, Y. Yang, A.J. McCullogh, S. Marczewski, C. Bennett, S.C. Kalhan, Elevated fatty acid oxidation, high plasma fibroblast growth factor 21, and fasting bile acids in non-alcoholic steatohepatitis, Eur. J. Gastroenterol. Hepatol. 23 (2011) 382–388.
[45] F. Nassir, J.A. Ibdah, Role of mitochondria in nonalcoholic fatty liver disease, Int. J. Mol. Sci. 15 (2014) 8713–8742.
[46] G.H. Koek, P.R. Liedorp, A. Bast, The role of oxidative stress in non-alcoholic steatohepatitis, Clin. Chim. Acta 412 (2011) 1297–1305.
[47] L.G. Guidotti, F.V. Chisari, Immunobiology and pathogenesis of viral hepatitis, Ann. Rev. Pathol. Mech. Dis. 1 (2006) 23–61.
[48] A.J. Czaja, Drug-induced autoimmune-like hepatitis, Dig. Dis. Sci. 56 (2011) 958–976.
[49] M. Chayanupatkul, S. Liangpunsakul, Alcoholic hepatitis: a comprehensive review of pathogenesis and treatment, World J. Gastroenterol. 20 (2014) 6279–6286.
[50] C.S. Lieber, L.M. De Carli, Hepatotoxicity of ethanol, J. Hepatol. 12 (1991) 394–401.
[51] K. Rasineni, C.A. Carol, Molecular mechanisms of alcoholic fatty liver, Indian J. Pharmacol. 44 (2012) 299–303.
[52] A. Berzigotti, Advances and challenges in cirrhosis and portal hypertension, BioMed Central Med 15 (2017), 200, 1–8, https://doi.org/10.1186/s12916-017-0966-6.
[53] G. D'Amico, L. Pasta, A. Morabito, M. D'Amico, M. Caltagirone, G. Malizia, F. Tinè, G. Giannuoli, M. Traina, G. Vizzini, F. Politi, A. Luca, R. Virdone, A. Licata, L. Pagliaro, Competing risks and prognostic stages of cirrhosis: a 25-year inception cohort study of 494 patients, Aliment. Pharmacol. Ther. 39 (2014) 1180–1193.
[54] G. D'Amigo, G. Garcia-Tsao, L. Pagliaro, Natural history and prognostic indicators of survival in cirrhosis: a systematic review of 118 studies, J. Hepatol. 44 (2006) 217–231.
[55] F. Feu, J.C. Garcia-Pagan, J. Bosch, A. Luca, J. Teres, A. Escorell, J. Rodes, Relation between portal pressure response to pharmacotherapy and risk of recurrent variceal haemorrhage in patients with cirrhosis, Lancet 346 (1995) 1056–1059.
[56] J.G. Abraldes, I. Tarantino, J. Turnes, J.C. Garcia-Pagan, J. Rodes, J. Bosch, Hemodynamic response to pharmacological treatment of portal hypertension and long-term prognosis of cirrhosis, Hepatology 37 (2003) 902–908.
[57] G. Garcia-Tsao, J.G. Abraldes, A. Bergizotti, J. Bosch, Portal hypertensive bleeding in cirrhosis: risk stratification, diagnosis, and management: 2016 practical guidance by the American Association for the study of liver diseases, Hepatology 65 (2017) 310–335.

[58] S. Dooley, J. Hamzavi, L. Ciuclan, P. Godoy, I. Ilkavets, S. Ehnert, E. Ueberham, R. Gebhardt, S. Kanzler, A. Geier, K. Breitkopf, H. Weng, P.R. Mertens, Hepatocyte-specific Smad7 expression attenuates TGF-beta-mediated fibrogenesis and protects against liver damage, Gastroenterology 135 (2008) 642–659.

[59] T. Nitta, J.S. Kim, D. Mohuczy, K.E. Behrns, Murine cirrhosis induces epithelial mesenchymal transition and alterations in survival signaling pathways, Hepatology 48 (2008) 909–919.

[60] S.L. Friedman, Seminars in medicine of the Beth Israel Hospital, Boston. The cellular basis of hepatic fibrosis. Mechanisms and treatment strategies, N. Engl. J. Med. 328 (1993) 1828–1835.

[61] A.M. Gressner, Cytokines and cellular crosstalk involved in the activation of fat-storing cells, J. Hepatol. 22 (1995) 28–36.

[62] N. Uyama, Y. Iimuro, N. Kawada, H. Reynaert, K. Suzumura, T. Hirano, N. Kuroda, J. Fujimoto, Fascin, a novel marker of human hepatic stellate cells, may regulate their proliferation, migration, and collagen gene expression through the FAK-PI3K-Akt pathway, Lab. Invest. 92 (2012) 57–71.

[63] N. Kinnman, C. Housset, Peribiliary myofibroblasts in biliary type liver fibrosis, Front. Biosci. 7 (2002) d496–d503.

[64] S.-J. Lee, K.-H. Kim, K.-K. Park, Mechanisms of fibrogenesis in the liver cirrhosis: the molecular aspects of epithelial-mesenchymal transition, World J. Hepatol. 6 (2014) 207–216.

[65] S. Mueller, L. Sandrin, Liver stiffness: a novel parameter for the diagnosis of liver disease, Hepat. Med. 2 (2010) 49–67.

[66] G. Millonig, S. Friedrich, S. Adolf, H. Fonouni, M. Golriz, A. Mehrabi, P. Stiefel, G. Pöschl, M.W. Büchner, H.K. Seitz, S. Mueller, Liver stiffness is directly influenced by central venous pressure, J. Hepatol. 52 (2010) 206–210.

[67] S. Mueller, G. Millonig, L. Sarovska, S. Friedrich, F.M. Reimann, M. Pritsch, S. Eisele, F. Stickel, T. Longerich, P. Schirmacher, H.K. Seitz, Increased liver stiffness in alcoholic liver disease: differentiating fibrosis from steatohepatitis, World J. Gastroenterol. 16 (2010) 966–972.

[68] C. Hézode, L. Castéra, F. Roudot-Thoraval, M. Bouvier-Alias, I. Rosa, D. Roulot, V. Leroy, A. Mallat, J.M. Pawlotsky, Liver stiffness diminishes with antiviral response in chronic hepatitis C, Aliment. Pharmacol. Ther. 34 (2011) 656–663.

[69] S. Mueller, Does pressure cause liver cirrhosis? The sinusoidal pressure hypothesis, World J. Gastroenterol. 22 (2016) 10482–10501.

[70] D.C. Balfour, T.B. Reynolds, D.C. Levinson, W.P. Mikkelsen, A.C. Pattison, Hepatic vein pressure studies for evaluation of intrahepatic portal hypertension, AMA Arch. Surg. 68 (1954) 442–447.

[71] G. Millonig, F.M. Reimann, S. Friedrich, H. Fonouni, A. Mehrabi, M.W. Büchler, H.K. Seitz, S. Mueller, Extrahepatic cholestasis increases liver stiffness (FibroScan) irrespective of fibrosis, Hepatology 48 (2008) 1718–1723.

[72] U. Arena, F. Vizzutti, G. Corti, S. Ambu, C. Stasi, S. Bresci, S. Moscarella, V. Boddi, A. Petrarca, G. Laffi, F. Marra, M. Pinzani, Acute viral hepatitis increases liver stiffness values measured by transient elastography, Hepatology 47 (2008) 380–384.

[73] F. Piecha, T. Peccerella, T. Bruckner, H.K. Seitz, V. Rausch, S. Mueller, Arterial pressure suffices to increase liver stiffness, Am. J. Physiol. Gastrointest. Liver Physiol. 311 (2016) G945–G953.

[74] G. Kleber, N. Steudel, C. Behrmann, A. Zipprich, G. Hübner, E. Lotterer, W.E. Fleig, Hepatic arterial flow volume and reserves in patients with cirrhosis: use of intra-arterial Doppler and adenosine infusion, Gastroenterology 116 (1999) 906–914.

[75] M. Dragoteanu, S.O. Cotul, C. Pîgleşan, S. Tamaş, Liver angioscintigraphy: clinical applications, Rom. J. Gastroenterol. 13 (2004) 55–63.

[76] J.M. Ryan, D.L. Shawcross, Hepatic encephalopathy, Medicine 39 (2011) 617–620.

[77] E.F. Wijdicks, Hepatic encephalopathy, N. Engl. J. Med. 375 (2016) 1660–1670.

[78] L. Zieve, G. Brunner, Encephalopathy due to mercaptans and phenols, in: D.W. McCandless (Ed.), Cerebral Energy Metabolism and Metabolic Encephalopathy, Springer, Boston, 1985, pp. 179–201.

[79] A.E. Gimson, Fulminant and late onset hepatic failure, Br. J. Anaesth. 77 (1996) 90–98.

[80] P.N. Trewby, R. Warren, S. Contini, W.A. Crosbie, S.P. Wilkinson, J.W. Laws, R. Williams, Incidence and pathophysiology of pulmonary edema in fulminant hepatic failure, Gastroenterology 74 (1978) 859–865.

[81] D. Bihari, A.E. Gimson, M. Waterson, R. Williams, Tissue hypoxia during fulminant hepatic failure, Crit. Care Med. 13 (1985) 1034–1039.

[82] R.E. Mebius, G. Kraal, Structure and function of the spleen, Nat. Rev. Immunol. 5 (2005) 606–616.

[83] K. Hansen, D.B. Singer, Asplenic-hyposplenic overwhelming sepsis: postsplenectomy sepsis revisited, Pediatr. Dev. Pathol. 4 (2001) 105–121.

[84] P.D. Sinwar, Overwhelming post splenectomy infection syndrome. Review study, Int. J. Surg. 12 (2014) 1314–1316.

[85] T.J. Marrie, G.J. Tyrrell, S.R. Majumdar, D.T. Eurich, Asplenic patients and invasive pneumococcal disease. How bad is it these days? Int. J. Infect. Dis. 51 (2016) 27–30.
[86] D.N. Williams, B. Kaur, Postsplenectomy care. Strategies to decrease the risk of infection, Postgrad. Med. 100 (1996) 195–205.
[87] R.N. Davidson, R.A. Wall, Prevention and management of infections in patients without a spleen, Clin. Microbiol. Infect. 7 (2001) 657–660.
[88] B.M. William, G.R. Corazza, Hyposplenism: a comprehensive review. Part I: basic concepts and causes, Hematology 12 (2007) 1–13.
[89] F.K. Swirski, M. Nahrendorf, M. Etzoldt, M. Wildgruber, V. Cortez-Retamozo, P. Panizzi, J.L. Figueiredo, R.H. Kohler, A. Chudnovkiy, P. Waterman, E. Aikawa, T.R. Mempel, P. Libby, R. Weissleder, M.J. Pittet, Identification of splenic reservoir monocytes and their deployment to inflammatory sites, Science 325 (2009) 612–616.
[90] M.A. El-Khishen, J.M. Henderson, W.R. Millikan Jr., M.H. Kutner, W.D. Warren, Splenectomy is contraindicated for thrombocytopenia secondary to portal hypertension, Surg. Gynecol. Obstet. 160 (1985) 233–238.
[91] M. Mejias, E. Garcia-Pras, J. Gallego, R. Mendez, J. Bosch, M. Fernandez, Relevance of the mTOR signaling pathway in the pathophysiology of splenomegaly in rats with chronic portal hypertension, J. Hepatol. 52 (2010) 529–539.
[92] Y. Chen, W. Wang, H. Wang, Y. Li, M. Shi, H. Li, J. Yan, Rapamycin attenuates splenomegaly in both intrahepatic and prehepatic portal hypertensive rats by blocking mTOR singling pathway, PLoS One 11 (2016), e0141159, 1–17, https://doi.org/10.1371/journal.pone.0141159.
[93] T. Akahoshi, M. Hashizume, K. Tanoue, R. Shimabukuro, N. Gotoh, M. Tomikawa, K. Sugimachi, Role of the spleen in liver fibrosis in rats may be mediated by transforming growth factor β-1, J. Gastroenterol. Hepatol. 17 (2002) 59–65.
[94] M. Asanoma, T. Ikemoto, H. Mori, T. Utsunomiya, S. Imura, Y. Morine, S. Iwahashi, Y. Saito, S. Yamada, M. Shimada, Cytokine expression in spleen affects progression of liver cirrhosis through liver-spleen cross-talk, Hepatol. Res. 44 (2014) 1217–1223.
[95] L. Li, M. Duan, W. Chen, A. Jiang, X. Li, J. Yang, Z. Li, The spleen in liver cirrhosis: revisiting an old enemy with novel targets, J. Transl. Med. 15 (2017), 111, 1–10, https://doi.org/10.1186/s12967-017-1214-8.
[96] R.C. Bone, R.A. Balk, F.B. Cerra, R.P. Dellinger, A.M. Fein, W.A. Knaus, R.M. Schein, W.J. Sibbald, Definitions for sepsis and organ failure and guidelines for the use of innovative therapies in sepsis. The ACCP/SCCM consensus conference committee. American college of chest physicians/society of critical care medicine, Chest 101 (1992) 1644–1655.
[97] M.M. Levy, M.P. Fink, J.C. Marshall, E. Abraham, D. Angus, D. Cook, J. Cohen, S.M. Opal, J.L. Vincent, G. Ramsay, International sepsis definition conference, 2001 SCCM/ESICM/ACCP/ATS/SIS international sepsis definitions conference, Intensive Care Med. 29 (2003) 530–538.
[98] R.A. Balk, Systemic inflammatory response syndrome (SIRS). Where did it come from and is it still relevant today? Virulence 5 (2014) 20–26.
[99] D.C. Angus, T. van der Poll, Severe sepsis and septic shock, N. Engl. J. Med. 369 (2013) 840–851.
[100] R.S. Hotchkiss, G. Monneret, D. Payen, Sepsis-induced immunosuppression: from cellular dysfunction to immunotherapy, Nat. Rev. Immunol. 13 (2013) 862–874.
[101] C.S. Deutschman, K.J. Tracey, Sepsis: current dogma and new perspectives, Immunity 40 (2014) 463–475.
[102] M. Singer, C.S. Deutschman, C.W. Seymour, M. Shankar-Hari, D. Annane, M. Bauer, R. Bellamo, G.R. Bernard, J.-D. Chiche, C.M. Coopersmith, R.S. Hotchkiss, M.M. Levy, J.C. Marshall, G.S. Martin, S.M. Opal, G.D. Rubenfeld, T. van der Poll, J.-L. Vincent, D.C. Angus, The third international consensus definitions for sepsis and septic shock (sepsis-3), J. Am. Med. Assoc. 23 (2016) 801–810.
[103] S. Fujishima, Organ dysfunction as a new standard for defining sepsis, Inflamm. Regen. 32 (2016), 24, 1–6, https://doi.org/10.1186/s41232-016-0028-y.
[104] J.L. Vincent, A. de Mendonca, F. Cantraine, R. Monero, J. Takala, P.M. Suter, C.L. Sprung, Use of the SOFA score to assess the incidence of organ dysfunction/failure in intensive care units: results of a multicenter, prospective study. Working group on "sepsis-related problems" of the European society of intensive care medicine, Crit. Care Med. 26 (1998) 1793–1800.
[105] Y.D. Podnos, J.C. Jiminez, S.E. Wilson, Intra-abdominal sepsis in elderly persons, Clin. Infect. Dis. 35 (2002) 62–68.
[106] M.D. Williams, L.A. Braun, I.M. Cooper, J. Johnston, R.V. Weiss, R.I. Qualy, W. Linde-Zwirble, Hospitalized cancer patients with severe sepsis: analysis of incidence, mortality, and associated costs of care, Crit. Care 8 (2004) R291–R298.

[107] B.A. Shah, J.F. Padbury, Neonatal sepsis. An old problem with new insights, Virulence 5 (2014) 170–178.

[108] D. Pittet, R.P. Wenzel, Nosocomial bloodstream infections. Secular trends in rates, mortality, and contribution to total hospital deaths, Arch. Intern. Med. 155 (1995) 1177–1184.

[109] M.J. Llewelyn, J. Cohen, Tracking the microbes in sepsis: advancements in treatment bring challenges for microbial epidemiology, Clin. Infect. Dis. 44 (2007) 1343–1348.

[110] K.E. Kollef, G.E. Schramm, A.R. Wills, R.M. Reichley, S.T. Mirek, M.H. Kollef, Predictors of 30-day mortality and hospital costs in patients with ventilator-associated pneumonia attributed to potentially antibiotic-resistant gram-negative bacteria, Chest 134 (2008) 281–287.

[111] M. Babu, V.P. Menon, P. Uma Devi, Prevalence of antimicrobial resistant pathogens in severe sepsis and septic shock patients, J. Young Pharm. 10 (2018) 358–361.

[112] C.E. Luyt, A. Combes, C. Deback, N.H. Aubriot-Lorton, A. Nieszkowska, J.L. Trouillet, F. Capron, H. Agut, C. Gibert, J. Chastre, Herpes simplex virus lung infection in patients undergoing prolonged mechanical ventilation, Am. J. Respir. Crit. Care Med. 175 (2007) 935–942.

[113] A.P. Limaye, K.A. Kirby, G.D. Rubenfeld, W.M. Leisenring, E.M. Bulger, M.J. Neff, N.S. Gibran, M.-L. Huang, T.K. Santo, L. Corey, M. Boeckh, Cytomegalovirus reactivation in critically ill immunocompetent patients, J. Am. Med. Assoc. 300 (2008) 413–422.

[114] G.-L. Lin, J.P. McGinley, S.B. Drysdale, A.J. Pollard, Epidemiology and immune pathogenesis of viral sepsis, Front. Immunol. 9 (2018), article 2147, 1–21, https://doi.org/10.3389/fimmu.2018.02147.

[115] L. Rózsa, P. Apari, V. Müller, The microbiome mutiny hypothesis: can our microbiome turn against us when we are old or seriously ill? Biol. Direct 10 (2015) 3, 1–9, https://doi.org/10.1186/s13062-014-0034-5.

[116] L. Rózsa, P. Apari, M. Sulyok, D. Tappe, I. Bodó, R. Hardi, V. Müller, The evolutionary logic of sepsis, Infect. Genet. Evol. 55 (2017) 135–141.

[117] E.A. Deitch, Rodent models of intra-abdominal infection, Shock 24 (Suppl. 1) (2005) 19–23.

[118] D. Rittirsch, L.M. Hoesel, P.A. Ward, The disconnect between animal models of sepsis and human sepsis, J. Leukoc. Biol. 81 (2007) 137–143.

[119] L.F. Poli-de-Figueiredo, A.G. Garrido, N. Nakagawa, P. Sannomiya, Experimental models of sepsis and their clinical relevance, Shock 30 (Suppl. 1) (2008) 53–59.

[120] J.Y. Chen, Y. Qiao, J.L. Komisar, W.B. Baze, I.C. Hsu, J. Tseng, Increased susceptibility to staphylococcal enterotoxin B intoxication in mice primed with actinomycin D, Infect. Immun. 62 (1994) 4626–4631.

[121] H.S. Warren, C. Fitting, E. Hoff, M. Adib-Conquy, L. Beasley-Topliffe, B. Tesini, X. Liang, C. Valentine, J. Hellman, D. Hayden, J.M. Cavaillon, Resilience to bacterial infection: difference between species could be due to proteins in serum, J. Infect. Dis. 201 (2010) 223–232.

[122] J. Zschaler, D. Schlorke, J. Arnhold, Differences in innate immune response between man and mouse, Crit. Rev. Immunol. 34 (2014) 433–454.

[123] G. Mathiak, D. Szewczyk, F. Abdullah, P. Ovadia, G. Feuerstein, R. Rabinovici, An improved clinically relevant sepsis model in the conscious rat, Crit. Care Med. 28 (2000) 1947–1952.

[124] C.C. Baker, I.H. Chaudry, H.O. Gaines, A.E. Baue, Evaluation of factors affecting mortality rate after sepsis in a murine cecal ligation and puncture model, Surgery 94 (1983) 331–335.

[125] A.J. Lewis, C.W. Seymour, M.R. Rosengart, Current murine models of sepsis, Surg. Infect. 17 (2016) 385–393.

[126] N. Zantl, A. Uebe, B. Neumann, H. Wagner, J.R. Siewert, B. Holzmann, C.D. Heidecke, K. Pfeffer, Essential role of gamma interferon in survival of colon ascendens stent peritonitis, a novel murine model of abdominal sepsis, Infect. Immun. 66 (1998) 2300–2309.

[127] L.M. Hoesel, T.A. Neff, S.B. Neff, J.G. Younger, E.W. Olle, H. Gao, M.J. Pianko, K.D. Bernacki, J.V. Sharma, P.A. Ward, Harmful and protective roles of neutrophils in sepsis, Shock 24 (2005) 40–47.

[128] R.S. Hotchkiss, G. Monneret, D. Payen, Immunosuppression in sepsis: a novel understanding of the disorder and a new therapeutic approach, Lancet Infect. Dis. 13 (2013) 260–268.

[129] R.S. Hotchkiss, I.E. Karl, The pathophysiology and treatment of sepsis, N. Engl. J. Med. 348 (2003) 138–150.

[130] A.J. Reber, T. Chirkova, J.H. Kim, W. Cao, R. Biber, D.K. Shay, S. Sambhara, Immunosenescence and challenges of vaccination against influenza in the aging population, Aging Dis. 3 (2012) 68–90.

[131] M.A. Cheever, Twelve immunotherapy drugs that could cure cancers, Immunol. Rev. 222 (2008) 357–368.

[132] F.S. Hodi, S.J. O'Day, D.F. McDermott, R.W. Weber, J.A. Sosman, J.B. Haanen, R. Gonzalez, C. Robert, D. Schadendorf, J.C. Hassel, W. Akerley, A.J. van den Eertwegh, J. Lutzky, P. Lorigan, J.M. Vaubel,

G.P. Linette, D. Hogg, C.H. Ottensmeier, C. Lebbé, C. Peschel, I. Quirt, J.I. Clark, J.D. Wolchok, J.S. Weber, J. Tian, M.J. Yellin, G.M. Nichol, A. Hoos, W.J. Urba, Improved survival with ipilimumab in patients with metastatic melanoma, N. Engl. J. Med. 363 (2010) 711–723.

[133] S.L. Topalian, F.S. Hodi, J.R. Brahmer, S.N. Gettinger, D.C. Smith, D.F. McDermott, J.D. Powerly, R.D. Carjaval, J.A. Sosman, M.B. Atkins, P.D. Leming, D.R. Spigel, S.J. Antonia, L. Horn, C.G. Drake, D.M. Pardoll, L. Chen, W.H. Sharfman, R.A. Anders, J.M. Taube, T.L. McMiller, H. Xu, A.J. Korman, M. Jure-Kunkel, S. Agrawal, D. McDonald, G.D. Kollia, A. Gupta, J.M. Wigginton, M. Sznol, Safety, activity, and immune correlates of anti-PD-1 antibody in cancer, N. Engl. J. Med. 366 (2012) 2443–2454.

[134] E.A. Said, F.P. Dupuy, L. Trautmann, Y. Zhang, Y. Shi, M. El-Far, B.J. Hill, A. Noto, P. Ancuta, Y. Peretz, S.G. Foncesa, J. van Grevenynghe, M.R. Boulassel, J. Brunneau, N.H. Shoukry, J.P. Routy, D.C. Douek, E.K. Haddad, R.P. Sekaly, Programmed death-1-induced interleukin-10 production by monocytes impairs $CD4^+$ T cell activation during HIV infection, Nat. Med. 16 (2010) 452–459.

[135] R.S. Hotchkiss, K.W. Tinsley, P.E. Swanson, R.E. Schmieg Jr., J.J. Hui, K.C. Chang, D.F. Osborne, B.D. Freeman, J.P. Cobb, T.G. Buchman, I.E. Karl, Sepsis-induced apoptosis causes progressive profound depletion of B and $CD4^+$ T lymphocytes in humans, J. Immunol. 166 (2001) 6952–6963.

[136] R.S. Hotchkiss, K.W. Tinsley, P.E. Swanson, M.H. Grayson, D.F. Osborne, T.H. Wagner, J.P. Cobb, C. Coppersmith, I.E. Karl, Depletion of dendritic cells, but not macrophages in patients with sepsis, J. Immunol. 168 (2002) 2493–2500.

[137] K.A. Felmet, M.W. Hall, R.S. Clark, R. Jaffe, J.A. Carcillo, Prolonged lymphopenia, lymphoid depletion, and hypoprolactinemia in children with nosocomial sepsis and multiple organ failure, J. Immunol. 174 (2005) 3765–3772.

[138] F. Venet, C.,S. Chung, G. Monneret, X. Huang, B. Horner, M. Garber, A. Ayala, Regulatory T cell populations in sepsis and trauma, J. Leukoc. Biol. 83 (2008) 523–535.

[139] B. Suetrong, K.R. Walley, Lactate acidosis in sepsis: it's not all anaerobic – implications for diagnosis and management, Chest 149 (2016) 252–261.

[140] B. Nolt, F. Tu, X. Wang, T. Ha, R. Winter, D.L. Williams, C. Li, Lactate and immunosuppression in sepsis, Shock 49 (2018) 120–125.

[141] H.B. Nguyen, E.P. Rivers, B.P. Knoblich, G. Jacobsen, A. Muzzin, J.A. Ressler, M.C. Tomlanovich, Early lactate clearance is associated with improved outcome in severe sepsis and septic shock, Crit. Care Med. 32 (2004) 1637–1642.

[142] A.H. Rishu, R. Khan, H.M. Al-Dorzi, H.M. Tamin, S. Al-Qahtami, G. Al-Ghamdi, Y.M. Arabi, Even mild hyperlactemia is associated with increased mortality in critically ill patients, Crit. Care 17 (2013), R197, 1–7, https://doi.org/10.1186/cc12891.

[143] M.A. Puskarich, S. Trzeciak, N.I. Shapiro, A.B. Albers, A.C. Heffner, J.A. Kline, A.E. Jones, Whole blood lactate kinetics in patients undergoing quantitative resuscitation for severe sepsis and septic shock, Chest 143 (2013) 1548–1553.

[144] E. Bolvardi, J. Malmir, H. Reitani, A.M. Hashemian, M. Bahramian, P. Khademhosseini, K. Ahmadi, The role of lactate clearance as a predictor of organ dysfunction and mortality in patients with severe sepsis, Mater. Sociomed. 28 (2016) 57–60.

[145] O.M. Pena, J. Pisolic, D. Raj, C.D. Fjell, R.E. Hancock, Endotoxin tolerance represents a distinctive state of alternative polarization (M2) in human mononuclear cells, J. Immunol. 186 (2011) 7243–7254.

[146] O.R. Colegio, N.Q. Chu, A.L. Szabo, T. Chu, A.M. Rhebergen, V. Jairma, N. Cyrus, C.E. Brokowski, S.C. Eisenbarth, G.M. Philips, G.W. Cline, A.J. Phillips, R. Medzhitov, Functional polarization of tumour-associated macrophages by tumour-derived lactic acid, Nature 513 (2014) 559–563.

[147] R.J. Arts, M.S. Gresnigt, L.A. Joosten, M.G. Neta, Cellular metabolism of myeloid cells in sepsis, J. Leukoc. Biol. 101 (2017) 151–164.

[148] X. Fan, Z. Liu, H. Jin, J. Yan, H.P. Liang, Alterations of dendritic cells in sepsis: featured role in immunoparalysis, BioMed Res. Int. 2015 (2015), article 903720, 1–10, https://doi.org/10.1155/2015/903720.

[149] K. Fischer, P. Hoffmann, S. Voelkl, N. Meidenbauer, J. Ammer, M. Edinger, E. Gottfried, S. Schwarz, G. Rothe, S. Hoves, K. Renner, B. Timischl, A. Mackensen, L. Kunz-Schughart, R. Andreesen, S.W. Krause, M. Kreutz, Inhibitory effect of tumor cell-derived lactic acid on human T cells, Blood 109 (2007) 3812–3819.

[150] M.F. Jiminez, R.W. Watson, J. Parodo, D. Evans, D. Foster, M. Steinberg, O.D. Rotstein, J.C. Marshall, Dysregulated expression of neutrophil apoptosis in the systemic inflammatory response syndrome, Arch. Surg. 132 (1997) 1263–1270.

[151] J. Demaret, F. Venet, A. Friggeri, M.-A. Cauzalis, J. Plassais, L. Jallades, C. Malcus, F. Poitevin-Later, J. Textoris, A. Lepape, G. Monneret, Marked alterations of neutrophil functions during sepsis-induced immunosuppression, J. Leukoc. Biol. 98 (2015) 1081–1090.

[152] K.A. Brown, S.D. Brain, J.D. Pearson, J.D. Edgeworth, S.M. Lewis, D.F. Treacher, Neutrophils in development of multiple organ failure in sepsis, Lancet 368 (2006) 157–169.

[153] M.A. Kovach, T.J. Standiford, The function of neutrophils in sepsis, Curr. Opin. Infect. Dis. 25 (2012) 321–327.

[154] B.M. Tavares-Murta, M. Zaparoli, R.B. Ferreira, M.L. Silva-Vergara, C.H. Oliveira, E.F. Murta, S.H. Ferreira, F.Q. Cunha, Failure of neutrophil chemotactic function in septic patients, Crit. Care Med. 30 (2002) 1056–1061.

[155] E. Crouser, M. Exline, M.D. Wewers, Sepsis: links between pathogen sensing and organ damage, Curr. Pharmaceut. Des. 14 (2008) 1840–1852.

[156] D. Rittirsch, M.A. Flierl, P.A. Ward, Harmful molecular mechanisms in sepsis, Nat. Rev. Immunol. 8 (2008) 776–787.

[157] D.C. Angus, The search for effective therapy for sepsis: back to the drawing board? J. Am. Med. Assoc. 306 (2011) 2614–2615.

[158] J. Cohen, S. Opal, T. Calandra, Sepsis studies need new direction, Lancet Infect. Dis. 12 (2012) 503–505.

[159] R.S. Hotchkiss, C.M. Coopersmith, J.E. McDunn, T.A. Ferguson, The sepsis seesaw: tilting toward immunosuppression, Nat. Med. 15 (2009) 496–497.

[160] J.S. Boomer, K. To, K.C. Chang, O. Takasu, D.F. Osborne, A.H. Walton, T.L. Bricker, S.D. Jarman 2nd, D. Kreisel, A.S. Krupnick, A. Srivastava, P.E. Swanson, J.M. Green, R.S. Hotchkiss, Immunosuppression in patients who die of sepsis and multiple organ failure, J. Am. Med. Assoc. 306 (2011) 2594–2605.

[161] F. Venet, A.P. Foray, A. Villars-Méchin, C. Malcus, F. Poitevin-Later, A. Lepape, G. Monneret, IL-7 restores lymphocyte functions in septic patients, J. Immunol. 189 (2012) 5073–5081.

[162] S. Inoue, J. Unsinger, C.G. Davis, J.T. Muenzer, T.A. Ferguson, K. Chang, D.F. Osborne, A.T. Clark, C.M. Coopersmith, J.E. McDunn, R.S. Hotchkiss, IL-15 prevents apoptosis, reverses innate and adaptive immune dysfunction, and improves survival in sepsis, J. Immunol. 187 (2010) 5310–5319.

[163] S. Inoue, L. Bo, J. Bian, J. Unsinger, K. Chang, R.S. Hotchkiss, Dose dependent effect of anti-CTLA-4 on survival in sepsis, Shock 36 (2011) 38–44.

[164] M. Varpula, M. Tallgren, K. Saukkonen, L.M. Voipio-Pulkki, V. Pettilä, Hemodynamic variables related to outcome in septic shock, Intensive Crit. Care 31 (2005) 1066–1071.

[165] E.M. Redl-Wenzl, C. Armbruster, G. Edelmann, E. Fischl, M. Kolacny, A. Wechsler-Fördös, P. Sporn, The effects of norepinephrine on hemodynamics and renal function in severe septic shock states, Intensive Care Med. 19 (1993) 151–154.

[166] E. Sorrentino, J. Diedler, M. Kasprowicz, K.P. Budohoski, C. Haubich, P. Smielewski, J.G. Outtrim, A. Manktelow, P.J. Hutchinson, J.D. Pickard, D.K. Menon, M. Czonnyka, Critical thresholds for cerebrovascular reactivity after traumatic brain injury, Neurocritical Care 16 (2012) 258–266.

[167] P. Asfar, F. Meziani, J.F. Hamel, F. Grelon, B. Megarbane, N. Anguel, J.P. Mira, P.F. Dequin, S. Gergaud, N. Weiss, F. Legay, Y. Le Tulzo, M. Conrad, R. Robert, F. Gonzalez, C. Guitton, F. Tamion, J.M. Tonnelier, P. Guezennec, T. Van der Linden, A. Vieillard-Baron, E. Mariotte, G. Pradel, O. Lesieur, J.D. Richard, F. Hervé, D. du Cheyron, C. Guerin, A. Mercat, J.L. Teboul, P. Radermacher, SEPSISPAM investigators, High versus low blood-pressure target in patients with septic shock, N. Engl. J. Med. 370 (2014) 1583–1593.

[168] R.S. Hotchkiss, L.L. Moldawer, S.M. Opal, K. Reinhart, I.R. Turnball, J.-L. Vincent, Sepsis and septic shock, Nat. Rev. Dis. Primers 2 (2016), 16045, 1–47, https://doi.org/10.1038/nrdp.2016.45.

[169] M. Levi, M. Schultz, T. van der Poll, Sepsis and thrombosis, Semin. Thromb. Hemost. 39 (2013) 559–566.

[170] P.A. Ward, The harmful role of C5a on innate immunity in sepsis, J. Innate Immun. 2 (2010) 439–445.

[171] H.D. Spapen, R. Jacobs, P.M. Honoré, Sepsis-induced multi-organ dysfunction syndrome. A mechanistic approach, J. Emerg. Crit. Care Med. 1 (2017), 27, 1–9, https://doi.org/10.21037/jeccm.2017.09.04.

[172] J.A. Clark, C.M. Coopersmith, Intestinal crosstalk: a new paradigm for understanding the gut as the "motor" of critical illness, Shock 28 (2007) 384–393.

[173] G. Piton, F. Belon, B. Cypriani, J. Regnard, M. Puyraveau, C. Manzon, J.C. Navellou, G. Capellier, Enterocyte damage in critically ill patients is associated with shock condition and 28-day mortality, Crit. Care Med. 41 (2013) 2169–2176.

[174] M. Gatt, B.S. Reddy, J. McFie, Bacterial translocation in the critically ill – evidence and methods of prevention, Aliment. Pharmacol. Ther. 25 (2007) 741–757.

[175] E.A. Deitch, Gut-origin sepsis: evolution of a concept, Surgeon 10 (2012) 350–356.
[176] M. Le Dorze, M. Legrand, D. Payen, C. Ince, The role of the microcirculation in acute kidney injury, Curr. Opin. Crit. Care 15 (2009) 503–508.
[177] A. Zarbock, H. Gomez, J.A. Kellum, Sepsis-induced AKI revisited: pathophysiology, prevention and future therapies, Curr. Opin. Crit. Care 20 (2014) 588–595.
[178] J. Yan, S. Li, S. Li, The role of the liver in sepsis, Int. Rev. Immunol. 33 (2014) 498–510.
[179] E.A. Woźnica, M. Inglot, R.K. Woźnica, L. Łysenko, Liver dysfunction in sepsis, Adv. Clin. Exp. Med. 27 (2018) 547–551.
[180] W.-Y. Kim, S.-B. Hong, Sepsis and acute respiratory distress syndrome: recent update, Thromb. Respir. Dis. 79 (2016) 53–57.
[181] K. Drosatos, A. Lymeropoulos, P.J. Kennel, N. Polak, P.C. Schulze, I.J. Goldberg, Pathophysiology of sepsis-related cardiac dysfunction: driven by inflammation, energy mismanagement, or both? Curr. Heart Fail. Rep. 12 (2015) 130–140.
[182] Y. Kakihana, T. Ito, M. Nakahara, K. Yamaguchi, T. Yasuda, Sepsis-induced myocardial dysfunction: pathophysiology and management, J. Intensive Care 4 (2016), 22, 1–10, https://doi.org/10.1186/s40560-016-0148-1.
[183] T. Suzuki, Y. Suzuki, J. Okada, T. Kurazumi, T. Suhara, T. Ueda, H. Nagata, H. Morisaki, Sepsis-induced cardiac dysfunction and β-adrenergic blockade therapy for sepsis, J. Intensive Care 5 (2017), 22, 1–10, https://doi.org/10.1186/s40560-017-0215-2.
[184] R. Sonneville, F. Verdonk, C. Rauturier, I.F. Klein, M. Wolff, D. Annana, F. Chretien, T. Sharshar, Understanding brain dysfunction in sepsis, Ann. Intensive Care 3 (2013), 15, 1–11, https://doi.org/10.1186/2110-5820-3-15.

CHAPTER

11

Conclusions

11.1 THE ALLIANCE OF HIGH COMPLEXITY AND DESTRUCTIONS IN LIVING SYSTEMS

The high order of complex biological structures and damage of this material are indispensable interrelated. Life appears to have Janus face—like properties. Synthesis of novel macromolecules and their arrangement into structural and functional units, on the one hand, and numerous destructions in cells and tissues, on the other hand, are in striking opposition to each other. To avoid an uncontrolled damage of their structures and to maintain their high structural and functional complexity, living organisms employ three main protective strategies:

i) inhibition of damaging reactions and repair of small defects,
ii) removal of deteriorated structural elements and replacement by novel ones, and
iii) augmentation of cell masses, i. e. growth in combination with cell divisions.

These strategies ensure the long-term survival of all living forms as species.

If we draw a closer look on individual human and animal organisms, we see a decrease in numerous physiological functions after the organism reached the optimum size at the beginning of adulthood. Then growth processes play only a minor role in defense against structural and functional deteriorations. This worsening becomes apparent during the second half of lifetime with increasing age. We learned from the analysis of destructive processes that growth in combination with cell divisions is the only strategy against destructions, which allows the maintenance of a low level of entropy per mass unit of biological structures in living systems.

11.2 DISTURBANCES OF HOMEOSTASIS

11.2.1 The Significance of a Regulated Internal Milieu in Multicellular Organisms

Higher animals and humans are multicellular organisms, which are characterized by the existence of a well-regulated internal milieu. Main physical properties of this milieu and concentrations of some constituents are held constant. The maintenance of these homeostatic parameters is of utmost importance for the long-term survival of complex living systems.

Cells, tissues, and organisms are open systems in a thermodynamic point of view. The energy consumed by these systems is used not only to create the highly ordered structural arrangement of macromolecules and to provide specific functional responses but also to correct any deviations in regulated homeostatic values. The existence of a well-defined internal milieu enables animals and humans much better to economize

Cell and Tissue Destruction
https://doi.org/10.1016/B978-0-12-816388-7.00011-5

energy and to survive unfavorable environmental conditions in comparison with an agglomeration of identical cells such as bacteria. Moreover, this milieu allows structural and functional differentiation of cells within complex organisms that greatly broadens the manifestation of living forms.

In humans and higher animals, the internal milieu is mainly presented by the circulating blood that delivers dioxygen and nutrients and provides them to all parts of the organisms. Removal of waste products from cells and tissues and secretion of wastes is another important function of blood. The nervous system is also an all-encompassing system, connecting the central regulatory unit, the brain, with all regions of the organism. General metabolic activities of the organism are regulated by endocrine hormones under control of the central nervous system as well. The immune system can additionally be activated by local and systemic disturbances to ensure and restore the integrity of affected tissues and organs.

11.2.2 Limits of Homeostatic Parameters

Concerning the endless fight against destructions, we have learned that defective dysfunctional cells can be driven into apoptosis, necrosis, or other kinds of cell death, removed by macrophages and replaced by novel synthesized cells. Providing the intactness of the underlying mechanisms, this recycling ensures a permanent rejuvenation of tissues. The same holds for many serum proteins, which are regularly synthesized in the liver and released into the blood. However, these mechanisms do not suffice for the maintenance of all homeostatic parameters of the internal milieu.

The analysis of underlying molecular, destructive mechanisms in the pathogenesis of diseases shows that disturbances in homeostasis of cellular and systemic parameters provide the basis of these diseases. The maintenance of a given parameter is regulated via regulatory circuits. Any deviation of the actual value from the nominal one of a homeostatic parameter induces an energy-consuming correction. If such deviations occur too often or they are too strong, the utilized energy increases followed by an energy deficit. Exhaustion of ATP diminishes the functional capacity of cells and tissues to respond adequately to disturbances of homeostasis. Under stress situations, more and more ATP is required for repair processes at the expense of normal functional activities. This further worsens the ability of organisms to provide useful work, to respond to unexpected events, and to resist against foreign microorganisms.

This general lack of energy provides the basis for deficiencies and exhaustion of natural protecting mechanisms, increasing impacts of cell and tissue destructions, and the development of multiple disease scenarios.

11.2.3 Pathogenesis of Chronic Inflammatory Diseases

On selected examples, we have seen the large variety of chronic disease forms that may affect human organism and disturb homeostatic parameters. Despite this diversity and the involvement of different organs in diseases, there are some common characteristics, which are valid for all these pathologies. They are summarized in Fig. 11.1.

At the beginning of disease, development should always be any impact that disturbs transiently the homeostasis in a certain region of the organism. Most of all, this is a local disruption of a barrier function of skin or mucous surfaces. Disturbance of homeostasis can result from bleeding and/or enhanced permeability for certain components through this barrier. Mucous and skin surfaces separate not only the outer milieu from inner compartments, but they are also important areas for exchange of gases and nutrients. Moreover, they prevent

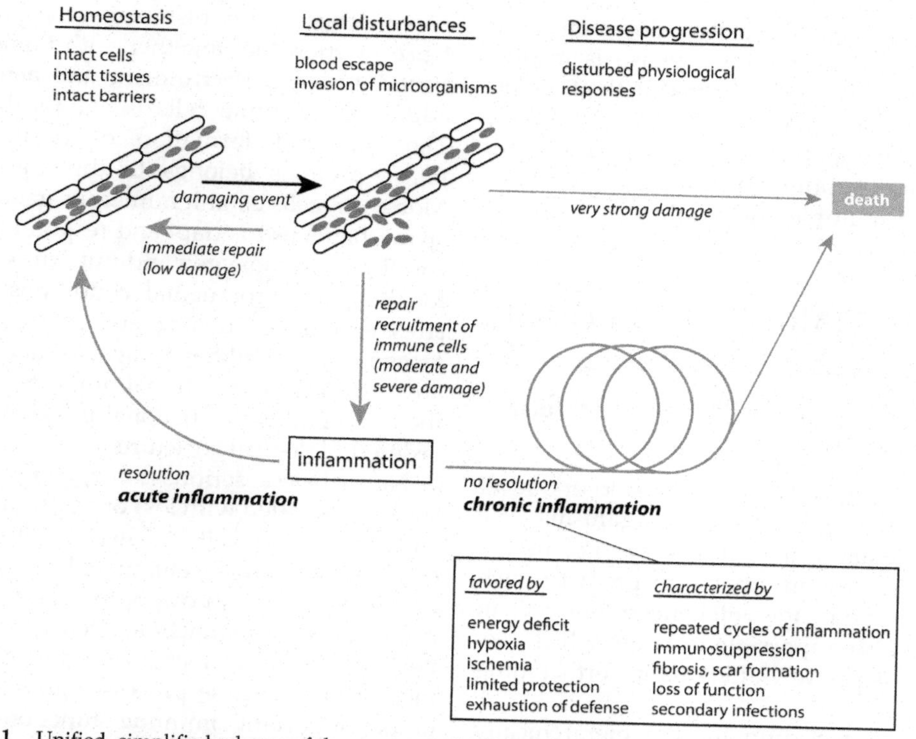

FIGURE 11.1 Unified, simplified scheme of the response of cells and tissues to local disturbances and the significance of chronic inflammatory diseases in these processes. Damaging events can immediately be repaired by preexisting protective mechanisms, cause an inflammatory response under activation of immune cells, or lead to a lethal outcome. In an acute inflammation, the inflammatory process is fully terminated, and the former homeostasis is restored. Numerous chronic disease scenarios are associated with insufficient resolution of inflammation. The imbalance between host's protective systems and immune cell–mediated tissue damage contributes to repeated cycles of inflammation and can further worsen the energy and health status of the host.

the invasion of external microorganisms in our body. Disruption of barrier functions initiates any response of protective systems.

Local disturbance of barrier functions results in blood escape and invasion of microorganisms. In adjacent tissue regions, the energy demand raises, and local ischemic and hypoxic states are favored by more severe impacts. There is an immediate activation of various defense mechanisms to stop the bleeding, to combat against invading microorganisms, and to remove any debris. Nevertheless, more or less pronounced necrosis areas can be formed. The overall degree of cell and tissue damage depends not only on the strength of the initiating impact but also significantly on the energy and health status of the affected tissues and the whole organism.

In many cases, the combined power of defense mechanisms is quite sufficient to restore the former homeostasis. Contrariwise, in some severe cases of damage, the organism passes away. Serious health problems arise from conditions, where despite the action of defense mechanisms, novel damage-associated molecular patterns are formed, microorganisms or cancer cells escape from the attacks of immune cells, host's defense mechanisms are weakened, and an immunosuppression status dominates too long. Then, an inflammatory phase is insufficiently terminated, and repeated activation cycles take place again and again. In worst cases, a chronic inflammation develops into a state, where immunosuppression dominates as seen in sepsis and multiple organ failure.

11.3 PECULIARITIES OF DEFENSE MECHANISMS

11.3.1 Destruction of Biological Material Versus Defense

Destruction of cells and tissues occurs at all stages of the development of organisms. We can speculate about the probability of some destructions, but we are unable to predict which certain kind of damage will concern our cells in future. It is also impossible to know which diseases and injuries will affect us in forthcoming time.

In humans and animals, the hierarchically organized protective systems are directed against most common disturbances of biological material. In the first line of defense reactions, already existing mechanisms contribute to inhibition, scavenging, and repair of unwanted alterations in biomolecules and toxic species. Example are the presence of antioxidants in aqueous and lipid phase, DNA repair systems, and serum proteins for scavenging of free heme proteins, neutrophil, and mast cell proteases. In addition, components of complement, coagulation, and contact systems circulate in blood, ready to be activated by appropriate disturbances. Immune cells residing in or patrolling through tissues also belong to immediately acting defense systems. By these mechanisms, the organism impedes any potential danger that may arise from low-level disturbances.

Under more severe stress situations, the activation of different inducible protective mechanisms contributes to increase the defense power. These mechanisms of second line of defense reactions act together with the aforementioned systems. Recruitment and activation of additional immune cells is a typical example. The enhanced formation of positive acute-phase proteins belongs to this category too. Moreover, cells exhibit numerous gene regulatory mechanisms to adapt and respond to varying conditions in dioxygen and nutrient supply and to altered environmental conditions including stress situations. Thus, organisms employ a great arsenal of possibilities to answer adequately to any external and internal impacts disturbing the structural and functional integrity of highly ordered biological structures.

Nevertheless, serious damage of cells and tissues is associated with loss of functions of cell organelles and cells. Then, defective cell constituents can be eliminated by autophagy, or cells are driven into apoptosis and other kinds of cell death. Dysfunctional cell debris and undergoing cells are recognized and digested by macrophages. These processes go in hand with other important immune functions such as detection, labeling, and elimination of antigens and unwanted microorganisms.

11.3.2 Exhaustion and Inactivation of Protective Systems

Despite the great significance and amazing variety of defense reactions, it again and again happens that organisms suffer from severe diseases and other threats. In other words, the normal physiological homeostasis is highly disturbed. In chronic diseases, it is very challenging or impossible to restore the former homeostasis.

Efficiency of protection highly depends on the energy status of cells and tissues and the intactness of defense systems. Moreover, when protective mechanisms are activated, there can be a decline in their capacity with increasing time. Under severe stress conditions, this can lead to complete exhaustion of certain defense reactions.

Of course, exhausting protective systems can be regenerated by time- and energy-consuming processes. An example for limited protection is the decline and exhaustion of haptoglobin and hemopexin after excessive intravascular hemolysis and/or rhabdomyolysis. A number of gene mutations and deficiencies are also associated with limited defense.

Long-term immunosuppression, which is observed in numerous patients with severe chronic diseases, is a very dangerous condition. This disturbs immune response against novel pathogens and facilitates infections with opportunistic microorganisms as well as the activation of silent viruses.

11.3.3 Potentiation of Damaging Reactions

In healthy organisms, cell and tissue damage results mainly not only from external events but also from chemical side reactions and physical disturbances. During inflammatory response, destructive reactions of biological material might also arise from activated immune cells. The latter process is favored by low efficient protective systems of the host as observed in chronically ill and immunocompromised persons. This is accompanied by an enhanced formation of necrotic cells, ischemic states, energy, and nutrient deficits, and by insufficient resolution of inflammatory reactions with scar formation and fibrosis induction. Finally, under conditions of immunodeficiency and limited protection, a loss of function is more often found in affected tissue areas than in healthy counterparts.

A decline and exhaustion of defense reactions potentiate cell and tissue damage. For example, this intensification can be caused by the prevalence of autoantibodies in different autoimmune disorders, the presence of cytotoxic free heme after excessive intravascular hemolysis and/or rhabdomyolysis, enhanced formation of reactive species due to redox recycling, or the unhindered growth of microbes or tumor cells. These amplification mechanisms are very dangerous to the organisms and lead to expanded dead tissue material, failure of single organs, sepsis, and, in extreme cases, multiple organ failure and lethal outcome.

11.4 WHAT CAN WE DO TO AVOID DREADFUL EFFECTS OF DAMAGING REACTIONS?

There does not exist any panacea against the adverse effects of damaging reactions, chronic inflammatory diseases, and functional worsening of physiological processes in advanced age. An active, healthy lifestyle, adequate physical and mental activities, as well as the avoidance of toxic and health-threatening substances are essential for well-consolidated protective systems. With these actions we cannot impede that we will be taken ill and that we will age, but we can postpone this inevitable period in our lifetime and can enhance the thresholds for the dominance of damaging reactions.

APPENDIX

Some Basics About Redox Reactions in Living Systems

A.1 CHEMISTRY AND PHYSICS OF REDOX REACTIONS

A.1.1 Oxidations and Reductions

In chemistry, reactions occurring under transfer of one or more electrons between reactants are called redox reactions. As these reactions are highly important for all chemical aspects of reactive species and processes of substrate oxidation, a short overview is given about redox chemistry in living systems.

In a redox reaction, the reductant is oxidized by liberating one or more electrons, whereas the oxidant is reduced under uptake of electrons. Oxidation and reduction are always closely linked to each other, as free electrons do virtually not exist. When oxidation and reduction exhibit the same rate in a redox reaction, a redox equilibrium exists.

Oxidation and reduction represent half reactions of the whole redox process. In a half reaction, oxidant and reductant form a redox couple. Two selected examples of biologically relevant redox reactions including the corresponding half reactions and redox couples are given in Fig. A.1.

A.1.2 Reduction Potentials

The tendency of the oxidant form of a redox couple to abstract one or more electrons from a substrate is described by the reduction potential E'. This value can be calculated by means of the Nernst equation:

$$E' = E'^{\circ} + (RT/nF)\ \ln(a_{ox}/a_{red}) \qquad (A.1)$$

The standard reduction potential E'° is referred to a concentration of 1 M of all reactants or a pressure of 101.3 kPa in case of gases. In chemistry, values for E'° of redox couples of interest are normally given at pH 0, whereas in biochemistry and other life sciences, these data are usually referred to pH 7. Standard values are conventionally referenced to the potential of the standard hydrogen electrode that was set to 0 V at pH 0. At pH 7, this value corresponds to −0.42 V.

In the Nernst Eq. (A.1), the gas constant R and the Faraday constant F are 8.31 J K^{-1} mol^{-1} and 96,485 As mol^{-1}, respectively. The absolute temperature T is usually set to 298 K. The factor n represents the number of electrons transferred in a single reaction step by this redox couple. The products of activities of all components

Example 1:

Cyt c(Fe^{3+}) + $O_2^{\bullet-}$ ⟶ Cyt c (Fe^{2+}) + O_2

Reduction: Cyt c(Fe^{3+}) + e^- ⟶ Cyt c (Fe^{2+})

Oxidation: $O_2^{\bullet-}$ ⟶ O_2 + e^-

Redox couples: Cyt c(Fe^{3+})/Cyt c (Fe^{2+}) and $O_2/O_2^{\bullet-}$

Example 2:

2 GSH + H_2O_2 ⟶ GS-SG + 2 H_2O

Reduction: H_2O_2 + 2 e^- + 2 H^+ ⟶ 2 H_2O

Oxidation: 2 GSH ⟶ GS-SG + 2 e^- + 2 H^+

Redox couples: H_2O_2/2 H_2O and GS-SG/2 GSH

FIGURE A.1 Two examples of biologically relevant redox reactions. Example 1 shows the reduction of the mitochondrial heme protein cytochrome *c* (Fe^{3+}) by superoxide anion radical (Section 4.5.5). Example 2 represents the enzyme-catalyzed oxidation of glutathione by hydrogen peroxide (Section 2.2.5). Half reactions (reduction, oxidation) and the involved redox couples are indicated. Abbreviations: *cyt c*, cytochrome *c*; *GSH*, glutathione (reduced form); *GS-SG*, glutathione (oxidized form).

involved in oxidation and reduction are given by a_{ox} and a_{red}.

Under conditions of redox equilibrium, the standard reduction potential for a redox couple can be measured or calculated from equilibrium activities of all participants. In most cases, the ratio a_{ox}/a_{red}, but not the total concentration of a redox component, determines the exact value of a reduction potential. Only if the stoichiometry of reduced and oxidized forms changes, the reduction potential depends also on reactant concentration. An example for this is the glutathione disulfide-glutathione redox couple GS-SG/2 GSH [1]. In this case, the concentration term in Eq. (A.1) corresponds to $\ln([GSH]^2/[GS\text{-}SG])$.

A.1.3 Differences to Chemicals Systems

Theory about redox reactions was primarily developed for chemical systems. There are some peculiarities in their description in biological systems.

i) These reactions are often embedded in a network of preceding, subsequent, and parallel reactions. Thus, it is unlikely that conditions exist for formation of chemical redox equilibria.
ii) Moreover, these reactions take mostly place in small microcompartments inside cells. Sometimes, redox components are localized on small segments of macromolecules, e.g., on an amino acid residue of a protein. This structural diversity impedes additionally the exact description of biological redox reactions.
iii) Redox components are more or less hydrated in cells. Hydrate shells can encumber redox reactions.
iv) In animals and humans, redox reactions occur mostly at 37°C. In chemistry, standard values of reduction potentials are referenced to 25°C. As usual, reduction

potentials are related here to 25°C. These potentials are higher by the factor 1.04 at 37°C.

v) In cells and tissues, majority of cytosolic reactions take place at pH values around 7.0–7.4 [2,3]. In some compartments, even higher (up to pH 8.2 inside mitochondria [4]) or lower (pH 4.0–4.5 in lysosomes [5]) pH values are possible. In contrast to chemical literature, standard reduction potentials of redox couples are related in life sciences to pH 7 but not to pH 0.

A.1.4 Dependencies of Reduction Potentials on pH Value

If protons are involved in a given reduction process, the corresponding reduction potential depends on pH value. Vice versa, reduction potentials are independent of pH, when no protons are embraced. Two examples for the latter case are

$Cl_2(aq) + 2e^- \rightarrow 2Cl^-$ with
$E'^\circ(Cl_2(aq)/2Cl^-) = 1.39$ V [6] and

$O_2(g) + e^- \rightarrow O_2^{\bullet -}$ with
$E^\circ(O_2(g)/O_2^{\bullet -}) = -0.33$ V [7].

How strong the pH effect on a reduction potential is expressed depends on the total number of electrons and protons in a reduction process. When equal numbers of electrons and protons are involved, the reduction potential decreases by 0.059 V per unit increasing pH value. This value results from Nernst equation inserting all constants and a temperature of $T = 298$ K. Other dependencies of reduction potentials and corresponding examples are given in Table A.1.

In description of redox processes, it is also necessary to consider pH-dependent changes in the protonation state of oxidants and reductants. Many of the above listed half reactions are only valid in a certain pH region due to the existence of pK_a values of reactants. For example, the pK_a value of superoxide is 4.81 [8]. Thus, in the one-electron reduction of dioxygen, the half reaction $O_2(g) + e^- + H^+ \rightarrow HO_2^\bullet$ dominates below pH 4.3, while above pH 5.4, the reduction potential is independent of pH and is described by $O_2(g) + e^- \rightarrow O_2^{\bullet -}$. Between pH 4.3 and 5.4, there is the intermediate region for the stepwise change of the slope in reduction potential. The dependence of reduction potential of this redox couple on pH is illustrated in Fig. A.2.

Considering the pK_a values of superoxide ($HO_2^\bullet$) and H_2O_2 being 4.81 [8] and 11.56 [9], respectively, the slope of −0.118 V per unit increasing pH value for the half reaction $O_2^{\bullet -} + e^- + 2H^+ \rightarrow H_2O_2$ is only valid between pH 5.4

TABLE A.1 Dependencies of Reduction Potentials on pH Value.

Number of Transferred Electrons	Number of Involved Protons	Change of Reduction Potential per Unit Increasing pH Value	Examples
1	0	0 V	$O_2 + e^- \rightarrow O_2^{\bullet -}$
2	0	0 V	$Cl_2 + 2e^- \rightarrow 2Cl^-$
1	1	−0.059 V	$OCl^- + e^- + H^+ \rightarrow Cl^- + H_2O$
2	1	−0.030 V	$HOCl + 2e^- + H^+ \rightarrow Cl^- + H_2O$
1	2	−0.118 V	$O_2^{\bullet -} + e^- + 2H^+ \rightarrow H_2O_2$
2	2	−0.059 V	$H_2O_2 + 2e^- + 2H^+ \rightarrow 2H_2O$

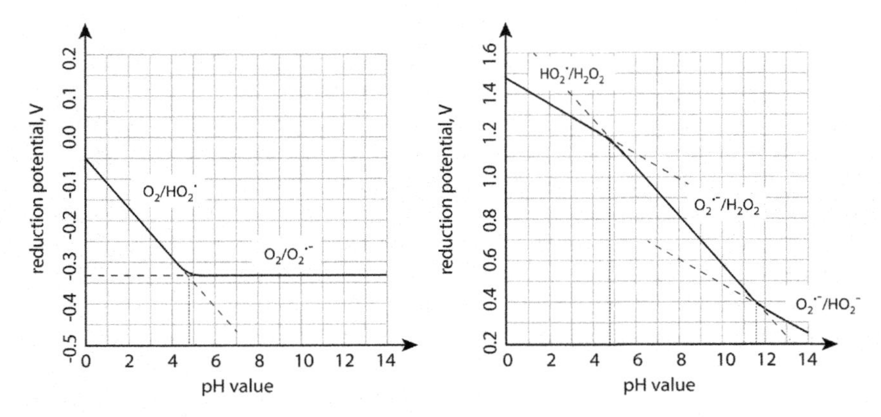

FIGURE A.2 Two examples of dependence of reduction potential on pH value. On the left side, the one-electron reduction of O_2(g) is presented. The dominating products are $HO_2^{\bullet}$ (below pH 4.81) and $O_2^{\bullet -}$ (above pH 4.81). At pH 7, the standard reduction potential of the couple O_2(g)/$O_2^{\bullet -}$ is equal to −0.33 V [7]. On the right side, the one-electron reduction of superoxide or superoxide anion radical to hydrogen peroxide or HO_2^- is given. The corresponding redox couples, which dominate below pH 4.81, in the pH range from 4.81 to 11.56 and above pH 11.56 are indicated in the figure. At pH 7, the standard reduction potential of the couple $O_2^{\bullet -}/H_2O_2$ is equal to 0.94 V [44]. Further explanations are given in the text.

and 11 (Fig. A.2). Below pH 4.3 and above pH 12, the reduction potential of this redox process changes with −0.059 V per unit increasing pH. Again, there are intermediate regions around pH values 4.81 and 11.56 for the change of the slope.

A special situation may occur when in a redox reaction the reduction potentials of both redox couples differ only slightly and these potentials exhibit different dependencies on pH values. Then, any pH changes can modulate this reaction. An example is the pH effect on the chlorinating and brominating activities of the heme proteins myeloperoxidase and eosinophil peroxidase [10,11]. At pH 7, the corresponding standard potentials are 1.16 and 1.10 V for the couple Compound I/ferric enzyme of myeloperoxidase and eosinophil peroxidase, respectively [12], 1.28 V for $HOCl/Cl^-$, H_2O [13] and 1.13 V for HOBr/Br^-, H_2O [14]. In the physiological relevant pH range (pH 5–8) for reactions of human heme peroxidases, the peroxidase potentials decline by −0.6 V per increasing pH unit, whereas the potentials for hypohalous acids decrease by −0.3 V per unit of increasing pH. As the pK_a value of HOCl is 7.53 at 37°C [15], at pH values higher 7, this decrease changes gradually to −0.6 V for HOCl. These different dependencies of reductions potentials on pH contribute to increasing halogenating activity of both peroxidases under slightly acidic conditions and the existence of different pH-dependent threshold values for chloride oxidation by both peroxidases [10].

A.2 THERMODYNAMICS OF REDOX REACTIONS

In a redox reaction, two redox couples interact with each other. The oxidized form of the redox couple with the higher reduction potential is reduced, whereas the reduced form of the second couple becomes oxidized. The reaction Gibbs energy Δ_rG' of a redox reaction can be calculated from the difference in reduction potentials $\Delta E'$ of both half reactions using

$$\Delta_r G' = -nF\Delta E' \quad (A.2)$$

A redox reaction proceeds spontaneously if $\Delta_r G'$ is negative.

$$\Delta_r G' < 0 \quad (A.3)$$

Alternatively, the equilibrium constant K' of all components in a redox equilibrium is linked with the difference of both standard reduction potentials $\Delta E'^{\circ}$ according to

$$\Delta E'^{\circ} = (RT/nF)\ln K' \quad (A.4)$$

However, not all redox reactions that are thermodynamically possible really occur with sufficient rate. Important for a high rate of a redox reaction is the degree of overlapping between electron orbitals from donor and acceptor [16,17]. If there is no overlapping between electronic orbitals of reactants, the activation energy is too high, and no redox reaction proceeds.

In addition, numerous other factors determine the rate of a redox reaction such as steric hindrance, availability of redox partners, or stability of hydrate shells. For example, redox reactions with reduction of H_2O_2 are comparable slow reactions because a transfer of electrons to an empty hardly accessible σ*-orbital is necessary [17]. Because of additional steric hindrance, reduction of alkyl hydroperoxides is yet slower [18,19].

A.3 STRONG OXIDANTS VERSUS SUBSTRATE OXIDATION

A.3.1 Easily Oxidizable Biological Material

In principle, it is possible to oxidize each kind of biological material. Degree of oxidative modification highly depends on strength of the applied oxidant, the amount of this oxidant, and the presence and capacity of antioxidative protecting systems.

Many kinds of biological material are very robust and resist efficiently to a high number of impacts. This property is fundamental for materials exposed to the outer milieu in cells and organisms giving living systems an additional protection. Materials of choice for surface covers are short-chain oligomeric carbohydrates and proteins with stable functional residues.

Other components of cells and tissues are more prone to oxidation. This concerns usually components inside cells and within regulatory units. Among easily oxidizable biological materials are unsaturated fatty acid residues in lipids, sulfur-containing and aromatic amino acids residues in proteins and small molecules, and some structural elements in lipids and carbohydrates characterized by weakened C—H bonds. Selected redox components are summarized in Table A.2 together with their reduction potentials.

In naturally occurring polyunsaturated fatty acids, all double bonds are in *cis* configuration, and two consecutive double bonds are always separated by two single bonds. H atoms are easily abstracted from carbons atoms at allylic sites to a double bond, especially at *bis*-allylic position between two double bonds. The standard reduction potentials of the couples allylic-C•/allylic-CH and *bis*-allylic-C•/*bis*-allylic-CH were calculated to be 0.7 and 0.6 V at pH 7, respectively [20]. More details are given in Section 3.2.1.

In some lipids and carbohydrates, abstraction of an H atom at carbons bearing a hydroxyl group and being in β-position to an ester or glycosidic bond is a further mechanism for initiating oxidative damage [21,22]. Data about standard reduction potentials for these oxidations are unknown. This issue is further described in Section 3.2.6.

In proteins and small molecules, sulfhydryl and methionyl groups are very susceptible to one- and two-electron oxidation. Their presence

TABLE A.2 Standard Reduction Potentials (E'°) of Selected Biological Substrates With Special Focus on Easily Oxidizable Materials.

Redox Couple	Half Reaction (Written as Reduction Process)	E'°, V (pH 7)	References
LH•/LH[a]	$-CH{=}CH{-}CH^{\bullet}{-}CH{=}CH{-} + e^- + H^+ \rightarrow -CH{=}CH{-}CH_2{-}CH{=}CH{-}$	0.6	[20]
	$-CH_2{-}CH{=}CH{-}CH^{\bullet}{-} + e^- + H^+ \rightarrow -CH_2{-}CH{=}CH{-}CH_2{-}$	0.7	[20]
	$-CH_2{-}CH{=}C^{\bullet}{-}CH_2{-} + e^- + H^+ \rightarrow -CH_2{-}CH{=}CH{-}CH_2{-}$	0.96	[20]
	$-CH_2{-}CH^{\bullet}{-}CH_2{-} + e^- + H^+ \rightarrow -CH_2{-}CH_2{-}CH_2{-}$	1.5–1.6	[20]
R-S•/R-SH	$R\text{-}S^{\bullet} + e^- + H^+ \rightarrow R\text{-}SH$	0.9–1.0	[23]
GS•/GSH	$GS^{\bullet} + e^- + H^+ \rightarrow GSH$	0.93	[24]
GS-SG/2 GSH	$GS\text{-}SG + 2\,e^- + 2\,H^+ \rightarrow 2\,GSH$	−0.24	[1]
Trx(-S-S-)/Trx(-SH, -SH)	$Trx(\text{-}S\text{-}S\text{-}) + 2\,e^- + 2\,H^+ \rightarrow Trx(\text{-}SH, \text{-}SH)$	−0.27 to −0.19	[25]
Met•/Met	$Met^{\bullet} + e^- + H^+ \rightarrow Met$	1.2–1.8	[26]
Met-O/Met, H_2O	$Met\text{-}O + 2\,e^- + 2\,H^+ \rightarrow Met + H_2O$	0.6–0.7	
Tyr•/Tyr	$Tyr^{\bullet} + e^- + H^+ \rightarrow Tyr$	0.88	[31]
		0.94	[30]
Trp•/Trp	$Trp^{\bullet} + e^- + H^+ \rightarrow Trp$	1.05	[30]
His•/His	$His^{\bullet} + e^- + H^+ \rightarrow His$	1.17	[32]
Guo•/Guo	$Guo^{\bullet} + e^- + H^+ \rightarrow Guo$	1.29	[33,34]
Ado•/Ado	$Ado^{\bullet} + e^- + H^+ \rightarrow Ado$	1.42	[33,34]
Comp I/ferric MPO	$Comp\ I + 2\,e^- + 2\,H^+ \rightarrow$ ferric MPO	1.16	[12]
Comp I/ferric EPO	$Comp\ I + 2\,e^- + 2\,H^+ \rightarrow$ ferric EPO	1.10	[12]
Comp I/ferric LPO	$Comp\ I + 2\,e^- + 2\,H^+ \rightarrow$ ferric LPO	1.09	[35]

[a] *Abbreviations:* Ado, *adenosine;* Comp, *Compound;* EPO, *eosinophil peroxidase;* GSH, *glutathione;* GS-SG, *oxidized glutathione;* Guo, *guanosine;* His, *histidine;* LH, *fatty acid molecule with abstractable H-atom;* LPO, *lactoperoxidase;* Met, *methionine;* Met-O, *methionine sulfoxide;* MPO, *myeloperoxidase;* Trp, *tryptophan;* Trx, *thioredoxin;* Tyr, *tyrosine.*

allows the existence of multiple redox-sensitive functional processes in cells. In addition, main oxidation products of thiol and thioether residues can often be reversed by cellular protective mechanisms. For proteins and peptides, standard reduction potentials of the couple R-S•/R-SH are in the range 0.9–1.0 V at pH 7 [23]. In case of glutathione, a potential of 0.93 V was determined [24]. Data about two-electron oxidation of sulfhydryls are known for glutathione and thioredoxin. At pH 7, the standard reduction potential of GS-SG/2 GSH is −0.24 V [1]. Values for various thioredoxins are in the range from −0.27 V to −0.19 V [25].

Data about redox chemistry of methionine and methionine residues in proteins are more complex. The corresponding value for Met•/Met for the standard reduction potential varies in a wide range from 1.2 to 1.8 V at pH 7 [26]. It highly depends on the near proximity of thioether function as the sulfur-based methionyl cation radical can interact with lone electron pairs of neighboring O-, N-, or S-atoms. Moreover, in some cases, even lower potentials are expected for one-electron methionine oxidation [27,28]. However, exact mechanism of one-electron methionine oxidation remains elusive. The exact value for the two-electron potential of the couple methionine sulfoxide/methionine is unknown. It should be somewhere in the range of 0.6–0.7 V, considering the fact that hypothiocyanite is unable to oxidize methionine in contrast to hypoiodous acid [29]. The corresponding reduction potentials for the couples HOI/I^-, H_2O and $^-OSCN/SCN^-$, H_2O are 0.78 and 0.56 V at pH 7, respectively [11].

One-electron oxidation of tyrosine, tryptophan, and histidine yields the corresponding radical species. For the couple Tyr•/Tyr, a standard reduction potential of 0.94 V at pH 7 was calculated [30]. Another determination of the Tyr•/Tyr couple yielded a reduction potential of 0.88 V [31]. The corresponding value for the couple Trp•/Trp is 1.05 V [30]. The standard reduction potential of the couple His•/His is 1.17 V at pH 7 [32].

Guanine has with 1.29 V at pH 7 the lowest one-electron reduction potential among DNA bases [33,34]. The values for the reduction potentials of the other DNA bases are about 0.1 –0.2 V higher [33,34].

At inflammatory sites, the activation of ferric heme peroxidases by hydrogen peroxides yields the powerful redox form Compound I (Section 2.5.1). The standard reduction potentials for the couple Compound I/ferric enzyme of human myeloperoxidase and eosinophil peroxidase are 1.16 and 1.10 V, respectively, at pH 7 [12]. The corresponding potential for bovine lactoperoxidase is 1.09 V [35]. The bovine isoform is often applied to model the behavior of human lactoperoxidase.

A.3.2 Which Reactive Species Are Responsible for Substrate Alterations?

It is impossible to give an unequivocal answer to this question. Important biologically relevant one- and two-electron oxidants and their standard reduction potentials at pH 7 are listed in Table A.3.

The highest standard reduction potential has the hydroxyl radical with 2.31 V [36]. Besides radiolysis of water, hydroxyl radicals are mainly formed as a result of the Fenton reaction on reaction of Fe^{2+} or Cu^+ with hydrogen peroxide (Section 2.2.6.2). Very reactive iron–oxygen complexes such as ferryl and perferryl species are also discussed as product of the Fenton reaction [37–41]. It is expected that these species have similar redox properties as hydroxyl radicals. However, data about reduction potentials of these species are not available.

Similar to hydrogen peroxide, lipid hydroperoxides (LOOH) are restricted in their reactivity because of completely filled electron orbitals [18,19], despite high values of reduction potentials for one- and two-electron reduction [20]. Thus, they react preferentially with transition metal ions, whereby their reduction dominates over oxidation. As these ions are usually under strict control, a significant decomposition of hydroperoxides by free Fe^{2+} or Cu^+ is only likely at strong inflammatory sites and hemorrhagic loci, where the control over these ions is limited or lost. The reduction potential for the one-electron reduction of lipid alkoxyl radicals (LO•) and peroxyl radicals (LOO•) are 1.7 and 1.0 V, respectively [20].

Several enzymes are able to catalyze the two-electron reduction of hydrogen peroxide and lipid hydroperoxides (Sections 2.2.5.4, 2.2.5.1 and 3.6.2). The corresponding reduction potentials at pH 7 are 1.32 V for $H_2O_2/2\ H_2O$ [36] and 1.8 V for LOOH/LOH, H_2O [20].

TABLE A.3 Standard Reduction Potentials (E'°) of Biologically Relevant One- and Two-Electron Oxidants.

Redox Couple	Half Reaction (Written as Reduction Process)	E'°, V (pH 7)	References
$HO^{\bullet}/H_2O$	$HO^{\bullet} + e^- + H^+ \rightarrow H_2O$	2.31	[36]
LOOH/LOH, H_2O[a]	$LOOH + 2\ e^- + 2\ H^+ \rightarrow LOH + H_2O$	1.8	[20]
LOOH/LO•, H_2O	$LOOH + e^- + H^+ \rightarrow LO^{\bullet} + H_2O$	1.9	[20]
LO•/LOH	$LO^{\bullet} + e^- + H^+ \rightarrow LOH$	1.7	[20]
$ONOO^-/NO_2^{\bullet}$, H_2O	$ONOO^- + e^- + 2\ H^+ \rightarrow NO_2^{\bullet} + H_2O$	1.6	[42]
$ONOO^-/NO_2^-$, H_2O	$ONOO^- + 2e^- + 2H^+ \rightarrow NO_2^- + H_2O$	1.3	[42]
$H_2O_2/2\ H_2O$	$H_2O_2 + 2\ e^- + 2\ H^+ \rightarrow 2\ H_2O$	1.32	[36]
$HOCl/Cl^-$, H_2O	$HOCl + 2\ e^- + H^+ \rightarrow Cl^- + H_2O$	1.28	[13]
$HOBr/Br^-$, H_2O	$HOBr + 2\ e^- + H^+ \rightarrow Br^- + H_2O$	1.13	[14]
LOO•/LOOH	$LOO^{\bullet} + e^- + H^+ \rightarrow LOOH$	1.0	[20]
$NO_2^{\bullet}/NO_2^-$	$NO_2^{\bullet} + e^- + H^+ \rightarrow NO_2^-$	1.04	[43]
$O_2^{\bullet -}/H_2O_2$	$O_2^{\bullet -} + e^- + 2\ H^+ \rightarrow H_2O_2$	0.94	[44]
Comp I/Comp II (MPO)	Comp I + e^- + H^+ → Comp II	1.35	[45]
Comp II/ferric MPO, H_2O	Comp II + e^- + H^+ → ferric MPO + H_2O	0.97	[45]
Comp I/Comp II (LPO)	Comp I + e^- + H^+ → Comp II	1.14	[35]
Comp II/ferric LPO, H_2O	Comp II + e^- + H^+ → ferric LPO + H_2O	1.04	[35]

[a] *Abbreviations:* Comp, *compound;* LOH, *fatty acid alcohol or lipid alcohol;* LOOH, *fatty acid hydroperoxide or lipid hydroperoxide;* LO•, *lipid alkoxyl radical;* LOO•, *lipid peroxyl radical;* LPO, *lactoperoxidase;* MPO, *myeloperoxidase.*

Peroxynitrite is a powerful one- and two-electron oxidant [42]. It results mainly from the rapid reaction between $O_2^{\bullet -}$ and NO. Another important nitrogen-based oxidant is nitrogen dioxide [43]. More details about peroxynitrite and nitrogen dioxide are given in Sections 2.3.2 and 2.3.4

Activation of heme peroxidases by hydrogen peroxide provides the basis for the production of strong two-electron oxidants such as HOCl and HOBr. Compound I and Compound II of myeloperoxidase, eosinophil peroxidase, and lactoperoxidase are involved in numerous substrate oxidations. The highest reduction potential exhibits the couple Compound I/Compound II of myeloperoxidase with 1.35 V at pH 7 [45]. The exact values for the reduction potentials of the couple Compound I/Compound II and

Compound II/ferric enzyme for eosinophil peroxidase are unknown. In more detail, reactions of heme peroxidases and hypohalous acids are described in Section 2.5.1

In Table A.3, the one-electron reduction of superoxide anion radical to hydrogen peroxide is listed with reduction potential of 0.94 V at pH 7 [44]. This conversion is part of the dismutation of two superoxide anion radicals to hydrogen peroxide and dioxygen.

Other strong oxidants are singlet oxygen and ozone. They are of particular interest in damaging reactions in light- and air-exposed tissues. Data about these oxidants are not included in Table A.3.

Fig. A.3 compares reduction potentials of selected biological important oxidants, which may contribute to damage of biological material, with reduction potentials of important targets susceptible to oxidation.

Data of Fig. A.3 and Table A.2 support the view that those substrate components are easily oxidized having standard reduction potentials lower than about 1.1 V at pH 7. The lowest limit for one-electron substrate oxidation is around 0.6 V. Otherwise, any oxidant should have a standard reduction potential of 1.0 V or higher to cause oxidation of these substrates. In addition, oxidants should not be restricted in their reactivity. Among biologically relevant oxidants, •OH, peroxynitrite, compounds I and II of heme peroxidases, HOCl, HOBr, singlet oxygen, $NO_2^{\bullet}$, LO•, and LOO• fulfill this condition.

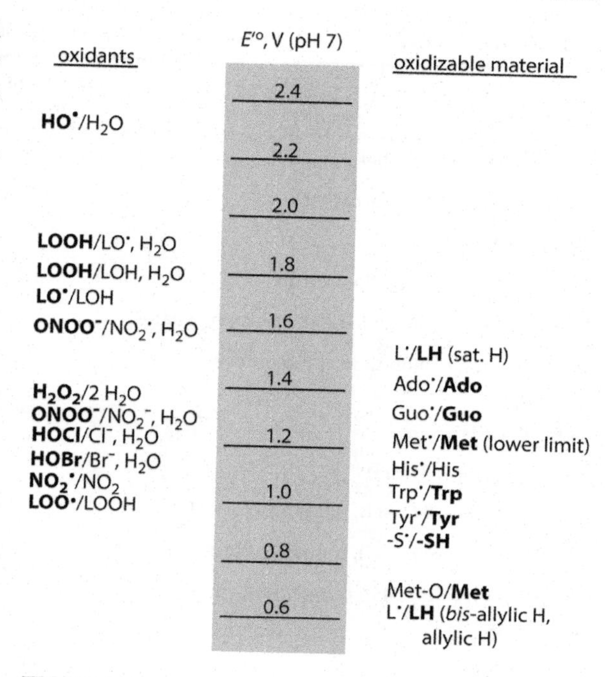

FIGURE A.3 Standard reduction potentials at pH 7 of biologically relevant oxidants in relation to selected reduction potentials of (easily) oxidizable material. In the indicated redox couples, either the oxidized form (for strong oxidants) or the reduced form (for oxidizable material) is given in bold. Abbreviations: *Ado*, adenosine; *Guo*, guanosine; *His*, histidine, *LH*, fatty acid molecule with abstractable H atom; *LOH*, fatty acid alcohol or lipid alcohol; *LOOH*, fatty acid hydroperoxide or lipid hydroperoxide; *LO•*, lipid alkoxyl radical; *LOO•*, lipid peroxyl radical; *Met*, methionine; *Trp*, tryptophan; *Tyr*, tyrosine. Further explanations are given in the text.

A.3.3 Defense Against Substrate Oxidations by Antioxidant Mechanisms

In stressed cells and tissues, all kinds of biological material can be subjected to damage. Defense against deteriorations takes place by numerous repair, replacement, and immune-based mechanisms. Antioxidant mechanisms act immediately, prevent excessive damage, and ensure redox homeostasis in cells.

Any ranging of chemical compounds by their reduction potentials reflects also on the ability of antioxidant systems to resist oxidative damage. Agents with very low reduction potential act as efficient antioxidants and sink for various reactive species and intermediary products during oxidative cascades. Standard reduction potentials for selected molecules known to act as antioxidant in lipid and aqueous phases of living systems are indicated in Table A.4. In addition to the low reduction potential, these agents

TABLE A.4 Standard Reduction Potentials of Selected Biologically Relevant Antioxidants.

Redox Couple	Transferred Electrons	E'°, V (pH 7)	Biological Significance	References
α-Tocopheryl•/α-tocopherol	1	0.48	Antioxidant	[46]
Semidehydroascorbyl•/ascorbate	1	0.28	Antioxidant	[47]
Urate•/urate	1	0.51	Antioxidant	[48]
Epigallocatechin•/epigallocatechin	1	0.43	Antioxidant	[49]
Epigallocatechin gallate•/epigallocatechin gallate	1	0.43	Antioxidant	[49]
Ubisemichinone anion radical/ubiquinone	1	−0.23	Electron transfer in mitochondria	[50]
GS-SG/2 GSH[a]	2	−0.24	Redox homeostasis	[1]
Trx(-S-S-)/Trx(-SH, -SH)	2	−0.27 to −0.19	Redox homeostasis	[25]
Grx(-S-S-)/Grx(-SH, -SH), human isoform 1	2	−0.23	Redox homeostasis	[51]
Grx(-S-S-)/Grx(-SH, -SH), human isoform 2	2	−0.22	Redox homeostasis	[51]
Prx(-S-S-)/Prx(-SH, -SH), human isoform 3	2	−0.29	Redox homeostasis	[52]
$NADP^+$/NADPH	2	−0.32	Redox homeostasis	[53]
NAD^+/NADH	2	−0.32	Redox homeostasis	[53]

[a] *Abbreviations:* Grx, *glutaredoxin;* GSH, *glutathione;* GS-SG, *oxidized glutathione;* Prx, *peroxiredoxin;* Trx, *thioredoxin.*

should also have a high reducing capacity, i. e. they should exist in a sufficient high concentration in the corresponding cellular compartments [1]. To maintain redox homeostasis in cells, these conditions are fulfilled for glutathione, thioredoxin, NADPH, and some other redox components (see Sections 3.7.1–3.7.4). Redox processes play also an important role in electron transfer in mitochondria. In Table A.4, only the reduction potential of the couple ubisemiquinone anion radical/ubiquinone is included.

References

[1] F.Q. Schafer, G.R. Buettner, Redox environment of the cell as viewed through the redox state of the glutathione disulfide/glutathione couple, Free Radic. Biol. Med. 30 (2001) 1191–1212.

[2] A. Roos, W.F. Boron, Intracellular pH, Physiol. Rev. 61 (1981) 296–434.

[3] G.R. Bright, G.W. Fisher, J. Rogowska, D.L. Taylor, Fluorescence ratio imaging microscopy: temporal and spatial measurements of cytoplasmic pH, J. Cell Biol. 104 (1987) 1019–1033.

[4] J. Santo-Domingo, N. Demaurex, The renaissance of mitochondrial pH, J. Gen. Physiol. 139 (2012) 415–423.

[5] J.R. Casey, S. Grinstein, J. Orlowski, Sensors and regulators of intracellular pH, Nat. Rev. Mol. Cell Biol. 11 (2010) 50–61.

[6] Gmelins Handbuch der Anorganischen Chemie, eighth ed., Chlor, Verlag Chemie, Weinheim, 1927, p. 62.

[7] Y.A. Ilan, G. Czapski, D. Meisel, The one-electron transfer redox potentials of free radicals. I. The oxygen/superoxide system, Biochim. Biophys. Acta 430 (1976) 209–224.

[8] B.H.J. Bielski, D.E. Cabelli, R.L. Arudi, Reactivity of $HO_2/O_2^{\bullet-}$ radicals in aqueous solution, J. Phys. Chem. Ref. Data 14 (1985) 1041–1100.

[9] W.H. Koppenol, Generation and thermodynamic properties of oxyradicals, in: C. Vigo-Alfredi (Ed.), CRC Critical Reviews in Membrane Lipid Oxidation, vol. 1, 1989, pp. 1–13. Boca Raton.

[10] H. Spalteholz, O.M. Panasenko, J. Arnhold, Formation of reactive halide species by myeloperoxidase and

eosinophil peroxidase, Arch. Biochem. Biophys. 445 (2006) 225–234.

[11] J. Arnhold, E. Monzani, P.G. Furtmüller, M. Zederbauer, L. Casella, C. Obinger, Kinetics and thermodynamics of halide and nitrite oxidation by mammalian heme peroxidases, Eur. J. Inorg. Chem. 2006 (2006) 3801–3811.

[12] J. Arnhold, P.G. Furtmüller, G. Regelsberger, C. Obinger, Redox properties of the couple compound I/native enzyme of myeloperoxidase and eosinophil peroxidase, Eur. J. Biochem. 268 (2001) 5142–5148.

[13] Gmelins Handbuch der Anorganischen Chemie, eighth ed., Chlor, Supplement A, Verlag Chemie, Weinheim, 1968, p. 244.

[14] Gmelins Handbuch der Anorganischen Chemie, eighth ed., Brom, Verlag Chemie, Weinheim, 1931, p. 290.

[15] J.C. Morris, The acid ionization constant of HOCl from 5 to 35°, J. Phys. Chem. 70 (1966) 3798–3805.

[16] R.G. Pearson, Orbital symmetry rules and the mechanism of inorganic reactions, Pure Appl. Chem. 27 (1991) 145–160.

[17] W.H. Koppenol, J. Butler, Mechanism of reaction involving singlet oxygen and the superoxide anion, FEBS Lett. 83 (1977) 1–6.

[18] J.D. Rush, Z. Maskos, W.H. Koppenol, Reactions of iron(II) nucleotide complexes with hydrogen peroxide, FEBS Lett. 261 (1990) 121–123.

[19] J.D. Rush, W.H. Koppenol, Reactions of Fe(II)-ATP and Fe(II)-citrate complexes with *t*-butyl hydroperoxide and cumyl hydroperoxide, FEBS Lett. 275 (1990) 114–116.

[20] W.H. Koppenol, Oxyradicals reactions: from bond-dissociation energies to reduction potentials, FEBS Lett. 264 (1990) 165–167.

[21] I.P. Edimicheva, M.A. Kisel, O.I. Shadyro, V.P. Vlasov, I.L. Yurkova, The damage of phospholipids caused by free radical attack on glycerol and sphingosine backbone, Int. J. Radiat. Biol. 71 (1997) 555–560.

[22] O. Shadyro, I. Yurkova, M. Kisel, O. Brede, J. Arnhold, Formation of phosphatidic acid, ceramide, and diglyceride on radiolysis of lipids: identification by MALDI-FOF mass spectrometry, Free Radic. Biol. Med. 36 (2004) 1612–1624.

[23] P.S. Surdhar, D.A. Armstrong, Redox potentials of some sulphur containing radicals, J. Phys. Chem. 90 (1986) 5915–5917.

[24] W.H. Koppenol, A thermodynamic appraisal of the radical sink hypothesis, Free Radic. Biol. Med. 14 (1993) 91–94.

[25] H. Follmann, I. Haberlein, Thioredoxins: universal, yet-specific thiol-sulfoxide redox cofactors, Biofactors 5 (1995-1996) 147–156.

[26] J. Bergès, P. de Oliviera, I. Fourré, C. Houée-Levin, The one-electron reduction potential of methionine-containing peptides depends on the sequence, J. Phys. Chem. B 116 (2012) 9352–9362.

[27] J. Kanski, M. Aksenova, C. Schöneich, D.A. Butterfield, Substitution of isoleucine-31 by helical-breaking proline abolishes oxidative stress and neurotoxic properties of Alzheimer's amyloid peptide (1-42), Free Radic. Biol. Med. 32 (2002) 1205–1211.

[28] V. Kadlcik, C. Sicard-Roselli, T.A. Mattioli, M. Kodicek, C. Houée-Levin, One-electron oxidation of β-amyloid peptide: sequence modulation of reactivity, Free Radic. Biol. Med. 37 (2004) 881–891.

[29] J.L. Beal, S.B. Foster, M.T. Ashby, Hypochlorous acid reacts with the N-terminal methionines of proteins to give dehydromethionine, a potential biomarker for neutrophil-induced oxidative stress, Biochemistry 48 (2009) 11142–11148.

[30] M.R. DeFilippis, C.P. Murthy, M. Faraggi, M.H. Klapper, Pulse radiolytic measurement of redox potentials: the tyrosine and tryptophan radicals, Biochemistry 28 (1989) 4847–4853.

[31] C. Giulivi, E. Cadenas, Heme protein radicals: formation, fate, and biological consequences, Free Radic. Biol. Med. 24 (1998) 269–279.

[32] S. Navaratnam, B.J. Parsons, Reduction potential of histidine free radicals: a pulse radiolysis study, J. Chem. Soc., Faraday Trans. 94 (1998) 2577–2581.

[33] S. Steenken, S.V. Jovanovic, How easily oxidizable is DNA? One electron reduction potentials of adenosine and guanosine radicals in aqueous solution, J. Am. Chem. Soc. 119 (1997) 617–618.

[34] S. Fukuzumi, H. Miyao, K. Ohkubo, T. Suenobu, Electron-transfer oxidation properties of DNA bases and DNA oligomers, J. Phys. Chem. A 109 (2005) 3285–3294.

[35] P.G. Furtmüller, J. Arnhold, W. Jantschko, M. Zederbauer, C. Jakopitsch, C. Obinger, Standard reduction potentials of the peroxidase cycle of lactoperoxidase, J. Inorg. Biochem. 99 (2005) 1220–1229.

[36] W.H. Koppenol, Thermodynamics of reactions involving oxyradicals and hydrogen peroxide, Bioelectrochem. Bioenerg 18 (1987) 3–11.

[37] J.D. Rush, W.H. Koppenol, Oxidizing intermediates in the reaction of ferrous EDTA with hydrogen peroxide, J. Biol. Chem. 261 (1986) 6730–6735.

[38] B. Halliwell, J.M.C. Gutteridge, Iron as a biological pro-oxidant, ISI Atlas Sci. Biochem. (1988) 48–52.

[39] S. Goldstein, D. Meyerstein, G. Czapski, The Fenton reagents, Free Radic. Biol. Med. 15 (1993) 435–445.

[40] W.H. Koppenol, The centennial of the Fenton reaction, Free Radic. Biol. Med. 15 (1993) 645–651.

[41] P. Wardman, L.P. Candeias, Fenton chemistry: an introduction, Radiat. Res. 145 (1996) 523–531.

[42] W.H. Koppenol, R. Kissner, Can 0=NOOH undergo hemolysis? Chem. Res. Toxicol. 11 (1998) 87–90.

[43] P. Wardman, Reduction potential of one-electron couples involving free radicals in aqueous solution, J. Phys. Chem. Ref. Data 18 (1989) 1637–1645.

[44] W.H. Koppenol, J. Butler, Energetics of interconversion reactions of oxyradicals, Adv. Free Radical Biol. Med. 1 (1985) 91–131.

[45] P.G. Furtmüller, J. Arnhold, W. Jantschko, H. Pichler, C. Obinger, Redox properties of the couples compound I/compound II and compound II/native enzyme of human myeloperoxidase, Biochem. Biophys. Res. Commun 301 (2003), 551–557.

[46] S. Steenken, P. Neta, One-electron redox potentials of phenols. Hydroxy- and aminophenols and related compounds of biological interest, J. Phys. Chem. 86 (1982), 3661–3667.

[47] N.H. Williams, J.K. Yandell, Outer-sphere electron-transfer reactions of ascorbate anions, Aust. J. Chem. 35 (1982) 1133–1144.

[48] M.G. Simic, S.V. Jovanovic, Antioxidation mechanism of uric acid, J. Am. Chem. Soc. 111 (1989) 5778–5782.

[49] S.V. Jovanovic, Y. Hara, S. Steenken, M.G. Simic, Antioxidant potential of gallocatechins. A pulse radiolysis and laser photolysis study, J. Am. Chem. Soc. 117 (1995) 9881–9888.

[50] A.J. Swallow, Physical chemistry of quinones, in: P.L. Trumpower (Ed.), Function of Quinones in Energy Conserving Systems, Academic Press, New York, 1982, pp. 59–72.

[51] J. Sagemark, T.H. Elgán, T.R. Bürglin, C. Johansson, A. Holmgren, K.D. Berndt, Redox properties and evolution of human glutaredoxins, Proteins 68 (2007) 879–892.

[52] A.G. Cox, A.V. Peskin, L.N. Paton, C.C. Winterbourn, M.B. Hampton, Redox potential and peroxide reactivity of human peroxiredoxin 3, Biochemistry 48 (2009) 6495–6501.

[53] F.L. Rodkey, Oxidation-reduction potentials of the triphosphopyridine nucleotide system, J. Biol. Chem. 213 (1955) 777–786.

Index

'*Note:* Page numbers followed by "f" indicate figures, "t" indicates tables and "b" indicate boxes.'

Printed in the United States
By Bookmasters